EAST MEADOW PUBLIC LIBRARY
3 1299 00978 3649

East Meadow Public Library
1886 Front Street, East Meadow, NY 11554
(516) 794-2570
www.eastmeadow.info

D1071129

BIRDS NEW TO SCIENCE

BIRDS NEW TO SCIENCE

FIFTY YEARS OF AVIAN DISCOVERIES

David Brewer

CHRISTOPHER HELM
LONDON

Christopher Helm
An imprint of Bloomsbury Publishing Plc

50 Bedford Square
London
WC1B 3DP
UK

1385 Broadway
New York
NY 10018
USA

www.bloomsbury.com

First published 2018

British Library Cataloguing-in-Publication Data
A catalogue record for this book is available from the British Library.

Library of Congress Cataloguing-in-Publication data has been applied for.

ISBN: HB: 978-1-4729-0628-1
ePDF: 978-1-4729-4589-1
ePub: 978-1-4729-0629-8

2 4 6 8 10 9 7 5 3 1

Typeset and designed by D & N Publishing, Baydon, Wiltshire
Printed and bound in China by C&C Offset Printing Co., Ltd

CONTENTS

ACKNOWLEDGEMENTS

This book would not have been possible without the generous help of numerous people, to all of whom I give my sincere thanks.

Ronald Orenstein, my co-author in a couple of previous publications, was as always unstinting in his help, from lending me books from his private library to offering advice based upon his encyclopaedic knowledge and positively elephantine memory. Janet Hinchcliffe has been hugely helpful in providing me access to the formidable resources of the Josselyn Van Tyne Library of the University of Michigan; I am also grateful to Irene Wu, of the Royal Ontario Museum Library, to Bernard Zonfrillo, for much diligent delving on my behalf in the library of the University of Glasgow, and to the staffs of the libraries of the Universities of Guelph and Waterloo, and of the County of Wellington Public Library system.

Numerous other people have provided help in a variety of ways, from sharing information, often unpublished, to helping me access some of the more obscure journal titles. These include Per Alström, David Ascanio, John Bates, David Beadle, Bruce Beehler, Marcos Bornschein, Bill Bourne, Nick Brickle, Donald Broom, Dante Buzzetti, Leah Pualaha'ole Caldeira, Chris Canaday, Leo Christidis, Martin Collinson, Thomas Donegan, Gordon Ellis, Lincoln Fishpool, Mary Gartshore, Carlos Gussoni, Paul Hebert, Sebastian Herzog, Jon Hornbuckle, Guy Kirwan, Niels Krabbe, Anne Lambert, Paul Leader, Alan Leishman, Bernabé López-Lanús, Jochen Martens, Ben Minteer, Jerry Olsen, Rodrigo Oliveira de Paiva, Ralph Parks, David Pearson, Leandra Pereira de Oliveira, Ben Phalan, Doug Pratt, Robert Prŷs-Jones, Pamela Rasmussen, Frank Rheindt, Phil Round, J. Rajeshkumar, Robert Ridgely, Paul Salaman, Paul Smith, Andy Symes, Dante Teixeira, Edson Vargas da Silva, George Wallace, Andrew Whitehouse, Bret Whitney and Andy Whittaker. I am particularly indebted to Tom Schulenberg for many helpful comments.

I am also most grateful to Katy Roper, who located photographs of all but a handful of the three hundred-odd species mentioned in the text; her painstaking efforts, along with the imaginative artistic design work of Namrita and David Price-Goodfellow, have been responsible for transforming my miserable scribblings into an attractive publication. I greatly appreciate the many constructive comments and meticulous editorial work of Nigel Redman, which freed me *frae mony a blunder an' foolish notion*. All residual errors and omissions are solely the responsibility of the author.

For secretarial and technical assistance I am most grateful to Hazel Brewer, Jennifer Brewer, Janice Crow, Mark Humphrey, Eva Schorer, Jenn Schorer, Danuta Szachanski and Bronwen Tregunno.

To all of these kind, generous people I extend my heartfelt appreciation.

SCOPE OF THE BOOK

This book is a study of all the species of birds described new to science since 1960 – in all a total of some 288 species. An account is given for each species, including details of how it was discovered, a brief description, details of habitat, food and feeding, breeding, voice, movements if any, the known range, its conservation status and a brief note on the etymology – that is, the origin of the specific (and generic, where appropriate) scientific name, and in some cases the vernacular name.

▲ Karthala Scops Owl. Grande Comore, Comoros, October 2012 (*Robert Hutchinson*).

We felt that a brief account of how the species was actually discovered would be of interest to general readers. While, in many cases, this is rather prosaic – the discoverer simply saw an unfamiliar bird and 'collected' (i.e. shot) it – in some cases the story is more interesting. (A short essay on the ethical controversy of collecting specimens appears later.) The discovery of the Karthala Scops Owl, for example, started with the British ornithologist C. W. Benson examining a sunbird's nest and finding, in the lining, the feather of a nocturnal bird, obviously an owl or a nightjar, on an island where no owl or nightjar had ever been recorded. So, of course, Mr Benson began nocturnal searches and very shortly uncovered a species of owl new to science. In other cases, new birds – that is, birds new to the scientific community – were, of course, not new at all – the local communities knew them very well indeed, and often even had individual names for them; it was simply that ornithologists were unaware of this local knowledge.

Following the section on discovery is a 'Description'. This is deliberately kept rather brief. Very frequently, an immensely detailed description, feather by feather, is given in the type description, the publication in which the discovery of the species is announced to the scientific world. We feel that this level of detail is unlikely to be of interest to the general reader, while lengthening the text substantially. Readers who do need such detail can

► Urrao Antpitta. Colibri del Sol, Colombia, May 2012 (*Barry Wright*).

◀ Rufous-winged Sunbird, female. Nyumbanitu, Udzungwa Mts, Tanzania, December 2002 (*Louis A. Hansen*).

usually find it in the type description, and these are highlighted in the reference section with the letters **TD** in bold.

After this, there is a brief summary of the species' habitat, the understanding of which is often of crucial importance in the conservation of the species. In many cases, the habitat requirements of a newly described species are very specific – indeed, this is often the reason that the species is only newly described, since a very specific habitat requirement often translates into a very restricted geographic range.

Following this are brief notes on food and feeding. In many cases, this brevity reflects a simple lack of knowledge on our part of the ecology of the species; knowledge of a species' diet is frequently only known from the examination of a few stomach contents of specimens.

The section on breeding is again very brief; indeed, in a majority of cases, there simply is no information, since the nest and eggs of the species concerned have never been described (a challenge, perhaps, to any red-blooded field ornithologist). In a lot of cases, the season of breeding can be inferred from the state of the sex organs of specimens, and this, more often than not, is all that we know.

In recent years field ornithologists have become much more aware of the importance of vocalisations in the differentiation of species and in speciation itself. This has often resulted in, especially in passerines described in the last 30 years or so, an exhaustive analysis of vocalisations, often accompanied by copious comparative sonograms. Again, we felt that these were outside the interest of the average reader, and our text section on the voice of the species is usually quite brief. Interested readers can again refer to the type description paper for more detail.

▲ White-headed Steamer Duck. Isla Escondida, Chubut, Argentina, November 2011 (*Nick Athanas*).

Most of the species described in this book are tropical, often confined to one mountain range or one island, and are essentially sedentary. Where migration or altitudinal movement does occur, we have noted it in the Movements section.

Readers of the section on Range should realise that a described range merely means that the species occurs within the geographic parameters given. It does not mean that the species is found everywhere in the indicated range, or even in all suitable habitat in that range, merely that all known occurrences have been within that range. Indeed, many newly described species are very uncommon, and patchy in their occurrence; in many cases, this is *why* they are only newly discovered, having been overlooked by previous ornithologists because of their sporadic occurrence.

We have included a section on the conservation status for each species, a paragraph that regrettably often makes for sad reading. Most statuses have been taken from the most recent analyses published by

BirdLife International, Cambridge, UK. In many cases the status is changing, lamentably only rarely for the better. A brief essay on conservation as it applies to newly discovered species appears later in this volume.

Finally, since we felt that the general reader would find it interesting, is a section on etymology. When a species is named, the describer has the privilege of choosing a 'specific epithet', that is, the second word in the binomial scientific name. (On those occasions when the new form is so unusual as to justify it, a new generic name also has to be coined.) The choice of the specific epithet is up to the describer, within the rules set out by the International Commission on Zoological Nomenclature (ICZN). The describer may use a descriptive word, usually either in Latin or Latinised Greek, such as *rufipennis*, red-winged, or *leucocephalus*, white-headed; or it can refer to a geographic location, or a habitat requirement, such as *amsterdamensis* from Amsterdam Island, or *calciatilis*, an inhabitant of limestone areas. Sometimes it can be a little whimsical, such as *perdita* for a long-overlooked species, or *solala*, meaning one-winged (the Nechisar Nightjar was described from a single wing found on an Ethiopian road; there does, however, seem to be fairly universal agreement among ornithologists that actual live examples of this species probably have two!) Very frequently, however, the specific epithet honours a person (if male, ending in '*i*'; if female, '*ae*'; and if more than one person, '*orum*'). The people so honoured include some of the most accomplished and distinguished ornithologists of our time, who have devoted their lives (and in several cases, given them) to the study and conservation of birds. We have included very brief biographical details of some of these people in the etymology section, but further details can be found in the excellent *Eponym Dictionary of Birds* (Beolens *et al.* 2014).

▲ Amsterdam Albatross with a chick (probably a couple of weeks old). Tourbières Plateau, Amsterdam Island, Indian Ocean, May 2013 (*Rémi Bigonneau*).

▲ Nava's Wren, male. Armando Zebadua, Mexico, March 2015 (*Nigel Voaden*).

The species covered by this work are those which have been described in publications dated 1960 or later. The literature has been searched up to June 2015, with later publications included as and when they became available. Some earlier literature in more obscure journals that are slow in being abstracted may have been missed. 'Splits', that is taxa which were previously classified as subspecies, but which subsequent work or data have shown to deserve full specific status, are not covered. In fact, the number of new species which have been, or are very likely to become, new full species by 'splitting' is very large and growing rapidly. The only exceptions to the 'splits' rule are taxa that were first described as a subspecies after 1960, and which have later been elevated to the status of full species. Thus, for example, Nava's Wren *Hylorchilus navai* was first described, in 1973, as a race of Sumichrast's Wren *H. sumichrasti navai*; in 1996, careful studies of vocalisations and other issues showed it to be a full species.

Very many extinct species have been described since 1960. These include species based on palaeontological evidence going back to the early evolution of birds as a separate class, and more recently extinct

◀ Limestone Leaf Warbler. Ba Be NP, Vietnam, March 2016 (*James Eaton*).

species, where the evidence might be more accurately described as archaeological rather than palaeontological. For example, the work of Olson and colleagues has revealed a treasure trove (or, perhaps more truthfully, a charnel house) of species which apparently became extinct in the Hawaiian Islands after the colonisation by Polynesians but before the arrival of Europeans (68% of Hawaii's 109 endemic species are now extinct). None of these are included. The only exceptions are species which were extant when described, but which have become extinct since (i.e. since 1960).

As will be seen from the chapter on invalid species, the description of a new species may not stand up to subsequent scrutiny, for a variety of reasons. The individual on which the type description depended might have been an abnormally plumaged example, for instance a melanistic individual of a known species, or perhaps a hybrid of two known species. Sometimes the evidence used to support a new species simply does not stand up to rigorous examination; or, sometimes, the new taxon should more properly be ascribed as a race of a known species. The acceptability of a claimed new species is frequently a moving target, subject to criticism based on new field studies or new genetic evidence. Thus it is entirely possible – indeed, highly probable – that some of the species covered by this book will, in the future, be demoted or even totally debunked. We have taken a rather liberal attitude to this problem. If a taxon is accepted by at least one major authority, such as the *IOC World Bird List* (version 7.1, 2017), the *eBird/Clements Checklist of Birds of the World* v2016, the *Howard & Moore Complete Checklist of the Birds of the World*, 4th Edition (2013, 2014), or the *HBW and BirdLife International Illustrated Checklist of the Birds of the World* (2014, 2016), we have included them, notwithstanding, in some cases, strong opinion in some quarters to the contrary. The views of the South American Checklist Committee (SACC) are frequently referred to for South American species. Our attitude is that we would prefer to be proven wrong for being too inclusive than for being insufficiently so.

GLOSSARY

(Most definitions taken from Erritzoe *et al.* 2007).

Allopatric. Living in different geographic areas. Antonym, sympatric.

Allotopic. Living in different habitats. Antonym, syntopic.

Caatinga. Semi-arid scrub forest, typical of NE Brazil.

Cere. A wax-like area at the base of the upper mandible, pierced by the nostrils, found in birds of prey, parrots etc.

Cerrado. Semi-xeric grassland, savanna, or woodland of interior SE Brazil.

Clade. A group of species including all descendants from a common ancestor and therefore constituting a monophyletic group.

Cloud-forest. Highly epiphytic montane forest situated usually between elevations of 1000 and 2500 m, sometimes higher, shrouded in fog most of the time.

Crissum. Undertail-coverts and area around vent.

Cryptic species. Two or more very similar species which do not interbreed.

Culmen. The ridge of the upper mandible, between the tip and skull, or cere if present.

Elfin forest. High-elevation, windswept, stunted humid forest, highly sculpted by prevailing winds; usually with heavy epiphytic growth.

Espinal. Dry lowland Argentinian savanna with scattered thorny bushes.

Genus, plural **genera**. A group of similar species considered more closely related mutually than any of them is with other species; in classification, a unit ranking between species and family, containing species sharing the first term of a binomial or trinomial scientific name.

Holotype. In taxonomy, the actual specimen designated to represent a new taxon when named. Synonym, type specimen.

IOC. The International Ornithological Congress, which, among other things, maintains the IOC World Bird List (current version 7.1, listing 10,672 extant species and 20,344 extant subspecies). http://dx.doi.org/10.14344/IOC.ML.7.1

Lores. The area between the eye and the upper mandible.

Malar stripe. A dark streak extending from the base of the bill down the side of the jaws to the side of the throat.

Monophyletic. A group of taxa containing all extant descendants of their most recent common ancestor, in contrast to polyphyletic.

Monotypic. Applied to a species, having no different subspecies.

Nomen dubium. A dubious application of a name.

Nomen nudum. A name given to a taxon with insufficient information provided to allow a proper assessment of its validity.

Nomen novum. A new name applied to a taxon.

Parapatric. Two or more species occurring in geographically adjacent areas, such that the ranges abut each other but do not overlap.

Paratype. Specimens, other than the holotype, used in the description of a new taxon in a type series.

Primaries. The outermost flight feathers of the wing, usually nine or ten in number, in some families up to twelve.

Polyphyletic. A group of organisms having different evolutionary origins, meaning that the last common ancestor is outside the assemblage considered.

Rectrix, plural **rectrices**. Tail feathers, excluding tail-coverts.

Remex, plural **remiges**. The flight feathers of the wing, ie primaries, secondaries and tertials, but excluding all wing-coverts.

Restinga. A moist tropical or subtropical coastal broadleaf forest, often on sandy soils, found in eastern Brazil.

SACC. The South American Classification Committee, acting under the auspices of the American Ornithologists' Union, a body comprising some of the top ranking contemporary experts on the avifauna of South America, whose duties include adjudicating on the taxonomy, nomenclature and, especially relevant to the current study, the validity of newly described species, of South American birds.

Secondaries. The inner flight feathers of the wing, located between the primaries and the tertials.

Sensu lato. In a broad sense, often applied to a group of species previously regarded as one.

Sensu stricto. The converse of *sensu lato*, meaning in a strict, narrow or restricted sense.

Sibling species. Two or more species that are morphologically very similar, often sympatric, but not interbreeding.

Sister species. A pair of species that are the closest relatives of each other, having a common ancestor not shared with any other species.

Superspecies. A closely related group of largely allopatric species, too distinct to be considered subspecies but more closely related mutually than to other species of the same genus or subgenus.

Sympatric. Occurring in the same geographical area, although not necessarily in the same habitat. Antonym, allopatric.

Syntopic. Living in the same habitats. Antonym, allotopic.

Syntype. A series of specimens used to describe a new taxon.

Taxon, plural taxa. Any group of organisms, irrespective of taxonomic rank. Frequently used to refer to species or subspecies.

Taxonomy. The theory and practice of classifying organisms into taxa.

Terra firme. Usually used to describe lowland Amazonian forest that is not subject to annual inundations, in contrast to *várzea* (q.v.)

Tertials. In passerines, the innermost secondaries, three in number; usually distinguished from the outer secondaries by being typically longer and of a more symmetrical shape and often of different colouring.

Type locality. The geographic location where the type specimen (q.v.) was obtained.

Type specimen. The actual specimen used to designate a new taxon. Synonym, holotype.

Várzea. Lowland Amazonian forest that is subject to annual flooding, often to a considerable depth.

Yungas. A dense Andean epiphytic cloud forest, typically with heavy tangled understorey, from about 1,000 to 2,500 m.

Zygodactyl. Having two toes (numbers 2 and 3) pointing forwards, numbers 1 and 4 backwards, as in woodpeckers and parrots.

THE CONCEPT OF SPECIES

No one definition has as yet satisfied all naturalists; yet every naturalist knows vaguely what he means when he speaks of a species.

Charles Darwin, 1859.

This work is an account of the species of birds described new to science over the last fifty years. However, we should at this point stop, and ask ourselves exactly what it is that we understand by the word 'species'. As Darwin pointed out, it is a noun widely used by the general public as well as by biologists and naturalists, but perhaps we should look a little more closely and define what it is that we mean by this word. Mankind, living in a natural environment, has always been aware of the different types of wildlife that he encountered in his day-to-day existence. The illiterate medieval English peasant, wallowing around in the mud of his lord's landholdings, was well aware of the difference between a Crow and a Rook, and both had distinct names, little different from their present-day ones, (cräwe and hrōc) in the Anglo-Saxon language of King Alfred the Great. However, the true definition and classification of species in a scientific way came much later. Widely recognised as the father of systematic taxonomic biology is the Swedish naturalist Carl von Linné (1707–1778), more usually recognised by the Latinised form of his name, Carolus Linnaeus. Linnaeus, son of a pastor, trained in medicine at the University of Uppsala, developing an early interest in botany. His talents was soon recognised, and in 1732 he was commissioned by the Uppsala Scientific Society to conduct a botanical expedition to Lapland, at that time no mean undertaking. Following this he went to Holland, where he was granted a medical degree and became the personal physician of a wealthy burgomaster who maintained a garden of exotic plants. During this period he wrote some of his most influential works, including *Systema Naturae* (1736), in which he organised plants, animals and minerals into classes, orders, genera and species. His most lasting contribution, with a huge influence on biology to this day, was the introduction of the binomial system, where a species is given a generic and a specific name, the generic being shared by related species but the specific being unique to the individual species. Thus he named Eurasian Blackbird *Turdus merula*, Ring Ouzel *Turdus torquatus*, Redwing *Turdus iliacus* and American Robin *Turdus migratorius*, all of which, found in the monumental 10th (1758) and 12th (1766) editions of *Systema Naturae*, are used to this day.

For Linnaeus, a species was a fixed entity, unchanging, an original creation. His concept of species was what we would now call the **Typological Species Concept**; in short, if a form *looked* different, anatomically and morphologically, it *was* different, and merited status as a species. However, with the recognition of the evolutionary process, a dilemma is encountered. If, as Darwin theorised, species are constantly diverging over time, at some stage in the process one should encounter a fluid range of intermediate forms; indeed, Darwin himself stated that "It will be seen that I look at the term species, as one arbitrarily given for the sake of convenience to a set of individuals closely resembling each other"; or in other words, there is an element of arbitrariness in the assignment of the term 'species' to a varied population of organisms "that does not essentially differ from the term variety, which is given to less distinct forms..." (*On the Origin of Species*, 1859, chapter 2). Notwithstanding the intellectual challenges inherent in an evolving natural world, most of the ornithological taxonomists of the remainder of the 19th century used, in effect, the Typological Species Concept. A more fluid recognition of the true nature of species came from the publications of, especially, two influential zoologists of the 20th century, Theodosius Dobzhansky and Ernst Mayr, leading to the **Biological Species Concept**, the one which will probably be most familiar to the majority of today's naturalists. Mayr's definition of a species was "groups of interbreeding natural populations that are reproductively isolated from other such groups" (Mayr 1969). To most people

▲ Chusquea Tapaculo. Tapichalaca Reserve, Ecuador, December 2009 (*Marc Guyt/AGAMI*).

this means, essentially, that a species does not interbreed with other species, and hence maintains itself as a separate entity, developing morphological and behavioural peculiarities which distinguish it from other, possibly very similar, species in the same habitat. In recent years a greater realisation has come to field ornithologists of the importance of vocalisations, aided by the development of sonogram technology, that were essentially unknown to the great taxonomic ornithologists of the 19th century; this has been a major factor in the detection of cryptic new species, especially in morphologically conservative groups such as the tapaculos.

In practical terms, some interbreeding between 'good' species does occur, especially in abnormal situations, particularly in captivity (the Mallard *Anas platyrhynchos* has been artificially crossed with almost 60 other species, including such totally implausible partners as the Canada Goose *Branta canadensis*). Nobody, however, would seriously suggest that this indicates that Mallards and Canada Geese are conspecific. In wild situations, on occasion an individual bird, well beyond its natural range as a vagrant, may in the absence of its own species pair with an individual of another species and even produce young (this is a fertile source of invalid species descriptions, as noted in the appendix on Invalid Species). Hybrids between different species – sometimes between different genera – of North American warblers (Parulidae) crop up with some regularity. But generally speaking, interbreeding between species is unusual and selected against.

A consequence of the evolutionary process is that species can be seen to be diverging from a common ancestor. If this has proceeded far enough for the two new forms to become essentially isolated genetically, we have what are frequently termed 'sibling species'. 'Sister species' are two or more species which are obviously more closely related to each other than to other species.

Another way of looking at species is the **Phylogenetic Species Concept**. This defines a species as any morphologically, behaviourally or genetically diagnosable population. Using this concept, Nelson & Platnick (1981) defined species as "simply the smallest detected samples of self-perpetuating organisms that have unique sets of characters". Cracraft (1983) defines a species as "the smallest diagnosable cluster of individual organisms within which there is a parental pattern of ancestry and descent". However, this concept has been criticised from several points of view; among other things, in its purest form it does not recognise the concept of subspecies, treating all diagnosable forms as full species within the concept (Erritzoe *et al.* 2007). It also has the potential to greatly increase the number of species, which might raise conservation concerns when scarce resources have to be allocated (Isaac *et al.* 2004). In fact, in a recently published paper, Barrowclough *et al.* (2016) suggest that the number of avian species currently recognised using the Biological Species Concept is a gross underestimate, the true number using a phylogenetic approach being at least twice, or even more, the figures listed using the traditional concept.

For the great majority of amateur field naturalists (which in practice means the great majority of field naturalists) the underlying intellectual arguments are confusing in the extreme. Zink & McKittrick (1995) give a very comprehensive discussion of the entire issue which is required reading for an understanding of the differing opposing concepts and their implications. To summarise by quoting from this paper, "emphasis on interbreeding or reproductive isolation typifies the biological species concept, whereas phylogenetic species concepts focus on diagnosability, patterns of descent, and monophyly".

Which brings us to the issue of subspecies. Linnaeus devised a binomial system, with two words, one specifying a genus, the second defining a species. Modern ornithology employs a trinomial system, the third word referring to a subspecies (the older practice, used for example in A. C. Bent's *Life Histories of North American...Birds* series, of giving subspecies full English names, with some 21 'Song Sparrows' (Eastern, Dakota, Tucson, Heermann's etc.) is now frowned upon, since it has the potential to cause endless confusion). A subspecies is defined as one or more populations being morphologically distinct from other populations considered to be the same species. Subspecies in their geographical contact zone may interbreed. One proposed standard for defining a subspecies is the '75% Rule' which requires, for a subspecies to be valid, that 75% of individuals are clearly identifiable (Erritzoe *et al.* 2007). The term 'race' is often used as a synonym for subspecies. Given the fluid nature of the evolutionary process, there are frequently differences of opinion among taxonomic ornithologists as to whether a particular population should be considered a full species, or a subspecies. This may be especially difficult when no zone of contact exists between the two forms where interbreeding might occur (as in isolated island populations). For example, Banks (1964), speaking of the western geographical races of the White-crowned Sparrow *Zonotrichia leucophrys*, concluded that the question of whether coastal races are specifically separate from inland ones is unanswerable, since their geographical isolation prevents the acid test of whether they would interbreed if given the chance; however, "the birds are...quite similar in ecologic, behavioral and physiologic characters, and on this basis one could postulate that the two types would interbreed.....the forms are best considered to be subspecies". The issue of whether a newly discovered bird population is a valid species or a race of a known species is one which occurs with great frequency in this book.

Influence of Biochemical Data on the Species Concept

Probably the most fundamental advance of twentieth century biology was the discovery of the biochemical mechanism of inheritance by nucleic acid replication. Without going into great technical details that are certainly beyond the scope and intent of this book, inheritance is guided by the sequence of four different bases, each a nitrogen-containing cyclic molecule, on the enormously long spine of a nucleic acid molecule. The spine is two-part – the famous double helix – and is so designed to reproduce itself by the so-called base-pair mechanism. The coding of the different bases ultimately gives rise to the morphological features of the organism. Mutations – that is, changes in the sequence of bases on specific parts of the nucleic acid spine – do occur, spontaneously or in response to outside factors, and when they do, the changes become heritable and may induce morphological changes in the organism. If these changes are advantageous to the survival or reproductive potential of the organism, by the mechanisms of Darwinian evolution they will become embedded in the population. Thus, if a species is in the process of diverging into two distinguishable sibling species, in theory this fact should be verifiable by merely examining the sequence of base pairs in the respective nucleic acids.

The word 'merely' is distinctly inappropriate, given the astronomical number of base pairs that govern the inheritance of an organism and its morphological features. If one had to rely upon the entire genetic make-up of an organism the task would be clearly utterly impossible. What is needed is a pared-down method of using genetic information; essentially a short piece of DNA that will be long enough to contain information that would distinguish species, but short enough to be practically manageable, and fast and efficient in use. Hebert *et al.* (2004) and Stoeckle & Hebert (2008) give an example of such a technique. The particular gene segment chosen gives rise to an enzyme called cytochrome c oxidase subunit 1. The region of the gene responsible for coding for this enzyme is small enough for the sequence of its base-pairs to be read easily with modern technology, but long enough to capture enough variation to distinguish between species. To give an example, in primates each cell has about 3.5 billion base pairs. The selected site contains 648 base pairs. Among human populations, one or two of these pairs may differ; but we diverge from our closest relative, the Chimpanzee, by about 60 base pairs and from the Gorilla by about 70.

Mitochondrial DNA (mt DNA), that is, DNA passed through the maternal line in the cell's mitochondria (a mitochondrion is a tiny separate structure within a cell, responsible for, among other things, providing energy to the organism by the metabolism of carbohydrates, and which does not occur in the cell nucleus), is particularly useful in genetic studies. Base sequence differences between different species are much more frequent in mt DNA than in DNA from the cell nucleus. It is also more abundant and easier to work with and to recover, especially from small tissue samples. However, it has been pointed out that sometimes mt DNA and nuclear DNA data may suggest different relationships in groups of organisms (Patton & Smith 1994, Degnan 1993).

Thus, in modern taxonomic studies, DNA analysis has become an essential tool, and it is difficult to imagine an acceptable species type description which does not in some part involve DNA analysis. This may be applied not only to recent, fresh tissue samples, but in many cases to old, degraded museum specimens whose DNA is incomplete or damaged; a sequence of 100 to 200 base pairs may be adequate to provide useful information.

The data arising from DNA analysis can be a hugely helpful adjunct to conventional morphological methods. First and foremost, it can give a clear indication of whether a new form is truly a 'good' species, or merely an abnormal example of an existing one. This can be especially important when the specimen base is small, or even dependent on one example alone. (Indeed, in some examples highly significant information has been provided by small blood or feather samples from an endangered species without the necessity of taking an actual specimen.) It can also provide information that is sometimes unexpected; a new species may, on genetic examination, prove to be a closer relative to a quite different-looking relative than to an apparently similar-looking one. Several examples of this are mentioned in the species accounts of our text. Sometimes organisms, while diverging genetically, remain very conservative in outward morphology; it has been estimated (Stoeckle & Hebert 2008) that up to 4% of named species of North American birds contain genetically distinct lineages that are likely to be separate species. Thus there may be numerous yet-to-be-described species lurking in museum drawers around the world; DNA analysis might unlock some of them.

DESCRIPTION OF THE WORLD'S BIRDS: AN HISTORICAL PERSPECTIVE

As noted in the chapter on the meaning of the word species, the systematic classification and description of the world's birds can be said to begin with the publication, in 1758, of Linnaeus' *Systema Naturae* (10th edition), a work whose influence extends to this day. In fact, a century before Linnaeus, John Ray, first of the great tradition of English parson-naturalists, attempted a similar venture, but his system, based on short descriptive phrases in Latin, was hopelessly cumbersome and totally impractical. He described the Northern Shoveler as "*Anas platyrhynchos altera sive clypeata Germanis dicta*" (another broad-billed duck, or, as the Germans say, bearing a shield). Linnaeus reduced this to a catchy *Anas clypeata*, which we use today, while appropriating *platyrhynchos* for the Mallard. Linnaeus, in his 1758 work, gave names to about 450 species of birds, more than half of them, predictably, from Europe. The remainder were from parts of the world at that time accessible; for example, more than 50 from North America, almost all from the Eastern seaboard; and a similar number from South America, concentrated mostly along the Caribbean coast and also in the West Indies – in short, wherever a sea captain, voyaging out of a European port, was likely to visit. Occasionally this gave rise to amusing errors – the type location of *Donacobius*, a monotypic family found in lowland South American swamps, is given as the Cape of Good Hope, South Africa – obviously the labels on packages were occasionally misplaced. But more impressively, Linnaeus described, in one fell swoop, more than half of the known species of the Western Palaearctic, and an even greater proportion of purely European ones. No zoologist since has achieved as much; and it should be noted that Linnaeus' primary interest was botany, not zoology. It is a little ironic that poor Linnaeus himself was not recognised in a bird's name for 200 years (until 1966), and then only by a subspecies, when Allan Phillips named an obscure race of the Clay-coloured Robin *Turdus grayi linnaei*, demonstrating a dreadful ingratitude among taxonomic ornithologists.

Linnaeus' pioneering example stimulated a substantial number of naturalists and savants to start the massive work of world avian classification. In the decade after the publication of the 10th edition of *Systema Naturae* some of the great names of 18th century zoology used the Linnaean system of nomenclature to classify ever-increasing numbers of species; by the end of the 1770s, more than two thirds of the species of the Western Palearctic and an even greater proportion from Europe proper, had been described, often with names used to this day. This period provided a series of extraordinarily productive and talented naturalists whose names are linked, as the describing authority, to many familiar species in an increasingly wide geographical context. A surprising number, like Linnaeus, had medical qualifications. The German Johann Friedrich Gmelin (1748–1804) was an astonishingly versatile person, who published significant works on medical science, but was also responsible for major contributions to ornithology, herpetology, entomology, malacology and botany. He published the 13th edition of *Systema Naturae*, which was the vehicle for many new descriptions. Pieter Boddaert, a Dutchman also qualified as a medical doctor, was responsible for a large number of descriptions of European species, as was the German Johann Bechstein (1757–1822). Many of the zoologists of this period relied on collectors scattered around the globe to provide the material which they then described. A notable exception was the great German zoologist Peter Simon Pallas (1741–1811), a hugely talented person who obtained his doctorate at the age of 19. He became a member of the Academy of Sciences of St Petersburg and made some of the first scientific explorations, on behalf of the Academy, into the expanding Russian possessions in Siberia and the Far East. No fewer than 13 taxa, species or races, have 'Pallas' in their vernacular names, while he was the describing authority of very many more. It should be noted that expeditions of this type were not without hazard; J. F. Gmelin's kinsman, S. Gmelin, while on a botanical trip to the Caucasus, was kidnapped and held to ransom by a local potentate, and in fact died in captivity, an event that led to a punitive expedition by the Tsarist government.

In the four decades after the publication of *Systema Naturae* the number of bird species described increased exponentially. However, this came to a crashing halt during the 20 years of the Napoleonic Wars, during which time the Royal Navy owned the world's oceans, inhibiting, among others, the efforts of the very productive French school of zoology from much effective exploration. This can be seen, very clearly, by looking at Figure 1 (page 18). One notable exception to this was the cataloguing of the Australian avifauna, in no small measure by the efforts of Dr John Latham (1740–1837). Latham, yet again a medical doctor, wrote among other things the descriptions of birds collected by Arthur Phillip, the Governor of the Botany Bay settlement in Australia. Latham also described several species from the Far East. At the same time, in North America, the work of the Scots-American Alexander Wilson led to the description of numerous species.

With the cessation of hostilities, there was a resurgence of ornithological exploration, involving several enormously productive ornithologists. One of these was the Frenchman Louis Jean Pierre Vieillot (1748–1831) who described a huge number of species in the late 1810s. Vieillot was perhaps overshadowed in the popular view by his fellow countrymen, the Compte de Buffon (1707–1788) and Baron Georges Leopold Cuvier (1769–1832), but his contributions were very substantial. He was perhaps a little unusual, for his time, in suggesting that there were things to be learned from studying live birds, instead of only museum specimens. Vieillot named some species for the Dutch ornithologist Coenraad Temminck (1778–1858), who in his own right named numerous species, and is commemorated in the common names of 17 species, from Europe, Asia, Africa and South America. Temminck was the first director of the prestigious Rijkmuseum van Natuurlijke Historie in Leiden, Holland, writing in 1815 the influential *Manuel d'Ornithologie,* a systematic study of European birds.

The half-century from 1822 to 1870 might be described as the Golden Age of ornithological description, with almost 50% of the world's bird species being described. There are several reasons for this. In the Old World, the 'Scramble for Africa' was in full swing, as the major European powers divided up that continent into their own colonial empires, with similar events occurring in southern Asia, Indonesia and the Far East. As the various colonial powers penetrated previously unknown areas and established local administrations, a variety of (often) amateur ornithologists began cataloguing the avifauna of hitherto unexplored areas, resulting in an explosion of new species descriptions. The peak decade was in the 1840s, when over 1200 new species were described. As "trade follows the flag", so did ornithological exploration, with naturalists from the great colonial powers – Britain, France and Germany in particular – describing the avifaunas of their countries' colonial holdings. This is illustrated by Figure 2 (pages 18–19); there is a very obvious bulge in the numbers of species described, especially, from Africa, and primarily by British, French and German naturalists, between 1870 and 1909. Thus the British had something of a monopoly on Indian birds. Edward Blyth (1810–1873), apparently a much put-upon gentleman, was Curator of the Museum of the Asiatic Society of Bengal. Some 16 species bear his name in the vernacular, and he described others. A most versatile person was Allan Octavian Hume (1829–1912), who in addition to writing extensively (and having 13 species named after him), was responsible for the introduction of free primary education in India, the founder of a vernacular newspaper and a co-founder of the Indian National Congress, a political party dedicated to Indian Home Rule. Brian Hodgson (1800–1894) was, initially, employed by the East India Company, and spent more than 30 years in Nepal and Himalayan India, during which time he amassed a huge collection of bird specimens. He described 79 as new to science, but was forestalled by others in his description of almost 50 more. Thomas Claverhill Jerdon (1811–1872) was also employed by the Company, as an assistant surgeon. His *Birds of India,* published 1862–1864, was a groundbreaking work; 13 species of birds bear his surname in their common names.

The German influence on African ornithology is, not surprisingly, mainly concentrated in East Africa in what were former German possessions, and many of the species of Kenya and Tanzania recognise the contribution of German naturalists. Wilhelm Rüppell (1794–1884) made two expeditions, the second with the Freiherr Friedrich von Kittlitz (1799–1874), to Africa. Between them they have 14 species named after them, although von Kittlitz was also active in the North Pacific. Rüppell published some of his work in 1845 in the *Systemische Übersicht der Vogel Nord-ost Afrikas* (a Systematic Survey of North-Eastern African Birds). His contemporary, Theodor von Heuglin was by profession a mining engineer who explored in Ethiopia and central Africa, publishing his work in 1869. Curiously, he was adamantly opposed to Darwin's theory of evolution, an opinion perhaps inherited from his pastor father. Numerous French and British naturalists, again many amateurs, made enormous contributions, so that by 1850 almost half of Africa's species had been described (although substantial areas of the continent were to remain unexplored for at least another 20 years). The Portuguese, in contrast, made only very minor contributions to African ornithology, despite their acquisition of huge tracts of territory in Angola and Mozambique.

The ornithology of Indonesia, New Guinea and related archipelagos spawned an impressive cast of zoologists, primarily British, Dutch and French, with later contributions by Germans. Sir Stamford Raffles (1781–1826) was a distinctly larger-than-life character; besides founding Singapore and creating the basis for British colonial rule in Malaysia, he fought several successful wars against the French and their Dutch allies, suppressed slavery and piracy, wrote a history of Java and still found time to name numerous species of birds, as well as *Rafflesia,* the largest flower in the world. The English scientist, Alfred Russell Wallace (1823–1913), perhaps best known for having conceived the theory of evolution independently from Charles Darwin, was also a major describer of numerous new species, not only of birds but of the entire fauna and flora, as well as making observations of the local people. However, some of most important contributions to the zoology of the region are due to the Italian Conte Adelardo Salvadori (1835–1923), whose distinguished career included serving as a medical officer in Garibaldi's army. He has 12 species named after him, and is the naming authority on a much larger number. The spike in the number of species described from Indonesia and New Guinea in the 1870s and subsequently is due in no small measure to Conte Salvadori.

For the most part, with a few notable exceptions, the species of South American birds were described by outsiders, Europeans and North Americans, rather than by local South American zoologists. The number of species described exploded at the end of the Napoleonic Wars. The Austrian Joseph Natterer (1787–1843) and the German, Johann von Spix (1781–1826) spent three very productive years (1817–1820) in Brazil, resulting in the description of a large number of species. The earliest serious describer of the South American fauna was probably the German, or rather, Prussian, Baron Friedrich von Humboldt (1769–1859), a man of extraordinary and eclectic accomplishments, who spent the years 1799–1805 travelling and collecting in Mexico and over much of northern South America, publishing a 'Personal Narrative' of his adventures which acted as an inspiration to Charles Darwin. Darwin himself, of course, added considerably to the ornithology of mainland South America as well as, more famously, the Galápagos Islands, before retiring to rural Kent to work on certain other matters.

There is a certain divide in the history of description of South American birds between those who travelled in the field, an activity certainly not without its hazards, and those who worked in (mainly) European museums, classifying specimen material sent to them by a variety of field collectors. In terms of nomenclatural fame, either as a naming authority or by being commemorated in the scientific or vernacular names of species, the latter course was by far the most rewarding in terms of eponymy. P. L. Sclater (1829–1913) and his son W. L. Sclater (1863–1944) named, between them, more taxa of birds than any other ornithologist – almost one thousand. Twenty-two species have the name Sclater in the vernacular, and fifty-two species or races are called *sclateri* or variants of it. The majority of these commemorate Sclater senior, and include forms from all continents except Europe. The elder Sclater travelled to North America and the Caribbean, but so far as can be told most of the type specimens bearing his name did not come from his own collecting activities. W. L. Sclater lived in India and in South Africa before succeeding his father as editor of *The Ibis* (which the elder Sclater had founded). The younger Sclater was, in his old age, the victim of enemy action, being killed by a flying bomb in London in 1944.

Another person who made enormous contributions to ornithology from within the walls of a museum was Richard Bowdler Sharpe (1847–1909), Assistant Keeper of the British Museum's Zoology Department from 1895 to 1909. He was a man of enormous energy, "a tonic to all zealous ornithologists" (Manson-Bahr, 1959) who, among other achievements, personally wrote over 200 descriptions and the 27-volume *Catalogue of the Birds of the British Museum* while rearing a family of 10 daughters. 'Sharpe's' appears as the vernacular name of 12 species from three continents, while the specific or subspecific epithet *sharpei* occurs 28 times.

Nineteenth-century German institutions also produced more than their fair share of highly productive taxonomic ornithologists, perhaps none more so than J. L. Cabanis (1816–1906). He founded, and edited for 40 years, the influential *Journal für Ornithologie*, still one of the most important ornithological periodicals today; his son-in-law, A. Reichenow (1847–1941), who was one of the leading authorities of his time on African birds (with 18 African and Melanesian species bearing his name), succeeded him as editor.

Field ornithology in South and Central America involved a large number of ornithologists, including some colourful characters. The Andes provided a fruitful area of study; the German Hans Graf von Berlepsch (1850–1915), the Swiss Baron Johann von Tschudi (1818–1889) (the influence of the aristocracy, of several nations, on ornithology was remarkable), and the Pole Wladislaw Taczanowski (1819–1890) all travelled extensively in South America.

In Central America, probably the most influential 19th century figures were Frederick Godman (1834–1919), who travelled enormously on several continents and his close friend Osbert Salvin (1837–1898). Together they published the monumental *Biologia Centrali Americana* which appeared in many volumes from 1888 to 1904. One of the (at that time) rather uncommon local zoologists who made major contributions to the avifauna of his own country was the Mexican Francois Sumichrast (1828–1882), for whom a number of birds and mammals are named.

In North America, the avifauna was comprehensively catalogued by the middle of the 19th century; 90% of all North American species had been described by 1870, and a mere eight new species between 1900 and the present. In Europe proper, the last truly new species (a few forms, like the Scottish Crossbill and the Iberian Chiffchaff were named later, and have since been 'split') was the Corsican Nuthatch, which remained undetected until named, by the indefatigable Richard Bowdler Sharpe, in 1884. By contrast, Africa, Asia and South America in the first three decades of the 20th century still produced over 500 new species. The majority of eastern North American species were documented by 1820; as the wave of European settlement surged westwards, so did the discovery of many characteristic western species. Not surprisingly, a considerable proportion of the naturalists responsible for the documentation of the western American fauna were military men: James Abert (1820–1897), Charles Bendire (1836–1897), William Hammond (1828–1900), Elliott Coues (1842–1899) and William McCown (1815–1879) were all U.S. Army officers whose duties took them into the West, where they were often involved in Indian wars. (Bendire relates an incident when he was up a tree, undetected, looking at a nest, when some hostile Apaches walked underneath; had they looked up, the history of North American ornithology would have lost one of its most colourful characters.)

As the 20th century progressed, the rate of discovery of new species started to diminish quite rapidly, although the first decade was, in Africa, remarkably productive as German, British and French ornithologists explored their respective colonial possessions. The early years of the century also saw the description of more than 200 new species from South America. Nevertheless, for obvious reasons, it became more difficult to discover new species and the ten-year totals steadily decreased with each new decade, until the 1950s produced only 40 new birds, more than half of which came from Africa or South America. Figures 1 and 2 show, respectively, the decennial rates of species discoveries, worldwide and by geographic region.

Figure 1 Description of the world's birds, by decade, from pre-1759 to 1959.

Decade	Species described by decade – whole world
Up to 1759	453
1760-1769	320
1770-1779	185
1780-1789	670
1790-1799	102
1800-1809	158
1810-1819	548
1820-1829	1049
1830-1839	1102
1840-1849	1092
1850-1859	846
1860-1869	792
1870-1879	732
1880-1889	445
1890-1899	425
1900-1909	304
1910-1919	191
1920-1929	151
1930-1939	126
1940-1949	51
1950-1959	41

Figure 2 Description of the world's birds, from pre-1759 to 1959, broken down by geographic region.
Notes:
Western Palearctic. As defined in Snow *et al.* 1998, *Birds of the Western Palearctic*, i.e. Europe from Iceland east to the Urals, Caucasus, Middle East from Iranian border west, northern Saudi Arabia north of 28°N, North Africa north of 20°N, Atlantic Islands including Azores and Cape Verde Islands.
Africa. All areas except Western Palearctic and Madagascar.
Indian Ocean. All islands including Socotra, Madagascar, islands of southern Indian Ocean, excluding Sri Lanka.
Indonesia, Philippines, New Guinea: includes islands of Melanesia (New Britain, Solomon Islands, Vanuatu etc).
Oceania. All islands excluding Melanesia, but including Hawaiian group.

Decade	Western Palearctic	North America	Mexico, Central America, West Indies	South America	Sub-Saharan Africa
Up to 1759	232	57	28	53	27
1760-1769	63	41	19	73	51
1770-1779	15	11	9	37	20
1780-1789	19	57	32	185	78
1790-1799	6	9	5	21	24
1800-1809	5	17	6	12	21
1810-1819	32	42	16	224	108
1820-1829	21	48	79	317	117
1830-1839	10	43	82	336	177
1840-1849	8	31	89	329	144
1850-1859	5	34	107	302	165
1860-1869	3	22	107	299	101
1870-1879	3	7	32	149	94
1880-1889	5	7	26	119	114
1890-1899	1	3	17	72	121
1900-1909	1	2	11	76	114
1910-1919	2	1	10	71	28
1920-1929	1	0	12	53	32
1930-1939	2	1	6	34	21
1940-1949	0	0	7	23	10
1950-1959	0	0	1	15	13

This brings us to the period covered by this book. In the table overleaf we have listed the geographic origins of the 288 species described since 1960. The divisions are geographic rather than political; thus, West Papua (Irian Jaya) is included with New Guinea rather than Indonesia, and Hawaii is classified as Oceania rather than the United States.

Predictably, more than half of the new species come from South America, with Brazil and Peru taking the lion's share. Asia contributes another 20% of the total, two thirds of which come from the island complexes of Indonesia and the Philippines, while some 40 come from Africa. Not surprisingly, there are no new species from mainland Europe, with only two, one of them cryptic, from the western Palearctic; what, perhaps, is rather surprising are the five new species from Australia, normally regarded as a very well known area ornithologically. As might be expected, there is a high level of endemism in newly described species; more than two thirds (194) are found in one country alone, 92 are found in more than one country (often rather narrowly, sometimes crossing a border by just a few kilometres) and two are of imprecise origin. The concept of endemism, when applied across an area with illogical human-drawn political boundaries, is of rather limited biological significance anyway; the study of avian distribution is far more meaningful when applied to geographic features and habitat types.

Especially in the earlier years of our study period, the major advances in South American taxonomic ornithology came from the efforts of, mostly, North American academic institutions; pre-eminent were, and still are, the Louisiana State University Museum of Zoology, the Field Museum of Chicago, the Academy of Natural Sciences of Philadelphia, the Smithsonian Institution, and the American Museum of Natural History in New York. A look at the Etymology sections of the species accounts will give a very good idea of the contributions of the extraordinary number of distinguished field ornithologists from these and other institutions. In more recent years, the work of local ornithologists (and even bird tour leaders) has become increasingly important, with, for example, major contributions from universities and other institutions in, especially, Brazil, Argentina and Colombia. In East and South-east Asia, the studies of German and Scandinavian scholars have resulted in a major series of discoveries in China and Indochina, while co-operations between American and Filipino universities have had similar results in the Philippines. In Indonesia, the collaboration of local ornithologists and western colleagues has produced a number of new species, a process which is continuing. We look forward to seeing, in future years, an ever-increasing participation in field ornithology by scientists from outside Europe and North America.

Decade	Madagascar, Indian Ocean	Mainland Asia	Indonesia, Philippines, New Guinea	Australia, New Zealand	Oceania, Antarctica
Up to 1759	2	42	9	1	2
1760-1769	14	36	22	0	1
1770-1779	11	37	11	2	5
1780-1789	20	120	46	64	13
1790-1799	0	11	4	18	1
1800-1809	1	21	10	63	2
1810-1819	4	49	31	39	3
1820-1829	2	132	150	70	13
1830-1839	21	270	61	84	18
1840-1849	18	214	108	107	44
1850-1859	21	82	79	20	31
1860-1869	35	102	86	16	21
1870-1879	29	127	240	14	37
1880-1889	14	29	103	8	20
1890-1899	6	39	122	16	28
1900-1909	3	30	41	15	11
1910-1919	3	20	31	14	11
1920-1929	3	16	22	0	12
1930-1939	5	14	30	3	10
1940-1949	0	2	7	2	0
1950-1959	0	2	7	1	1

Distribution of newly described species, 1960–2016, by country of type description.

Azores 1
Algeria 1
Total: Western Palearctic 2

United States 1
Total: North America 1

Mexico 4
Panama 2
Caribbean (Puerto Rico) 1
Total: Mexico, Central America, West Indies 7

Colombia 17
Venezuela 5
Guyana 1
French Guiana 1
Ecuador 9
Peru 49
Brazil 50
Bolivia 4
Argentina 8
Chile 2
Unknown 1
Total: South America 147

Liberia 4
Ivory Coast 1
Nigeria 3
Cameroon 1
Gabon 1
Chad 1
Democratic Republic of the Congo 5
Sudan 1
Ethiopia 2
Somalia 1
Uganda 1
Kenya 2
Tanzania 8
Zambia 1
Mozambique 1
Total: sub-Saharan Africa 33

Madagascar 4
Indian Ocean (excluding Madagascar) 10
Total: Madagascar/Indian Ocean 14

Palestine 1
Tibet 1
Nepal 1
India 2
Myanmar (Burma) 1
Japan 1
Sri Lanka 1
China 9
Taiwan 1
Laos 1
Cambodia 2
Vietnam 4
Thailand 1
Russian Federation 1
Total: mainland Asia 27

Indonesia 13
Philippines 20
New Guinea and Melanesia 10
Total: Indonesia, Philippines, New Guinea 43

Australia 5
New Zealand and dependencies 2
Total: Australia, New Zealand 7

Oceania (including Hawaii) 7
Total: Oceania, Antarctica 7

World Total: 288

SPECIES ACCOUNTS

In this, the major part of the book, we give accounts of the 288 species that have been described and published since 1960. The details of how we have dealt with these species, and what has been included and omitted, are explained in the 'Scope of the Book' (pp.7–10).

*A*raripe Manakin. Arajara Park, Barbalha, Ceara, Brazil, August 2016 (*Robson Czaban*).

OKARITO KIWI
Apteryx rowi

A recently recognised and highly restricted species.

Type information *Apteryx rowi* Tennyson *et al.* 2003, South Okarito Forest, West Coast, South Island, New Zealand.

Alternative names Okarito Brown Kiwi, Rowi.

Discovery Kiwis have been well known from the west coast of New Zealand for more than a century and a half; however, the taxonomy has been confused, the Okarito Forest population being lumped with the more widespread Brown Kiwi *A. australis*. Recent DNA work has shown that the Okarito population is in fact more closely related to North Island Brown Kiwi *A. mantelli* than it is to its closest neighbours; further evidence comes from an examination of the feather lice (*Apterygon*), which indicate the separation of *rowi* and *australis*. Molecular evidence suggests that *A. rowi* diverged from *A. mantelli* about 6.2 million years ago, and both diverged from *A. australis* about 8.2 million years ago (Burbidge *et al.* 2003). There is a dearth of good specimens of the new species, and of course no further ones can be collected; the holotype, an immature female, presently in the Canterbury Museum, was banded as a juvenile in September 1999 and accidentally killed on a road in August 2002. The species was named to forestall the repetition of any *nomen nudum* (i. e. a description based on insufficient data) (Tennyson *et al.* 2003), and to provide a valid scientific name for a critically endangered population.

Description 55 cm. Monotypic. As a kiwi, unmistakable and totally unlike any other bird. Differs from *A. australis* in having a greyer head and belly, outer wing transversely barred pale and darker, and shorter, pink (never creamy or horn-coloured) bill.

Habitat Coastal podocarp-hardwood forest.

Food and feeding Nocturnal feeder; specific diet items not recorded, presumably similar to those of other kiwi species (i. e. soil and litter-dwelling invertebrates, including earthworms, millipedes, larval beetles, and moths and spiders; also some vegetable matter).

Breeding Nests in burrows excavated by the birds themselves or occasionally in natural cavities. One large white or greenish-white egg, up to 20% of the weight of the laying female; incubation by both sexes (unlike some other kiwi species, where the male alone incubates); in other kiwis incubation period may be 63–92 days (F. Jutglar 1992 HBW 1). Female may lay up to three eggs in a year. Young self-sufficient immediately on hatching, but may remain with parents for several years. Monogamous; usually pairs for life. Birds are exceptionally long-lived, possibly 80–100 years.

Voice Male gives a high-pitched ascending whistle repeated 15–25 times; female gives a slower and lower-pitched hoarse guttural call repeated 10–20 times (Robertson, H.A. in Miskelly, Ed., New Zealand Birds Online 2013). Pairs often duet.

Movements Sedentary.

◄ Okarito Kiwi female. Okarito Forest, Westland, South Island, New Zealand, February 1999 (*Tui de Roy/ Nature Picture Library*).

Range Endemic. Now restricted to an area of about 10,000 ha on the west coast of New Zealand, bordered by the Okarito River to the south and the Waiho River to the north. Recent relocation programme involved the release of birds on the predator-free Blumine and Mana Islands in Queen Charlotte Sound, Marlborough, South Island. DNA evidence suggests that *A. rowi* was formerly much more widespread on South Island and may even have extended onto North Island (2013 HBW SV).

Conservation status Not currently classified by BirdLife International, but certainly Endangered or higher. Restricted (apart from artificial relocations) to the 11,000 ha Okarito Kiwi Sanctuary. Main threat is predation by introduced predators, specifically Stoat (*Mustela erminea*) and Common Brushtail Possum (*Trichosurus vulpecula*), as well as feral dogs and cats. Control of these predators is not easy. Removing eggs from the wild and incubating them artificially can increase hatching rates dramatically (from as low as 5% to 90%); chicks are then kept in an enclosure until they have developed their nocturnal instincts, then released on Blumine Island until they are large enough to defend themselves against predators, at which point they are returned to Okarito (2013 HBW SV). World population was estimated at 160 birds in 1995, 200 in 2000 (Robertson & de Monchy 2012) and 375 in 2011 and 2012 (2013 HBW SV; New Zealand Birds Online).

Etymology The specific name *rowi* is derived from the Maori name for the species (also rendered as 'roa' and 'rohi', though there is confusion as to which species of kiwi is referred to in some circumstances). The form 'rowi' was chosen in consultation with the Ngai Tahu, the principle iwi (tribe) of South Island (Tennyson *et al.* 2003).

TACARCUNA WOOD QUAIL
Odontophorus dialeucos

A distinctively marked wood quail with a very restricted range in southern Panama and northern Colombia.

Type information *Odontophorus dialeucos* Wetmore 1963, Darién, Panama.

Alternative names Spanish: Corcovado del Tacarcuna.

Discovery The type specimen was obtained on 7 June 1963 by a collecting expedition of the Gorgas Memorial Laboratory, directed by Dr Pedro Galindo, 6.5 km west of the summit of Cerro Malí, Darién, at an altitude of 1,450 m.

Description 22–25 cm. Monotypic. A typical wood quail, very dark brownish-black over most of the body, with a diagnostic strikingly white upper chest, face and supercilium, and warm buffy on the nape. Female brighter tawny-brown on underparts; juvenile similar to female but with less white on chin. Not confusable with any other *Odontophorus* species in its limited range.

Habitat Humid subtropical forest, mostly 1,200–1,450 m (Wetmore 1963), sometimes down to 1,050 m (Ridgely & Gwynne 1989).

Food and feeding Largely terrestrial, occasionally perches low in groups up to 6–8 birds in vegetation. No data on food items.

Breeding Nest and eggs undescribed; birds in immature plumage in early June (Madge & McGowan 2002).

Movements Probably sedentary.

Voice Not recorded.

Range Very restricted geographically, the total range being probably no more than 100 km², confined to the Cerro Malí and Cerro Tacarcuna at the southern end of the Serranía del Darién, Panama and in immediately adjacent Colombia (Wege 1996).

Conservation status Restricted-range species, described as fairly common in suitable habitat. Population estimates 6,000–15,000 mature individuals (BirdLife International 2013), with a classification of Vulnerable due to small total range. Panamanian range entirely within Darién National Park (though habitat destruction occurs at lower altitudes); small Colombian range is within Parque Nacional Los Katios (Rodríguez 1982; Pearman 1993).

Etymology *Dialeucos*, from the Greek *dialeukos*, 'marked with white'.

▼ Tacarcuna Wood Quail, adult male. Flank of Tacarcuna, Darién, Panama, September 2016 (*Alexis Guevara and Euclides Campos*).

GUNNISON SAGE GROUSE
Centrocercus minimus

A cryptic species, the first new species to be described in the continental USA since the 19th century.

Type information *Centrocercus minimus* Young *et al.* 2000, 23 km south-east of Gunnison, Gunnison County, Colorado, USA.
Alternative names Gunnison Grouse.
Discovery As part of a management programme of the Sage Grouse conducted by the Colorado Division of Wildlife, starting in the 1970s, wings were collected from hunters. In 1977 it was noted that wings from the Gunnison Basin in SW Colorado were substantially smaller than examples from other parts of the state. Subsequent studies in the 1980s and 1990s showed significant differences in plumage, morphometrics and breeding behaviour from other populations. In addition, genetic studies have shown additional significant differences from other Sage Grouse populations (Young *et al.* 2000; Oyler-McCance *et al.* 2015), the remaining populations gaining the suggested English names of Greater Sage Grouse or Mountain Sage Grouse.
Description Male 44–51 cm; female 32–38 cm. Monotypic. Sage Grouse are unmistakable, with a long barred tail. Males have striking black-and-white facial markings, and in display engorged yellow patches above the eye and two inflatable yellow pouches set in a mass of white feathers from the lower chest to the neck. Females lack these but have, as do the males, a black belly. Gunnison Sage Grouse differs from Greater Sage Grouse in its smaller size (body weight about 30% less), wing length and smaller bill. In plumage, the tail is paler and the ornamental feathers on the neck are longer and broader (Young *et al.* 2000; del Hoyo *et al.* 2013).
Habitat Associated with sage-brush (*Artemisia*) usually with forb and grass undercover; in winter, in more sheltered areas of large sage-brush with less snow-cover and deciduous shrubs (Young *et al.* 2000); usually below 3,000 m, and in winter typically 1,800–2,800 m.
Food and feeding Forages on the ground. In winter predominantly the leaves of sage-brush (*Artemisia*); in summer also forbs and insects. In disturbed and fragmented habitats, forages in fields of alfalfa, wheat and beans (Young *et al.* 2000).
Breeding A lekking species; males gather at selected leks from late March to May; only 10–15% of males, usually older experienced birds, successfully mate each year. Lekking grounds have low vegetation with sparse shrubs, but are surrounded by higher bushes. Nest is a

◄ Gunnison Sage Grouse, adult male displaying at a lek. Gunnison County, Colorado, USA, April 2009 (*Gerrit Vyn*).

► Gunnison Sage Grouse, adult female. Gunnison County, Colorado, USA, April 2009 (*Gerrit Vyn*).

shallow depression in the ground, concealed by higher vegetation; clutch 6–10, incubation and chick care by female alone.

Voice Cackling and clucking notes. In display, male vocalisations differ from the corresponding ones of Greater Sage Grouse and probably act as an isolating mechanism. Display calls include a series of nine popping or hooting noises, produced by the air-sacs (in contrast to two, higher-pitched, in Greater Sage Grouse), followed by three (as opposed to two) mechanically produced wing-swish sounds. Additionally, some of the visual displays of lekking birds differ between the two species.

Movements Largely sedentary, with minor movements into more suitable habitat in winter.

Range Endemic. Now restricted to the Gunnison Basin of SW Colorado, and in a very small area of adjacent eastern Utah.

Conservation status Classified by BirdLife International as Endangered. The population and range have declined catastrophically since the time of European settlement; historically it was found from SW Colorado, SE Utah, NE Arizona and NW New Mexico and possibly also SW Kansas and NW Oklahoma (Young *et al.* 2000, US Fish and Wildlife Service Bulletin 2011). Formerly it occurred in no fewer than 20 counties in Colorado; it is now restricted to eight counties in SW Colorado and one county in eastern Utah, with local extirpations continuing into the 1990s (Braun *et al.* 2014, who consider, not too optimistically, the possibilities of restoration of locally extirpated populations). All populations are isolated and fragmented, with a total area of less than 500 km^2. Current population estimate is 2,500–2,600 individuals, equating to about 1,700 mature birds. In a recent study in the western portion of the species' range, nest survival over a 38-day period (ie, laying and incubation) was about 50%, which is higher than previous estimates and generally higher than corresponding figures for Greater Sage Grouse (Stanley *et al.* 2015). Main causes of the decline are habitat loss and degradation due to agricultural practices, road and home developments, power lines and other industrial developments, and possibly increased deer populations. Loss of genetic diversity, exacerbated by shrinking gene pools in fragmented populations is a major concern, as is the appearance of the West Nile virus in the species' range. Actual hunting has now ceased and lek sites are now protected; conservation plans are being implemented, including among other things cooperative measures with local landowners. Recent captive-breeding efforts have been quite successful and would appear to offer a route to supplementing wild populations (Apa & Wiechman 2015).

Etymology *Minimus* Latin smallest, in view of its size difference from *C. urophasianus*.

UDZUNGWA FOREST PARTRIDGE
Xenoperdix udzungwensis

One of the most interesting and unusual recent discoveries in Africa; originally thought to be an aberrant species of forest francolin but actually most closely related to the Asian genus *Arborophila*, meriting the erection of a new genus (Madge & McGowan 2002).

Type information *Xenoperdix udzungwensis* Dinesen *et al.* 1994, Ndundula Mountains, Tanzania.

Discovery In early July 1991 a flock of four or five peculiar francolin-shaped birds which did not correspond to any known species was observed by a group of Danish ornithologists who were conducting field studies in the remote and pristine Udzungwa Mountains of southern Tanzania. In the following few days several more observations were made and careful notes of plumage and behaviour were taken. Although it was suspected that an unknown species was involved, this was not confirmed until a return to Dar es Salaam allowed consultation with relevant literature. The group returned to the site a few weeks later to collect specimen material, using mist-nets. This proved unsuccessful; fortunately it was possible to obtain three specimens which were collected with the help of local people who used snares (Dinesen *et al.* 1994).

The widely disseminated story of the species having been discovered as a set of unusual legs projecting from a native cooking-pot is entirely apocryphal (L. A. Hansen, pers. comm.).

Description 29 cm. Monotypic. Upperparts olive-brown, with prominent black-margined rufous bars; face and supercilium rufous-brown; underparts grey with large blackish oval spots on flanks; bill bright red; tarsi yellow. Females apparently similar but with darker crown and more extensive blotching on underparts. Juveniles have darker bills, black above with coral-red cutting-edges, and yellow-brown legs.

Habitat Exclusively in montane and submontane evergreen forest, 1,350–1,900 m, with *Podocarpus* sp., *Ficus* sp. and *Cyperus* sp. Absent from more open habitat and swamps (Hansen 2007).

Food and feeding Forages on the ground, flicking over dead leaves; food items include invertebrates (beetles, ants, flies, woodlice); also seeds (McGowan 1994 HBW 2).

Breeding Nest and eggs undescribed; chicks seen in late November and early December.

Voice Subdued high-pitched peeping notes *djuii*, and a whistled song have been reported (Stevenson & Fanshawe 2002).

Movements Probably sedentary.

Range Endemic. Restricted to Ndundulu and Nyumbanito Mountains, both in the Udzungwa Mountain Range.

Conservation status Listed by BirdLife International as Endangered since it has a very small range (ca 190 km^2), where it may be locally common, with densities of 15–25 birds per km^2. Total population estimated at 3,700 individuals though there appear to be significant fluctuations. Population probably stable. Threats include snaring for food by local people, which may be significant; the construction of access trails may also increase predation by lions, hyenas etc.

Etymology The newly erected generic name, *Xenoperdix*, from Greek, means strange or foreign partridge; the specific name refers to the Udzungwa Mountain Range.

▼Udzungwa Forest Partridge, adult (left) and immature (right). Ndundulu Forest, Udzungwa Mountains, Tanzania, 2006 (*Louis A. Hansen*).

RUBEHO FOREST PARTRIDGE
Xenoperdix obscuratus

An isolated population of *Xenoperdix*, separated by 100 km of unsuitable habitat from the main localities in the Udzungwa Mountains.

Type information *Xenoperdix udzungwensis obscuratus*, Fjeldså & Kiure 2003, Mafwemiro Forest, Rubeho Mountains, Tanzania.

Discovery In March 2000 one bird was seen, and then in December 2000 and January 2001 three specimens were obtained, of a *Xenoperdix* which was named as a new race of *X. udzungwensis* (Fjeldså & Kiure 2003). Based on genetic and morphological criteria, Bowie & Fjeldså elevated this population to full specific status in 2005. Evidence was presented suggesting that the two populations had had no significant gene flow for about 200,000 years.

Description 25–28cm. Monotypic. Differs from *udzungwensis* in its smaller size, speckled face, absence of a necklace of white spots, lack of ochraceous on the undertail-coverts, less marked barring on the secondaries, scaly appearance of the secondary wing-coverts and narrower rectrices.

Habitat So far found only in the Mafwemiro Forest in mostly mature forest with *Podocarpus* trees and open understorey, with a canopy height of 25–30 m.

Food and feeding Nothing recorded.

Breeding No data.

Voice Not recorded.

▲ Rubeho Forest Partridge. Mafwomero Forest, Rubeho Mountains, Tanzania, June 2007 (*Francesco Rovero/MUSE-Museo delle Scienze*).

Movements Probably sedentary.

Range Endemic. So far seen only on Chugu Hill, in the north of the forest reserve; seemingly absent from other areas.

Conservation status Not assessed, but total range is very small (the Mafwemiro Forest is 32 km^2 and the bird is absent from substantial parts of this). Total population may be in the range of a few hundred individuals (Doggart *et al.* 2006).

Etymology *Obscuratus*, Latin, 'obscured' or 'concealed'.

WHITE-HEADED STEAMER DUCK
Tachyeres leucocephalus

A large flightless duck with a limited range in southern Argentina.

Type information *Tachyeres leucocephalus* Humphrey & Thomson 1981, Puerto Melo, Chubut, Argentina.

Alternative names Chubut Steamer Duck. Spanish: Patovapor Cabeciblanco, Patovapor Cabeza Blanca.

Discovery Prior to 1979, flightless steamer ducks had been seen by numerous observers in the range of this species; however, these were usually ascribed to other species, either Magellanic Steamer Duck *Tachyeres pterneres* or Flying Steamer Duck *T. patachonicus*. In 1979, Todd, basing his opinions on those of the Argentinian ornithologist Maurice Rumboll, suggested that the Chubut population might refer either to Falklands Steamer Duck *T. brachypterus* or an unknown taxon. Humphrey and Thomson collected a long series of specimens in 1979 and described the species in 1981; further confirmation of the separateness of this taxon was provided by electrophoretic protein studies (Corbin *et al.* 1988).

Description 61–74 cm. Monotypic. A heavy sexually dimorphic flightless duck, only confusable with other species of the genus. It differs from the apparently allopatric Magellanic Steamer Duck (also flightless) in having a much whiter head and, in the female, a much more prominent post-ocular stripe; Flying Steamer Duck, which does overlap in small numbers outside the breeding season, has a greyer head in the male and much restricted facial markings in the female.

◀ White-headed Steamer Duck, male (top) and female (bottom). Punta Tombo, Chubut province, Argentina, October 2014 (*Kurt Hennige*).

Juvenile White-headed is similar to the female but facial markings are more diffuse. There are numerous osteological differences from others members of the genus. The downy young is also distinguishable, with a whitish upper eyelid and the crown darker than the cheeks, features not found in the other three species (Humphrey & Livezey 1985).

Habitat Coastal. Tends to occur on rocky rather than sandy shores, but usually in sheltered waters protected from full oceanic wave action.

Food and feeding Feeds mostly by diving, also by upending, typically remaining underwater for 30 seconds or so; based on stomach contents, molluscs including snails and bivalves predominate; also crustaceans, occasionally fish and rarely plant matter (Livezey 1989).

Breeding Breeding season October to February. Nests almost invariably on islands; in one study, 170 nests were located on islands of 0.5–55 ha in size, from 0.01–5.6 km offshore and from 1–159 m from the high water mark (Agüero *et al.* 2010, 2011). Nests were in bays, protected against waves, and usually located under small bushes, often well concealed, in a simple depression, copiously lined with down and often close to other nests. Eggs creamy-tan, clutch size 3–6. Incubation by female alone, but both sexes attend and protect the ducklings (Humphrey & Livezey 1985).

Voice Not recorded. Steamer ducks as a group are not particularly vocal.

Movements Probably sedentary.

Range Endemic. Coast of southern Argentina in Chubut province, from the mouth of the Chubut river (43° 21' S) to the borders of Chubut and Santa Cruz provinces (46° 00' S) (Agüero *et al.* 2010).

Conservation Status Currently classified as Near Threatened (IUCN 2010). Man-made threats include disturbance by seaweed collection and oil exploration; collection of eggs for food minimal (Agüero *et al.* 2010). Given the restricted range, and restricted habitat within that range, the biggest potential threat would come from a major oil spill, a not impossible circumstance.

Etymology *Leucocephalus*, from the Greek *leukocephalus*, white-headed.

ANTIPODEAN ALBATROSS
Diomedea antipodensis

One of the complex group of giant albatrosses of the Southern Ocean whose taxonomy is still the subject of conflicting opinions. For the purposes of this book we have followed the treatment of del Hoyo *et al.* (2013) and del Hoyo *et al.* (2014).

Type information *Diomedia exulans antipodensis* Robertson & Wareham 1992, Central Plateau, Antipodes Island, New Zealand.
Alternative names Gibson's Albatross (*gibsoni*).
Discovery The presence of giant albatrosses on the subantarctic islands south of New Zealand has been known since the early 19th century – indeed, the SE promontory of Antipodes Island is called Albatross Point; however, their taxonomy has remained obscure until recently. In 1992 Robertson & Wareham described two new subspecies of the Wandering Albatross *Diomedea exulans*: *D. e. antipodensis* from Antipodes Island and *D. e. gibsoni* from the Auckland Islands group. Based on ecological differences and morphology, Brooke (2004) elevated *antipodensis* to full specific status, treating *gibsoni* as its subspecies – a treatment followed by del Hoyo *et al.* (2013), although other opinions exist (e.g. Robertson & Nunn 1998). Currently, *gibsoni* is given full specific status by the SACC (Proposal 388, January 2009).
Description 110 cm. A giant albatross, males weighing up to 7.5 kg, with a wingspan of 3 m. Male: head white, crown solid brown; body mostly white with some variable brown vermiculations; upper surface of wings plain dark brown; underside white with dark brown tips to remiges, brown wing-tips and carpal joint; rectrices dark brown; iris dark brown; bill pink, tipped yellowish-horn above, bluish-grey below; tarsi and webs greyish to pink. Female has extensive chocolate-brown on chest and flanks. Race *gibsoni* whiter than nominate, with some white on inner section of upperwing. Somewhat smaller than Wandering Albatross *D. exulans*.
Habitat Open oceans; nests on remote islands. Forages over pelagic waters, in oceans of depth up to 6,000 m, but more concentrated in areas of 1,000 m depth.
Food and feeding Feeds mostly from surface or in shallow dives (up to 1 m). Food items comprise squid, fish and crustaceans. During the breeding season, trips are longer (7–13 days) during incubation period than in chick-rearing stage (average four days).

▼ Antipodean Albatross *D. a. gibsoni*. Off Kaikoura, New Zealand, November 2015 (*Ray Plowman*).

◀ Antipodean Albatross *D. a. antipodensis*. South of Guafo Island, Chile, March 2016 (*Fabrice Schmitt*).

◀ Antipodean Albatross, immature. Mayor Island, Bay of Plenty, New Zealand, March 2008 (*Kim Westerskov/Getty Images*).

Breeding First breeding at age of seven for Antipodes Islands birds and eight for the Auckland Island population. Successful breeders breed every second year. One egg, white, with some reddish spots at blunt end; incubation by both sexes in long shifts, totalling 75–85 days. Young fed by both sexes, leaving nest at about nine months. Egg-laying from late December to late January for Auckland Island birds (race *gibsoni*), early January–early February for Antipodes birds (nominate race).

Voice Apparently silent except on breeding grounds; guttural croaks.

Movements Ranges widely across the Southern Ocean, in SW Pacific, and southern Indian Ocean (Walker & Elliott 2006). Non-breeding juvenile males of the nominate race migrate eastwards to waters off western Chile; females, and juvenile males of *gibsoni* moved westwards to areas in SE Indian Ocean (Walker *et al.* 1995, Walker & Elliott 2006). Rarely enters iceberg belt north of Antarctica.

Range Pelagic, endemic breeder. The nominate race nests on Antipodes Island with small numbers on Campbell Island and Pitt Island in the Chatham Islands group; race *gibsoni* nests on Disappointment Island, Adams Island and Auckland Island in the Auckland group.

Conservation status Currently listed by BirdLife International as Vulnerable. All of the nesting areas are secure and of restricted access. The main threat is accidental by-catch in long-line fishing, especially from the fishing fleets from Taiwan and South Korea, with further mortality from the Chilean swordfish fishery. In 2006 58 birds caught during a single fishing trip. With a long-lived, slowly-reproducing species with a delayed onset of sexual maturity, losses of adult birds that would be inconsequential in a small passerine assume a major significance; in 2005 and 2006, male and female annual survival rates in the Auckland Islands population were 88% and 80% respectively (BLI SF 2014), an attrition rate that if maintained is uncomfortably high. Clearly, major efforts to reduce by-catch from fishing operations are essential. Counts of breeding pairs from 2006 to 2009 gave an average of 3,277 pairs in the Auckland Islands (with trivial numbers elsewhere); on Antipodes Island, 4,565 pairs between 2007 and 2009, giving an overall breeding population of about 16,000 individuals (ACAP 2009).

Etymology The specific epithet refers to Antipodes Island; the subspecific name for the Auckland population is named for J. D (Doug) Gibson, "who, over many years, helped unravel the colour morphs of the thousands of Wandering Albatrosses that he banded off New South Wales" (Robertson & Wareham 1992).

AMSTERDAM ALBATROSS
Diomedea amsterdamensis

An albatross demonstrating the most extreme example of neoteny (i.e. the retention into adulthood of immature plumage) found in the genus.

Type information *Diomedea amsterdamensis* Roux *et al.* 1983, Amsterdam Island, southern Indian Ocean.
Alternative names French: Albatros d'Amsterdam.
Discovery Although numerous observers, dating from the early part of the nineteenth century, had seen giant albatrosses at sea in the vicinity of Amsterdam Island (Paulian 1953; Bourne & David 1995) it was not until 1951 that P. Paulian found and photographed a population nesting on the Plateau des Tourbières, at an altitude of 600 m, on the island (Paulian 1960). Paulian himself was not able to determine whether the Amsterdam population was differentiated from that of Tristan da Cunha (presently treated by most authorities as a full species, *D. dabbena*). In a careful study published in 1983, Roux *et al.* classified the Amsterdam population as a full species, *D. amsterdamensis*, based on plumage characteristics and its apparent reproductive isolation from other groups. This treatment was not universally accepted, with many authorities treating the taxon as a race of Wandering Albatross *D. exulans*; however, recent biochemical studies (e.g. Rains *et al.* 2011; Milot *et al.* 2007) strongly support full species status of the Amsterdam Island population. The type specimen is a bird found dead at the breeding site in 1982 and currently housed in Paris; no further specimens can of course be taken.
Description 107–122 cm. Monotypic. A huge albatross (though slightly smaller than Wandering Albatross), males weighing as much as 8 kg with a wingspan of 3 m. Adult has entire upperparts, apart from white face-mask, dark chocolate-brown; underparts largely white, with a variable brown breast-band and brown undertail-coverts; underwings mainly white with dark tip and leading edge; bill pink with dark tip and dark cutting edges, which is probably the best distinction from immatures of *D. exulans*, which lack these.
Habitat Open oceans; nests at 417–640 m in peat bog with much moss.
Food and feeding Cephalopods, fish, crustaceans, taken from or just below the surface.

▼ Amsterdam Albatrosses displaying (male on the left and female on the right). Tourbières Plateau, Amsterdam Island, Indian Ocean, February 2013 (*Rémi Bigonneau*).

▲ Amsterdam Albatross, adult (probably female) in flight over D'Entrecasteaux Cliffs, Amsterdam Island, Indian Ocean, March 2013 (*Rémi Bigonneau*).

Breeding Breeds only every second year. Males arrive at the breeding site in late January to mid-February, females about 10 days later; pairs are stable from year to year. The single white egg is laid from mid-February to early March; incubated by both sexes, in shifts ranging from 5 to 7 days, for 46 to 83 days in total; fledging period about 235 days, young fed by both sexes. Young may return to colony as early as four years old, but at least one bird did not breed for the first time until the age of nine (Jouventin *et al.* 1989).

Voice No recorded data, but presumably silent at sea; calls on land presumably similar to those of Wandering Albatross.

Range and Movements Widely pelagic, endemic breeder. Breeds only on Amsterdam Island, southern Indian Ocean. Ranges widely from the breeding island, possibly as far as New Zealand with a confirmed sighting off Western Cape, South Africa. Satellite tracking shows a feeding range for adults from eastern South Africa to south of Western Australia.

Conservation status Critically Endangered. Current population is estimated at 170 individuals, of which 80 are mature adults, with most recently 26 breeding pairs (Rains *et al.* 2011) (at the time of discovery, there appeared to be in the region of only 5–6 pairs). Palaeontological findings suggest that at one point the breeding area on Amsterdam Island was considerably more extensive than today, the contraction probably due to the effects of a herd of feral cattle, which increased from the five introduced in 1871 to about 2,000 in 1988. Cattle were eradicated 2011. From 1984 onwards there has been a slow increase in the albatross breeding population. Major threats are losses of birds, both adults and immatures, due to by-catch from fishing operations (Inchausti & Weimerskirch 2001; Lewison & Crowder 2003; Mills & Ryan 2005; Rivalan *et al.* 2010; Thiebot *et al.* 2014a), the control of which is essential to the survival of the species. Although Black Rats *Rattus rattus* are present, there seems to be no evidence of their actually causing mortality of chicks, at least of healthy chicks (Thiebot *et al.* 2014b). A more recent concern is the appearance of avian diseases in the Amsterdam population of the Yellow-nosed Albatross *Thalassarche chlororhynchos*, with evidence that chicks of the Amsterdam Albatross may now be affected, causing higher chick mortality.

Etymology *Amsterdamensis,* Latinised form of Amsterdam. Although the island was discovered by the Portuguese in 1522, it was not actually named (by the Dutchman Anthony Van Diemen) until 1633.

VANUATU PETREL
Pterodroma occulta

A medium-sized, little-known, gadfly petrel, closely similar to several other species. Its full specific status is not currently accepted by BirdLife International (2013).

Type information *Pterodroma occulta* Imber & Tennyson 2001, based on specimens taken in 1927 at sea, 30 nautical miles east of Mera Lava, Banks Islands, Vanuatu.
Alternative names Falla's Petrel
Discovery Six specimens were taken by R. H. Beck when the Whitney South Seas Expedition was becalmed "30 miles E of Manelav Islands", on 28 January 1927. The location is spelled variously on specimen labels and in logs as Manelav, Melapav, Melapao, Meralav and Meralov, but refers to the island now called Mera Lava. The specimens were catalogued as Juan Fernandez Petrels *Pterodroma externa*, at that time regarded as conspecific with White-necked Petrel *P. cervicalis* (Peters 1931). In 1976, Falla identified them as *cervicalis*, albeit "uniformly smaller" (Falla 1976). In April 1983 a road-killed gadfly petrel was found in northern New South Wales, which was identified as *cervicalis*, again small (Boles *et al.* 1985). In 1992 and 1993 four of the original Beck specimens were examined critically, alongside the New South Wales specimen, and consistent differences from authentic *cervicalis* were noted, prompting the description of the Vanuatu and the New South Wales specimens as a new species, *P. occulta*.
Description 35 cm. Monotypic. A typical gadfly petrel, black-capped, white-naped, with a white forehead, face and underparts; underwing white, with dark leading edge and largely dark primaries, giving a generally dark tip; very difficult to separate from *cervicalis* in the field, the smaller size and smaller bill being extremely fine distinctions, further complicated by individual variation (Shirihai & Bretagnolle 2010). The greater extent of dark on the underwing tip may be of value, though some specimens of *cervicalis* also show more dark than others.
Habitat Pelagic. Little information.
Food and feeding Flying fish, captured in the air, and squid, taken from the surface. A solitary species, usually found on the edges of flocks of boobies (*Sula*) and noddies (Sternidae) (2013 HBW SV).
Breeding Nest only recently discovered, at an altitude of about 590 m on Mt Suretamatai, Vanua Lava. Nests are located in burrows, tunnelled under large rocks and boulders and about 70 cm long. Birds are present at the breeding site in late January to early March and one egg was found on 21 February; the breeding season is undoubtedly more protracted but no data are available (Totterman 2009). No information on clutch size or egg colour, but all related *Pterodroma* petrels lay one white egg.

▲ Vanuatu Petrel. Vanua Lava, Banks Island, Vanuatu, April 2014 (*Kirk Zufelt*).

Voice Apparently only calls at the nesting site. Two basic calls; a rapid decelerating *kok-kok-kok-kok-kok* and a drawn-out *toooooo-wit* (2013 HBW SV). There is a difference in male and female calls from the burrow, the males sounding "clear" and the females "hoarse" (Totterman 2012).
Movements Widely ranging, but data scanty due to difficulties in field separation from *cervicalis*.
Range Pelagic, endemic breeder. So far only known to breed on Vanua Lava, Banks Islands, Vanuatu. Reports of nesting on Mera Lava seem to be the result of confusion with Audubon's Shearwater *Puffinus lherminieri*. The type specimen was taken at sea 30 nautical miles (56 km) east of the island of Mera Lava in the same group. Other observations at sea have been made between Vanuatu and New Caledonia, off southern Vanuatu and off northern Vanuatu. Doubtless more widely spread than current sparse identifications suggest.
Conservations status Not currently assessed by BirdLife International, since not recognised by them as a full species. However, undoubtedly warrants some classification in the range of Vulnerable or Near Threatened. The population, while not accurately known, seems to be very small – possibly in the hundreds – and the known breeding area is very small, vulnerable to introduced predators and stochastic events. Formerly hunted for food by local people, though this seems to have ceased.
Etymology *Occulta*, Latin, 'hidden' or 'concealed', referring to the fact that the original specimens languished undetected for 60 years. Vanuatu, from the name of the country (formerly New Hebrides). Falla, honouring Sir Robert Alexander Falla (1901–1979), the highly distinguished New Zealand ornithologist.

BARAU'S PETREL
Pterodroma baraui

A typical gadfly petrel with a very restricted distribution.

Type information *Pterodroma baraui* Jouanin 1964, Saint-Denis, Réunion, Indian Ocean.

Alternative names French: Pétrel de Barau. Local name: Taille-vent.

Discovery This species appears to be quite well known to the inhabitants of Réunion from time immemorial, having a local name under which it may have been described as early as 1804 (Bory de St-Vincent, 1804). However, it was not formally described until 1964; originally ascribed to the genus *Bulweria*.

Description 38 cm. Monotypic. Similar to a number of other members of the genus; however, there is no other confusable species in the western Indian Ocean. Dark brown above, the back and secondary coverts paler; black cap, contrasting with pale cheeks and forehead; underparts and underwing white, with dark leading edge to outer wing and dark tips to secondaries, more extensive on primaries. Some individuals show a darker 'M' mark across the upperwing (Onley & Scofield 2007). Bill black, eyes dark brown, legs pinkish.

Habitat A pelagic and offshore species, coming to land only to breed. It favours deep, warm pelagic waters, typically with strong winds.

Food and feeding Main prey items are small fish and squid. In southern subtropical waters, usually solitary, feeding by seizing prey from the surface. Elsewhere it is frequently found in association with other species (especially Tropical Shearwater *Puffinus (lherminieri) bailloni*), feeding on schooling fish, when other fishing techniques such as dipping and surface-plunging are employed (Stahl & Bartle 1991). During the early part of the nesting season males forage further offshore than females, but later on this difference disappears (Pinet *et al.* 2012).

Breeding Nests in tunnels burrowed out of volcanic ash in sea cliffs, in areas with a sparse cover of vegetation such as *Philippia montana* and *Sophora denudata*; an undisturbed humus layer appears to be necessary. Burrows may be quite densely packed; in one 100 m^2 quadrant there were 62 active burrows (Probst *et al.* 2000). Barau's Petrel is one the highest nesting procellarids; nest sites in Réunion were located at 2,700–2,900 m, an altitude at which temperatures sometimes drop to 0°C. The lone nest on Rodrigues Island was at 320 m. Barau's Petrel is also unusual in its family by coming to the nest in daylight (Bretagnolle & Attié 1991). Burrows average 98 cm in length. One egg, white; incubation probably about 55 days by both sexes. Chicks

◄ Barau's Petrel. Gol Estuary, Réunion, December 2015 (*Ken Behrens*).

► Barau's Petrel. Gol Estuary, Réunion, December 2015 (*Ken Behrens*).

fledge between November and February, after a total breeding period of 100–120 days.

Voice Very vocal at the colony; a hoarse, throaty *oaou* followed by a rapid *kekekeke*, repeated two or three times; also *gor-wick* and variants (Jouanin & Gill 1967); silent at sea.

Movements Ranges widely across the Indian Ocean; one vagrant record off Victoria, Australia (Carter *et al.* 1989); one record from Enganno, Indonesia (Eaton *et al.* 2016b), three records from eastern South Africa (Richards Bay area), one from the Bay of Maputo, Mozambique (Sutherland 2005) and one Atlantic record, from Namibia (de Boer 2015). Geolocator studies show it to be highly mobile but also with much individual variation, with post-breeding birds (March–April) ranging widely from south of Madagascar, across the southern Indian Ocean to west of Australia, sometimes into Western Australian waters, in one case as far south as 45°S. From April to August most geolocator records are further north with the bulk of birds from May to July in the central Indian Ocean, from about 12°S to 25°S and 65°E to 95°E, again with some variation between the two study years (2008 and 2009) (Pinet *et al.* 2011). It has, however, also been seen in small numbers as far north as 11°N (van den Berg *et al.* 1991). Birds seem to favour areas with consistently strong winds and avoid areas with weaker easterly winds.

Range Pelagic, endemic breeder. Currently breeds only on Réunion with one single record from Rodrigues (Cheke & Hume 2008). Fossil evidence suggests that it may also have bred on Amsterdam Island before the habitat there was devastated by human activity (In 1799, hogs were introduced onto Amsterdam, "which… were very dextrous at plundering the eggs of seabirds and capturing the young ones, and even the old ones if they can get hold of them" (Bourne & David 1995), an account which might well apply to the native diurnal Barau's Petrel).

Conservation status Currently listed by BirdLife International as Endangered. Several serious threats exist. Previously taken apparently in large quantities as food, causing significant stress on the entire population, though this now appears to be minor or discontinued.

Major threats are predation by feral cats (Barau's Petrel appears to be the preferred prey; 58% of petrel remains were of adult birds (Faulquier *et al.* 2009) and by rats. Unlike cats, rats cannot take adults, but prey on nestlings and eggs. Trampling of nests by ungulates (or even humans) could have deleterious effects. A second major cause of concern is the disorientation and subsequent stranding of birds by artificial light (Le Corre *et al.* 2002). Between 1995 and 2004, 3,762 stranded petrels were retrieved. The local population has proven most helpful in retrieving stranded birds and releasing them with a 90% success rate; amelioration measures to reduce the impact of artificial light have been proposed (Salamolard *et al.* 2007).

In 2001, there were an estimated 4,000 to 6,500 breeding pairs, in 10 colonies; a more recent estimate (2007) is 3,000–4,000 pairs, equating to 9,000–12,000 individuals in total (BLI SF 2013)

Clearly vigorous and prompt conservation action, especially the removal of introduced predators, is essential; Pinet *et al.* (2001) suggest that without them the species is in danger of extinction.

Etymology The scientific and vernacular names commemorate Armand Barau (1921–1989), ornithologist and agricultural engineer. He was one of the authors of *Oiseaux de la Réunion*.

BRYAN'S SHEARWATER
Puffinus bryani

A black-and-white shearwater, possibly the smallest member of its genus, previously confused with other small *Puffinus* species.

Type information *Puffinus bryani* Pyle *et al.* 2011, from a specimen collected in 1963 on Sand Island, Midway Atoll, Pacific Ocean.

Discovery The type specimen was taken in 1963 from a burrow within a colony of Bonin Petrels *Pterodroma hypoleuca* on Sand Island, Midway Atoll. Due to its small size, it was identified as a Little Shearwater *Puffinus assimilis,* which would have been the first record of the species from the Hawaiian Islands and for the north Pacific. Later examination of the specimen by Pyle revealed blacker undertail-coverts, more typical of Audubon's Shearwater *Puffinus lherminieri,* than *assimilis.* This prompted a full mensural and molecular analysis that resulted in its description as a new species.

Description 25 cm. Monotypic. A very small shearwater, black above and white below, bill blackish with a grey base, tarsi largely bluish. Differs from Little Shearwater in having blackish undertail-coverts and a longer tail; from Audubon's Shearwater in having more white on the face and smaller size. Sexes presumably similar; juvenile undescribed.

Habitat Open oceans, nesting on small oceanic islands.

Food and feeding No data.

Breeding Eggs recorded on 25 February; no other details published (Platt 2015).

Voice Described as a high-pitched cry (Platt 2015).

Movements No data.

Range The original specimen, and further observations in the winter of 1991–1992 (Pyle & Pyle 2009) do not involve birds that were obviously nesting. Six specimens from the Bonin (Ogasawara) Islands, south of Japan, obtained since 1997 but before Pyle *et al.*'s description, were provisionally identified as Little Shearwaters, but subsequent examination and DNA analysis showed them to be *bryani* (Kawakami *et al.* 2012). Recently, a small number of breeding birds have been found on Higashijima Island, about 3 km east of Chichijima Island in the central Bonins (Platt 2015). Some ten birds were observed, though no comprehensive population study has been published. Higashijima, which is uninhabited, is about 400 m wide and 900 m long. Whether other breeding populations are found on other remote islets is not known. It also seems probable that small shearwaters observed from the ferry between the Japanese mainland and the Bonins are of the new species (Chikara 2011); to complicate matters further, Chikara also suggests that the birds illustrated and described in Onley & Scofield (2007) as Bannerman's Shearwater *P. bannermani,* whose breeding grounds are believed to be in the Bonins, in fact more resemble *bryani.*

Conservation status Not presently classified by BirdLife International but probably deserves a rating of Endangered or higher. The population is unknown but appears to be small. Predation of seabirds by Black Rats (*Rattus rattus*) is severe on the Bonins; of the six specimens from the islands, three were killed by rats. Rats had already been exterminated on Higashijima, where the three corpses were found, prior to the discovery of the breeding population. The survival of Bryan's Shearwater is dependent on maintaining its breeding islands free from introduced predators, notably rats and cats. Some twenty of the islands and islets in the Bonin group still harbour rats.

Etymology Named in honour of Edwin H. Bryan Jr (1898–1985), who participated in the Whiting South Seas Expeditions in 1920–1923 and was Curator of Collections at the Bernice Bishop Museum, Honolulu, for almost half a century.

◄ Bryan's Shearwater. Higashijima Island, Bonin Islands, Japan, February 2015 (*Kazuto Kawakami*).

PINCOYA STORM PETREL
Oceanites pincoyae

A puzzling storm petrel, observed, photographed (and recognised as "something different") in 2009.

Type information *Oceanites pincoyae* Harrison *et al.* 2013, at sea, south of Puerto Montt, Chile.

Alternative names Puerto Montt Storm Petrel. Spanish: Paíño Pincoya.

Discovery A group of Irish and American birders made a 12-day voyage from Valparaiso, Chile, to Buenos Aires, Argentina, in February 2009. A few days previously, on a pelagic trip out of Valparaiso, a number of unfamiliar storm petrels were encountered, alerting the participants to the necessity of careful examination of all storm petrels observed. Within a day of leaving Valparaiso, the new birds were encountered. Careful notes and photographs were made, leading to a publication suggesting that a mysterious unidentified species was involved (O'Keefe *et al.* 2009). Shortly thereafter, Harrison and others undertook five voyages south of Puerto Montt, during which some 3,000 observations were made, 2,000 photographs taken, 14 birds were captured, measured and released, and the type specimen taken. In fact, this species may have been observed as early as 1983, and two specimens were also collected in 1972 and 1983 at El Bolsón, Argentina, just 80 km east of the site of the subsequent Chilean observations. These were originally identified as Wilson's Storm Petrels *Oceanites oceanicus*, then on re-examination 20 years later as White-faced Storm Petrels *O. gracilis* before being finally recognised as an undescribed taxon.

Description 16 cm. Monotypic. A typical *Oceanites* storm petrel, with rounded wings, square-cut tail, conspicuous white rump and protruding feet with yellow webs. Distinguished from all other *Oceanites* species by bold white ulnar bars and distinctive white vanes to outermost rectrices, distinctive underwing pattern and white on lower belly and ventral region. Sexes similar. Juvenile similar to adult but dorsal plumage blacker and underparts greyer; often with white loral spot.

Habitat Oceanic; so far, only observed in sheltered waters of depths 100–200 m.

Food and feeding Feeding techniques unique among members of the genus, including "mouse-runs" with folded wings and tarsi semi-submerged; also dives repeatedly (2013 HBW SV).

Breeding Nest and eggs undescribed. Based on moult data, breeding season hypothesised to involve egg-laying in mid-November and fledging from middle to late February.

Voice Feeding groups make "an incessant sparrow-like chatter" (Fjeldså & Sharpe 2016).

Movement Appears to be fairly sedentary.

Conservation status Current population size estimated at ca 3,000 individuals, with up to 1,000 individuals seen at some good feeding sites. Discovery and protection of breeding site(s) are a priority; infestation of them by rats or cats would be very damaging. The use of polystyrene fishing floats, which on being damaged break up into little, ingestible particles, is a potential threat.

Etymology *Pincoyae*, genitive feminine for Pincoya, in Chilean mythology the spirit of the Chilotan Sea, a benevolent deity, good and helpful to fishermen, coming to the aid of distressed mariners (Harrison *et al.* 2013; Quintana 1987).

▼ Pincoya Storm Petrel. Reloncavi Sound, Los Lagos, Chile, February 2016 (*Rodrigo Tapia Jiménez*).

MONTEIRO'S STORM PETREL
Oceanodroma monteiroi

A cryptic species, extremely difficult to distinguish in the field from the sympatric Band-rumped Storm Petrel *O. castro*, but genetically isolated.

▲ Monteiro's Storm Petrel. Near the Azores, Portugal, September 2016 (*Marc Guyt/Agami*).

Type information *Oceanodroma monteiroi* Bolton *et al.* 2008, Praia islet, off Graciosa, Azores, Portugal.

Discovery In 1996 Luis Monteiro demonstrated that there were two populations of small *Oceanodroma* petrels nesting in the Azores, completely separated by asynchronous breeding; in fact, the two groups overlapped at the breeding sites only in August and early September (Monteiro & Furness 1998). Further work showed that the late-breeding birds ('hot-season') did not respond to 'cool-season' burrow calls although the converse was not demonstrated (Bolton 2007). The differences in the calls of the two groups are obvious to human ears. Later investigations showed that major differences in the two populations, in timing of moult, in foraging areas as shown by isotopic analysis of feathers, and in a number of subtle but significant morphological features were consistent and supported full specific status for the 'hot-season' birds. DNA analysis suggests that the divergence between the two populations in the Azores goes back between 70,000 and 150,000 years (Bolton *et al.* 2008).

Description 19 cm. Monotypic. A typical *Oceanodroma* storm petrel, with all-dark plumage relieved by a white rump and pale edgings on the wing-coverts. Differs from *O. castro* in the measurements of wing, tail and tarsus, dimensions of bill and depth of tail fork; only the last of these has any potential use in the field. Possibly more useful is the state of wing moult; in August *O. monteiroi* has obviously faded and worn plumage with irregular inner primaries caused by moult. Fortuitously, this is the time of year when both are present at the breeding islands.

Habitat Open ocean.

Food and feeding Few data; thought to be squid and small fish.

Breeding Nests in burrows excavated in soil, but readily accepts artificial sites. Egg-laying in late April or early May, totally asynchrononous from *O. castro*. Clutch one white egg, incubation period about one month, fledging period two months. Nesting success very low, about 0.16 young per pair (2013 HBW SV). Competition for nest sites apparently a significant cause.

Voice Only heard around nesting sites. Burrow call, a throaty purring, punctuated at regular intervals by short sharp *wicha* notes (2013 HBW SV).

Movements Few data. Isotopic feather analysis shows that *O. monteiroi* appears to maintain the same foraging environment summer and winter (in contrast to *O. castro*) and hence, unlike that species which regularly occurs in North American waters, is likely to be relatively sedentary.

Range Pelagic, endemic breeder. So far, breeding only proven from two islets, Praia and Baixo, off Graciosa in the western Azores.

Conservation status Classified as Vulnerable because of small total numbers (estimated in 1999 at 250–300 pairs) (Bolton *et al.* 2008), low reproduction success and breeding sites being confined to two small islets, each no more than 500 m in length. Storm petrels are extremely vulnerable to predation by rats; the two islets are currently rat-free, but one of them is frequently visited by local people from Graciosa, with possibly some risk of rodent introduction. Predation by locally resident Long-eared Owls *Asio otus* is a significant drain on the population.

Etymology Named after Luis Monteiro (1962–1999), whose pioneering work originally established the asynchrony of breeding of the two *Oceanodroma* populations. Dr Monteiro was killed in a plane crash on the island of San Jorge; his final publication, on Fea's Petrel *Pterodroma feae*, appeared posthumously.

HOODED GREBE
Podiceps gallardoi

One of the most spectacular species discovered in South America in the 20th century, the Hooded Grebe has declined alarmingly in the last 20 years, due, mainly, to a series of ill-advised human activities.

Type information *Podiceps gallardoi* Rumboll 1974, Laguna Las Escarchadas, east of Calafate, Santa Cruz, Argentina.

Alternative names Mitred Grebe. Spanish: Zampullín Tobiano, Macá Tobiano.

Discovery The type specimen was collected at a location about 50 km east of Calafate, Santa Cruz, in April 1974 by E. Shaw and described by M. Rumboll. It is somewhat remarkable that such a spectacular, conspicuous bird could remain unknown, in an area quite reasonably accessible, for so long. In fact, in April and May 1834, Charles Darwin must have come very close to some of the breeding habitat (Darwin, 1839), while in 1902 H. Hesket Prichard, while travelling in Patagonia, observed "many divers (sic) which I could not identify" (Hesket Prichard 1902).

Description 32 cm. Monotypic. Unmistakable. Forehead white, front of crown consisting of fine hair-like erectile feathers of an orange-rufous colour; remainder of head, ear-coverts and upper throat jet black; yellow wattle (unique among grebes) below eye; back of neck with thin black median stripe; upperparts dark slaty-grey; neck and remainder of underparts silvery-white with blackish spots and smudges on flanks; wings largely white, inner and lesser coverts and tips to outer primaries black; eye bright scarlet-red; bill bluish-grey; legs and feet blackish. Sexes similar; juveniles have black hood on top and sides of head, with a white throat.

Habitat Breeds on windswept upland lakes, often volcanic crater lakes, usually at 500–1,500 m. Winters in saltwater estuaries.

Food and feeding Feeds gregariously. Feeds mostly by dives of up to 30 seconds duration, usually near the shoreline. Food items (in breeding lakes) include small snails, copepods, midge larvae, bugs, beetles and larvae (Fjeldså 1986). Few data on diet in wintering areas; one bird found dead in the Río Gallegos estuary in May had stomach contents of mainly fish, especially Falkland sprat *Sparttus fuegensis,* and spider crabs *Halicarcinus planatus,* with lesser amounts of algae and molluscs (Imberti *et al.* 2011).

Breeding A colonial nester, without a territory except for the immediate vicinity of the nest; colonies vary (or did) in number up to 100 pairs. Birds nest relatively late in the season to take advantage of emergent vegetation on which to anchor their floating nests. Courtship is very elaborate, involving a variety of visual displays (Storer 1981). Two eggs are laid, from December to February, but only one chick is raised (Fjeldså & Krabbe 1990). Incubation 20–21 days. Chicks are fed initially with very small items, changing to snails after about two weeks (Imberti *et al.* 2011). Food supply seems to be irregular and chicks may starve on occasions (Fjeldså 1986). One hybrid with Silvery Grebe *P. occipitalis* reported (Storer 1982).

► Hooded Grebes. Strobel Plateau, Argentina, December 2008 (*Pete Morris*).

Voice Very vocal. Call is a characteristic whistled *terr-wheee-err* or *teete-wheee-err*; also mechanical ticking calls, used in display and aggressively, and a loud *ki-wee, ki-wee*, also during display (Fjeldså 2004; Straneck & Johnson 1984).

Range Endemic breeder. Breeds in interior lakes in Santa Cruz, Argentina (Lancelotti *et al.* 2009); several occurrences in Magallanes, Chile. The wintering grounds were not discovered until 1994, initially in the Coyle estuary on the Atlantic coast of Santa Cruz, subsequently in several other locations, notably the Río Gallegos and Río Santa Cruz estuaries (Imberti *et al.* 2004). However, wintering numbers do not account for all the breeding population, leading to speculation that some birds may winter on ice-free lakes (Imberti *et al.* 2011).

Conservation status Currently classified by BirdLife International as Critically Endangered. The species has undergone a catastrophic decline since its discovery, leading to a very real anxiety as to its very survival as a species (Roesler *et al.* 2011; Imberti & Casañas 2011; O'Donnell & Fjeldså 1997). Earlier population estimates were in the region of 3,000–5,000 individuals, leading to a classification of Low Risk (Konter 2008); however, a thorough study in 2009–2011 gave a population of only about 800–900 birds, with an estimated decline of 80% over 25 years (Roesler *et al.* 2012). There are several obvious causes and possibly some more subtle ones for this decline. Introduced North American Mink *Neovison vison* are devastating predators; one individual mink is known to have killed 15 adults and 7 juveniles within a week, while at a second location, one mink killed ten adults and five chicks (Anon. 2013). In its original state, the Hooded Grebe had apparently no predators (Fjeldså 1986), leaving it very vulnerable to introduced ones. A second serious threat is that of Kelp Gull *Larus dominicanus*, which is an effective predator on nests. Although Kelp Gulls are native to Argentina, until recently they were not found in the barren interior lakes favoured by the grebe; however, increased food supplies provided by expanded agriculture has apparently caused an increase and an expansion of range. A further cause of concern is the introduction of alien game fish such as trout and salmon, which not only change the ecology of the lakes but also predate small chicks. Further concerns involve reduced rainfall, which may cause some breeding lakes to dry up or become unsuitable.

The survival of this charismatic species will depend upon the extirpation of mink, the deterrence of Kelp Gulls and the prohibition of further game fish introductions. Fortunately, vigorous measures, involving 'colony guardians' on 24-hour watch, are being implemented (Anon. 2013a). These appear to be effective; in the 2013–2014 breeding season, there were zero instances of mink predation, and somewhere in the region of 85–105 chicks survived to fledging, a huge improvement on previous years, with evidence that the overall population is at least stabilising (Anon. 2015a). Most recent data for the 2014–2015 season are also very encouraging. Mink numbers have been further reduced; 771 adult birds produced 138 surviving juveniles from 357 nests in 12 colonies (A. Lambert, pers. comm.)

Etymology Named in honour of the Gallardo family, whose name has been associated with the Museo Argentino de Ciencias Naturales 'Bernardino Rivadavia' (Rumboll 1974).

GREATER YELLOW-HEADED VULTURE
Cathartes melambrotus

A widely distributed vulture which was, nonetheless, not described until 1964.

Type information *Cathartes melambrotus* Wetmore 1964, Kartabo, British Guiana (Guyana), based on a specimen taken in 1930.

Alternative names Forest Vulture. Spanish: Aura Selvática; Portuguese: Urubu-da-mata.

Discovery The taxonomy of the yellow-headed vultures remained extremely confused until Alexander Wetmore finally distinguished this species and the races of Lesser Yellow-headed Vulture *C. burrovianus*. This involved examination of numerous specimens, some going back 150 years.

Description 74–81 cm. Monotypic. A very large vulture, iridescent on the back, with the bare skin of the face bright yellow, becoming bright orange on neck and bluish on top of head. Somewhat larger than *burrovianius*, with darker glossy black plumage and contrasting dark inner primaries; easily distinguished in the field, even at great distances, by its characteristic flat-winged 'jizz', with almost no dihedral when soaring.

Habitat Lowland humid forests, rarely over forest-edge grassland, sea level to 700 m (Hilty & Brown 1986).

Food and feeding Carrion, located, even when not visible from above, by scent (Graves 1992). At food, subordinate to King Vulture *Sarcorhamphus papa* (Gómez *et al.* 1994). Needs other animals to open up large carcasses.

Breeding Nests in hollows in stumps or on surfaces like cave floors. Young fed by regurgitation. Eggs cream-coloured, blotched with brown on blunt end, 1–3, usually 2.
Voice Essentially silent, having no syrinx. Makes low hissing sounds.
Movements Nothing known; probably sedentary.
Range Occurs over much of the Amazon Basin, from S & E Venezuela, SE Colombia, the Guianas, N & W Brazil (east to Maranhão and south to Mato Grosso), N Bolivia, E Peru and E Ecuador.
Conservation status Least Concern. It has a very large distribution, over 6,000,000 km² and in some areas (away from habitation) is the most common vulture. Nevertheless, population may be declining. It seems to be dependent on large areas of lowland forest and to be sensitive to significant disturbances (Lees & Peres 2006), and also possibly to diminished food resources due to overhunting (Houston 1994 HBW 2).
Etymology *Melambrotus*, from two Greek roots meaning 'black, of the dead'.

▲ Greater Yellow-headed Vulture. Caxiuanã, Pará, Brazil, November 2005 (*Arthur Grosset*).

CRYPTIC FOREST FALCON
Micrastur mintoni

A widespread but cryptic species, only detected in 1997.

Type information *Micrastur mintoni* Whittaker 2003, Caxiuna, Pará, Brazil.
Alternative names Portuguese: Falcão Cryptico; Spanish: Halcon-montés Críptico.
Discovery In October 1997, while conducting field observations in Pará State, Brazil, Andrew Whittaker encountered a bird superficially resembling a Lined Forest Falcon *Micrastur gilvicollis* but with very distinctive vocalisations. Subsequent work showed that birds of this vocal-type differed from *gilvicollis* in a number of minor, but nevertheless consistent, morphological features, in plumage and in measurements, most notably in wing/tail ratios. Armed with these distinctions, a review of specimens in a variety of institutions revealed numerous examples, in some cases more than a century old, which had been misidentified as *gilvicollis*, showing that the new taxon has, or had, a widespread distribution across Amazonia.
Description 30–35 cm. In form, a typical *Micrastur* forest falcon, with a dark grey crown, grey back and wings; white underparts, the chest and upper flanks with fine blackish barring (largely absent in immatures); the tail dark grey with one obvious white band about half way down, usually one narrower band nearer the base, concealed by primary tips; bill black-tipped

▼ Cryptic Forest Falcon. Rio Azul Jungle Lodge, Novo Progresso, Pará, Brazil, July 2016 (*Bruno Rennó*).

with orange base, bare skin around eye orange, legs orange-yellow; eye white in adults, grey in immatures. It differs from *gilvicollis* in having a more extensive patch of bare orange skin around the eye, and different tail pattern: one broad white central tail band in adults (in *gilvicollis*, two narrower white bands), and in immatures, two broad white or buff bands (in *gilvicollis*, two or three narrower white or buff bands); and in having the dark bars on the underparts less extensive. Possibly two subspecies (see under Conservation status).

Habitat Lowland *terra firme* forests, with tall trees and dense understorey; also seasonally flooded *várzea* forest; sometimes in second-growth (Whittaker 2004).

Food and feeding Diet includes invertebrates and reptiles; follows army-ant swarms, opportunistically snapping-up fleeing prey. On one occasion attracted aggressively by a tape-recording of a Rufous-necked Puffbird *Malacoptila rufa* (Whittaker 2009).

Breeding Nest and eggs undescribed. Presumably nests in cavities as do other members of the genus (a female specimen collected on 4 June showed heavily abraded rectrices, consistent with cavity-nesting). Probably nests December–May (Whittaker 2003); food-begging recorded (in Mato Grosso) in September.

Voice Quite distinctive from that of *gilvicollis*. A territorial advertisement call, *uk-uk-uk-uk*; a series of short, fast cackles, *ca-ca-ca-ca*... and a quacking song, consisting of three loud, lamenting notes, *uuk, qui, qua-qua* (2013 HBW SV). Especially vocal just before dawn.

Movements No data; probably sedentary.

Range Widespread in Amazonia, in Brazil from Pará and W Maranhão south to Rondônia, and into lowland Eastern Bolivia (north Santa Cruz).

A disjunct population exists in SE Brazil (two specimens from southern Bahia, three from Espírito Santo, the most recent in 1972); two recent sightings in Espírito Santo, and possible tape recordings in 2011 (del Hoyo *et al.* 2013). An occurrence, documented by both photographs and recordings, in the Reserva Natural Vale, Espírito Santo, in July 2012 (Simon & Magnago 2013).

Conservation status Currently classified as Least Concern by BirdLife International, on the basis of large geographical range and being, at least in parts of its range, apparently fairly common if not always easy to detect (one transect in Santa Cruz, Bolivia, gave 3–4 pairs/km). May be a slow decline in population but not currently enough to trigger a change to Near Threatened. The population in Atlantic Brazil is quite another matter. The last specimen was taken in 1972, with the only documented recent record, of a single bird, in 2012 (Simon & Magnago 2013). This population is certainly very small, since most observers have failed to detect it in apparently suitable remaining habitat; it could well be Critically Endangered. Based on three of the existing museum specimens, this disjunct population might be subspecifically distinct (Whittaker 2003).

Etymology Named in honour of Clive D. T. Minton. Dr Minton, a metallurgist by profession, was responsible for initiating the Wash Wader Ringing Group in eastern England in the early 1960s. By, for the first time, capturing large numbers of waders by cannon-netting, he hugely expanded our knowledge of wader migration, demography and ecology. After emigration to Australia in 1978, he then did precisely the same thing with the migrant wading birds of that continent, founding the Australasian Wader Study Group and the Broome Bird Observatory. Recipient of the Order of Australia (Member) for "services to ornithology", the Bernard Tucker and Eisenmann medals, and a Fellowship of the Royal Australasian Ornithologists' Union.

SOCOTRA BUZZARD
Buteo socotraensis

An endemic raptor of Socotra, an island sometimes described as the "Galápagos of the Indian Ocean" with a very high level of endemism.

Type information *Buteo socotraensis* Porter & Kirwan 2010, based on a specimen taken in 1899, Hadibu Plain, Socotra, Yemen.

Discovery That large *Buteo*-type buzzards occur on Socotra has been well known for over 100 years. However, the taxonomic status of these birds has been a source of much confusion. For a detailed account of the complexities that led to its final description and its relationships with other *Buteo* taxa, see Porter & Kirwan (2010); see also Clouet & Wink (2000). Previous workers often assumed that the Socotran population referred to *Buteo buteo*, complicated by the fact that the 'Steppe Buzzard', *B. b. vulpinus*, occurs on the island as a vagrant.

Description 50 cm. Monotypic. A relatively small buzzard, upperparts and crown brownish, the upperwings brown with pale bases to the primaries, rectrices pale greyish, finely barred with narrow darker bars; chin, breast and upper belly white with burnt umber speckling, becoming heavier on lower belly; underwing pale with dark tip and trailing edge and with a solid brown carpal patch. Sexes similar. Essentially

▲ Socotra Buzzard. Socotra, Yemen, October 2008 (left) and February 2011 (right) (*Richard Porter*).

not confusable with any other local species, except vagrant *vulpinus*.

Habitat Open rocky country, often in areas of steep ravines, sea level to 1,370 m, mostly 150–800 m. Not dependent on trees.

Food and feeding Soars with wings in a slight dihedral; sometimes hovers. Most prey seems to be taken by launching from perches on outcrops and rocks. Prey items include reptiles and invertebrates.

Breeding Nests on cliff ledges or in rock crevices, in shaded sites, at altitudes of 150–650 m; nest is a bulky assemblage of twigs and sticks sometimes supported by vegetation. Nesting season October to April/May during the rainy season (Clouet *et al.* 1994). Eggs 1–3, usually 1–2 (Porter & Kirwan 2010).

Voice High-pitched mewing calls, *peeeooo* (Porter & Aspinall 2010), similar to those of other members of the genus, but with shorter gaps between calls.

Movements Apparently sedentary.

Range Endemic. Confined to Socotra Island.

Conservation status Classified by BirdLife International as Vulnerable, mainly because of its limited range. Population estimated at 250 pairs; no evidence as to whether this is increasing or decreasing. Some 75% of Socotra is protected, although this does not always prevent unwise development (BirdLife Data Zone, case study 27 2013). Other threats include the taking of young from the nest by local people under the (mistaken) belief that these are saleable for falconry purposes, and possible competition for nest sites by Peregrine Falcons *Falco peregrinus*, Egyptian Vultures *Neophron percnopterus* and Brown-necked Ravens *Corvus ruficollis*.

Etymology Both common and specific names refer to the island; Socotra is preferred to Socotran as a qualifying adjective.

PINSKER'S HAWK-EAGLE
Nisaetus pinskeri

Originally described as a putative subspecies of (Northern) Philippine Hawk-Eagle *N. philippensis;* genetic evidence suggests that the two forms are specifically distinct (Gamauf *et al.* 2005; Haring *et al.* 2007). All the Old World species of *Spizaetus*, about nine species in all, have been moved to the genus *Nisaetus* (Helbig *et al.* 2005).

Type information *Spizaetus philippensis pinskeri* Preleuthner & Gamauf 1998, Carcanmadlan area, Surigao del Sur, Mindanao, Philippines.

Alternative names Visayan Hawk-Eagle. Spanish: Águila-azor de Pinsker.

▼ Pinsker's Hawk-Eagle. Mindanao, Philippines, November 2006 (*Robert Hutchinson*).

Discovery As part of a study of hawk-eagles of the (then) genus *Spizaetus*, M. Preleuthner and A. Gamauf studied a series of specimens from eight different museums as well as five live birds in captivity, backed up by field observations. On this basis they described the form as a "possible new subspecies of the Philippine Hawk-Eagle". Subsequent genetic work suggested that, in fact, full specific status is appropriate, though this is not universally accepted.

Description 54–61 cm, females larger than males. Monotypic. In form a typical hawk-eagle, large, quite long-tailed, with feathered tarsi and prominent crest. Head brown or olive-buff with darker streaking; upperparts dark brown; tail dark brown with four or five narrow darker bars; throat white with dark central stripe; upper chest white and lower chest and belly rusty-brown, both with heavy lateral streaking; lower belly, vent and thighs whitish with dense brown or black barring; iris bright chrome-yellow; bill and cere blackish; tarsi yellow.

Habitat Mature forest, including some areas selectively logged, but not apparently tolerant of heavy disturbance; sea level to 1,900 m, mostly below 1,000 m.

Food and feeding No data on diet, but probably birds. Perches in forest or forest-edge, or soars above canopy.

Breeding Nest and eggs undescribed; other members of the genus build substantial nests high up in large trees.

Voice Call, a disyllabic *whee-whit.*

Movements Presumably sedentary.

Range Endemic. Central and south Philippines; Mindanao, Samar, Biliran, Negros, Basilan, Siquijor and Bohol.

Conservation status Currently classified by BirdLife International as Endangered, and apparently declining rapidly. Forest destruction in the Philippines is major and ongoing; between 1970 and 1990 40% of total forest cover was lost (Uitamo 1999); by another estimate, there was a 44% decrease in closed-canopy forest between 1987 and 2002 (Walpole 2010) (The human population of the Philippines has increased from 21 million in 1950 to 100 million in 2014.) Most common on Mindanao, where 320–340 pairs were estimated in the late 1990s (Preleuthner & Gamauf 1998); total population estimate (BLI SF 2014) is 600–800 mature individuals, with 900–1,200 individuals in total.

Etymology Named in honour of Prof. Dr Wilhelm Pinsker (b. 1945) of the University of Vienna, for "his eminent skill in the guidance of his students" (Preleuthner & Gamauf 1998).

TSINGY WOOD RAIL
Canirallus beankaensis

A rail of very restricted distribution; the taxonomy is currently under review, and it is not accepted as a full species by all authorities. The endemic Malagasy genus *Mentocrex*, which was erected to accommodate the more widely distributed Madagascar Wood Rail as *M. kioloides*, on the basis of its imperforate nostrils, is presently subsumed into the African genus *Canirallus* (Taylor & van Perlo 1998). *Canirallus* is generally regarded as a primitive genus, with patterned natal down in the nidifugous young (Taylor, HBW 3 1996). *Canirallus* may be closer to *Sarothrura* than to other genera.

Type information *Mentocrex beankaensis* Goodman *et al.* 2011, Forêt de Beanka, Maintirano, western Madagascar.

Alternative names French: Râle des Tsingy; Malagasy: Tsikozanalan'i Tsingy.

Discovery Madagascar Rail, *C. kioloides*, widespread if uncommon in eastern Madagascar, is presently treated as two races, the nominate covering the eastern half of Madagascar, including the High Plateau, and *C. k. berliozi*, restricted to Sambirano in the extreme north-west of the island, and occurring in lowland transitional humid-dry deciduous forest. A third population was discovered in July 1987 in limestone karst habitat in central western Madagascar (Taylor 1996; Langrand 1990). Differing opinions as to the taxonomy of this distinctively coloured population have been advanced; Taylor (1996) and Morris & Hawkins (1998) suggested a third subspecies of *kioloides*, while Fishpool & Evans (2001) suggested full specific status. Finally, Goodman *et al.* (2011), having collected a holotype and a paratype, described it as a full species *C. beankaensis*, based on plumage, measurements and molecular genetics.

Description 30 cm. Monotypic. A large rail, overall cinnamon-rufous below; crown dull smoke-grey, slightly mottled; back olivaceous-umber; lower flanks obscurely barred; rump, uppertail-coverts and uppertail ferruginous; iris reddish-chestnut; bill light bluish-grey with pale horn or ivory tip; tarsi blackish. Sexes probably similar. Differs from *kioloides* in larger size, darker overall coloration and in facial pattern. The paratype was a downy female; the iris was dark brown, the bill black with a pale tip.

Habitat Dry deciduous forest in limestone karst country, with rocky pinnacles, at 100–320 m (2013 HBW SV). Secretive in habits.

Food and feeding No data. *Kioloides* eats insects, amphibians and seeds (Taylor 1996).

Breeding No data. The nest of *kioloides* is a roughly-made bowl of grass, ferns and leaves, 2–3 m above the ground in bushes or conifers; eggs two, pinkish-white with rufous and grey speckling (Taylor 1998).

Voice Not recorded.

Movements Probably sedentary.

Range Endemic. Confined to the Bemaraha Massif from Beanka in the north to Bemaraha in the south in central-west Madagascar, in an area probably 125 km long and usually no more than 5 km wide.

Conservation status Classified by BirdLife International as Near Threatened, in view of its highly limited range and probable small population. Two areas of the range are protected, but activities such as firewood-gathering, setting of fires and hunting (in the north by bandits) occur (2013 HBW SV).

Etymology The name 'Tsingy' comes from the Malagasy word for a rock pinnacle. The specific name *beankaensis* is derived from the type locality, the Forêt de Beanka; 'beanka' in Malagasy means "the place of many owls" (in this instance, apparently, the Madagascar Long-eared Owl *Asio madagascariensis*).

▼ Tsingy Wood Rail. Tsingy de Bemaraha National Park, Madagascar, October 2014 (*Ken Behrens*).

CALAYAN RAIL
Gallirallus calayanensis

A very distinctive flightless rail from the Babuyan Islands, north of Luzon in the Philippines. Probably most closely related to Okinawa Rail *G. okinawae.*

Type information *Gallirallus calayanensis* Allen *et al.* 2004, Longog, Barangay Magsidel, Calayan Island, Philippines.

Alternative names Local name: Piding.

Discovery Calayan Island was until very recently remarkably unexplored ornithologically; apparently no investigations occurred for 100 years prior to 2004. In May 2004 an expedition was mounted to conduct a faunal inventory. On 11 May a group of rails was seen and heard on a path in a coconut grove, with plumage totally different from the only other rail on the island, the Buff-banded Rail *G. philippensis.* Over the following days, a totally unfamiliar rail with dark, unbarred plumage and bright red bill and legs, and clearly an undescribed species, was seen very well by several expedition members. One bird, a subadult female, was caught on 14 May and became the holotype, now deposited in the National Museum of the Philippines, Manila.

Description 30 cm. Monotypic. Essentially all dark, the upperparts dark olive, rump olive-brown; chin white; wings blackish-brown; rectrices very short and hair-like; iris orange-brown to deep brown, narrow orbital ring dull orange; bill heavy, scarlet with yellow-orange tip; tarsi orange-red. Juvenile similar to adult but may have whitish spot on rear ear-coverts (Allen *et al.* 2004, del Hoyo *et al.* 2013). Discussion with local people revealed that the species was quite well known to them, under the name 'Piding'. Probably flightless or nearly so.

Habitat Forest, both primary and secondary. Not restricted, as first thought, to forest on a coralline base (BLI SF 2013); also occurs in quite degraded habitat and in coconut plantations with thick fern undergrowth. Appears to have some preference for the vicinity of streams.

Food and feeding Terrestrial, feeds by pecking and overturning leaves. Stomach contents of the holotype included snails, and beetle and millipede fragments.

Breeding So far one nest only described, found on 2 June 2009. This nest was a loose construction of dried leaves and stems, forming a shallow cavity 12 cm in diameter, located on the ground near the buttress of a fig tree *Ficus congesta.* Eggs three, pale pink with blotches of reddish-brown and dark lilac. Fledged chicks have been seen as early as 6 April and juveniles encountered in May (Olivieros & Layusa 2011). Anecdotal information from local people indicates a clutch size of three to seven eggs.

Voice A series of hoarse, staccato calls *ngeck, ngeck, ngeck* repeated at the rate of about seven per second, given by both sexes and small groups. Also alarm calls, *ngreck, skeet* etc, sometimes extending into a trumpeting scream (Allen *et al.* 2004).

Movements No evidence of movements.

Range Entirely confined to suitable habitat on Calayan Island, Babuyan Islands, Cagayan Province, Philippines.

Conservation status Currently classified by BirdLife International as Vulnerable. Population was estimated in 2004 as 100–200 pairs; however, later studies suggest a range of about 36 km², where in some places it is locally common, with a population of 3,800 to 6,500 individuals, equating to about 2,500 to 4,300 mature birds. Population is suspected to be stable in the absence of evidence for any substantial declines or threats. Much of the range is located on very poor soils, uninviting to agriculture, although some forest clearance and unregulated logging does occur. Predation by domestic dogs may be a problem which could well increase if unwise road construction occurs. Any evidence of population decline would raise the conservation classification.

Etymology Both the scientific and the vernacular names refer to the only area of occurrence.

◄ Calayan Rail. Longog Wildlife Sanctuary, Calayan Island, Philippines, June 2016 (*Don Geoff Tabaranza*).

OKINAWA RAIL
Gallirallus okinawae

The description of a new, large, spectacular and nearly flightless species of rail from a well-populated prefecture of an advanced country caused a major sensation in ornithological circles at the time. Presently placed in the genus *Gallirallus*.

Type information *Rallus okinawae* Yamashiwa & Mano 1981, Mt Fuenchiji, Kunigami-gan, Okinawa.
Alternative names Japanese: Yambaru-Kuina.
Discovery The junior author of the type description, T. Mano, saw an unknown bird on a woodland path near Mt Yonaha, Kunigami-gan, in June 1978 and again in 1979 and 1980. A carcass of an adult female (which became the holotype) was obtained from the roadside at Mt Fuenchiji in early June 1981; an expedition was mounted later that month, resulting in the capture of a juvenile on 28 June and an adult on 4 July. These, which became paratypes, were banded and released back at the points of capture. In fact, reports of an unknown rail in the mountainous region of northern Okinawa date back to the early 1970s, and there were rumours of a forest bird, known to locals as *'Agachi Kumira'* before that (Brazil 1991).

Description 30 cm. Monotypic. A strikingly marked rail with short wings, almost no tail, large heavy red legs and bill, olive back, black face with a conspicuous white cheek mark; iris blood-red. Juvenile has paler, olive-tinged upperparts and head, mottled underparts, brownish bill and fleshy yellow-ochre legs (Taylor 1996 HBW 3). Not unlike Barred Rail (*Gallirallus torquatus*) of the Philippines, and obviously closely related, but in its limited range unmistakable.
Habitat Subtropical broadleaved evergreen forest, both primary and secondary, with patches of standing water; also in scrub and around cultivated areas with damp ground. Sea level to 498 m (highest mountain peak in area).
Food and feeding Feeds on the ground and in shallow water; prey items include insects (especially locusts) and amphibians, snails and lizards.
Breeding Breeds in May to July; apparently monogamous with long pair-bonds. Nest is on the ground, clutch size 2–3 eggs, pale pink with blotches of reddish-brown and dark lilac (Olivieros & Layusa 2011). Incubation period not recorded. Nidifugous young are covered in black down.

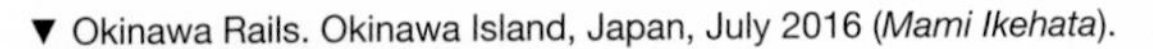
▼ Okinawa Rails. Okinawa Island, Japan, July 2016 (*Mami Ikehata*).

◄ Okinawa Rail. Okinawa Island, Japan, August 2016 (*Ted Shimba*).

Voice Extremely vocal; calls individually or in duets (Ikenaga & Gima 1993), mostly in early morning or evening. Calls lightly varied, including pig-like squeals, a deep bubbling of *gu-gu-gu-gu*, a high *kwi-kwi-kwi* etc.
Movements Largely sedentary, but in winter some individuals may wander, reaching areas south of the main distribution (Brazil 1991).
Range Endemic. Confined entirely to the northern quarter of Okinawa prefecture, known as Yambaru.
Conservation status Currently classified as Endangered. Originally thought to be very rare, but subsequent investigations showed it to be more widespread and catholic in habitat requirements than originally believed. In the early 1990s, the population was believed to be in the probable range of 1,000–2,000 birds. However, between 1996 and 2004 the population declined seriously to an estimated 720 birds, with a contraction of range to the north. Since 2006 the population appears to have stabilised (BLI SF 2013).

Main threats include predation by the mongoose *Herpestes javanicus* and weasels (*Mustela*), both unwisely introduced. Rail populations are significantly reduced in areas where mongooses are present. Between 2000 and 2004 the local Ministry of the Environment conducted an intensive trapping programme for mongooses, some 5,000 being caught. Correlation of rail numbers and mongoose trapping clearly indicates that it is mongoose predation which has caused the recent range contraction of the rail (feral cats are also predators). Further predator control is now needed urgently if the extinction of Okinawa Rail is to be averted (Ozaki *et al.* 2006).

Road casualties are also significant (in a 5-year period, 22 dead birds were found, mostly in May and June, which is the breeding season (Kotaka & Sawashi 2004). Prevention of these involves traffic-calming on the higher speed sections of the highways in the relevant habitat. Fragmentation of habitat may lead to genetic interbreeding; work on genetic diversity and phylogeny indicates that the species may have passed through a recent bottleneck (Ozaki *et al.* 2010). Habitat destruction for golf courses, pineapple plantations and dam construction has reduced the species' range since its discovery. Predation of eggs and young by the terrestrial native poisonous snake *Trimeresurus* occurs, which may account for the rail's habit of roosting in trees (Harato & Ozaki 1993); there is one observation of an adult bird being taken by this species (Hiragi & Ikeuchi 2015). However, the potential introduction of the arboreal and nocturnal Brown Tree Snake *Boija* would be disastrous. Other conservation measures include captive breeding. The species, along with the endemic Okinawa Woodpecker *Sapheopipo noguchii*, has been designated a National Monument (Taylor & van Perlo 1998). However, while prohibiting hunting and trapping, it does not necessarily bring with it more habitat preservation, a far more important issue.
Etymology Both scientific and vernacular names refer to the unique range of the species.

ROVIANA RAIL
Gallirallus rovianae

A flightless rail found on several islands in the Solomon Islands, apparently well known to local inhabitants but almost unknown to ornithologists.

Type information *Gallirallus rovianae* Diamond 1991, Munda, New Georgia, Solomon Islands.
Alternative names Kitikete (Roviana language).
Discovery Between 1972 and 1976 Jared Diamond was surveying local birds in the Solomon Islands. Local residents on the island of New Georgia described a bird, chicken-sized and flightless, and common enough to have acquired a name in the Roviana language of 'Kitikete'; on the island of Kolombangara the same bird was called 'Keremete' in the local language (both names are onomatopoeic). Another onomatopoeic name is 'Pio-piu' (Dutson 2011). The islanders were very familiar with the bird, and gave entirely plausible and credible accounts of its habits – flightless, a very fast runner only to be caught by dogs, diet, nest etc. In fact, in a dictionary of the Roviana language, published in 1949 but based on research dating back to 1928, the word 'kitikete' was defined as a 'dark, nimble bird' (Waterhouse 1949). Nevertheless, the species remained unknown to Western science until described in 1991 by Diamond, based on a specimen, initially mummified and subsequently turned into a study skin currently housed in the American Museum of Natural History, New York, and collected in June 1977 by Alisasa Bisili, a retired government officer (Diamond 1991). Apart from the observations of Gibbs (1996) on Kolombangara Island of what is almost certainly an undescribed race, virtually nothing is known about this species.
Description 30 cm. Based solely on the holotype, a medium-sized flightless rail, chestnut-brown above, nape rufous, grey post-ocular eye-stripe, chin and throat whitish, remainder of underparts grey with black bars extending to undertail-coverts; pinkish-tan wash on tips of the neck and breast feathers, forming a breast band; colour of bare parts not known. The birds observed on Kolombangara were quite distinct, lacking a breast-band and being generally darker (Gibbs 1996).
Habitat Forest, including second growth, abandoned gardens, coconut plantations and other degraded habitats (Taylor 1996).
Food and feeding Anecdotal reports from local people on Kolombangara Island describe the diet as omnivorous (worms, seeds, coconut shoots, potatoes, taro, and small crabs) (Diamond 1991).
Breeding Details from local informants. Nest is described as a depression on the ground lined with debris, built in June (dry season). Clutch 2–3 (Diamond 1991).
Voice Very vocal. Call a clattering *kik-kik-kitikek-kitikek*, a loud nasal *teku-ku* etc (Dutson 2011).
Movements Probably sedentary.
Range Endemic. Present, at least according to local reports, on New Georgia, Wana Wana, Kohinggo Rendova and Kolombangara Islands. Apparently absent from the nearby islands of Gatukai, Simbo, Ganonga, Villa Lavella and Gizo.
Conservation status Currently classified by BirdLife International as Near Threatened. Most common on Kolombangara Island, apparently less common or rare on the other islands of its range. Predation by domestic dogs is apparently a serious problem. Total population fewer than 10,000 individuals, with 1,500–7,000 mature birds (BLI SF 2013).
Etymology The English and scientific names are derived from Roviana, the name of the language spoken on New Georgia and neighbouring islands.

► Roviana Rail. Kolombangara, Solomon Islands, April 2013 (*Kirk Zufelt*).

TALAUD RAIL
Gymnocrex talaudensis

One of two new rallids from Karakelong, the Talaud Rail is essentially an unknown species to ornithologists, though possibly well known to local people.

Type information *Gymnocrex talaudensis* Lambert 1998a, Karakelong, Talaud Islands, northern Indonesia.

Alternative names Local: Tuu-a or Tu'a.

Discovery In August 1996, Frank R. Lambert had a clear if brief view of a distinctive, unfamiliar rallid walking casually across a tarmac road on the island of Karakelong in the Talaud Archipelago, Indonesia. The bird was not encountered again, but enquiry among local villagers confirmed that such a bird, with a "chocolate-red" head, long yellowish bill and white around the eye was indeed quite well known locally, under the name 'Tuu-a'. Reportedly it was occasionally caught and eaten by local people, using snares or dogs. In early September 1996 a bird-catcher from the village of Rainis was encountered in Beo, Karakelong, selling various birds caught near his village, among them a live example of the rallid seen in August. This individual, which was missing a tail and numerous primaries, died shortly after; with the agreement of local conservation officers, it was purchased and made into a museum specimen, becoming the holotype. In May 2000 Jim C. Wardill, who was conducting fieldwork for a conservation organisation, found a live specimen of Talaud Rail in a villager's home in Rae, near Beo. The bird had only recently been captured (by a snare) and was in good condition. It was purchased from the villager for a small sum, measured, photographed and described in detail, and then released back into the wild (Wardill 2001). His observations, along with those of Lambert and his colleagues, are to date the only ones made of this species by ornithologists.

Description 33–35 cm. Monotypic. A robust, possibly flightless or near-flightless rail with a chestnut head, neck and breast, bright yellow, dark-tipped bill, a diagnostic pale oval of bare skin around and behind the eye, which is scarlet-red with a cerise-pink eye-ring; back olive-green; rump and uppertail-coverts mid-brown; tail black, tarsi dull pink in front, dull yellow at rear (Wardill 2001) or yellowish with pinkish feet (Lambert 1998a).

Habitat Patches of long wet grass and rank vegetation at the edge of lowland forest; may prefer a mosaic of wetland habitats and riverine forest (del Hoyo *et al.* 2013).

Food and feeding The stomach of the holotype contained snail and beetle remains (Lambert 1998a). No other data.

Breeding No data.

Voice Local residents ascribe a deep *ump-ump-ump* call as coming from this species; a series of about 15 high-pitched *peet-peet-peet* calls was heard by Lambert on his original August 1996 observation and most probably came from the bird in question.

Movements Probably sedentary.

Range Endemic; apparently confined to Karakelong Island.

Conservation status Currently classified by BirdLife International as Endangered. Although local people seem to know the species quite well, describing it accurately, only one authenticated observation, that of Wardill in 2001, has been made since the original

◀ Talaud Rails, taken from a villager's house where they had been trapped for consumption. Beo, Karakelang Island, Talaud Islands, Indonesia, September 2008 (*Bram Demeulemeester*).

discovery. In some cases local knowledge may be compromised by confusion with one or other of the bush-hens (*Amauronis*); the species seems to be scarce since in four months fieldwork on the island it was neither encountered in the field nor found in trappers' catches (Wardill 2001). Threats include direct human predation for food, using snares and dogs and possibly predation by rats (*Rattus*).

Etymology Both vernacular and scientific names refer to the Talaud Archipelago, Indonesia.

TALAUD BUSH-HEN
Amaurornis magnirostris

A virtually unknown species, with almost no observations from ornithologists, although it appears to be quite well known to local trappers.

Type information *Amaurornis magnirostris* Lambert 1998b, from a specimen purchased in Beo, Karakelong Island, Talaud, northern Indonesia, from a native hunter.

Discovery In August 1996 a puzzling rallid was heard, and very briefly seen, in primary forest on Mount Manuk, Karakelong Island, Indonesia. It appeared to be a species of bush-hen (*Amaurornis*) but seemed to be larger and darker than Rufous-tailed or Pale-vented Bush-hen *A. (olivaceus) moluccanus*. On subsequent days the bird was encountered several times; it was always excessively secretive, and even though it responded to playback of its call, views were always brief. Nevertheless, it was obvious to the observer that an unknown taxon was involved. Finally, in early September 1996 a man was encountered in Beo, selling rallids that had been caught near his home village of Rainis. His stock consisted of three live and one dead Rufous-tailed Bush-hens, a *Gymnocrex* rail (see species account of Talaud Rail) and one dead specimen of the unknown bush-hen. This was, after discussion with local conservation officers, purchased and prepared as the holotype (Lambert 1998b). At the time of writing, the holotype appears to be a unique specimen.

Description 30 cm. Monotypic. A very robust bush-hen, with heavy, broad and arched greenish bill; iris bright red; legs dark olive-brown, yellow in front; plumage generally unmarked dusky brown. Resembles the sympatric *A (olivaceus) moluccanus* but darker below with no contrasting undertail-coverts and a longer, broader bill (Lambert 1998b). It should be noted that the taxonomy and nomenclature of the *Amaurornis* bush-hens is currently very fluid; Lambert suggests common names of Philippine Bush-Hen for *A. olivaceus* and Pale-vented Bush Hen for *A. moluccanus*, in replacement for Rufous-tailed Bush Hen or Plain Bush Hen.

Habitat Primary forest, second growth, plantations and remnant forest patches up to 290 m, including scrubby grasslands up to at least 3 km from the forest edge (Riley 2003). Also wet swampy habitat (BLI SF 2013).

▲ Talaud Bush-hen, adult, one of two individuals confiscated from Melonguane Market. Beo, Karakelang, Talaud Islands, Indonesia, November 2014 (*Robert Martin*).

Food and feeding No data.

Breeding No data.

Voice A monotonous series of loud frog-like notes, sounding like low-pitched barks (2013 HBW SV).

Movements Probably sedentary.

Range Endemic. So far only known from Karakelong (or Karakelang) Island in the Talaud Islands, northern Indonesia. Potentially occurs on nearby islands such as Salebabu and Kabaruang, but fieldwork needed (Lambert 1998b).

Conservation status Currently classified by BirdLife International as Vulnerable. Riley (2003) estimated the total population in the range of 2,350 to 9,560 individuals. Current threats include further forest destruction, including illegal logging, trapping for food by local people and predation by introduced Rice-field Rats (*Rattus argentiventer*). Some 350 km² of forest remains on Karakalong, with much of it receiving nominal protection as a wildlife reserve; however, actual enforcement is ineffective (BLI SF 2013).

Etymology The common name celebrates the Talaud Islands, home to several endemic species and races. The specific name *magnirostris* emphasises the large bill of the species.

BUKIDNON WOODCOCK
Scolopax bukidnonensis

The discovery of this endemic species was complicated, and delayed by confusion with other woodcock species, until its unequivocal description by Kennedy *et al.* in 2001.

Type information *Scolopax bukidnonensis* Kennedy *et al.* 2001, Mt Kitanglad, 1,530 m, Mindanao, Philippines.
Alternative names Philippine Woodcock.
Discovery In February 1993 two of the authors of the type description flushed a woodcock from a patch of relict forest at 1,600 m on Mt Kitanglad in north-central Mindanao. On subsequent days woodcocks were heard and seen performing a characteristic crepuscular territorial flight ('roding'), clearly establishing that they were breeding birds, not winter visitors from further north. Subsequently more detailed observations were made and the calls were recorded (Harrap & Fisher 1994). In January 1995 a displaying male bird was mist-netted at the same location (this became the holotype). This specimen was clearly different in plumage from other woodcock species and was described as a new taxon. Examination of specimens already in collections revealed several other individuals, previously misidentified as Eurasian Woodcock *S. rusticola*, indicating that the new taxon was geographically quite widely spread, occurring on the island of Luzon as well as Mindanao. In 1999, two further specimens, a male and a female, were taken on Mt Kitanglad. Whether *rusticola* does actually occur as a winter visitor in the Philippines is not totally clarified; some key specimens in the National Museum of the Philippines were destroyed by bombing during World War II.

Description 30 cm. Monotypic (but see below). Very similar to all other woodcock species; a dumpy, cryptically-plumaged shorebird, the back dusky-brown, shading to olive-brown on the rump, heavily barred, the underparts cinnamon-buff with dense dusky-brown bars. Differs from Eurasian Woodcock *S. rusticola* in having darker and narrowly patterned upperparts and more obvious lateral lines on the scapulars (Kennedy *et al.* 2001).
Habitat Mountain forest, 900–2,760 m, mostly ca 1260–1650 m on Mt Kitanglad. On Mindanao, in bracken (*Pteridium*) and scrub and forest remnants in areas with many clearings, but on Luzon mainly in low mossy cloud-forest. On the Babuyan Islands (see below), recorded in lowlands (2013 HBW SV).
Food and feeding Crepuscular or nocturnal. No data on food items.
Breeding One nest ascribed to this species was found at 2,600 m on Mt Kitanglad in early September 1996. The nest was a slight depression on the ground lined with a layer of dead and live grasses and ferns. The nest held two nestlings. Downy young most resemble those of Javan Woodcock *S. saturata* but with a white patch on the side of the neck. Gonadal data from specimens suggests a protracted breeding season, from January to September; eggs found in May (Kennedy *et al.* 2001).
Voice The calls during 'roding' are loud and distinctive, a hard metallic rattle '*ti-ti-ti-ti*', interspersed with a series of low grunts. Peak display apparently occurs December–February (Peterson *et al.* 2008).
Movements No evidence of movement.

◄ Bukidnon Woodcock. Mt Kitanglad, Mindanao, Philippines, March 2016 (*Bram Demeulemeester*).

Range Endemic. Now realised to be much more widespread than previously thought; has been found in four provinces of Luzon and at least four locations on Mindanao. In addition, has been recorded at lower elevations in the Babuyan Islands to the north of Luzon; Calayan (2013 HBW SV) and Babuyan Claro (Allen *et al.* 2006). Searches in suitable habitat on Negros have so far not detected the species there (Kennedy *et al.* 2001). The presence at lower altitudes on the Babuyan Islands suggests the possibility of the existence of a separate subspecies, though currently there is no morphological evidence for this.

Conservation status Currently classified by BirdLife International as Least Concern, due to its wide geographical distribution and the fact that in some areas it appears quite common. Nonetheless, habitat destruction on the Philippines remains a severe problem.

Etymology The English and specific names are derived from the word 'bukidnon' in the Visayan language, meaning 'of the mountain'.

CAMIGUIN HANGING PARROT
Loriculus camiguinensis

A small hanging parrot, unique in its genus in lacking sexual dimorphism. The designation as a full species in not universally accepted.

Type information *Loriculus camiguinensis* Tello *et al.* 2006, Kasangsangan, Camiguin Island, Philippines.

Alternative names Colasisi (includes all members of the *L. philippensis* group).

Discovery The designation of the population of hanging parrots on the island of Camiguin, north of Mindanao, came from a careful examination, published in 2006, of a series of specimens taken in the 1960s. However, Austin Rand, an expert in Philippine ornithology, made an undated notation in pencil, "subsp. nov" on the label of a specimen taken on Camiguin at that time, but never followed the matter up further. Careful

▼ Camiguin Hanging Parrot, captive adult (left) and captive juvenile (right). Camiguin Island, Philippines, January 2006 (*Thomas Arndt*).

morphometric analysis of a series of 23 specimens led to the conclusion that a new species was involved; this has not, to date, been backed up by genetic analysis.

Description 14 cm. Monotypic. A small green parrot, tinged orange on upperparts; forehead and forecrown scarlet; lores, chin, cheeks and throat turquoise-blue; rump and uppertail-coverts scarlet; thighs turquoise-blue; rectrices green with some light blue on tips; bill orange to orange-red; legs yellowish-horn to fleshy-orange. Sexes similar. Juveniles largely or totally lack the red on the crown. Distinguished by lack of sexual dimorphism (both sexes resemble the females of the races of *L. philippinensis* from adjacent islands, among other things lacking a bright red throat in the male), and by its larger size.

Habitat Canopy in upland forest (2013 HBW SV).

Food and feeding Feeds singly, in pairs or in small groups. Diet includes nectar, seeds (especially wild bananas), soft fruit etc (Arndt 2006a).

Breeding Nest and eggs undescribed. Local knowledge states that breeding takes place in September to November. One informant stated that he had repeatedly found nests in dead tree-ferns (Arndt 2006a).

Voice A high-pitched quickly repeated *tziit-tziit-tziit* (Arndt 2006a).

Movements Probably sedentary.

Range Endemic. Confined to uplands of Camiguin Island, Philippines; formerly down to 300 m.

Conservation status Not assessed by BirdLife International, but undoubtedly worthy of a classification of Threatened or Endangered. Does not appear to be able to adapt to conversion of forest to coconut plantation, which has largely replaced the original forest at all altitudes below 600 m. The remaining habitat may be as little as 20 km^2. A further and obviously serious threat is trapping for the cagebird trade. In one visit to the island, Arndt (2006a) personally found 35 individuals in captivity and estimated that no fewer than 100, and possibly more, were taken from the wild every year, probably an unsustainable number. Most birds caught are sold to Filipino tourists. Inbreeding of a reduced population may be a further threat; a blue morph recently observed may indicate a very narrow gene pool.

Etymology Both English and scientific names derived from the island of Camiguin.

SULPHUR-BREASTED PARAKEET
Aratinga maculata

A spectacular parakeet which is suffering, in conservation terms, from that very fact.

Type information *Psittacus maculatus* Statius Müller, 1776, Monte Alegre, Pará, Brazil.

Alternative names Portuguese: Cacaué.

Discovery The species has a very confused and confusing taxonomic history. It was described in 2005 under the new name of *Aratinga pintoi*; however, the specific name *maculatus*, long dismissed as invalid, applies to this form and consequently has priority. Thus the holotype of *A. pintoi* has now been designated as the neotype of *Psittacus maculatus* (2013 HBW SV; Nemésio & Rasmussen 2009). The separation of this taxon from the more widespread Sun Parakeet *A. solstitialis* (under the name *pintoi*) came as a result of a very thorough analysis by Silveira and his colleagues of almost 400 specimens in 13 different collections in five countries (Silveira *et al.* 2005a). The species was long dismissed as a juvenile of *solstitialis* or a hybrid between that species and the allopatric Jandaya Parakeet *A. jandaya*. Not accepted as a valid species by all authorities.

Description 30 cm. Monotypic. Upperparts bright yellow, greenish-yellow on crown, the orbital area and lower forehead orange; lesser coverts yellow with green centres, greater coverts and tertials green; primaries deep blue with green basal half of outer webs; rectrices bright green with blue tips; underparts bright yellow with orange tinge on belly; eye dark grey; bare orbital skin bluish-grey; bill black; tarsi dark brownish. Sexes similar; juveniles with green head, mantle and wing-coverts. Differs from *solstitialis* in having much more extensive green centres to lesser coverts and feathers of back, and crown yellow, rather than golden, and lesser extent of orange on the underparts.

Habitat Open areas with sandy soils and scattered small trees and bushes; will enter orchards; also forages in gallery forest (2013 HBW SV). In the Sipaliwini Savanna of southern Suriname, occurs in areas where relatively large tracts of forest abut open savanna (Mittermeier *et al.* 2010).

Food and feeding Occurs in small flocks and singly. Food items include fruit, seeds and flowers.

Breeding One nest with one egg found in a *Hymenaea* tree in September (2013 HBW SV). No other data.

Voice Most vocal when flying; a *kew* or *screek-schkeek-sckreek* and a short weak *krek*, resembling the calls of *Brotogeris* parakeets, given when perched (da Costa *et al.* 2011).

Movements May show local movements according to food supply.

▲ Sulphur-breasted Parakeet, captive individual, June 2013 (*Marc Chrétien*).

Range Occurs in scattered locations in northern South America. In Brazil, known from several locations in Pará state and, more recently, a disjunct population in Amapá; southern Suriname (Mittermeier *et al.* 2010); probably also in French Guiana, although specimen confirmation is lacking. Reports from other Brazilian locations may be in error, due to confusion with *solstitialis*, or may involve escaped cagebirds (da Costa *et al.* 2011).

Conservation status Not classified by BirdLife International, but clearly warrants some classification from Vulnerable upwards. Along with other spectacular golden-yellow parakeets, under continuous pressure from the cagebird trade, the control of which is clearly vital. In some locations (e.g. Sipaliwini) appears to be still quite common. Captive specimens may not be pure genetically. Da Costa *et al.* (2011) suggest that the species may actually or potentially benefit from forest clearing.

Etymology The original scientific name honours Dr Olivério Mário de Oliviera Pinto (1896–1981), "one of the most outstanding Brazilian ornithologists" (Silveira *et al.* 2005a), who was the first to point out the differences between this species and *solstitialis*.

MADEIRA PARAKEET
Pyrrhura snethlageae

A recently described species from the legendarily complex genus *Pyrrhura*, the source of highly divergent opinions among ornithologists. Regarded by many as a race of Santarem (Hellmayr's) Parakeet *P. amazonum*, which itself is split from Painted Parakeet *P. picta*. The SACC voted to treat *snethlageae* as a race of *amazonum* (Proposal 306), but again, not unanimously. Treated as a valid species by HBW SV (2013).

Type information *Pyrrhura snethlageae* Joseph & J. M. Bates 2002, 4 km upstream from Río Itenez, West Bank of Río Paucerna, province of Velasco, Santa Cruz, Bolivia.

Alternative names Madeira Conure; Cristalino Parakeet (race *lucida*). Spanish: Cotorra del Madeira; Portuguese: Tiribi-do-madeira.

Discovery Described as part of a far-reaching study of the genus (Joseph 2000; Joseph & Bates 2002, in Joseph 2010), based largely on detailed examination of specimens.

Description 22 cm. A typical *Pyrrhura* parakeet, medium-sized with long, graduated reddish tail. Crown

◄ Madeira Parakeet. Alta Floresta, Mato Grosso, Brazil, May 2012 (*João Quental*).

dark grey-brown; mantle and scapulars green, becoming green tipped red on back and rump; rectrices dull green at base; neck and breast grey-brown with obvious darker centres to feathers; narrow green band across chest; lower chest and belly yellowish-green with broad deep red central area; iris orange to red or darkish brown, with grey orbital ring; bill blackish with grey cere; tarsi blackish-grey. Race *lucida* is smaller, paler, with pronounced bluish on forehead and anterior superciliaries (Arndt 2008).

Habitat Humid lowland forest and forest-edge, both *várzea* and *terra firme* (Olmos *et al.* 2011).

Food and feeding No specific data; *P. picta* feeds in flocks on a wide variety of fruits, flowers and seeds.

Breeding Nest and eggs undescribed, but presumably similar to *P. picta* which nests in tree cavities and (in captivity) lays 4–5 white eggs.

Voice No data.

Range Endemic. Precise limits uncertain. Nominate race: Amazonian Bolivia (Santa Cruz) and west-central Brazil (Acre, Rondônia) in the drainage of the Madeira River; race *lucida*: Mato Grosso, Brazil, in region of Rio Teles Pires and Rio Cristalino, and Rio Peixoto de Azuvedo (Arndt 2008).

Conservation status Classified by BirdLife International as Vulnerable. Although the overall range is quite extensive, deforestation for cattle ranching and soya bean production is extensive and ongoing, aided by recent changes in Brazilian law giving an amnesty for previously cleared areas. Capture of birds for the pet trade is another possible threat. Some parts of the range protected by national or state parks.

Etymology Named after Maria Emilie Snethlage (1868–1929), a German ornithologist who collected extensively in Amazonian Brazil. At a time when female field ornithologists were a distinct rarity, she was a pioneer, the first woman to direct a museum in Brazil, naming several new species of birds, and having at least three named after her. Dr Snethlage herself recognised that the Rio Madeira birds were undescribed, but did not go any further (Snethlage 1914).

GARLEPP'S PARAKEET
Pyrrhura parvifrons

A further species in the complex *Pyrrhura* genus, whose taxonomic validity is very disputed.

Type information *Pyrrhura parvifrons* Arndt 2008, Yurimaguas, Shanusi, Loreto, Peru.
Alternative names Amazon Red-fronted Parakeet. Spanish: Cotorra de Garlepp.
Discovery The type description came from the comprehensive study by T. Arndt (2008) of the genus *Pyrrhura*. Although recognised in HBW SV (2013), in the IOC World Bird List it is treated as a race of Rose-fronted Parakeet *P. roseifrons* (as is *P. snethlageae*); not currently accepted by the SACC (Proposal 484).
Description 22 cm. Monotypic. Very similar to other *Pyrrhura* parakeets, medium-sized with a long graduated tail and largely green plumage. Head dark brown, with a variable red forehead; mantle and scapulars green; back and rump largely maroon-red; tail maroon-red above with green base, dull red below; upper chest and bib brownish to dirty white with darker central streaks; narrow green band on lower chest; underparts yellowish-green, centre of belly with large maroon-red or deep red patch; iris dark brown with pale orbital ring; bill blackish with whitish cere; tarsi blackish-grey.
Habitat Lowland humid forest.
Food and feeding No data; diet presumably similar to other members of the genus, i.e. fruit, seeds and flowers.
Breeding Nest and eggs undescribed, but presumably nests in cavities.
Voice No data.
Movements No data.
Range Endemic to Peru. Two disjunct populations without, apparently, any connection between them (Arndt 2008, *contra* Joseph 2002). SW Loreto, NE San Martín and further north-east in Loreto, near the Río Orosa.
Conservation status Currently classified by BirdLife International as Least Concern. Although the population is suspected to be in decline, it is not believed to approach the threshold for Near Threatened.
Etymology Named for Gustav Garlepp (1862–1907), a German zoologist and collector who, with his brother Otto (1864–1959) worked extensively in Brazil, Peru and Bolivia. In 1901 he settled in Paraguay, where he was murdered in 1907. G. Garlepp collected what is now the type specimen of Garlepp's Parakeet in Peru in 1885; this is now housed in the Senckenberg Museum, Frankfurt am Main, Germany.

► Garlepp's Parakeets. Tarapoto–Yurimaguas road, Peru, July 2012 (*Thomas Arndt*).

WAVY-BREASTED PARAKEET
Pyrrhura peruviana

A third controversial taxon of *Pyrrhura* parakeet, recently described from Amazonia.

Type information *Pyrrhura peruviana* Hocking *et al.* 2002, Río Santiago, Puerto Galilea, Amazonas, Peru.

Alternative names Wavy-fronted Conure. Spanish: Cotorra peruviana.

Discovery Described by Hocking, Blake and Joseph, in Joseph's extensive 2002 study of *Pyrrhura* taxonomy. The type specimen was taken by P. Hocking in Dpto. Amazonas, Peru, in December 1965; some 15 other specimens were studied. In the IOC World Bird List treated as a race of Rosy-fronted Parakeet *P. roseifrons*, following the opinions of Ribas *et al.* (2006), based on mitochondrial DNA evidence. Current treatment by the SACC also classifies *peruviana* as a race of *roseifrons*; however, in HBW SV (2013) the treatment of Joseph (2002) is followed, according *peruviana* full species status.

Description 22 cm. Very similar to other *Pyrrhura* parakeets of western Amazonia; medium-sized, long-tailed, the head brownish with deep maroon band across forehead, with a greenish-blue forecrown; ear-coverts off-whitish; throat dark brown, the feathers with broad dull white or creamy-white to light yellow fringes; narrow green breast-band; centre of belly reddish-brown, sides of abdomen to vent green; back green; rump and rectrices reddish-brown; primaries, primary-coverts and alula bluish-green; iris orange; bill black; tarsi black. Race *dilutissima* (Arndt 2008) similar, but bluish on forecrown more restricted, face rusty-red, small whitish-cream or yellowish crescent in front of eye.

Habitat Humid lowland forest; type specimen taken at ca 250 m; up to 875 m in Ecuador (Loaiza *et al.* 2005).

Food and feeding No data. Frequents salt-licks (Loaiza *et al.* 2005).

Breeding Nest and eggs undescribed.

Voice No data.

Movements No data.

Range Two disjunct populations; nominate race: northern Peru (Amazonas and Loreto) (Schulenberg *et al.* 2007) and southern Ecuador (Zamora-Chinchipe and Morona-Santiago) (Loaiza *et al.* 2005); race *dilutissima:* central Peru (Junín and Cuzco) (Arndt 2008).

Conservation status Currently classified by BirdLife International as Least Concern, although the population appears to be decreasing (BLI SF 2014).

Etymology *Peruviana*, from Peru.

◄ *Pyrrhura* parakeets, probably Wavy-breasted Parakeet *P. peruviana*. Puerto Ocopa–Atalaya road, Dpto. Junín, Peru, June 2012 (*Daniel Lane*).

AZUERO PARAKEET
Pyrrhura eisenmanni

One of two Central American representatives of the essentially South American genus *Pyrrhura*.

Type information *Pyrrhura picta eisenmanni* Delgado 1985, Los Piraguales, El Cortezo de Tonosí, Los Santos province, Panama.
Alternative names Azuero Conure, Eisenmann's Parakeet. Spanish: Cotorra azuense.
Discovery The presence of a *Pyrrhura* parakeet in central Panama went undetected for a long time; neither Wetmore (1965 *et seq.*) nor Ridgely (1976) were aware of its existence. In 1985 F. Delgado described an isolated new race of Painted Parakeet *P. picta eisenmanni*, confined to a small area of the Azuero Peninsula, which projects southwards from central Panama into the Pacific Ocean. The Painted Parakeet complex has an essentially Amazonian distribution, with several curiously isolated populations in Colombia; the closest of these to the Panamanian population, *P. (p.) subandina*, is some 500 km distant. More recent authorities (e.g. Forshaw 2010; del Hoyo *et al.* 2016 HBW Alive) treat *eisenmanni* as a full species, along with several other former subspecies.
Description 22 cm. Monotypic. In form a typical *Pyrrhura* parakeet, medium-sized, with a long, dull red tail. Forehead, crown and nape dull sooty, indistinct blue on forecrown; narrow red frontal band, extending to lores and ocular region, with dusky red cheeks; ear-coverts buffy-white; chin white; throat and chest dull blackish, broadly edged white; breast dark greenish-blue, edged white in centre and yellow on flanks; centre of belly with dull red patch; lower back to uppertail-coverts green; shoulder very narrowly fringed red; iris pale ochre, the orbital skin dull sooty; bill dull black; tarsi greyish-black.
Habitat Humid forest in hills, sometimes at forest edge and in partially cleared areas; sea level to 1,660 m (Forshaw 2006; Juniper & Parr 1998).
Food and feeding Feeds in mid- and upper levels of forest, usually in small flocks. Few data on diet; visits bean fields.
Breeding Nest and eggs undescribed, but doubtless in holes in trees; birds in breeding condition January–February, egg-laying reported in March.
Voice Flight-note a short *eek*, a harsh, guttural *kleek-kleek* when perched; contact call *peea*.
Movements Apparently some local, probably altitudinal, movements, since local information indicates that it has a variable presence in some locations.
Range Endemic. Restricted to a fairly small area of the southern Azuero Peninsular, in Veraguas and Los Santos Provinces, Panama.
Conservation status Classified by BirdLife International as Endangered; the overall range is small (about 700 km^2), and even in 1985 had been substantially diminished by clearing for agriculture; some of the range is protected by the Cerro Hoya National Park. Estimated population in the region of 2,000 individuals (World Parrot Trust 2014).
Etymology The scientific name honours Eugene Eisenmann (1906–1981). Mr Eisenmann, a lawyer by training, was born in Panama. In 1956 he abandoned the legal profession and became a full-time ornithologist, making major contributions to the study of birds in Central and South America (Bull & Amadon 1983).

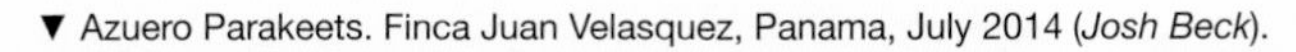
▼ Azuero Parakeets. Finca Juan Velasquez, Panama, July 2014 (*Josh Beck*).

EL ORO PARAKEET
Pyrrhura orcesi

A highly restricted parakeet of humid tropical forest on the western slope of the Ecuadorian Andes.

Type information *Pyrrhura orcesi* Ridgely & Robbins 1988, ca 9.5 km west of Piñas, El Oro province, Ecuador.
Alternative names El Oro Conure. Spanish: Cotorra de El Oro.
Discovery In August 1980 R. S. Ridgely, P. Greenfield and R. A. Rowlett were investigating remnant patches of cloud forest west of Piñas, El Oro province, Ecuador when a flock of nine unfamiliar parakeets was observed for 15 minutes. Although obviously members of the widespread genus *Pyrrhura*, the birds had unique plumage features not shared with any other members of the genus. This could not be further investigated until an expedition was mounted in June 1985, when a series of specimens was collected; a subsequent field investigation in 1986 found the new taxon a further 100 km north of the type locality. One specimen, taken in September 1939, lay unrecognised in the British Museum for 50 years.
Description 22 cm. Monotypic. In form a typical *Pyrrhura*, relatively small, with a long pointed dark maroon tail, overall bright green with bright red forehead; a reddish patch on the lower belly; greater primary-coverts bright red, conspicuous in flight; primaries with some blue, most conspicuous on closed wing; orbital ring pale pinkish; bill dusky. Female with reduced red on forehead, immature with less red on head and wing, red on belly absent (Ridgely & Greenfield 2001; Collar 1997 HBW 4). Distinguished from all other *Pyrrhura* parakeets (all of which are allopatric) by the virtual absence of conspicuous scaling on the chest and neck; much smaller than sympatric *Aratinga* parakeets.

▼ El Oro Parakeets. Buenaventura, El Oro, Ecuador, March 2016 (*Dušan Brinkhuizen*).

Habitat Very humid epiphyte-rich upper tropical forest, mostly 600 to 1,200 m (although the 1939 specimen was taken at 300 m; however, much of the potentially viable habitat at that altitude has now been destroyed). Occasionally up to 1,550 m (Ridgely & Greenfield 2001); also in gardens and orchards with tiny patches of wet forest (Best *et al.* 1993).
Food and feeding Feeds gregariously; main items fruits of fig (*Ficus*), *Heliocarpus*, euphorbias (*Hyeronima*), *Cecropia* and berries (Collar 1997 HBW 4).
Breeding Nests in cavities in trees, especially *Dacryodes peruviana*. Takes very readily to nest-boxes (BLI SF 2013). Nesting period probably variable; young being fed by adults in August (Ridgely & Robbins 1988), copulation observed in January (López-Lánus & Lowen 1999).
Voice Flight calls a rather trilling but quite harsh and metallic *tchreet-tchreeet*; also a quieter chirruping when perched (Juniper & Parr 1998).
Movements May show local or seasonal short-distance movements, since abundance of birds at any particular location is rather variable.
Range Endemic. Restricted to a very limited area in El Oro, Azuay, Cañar and Loja, the total range being a strip, only 100 km long and 5–10 km wide. Probably wider-ranging formerly, before extensive habitat destruction.
Conservation status Currently listed by BirdLife International as Endangered. The very limited and probably fragmented range, and the extensive and ongoing destruction of forest by logging, conversion to cattle pasture and mining are serious threats. The preferred nesting tree *Dacryodes* is frequently harvested. One reserve (Buenaventura, created by the Fundación Jocotoco with help from the World Land Trust), protects about 2,200 ha; the population here is about 60 birds year-round and ca 120 seasonally. More recently a further 280 ha has been added (Anon. 2015b). A successful nest-box programme has been initiated; local education in nearby schools emphasises the benefits of conservation (Waugh 2004).
Etymology The specific epithet honours Dr Gustavo Orcés (1902–1999) "in recognition of his many contributions to Ecuadorian ornithology and his continuing encouragement of younger generations of field biologists" (Ridgely & Robbins 1988).

AMAZONIAN PARROTLET
Nanopsittaca dachilleae

One of two members of the genus *Nanopsittaca*, both with very restricted geographic distributions; initially thought to be a member of the widespread genus *Forpus*.

Type information *Nanopsittaca dachilleae* O'Neill *et al.* 1991, 65 km ENE of Pucallapa, Dpto. Ucayala, Peru.
Alternative names Manu Parrotlet. Spanish: Cotorrita Amazónica; Portuguese: Periquito-da-amazónia.
Discovery In 1985 C.A. Munn saw a group of small green parrotlets in Manu National Park in eastern Peru. Their small size and shape recalled parrotlets of the genus *Forpus*, but unlike typical examples of that genus, there was no obvious sexual dimorphism; they could not be assigned to any known species. Similar birds were seen subsequently on numerous occasions. In July 1987 J. P. O'Neill and colleagues collected a number of specimens while conducting an ornithological survey of the hills on the Peru–Brazil border from a location on the banks of the Río Shesha. This confirmed that a new species was involved, and that it shared the genus *Nanopsittaca*, previously monotypic, with Tepui Parrotlet *N. panychlora* which occurs on several isolated mountain peaks in SE Venezuela, S Guyana and N Brazil. The holotype was obtained on 29 July at an altitude of 300 m.
Description 14 cm. Monotypic. A small, uniformly greenish parrotlet with a short, square-ended tail and rather pointed wings; the lores, forehead and anterior crown are pale powdery blue; underparts more yellowish-green; iris greyish-brown; bill and tarsi pinkish (O'Neill *et al.* 1991).
Habitat Lowland riverine forest, not found in closed canopy forest nor in areas disturbed by human activity, up to 300 m (Collar 1997 HBW 4).
Food and feeding Gregarious, usually in flocks of up to a dozen or more birds. Food items include the seeds of *Guadua* bamboo (Lebbin 2006), *Cecropia* catkins, *Vernonia* seeds and fruits of the arboreal epiphytic cactus *Rhipsalis* (Juniper & Parr 1998). Will congregate with other psittacids at clay-licks to obtain minerals.
Breeding Largely unknown; birds were seen inspecting a hole in a clump of epiphytes near the top of a 25 m tree in July and September (O'Neill *et al.* 1991).
Voice High-pitched piping calls, recalling poultry chicks; chirping, squeaking and chattering sounds when in a flock (Collar 1997 HBW 4).
Movements May possibly be a short-distance nomad, taking advantage of local fruiting of *Guadua* bamboo.
Range Confined to eastern Peru (Ucayali and Madre de Dios), NW Bolivia (La Paz) (Parker *et al.* 1991) and western Brazil (Acre) (Whitney & Oren 2001).

▲ Amazonian Parrotlet. Tambopata Research Center, Madre de Dios, Peru, July 2013 (*Andre Moncrieff*).

Conservation status Currently classified by BirdLife International as Near Threatened. The limited total range, which while largely protected by its remoteness, is still vulnerable to pressures from selective logging, pipeline and road construction, and human colonisation (Lloyd 2004). *Guadua* bamboo is in demand for use in dwelling construction. A considerable portion of the species' range is protected by Manu National Park. Current population estimates, admittedly with considerable uncertainty, are in the region of 10,000 birds, with 6,700 mature individuals (BLI SF 2013).
Etymology Named in honour of Barbara D'Achille, murdered by terrorists of the Sendero Luminoso movement on 31 May 1989 in Huando district, Peru. Mrs D'Achille was originally Latvian, but spent her adult life in western Europe and Latin America. After raising a family, she devoted the last few years of her life to conservation work, becoming one of the most respected environmental journalists on the continent, winning, among other awards, the Koepcke Prize for Environmental Journalism (O'Neill *et al.* 1991). Her party was ambushed by a group of Senderistas on a remote road in Huancavelica Department; when she refused to conduct an interview, doubtless laudatory, with the terrorists' leader, she was then killed by them (Anon. 2000). In addition to the parrotlet, a nature reserve in Peru has also been named in her honour.

BALD PARROT
Pionopsitta aurantiocephala

A large and conspicuously marked parrot, previously undetected due to confusion with Vulturine Parrot *P. vulturina*. Has been variously put in the genera *Pyrilia* and *Gypopsitta*.

Type information *Pionopsitta aurantiocephala* Gaban-Lima *et al.* 2002, left bank of Cururuaçu River, Pará, Brazil.

Alternative names Orange-headed Parrot. Portuguese: Papagaio-de-cabeça-laranja.

Discovery Large square-tailed parrots with featherless, orange heads were described by Helmut Sick as immatures of Vulturine Parrot *P. vulturina*, but juveniles of *vulturina* have heads fully covered with greenish feathers. Sick incorrectly described immatures as having orange or pumpkin-coloured bare heads, noting that they kept in separate flocks from the black-headed adults (Sick 1993). In fact, in *vulturina*, there is no orange-headed intermediate plumage between feathered juveniles and black-headed bald adults. This anomaly was resolved by Gaban-Lima *et al.* who collected the holotype and paratypes in 1999 – orange-headed birds, obviously adult, with completely ossified skulls and sexually mature gonads. Further comparison with authentic specimens of *vulturina* revealed that, quite apart from the differences in colour, the bare skin on orange-headed birds was greater in extent, including the nape; the new taxon also lacked the pronounced black and yellow collar of *vulturina*. In fact, examination of collections in several museums revealed a number of specimens attributable to the new species; in dried specimens, the head fades from orange to a paler colour, whereas in aged specimens of *vulturina* the black bare skin remains black.

Description 23 cm. Monotypic. A bulky, square-tailed, medium-sized parrot, predominantly green; orange shoulder-patch conspicuous on the closed wing; underwing-coverts bright red; a diffuse yellowish collar with fine blackish tips to the feathers; head bare, apart from some fine black bristles, the skin bright orange. Vulturine Parrot is the only confusable species; it has bare black skin on the face and a much more pronounced black and yellow collar.

Habitat Lowland gallery forest and white-sand forest.

Breeding No data.

Movements No evidence of movements, though short-range nomadic movements to take advantage of fruiting trees might be expected.

Range Endemic. Restricted to a few locations on the lower Madeira and upper Tapajós rivers in Pará state (Gaban-Lima *et al.* 2002).

Conservation status Currently classified by BirdLife International as Near Threatened, based on the limited range and an estimated total population of not more than 10,000 individuals. Although some of the range is economically managed through environmental tourism, forest destruction from logging is a constant threat (Gaban-Lima *et al.* 2002).

Etymology The specific name, *aurantiocephala* (an illegitimate combination of Latin and Greek roots) refers to the orange head of adult birds.

◄ Bald Parrot. Thaimaçu, Pará, Brazil, April 2003 (*Arthur Grosset*).

WHITE-FACED AMAZON
Amazona kawalli

A bird which, after an extraordinary history of confusion, is now recognised as a cryptic species separate from the widespread Mealy Parrot *A. farinosa*.

Type information *Amazona kawalli* Grantsau & Camargo 1989, Mato Piri, right bank of the Rio Juruá below Eirunepé, Amazonas, Brazil.

Alternative names Kawall's Amazon, White-faced Parrot. Portuguese: Papagaio-dos-Garbes, Papagaio-de-cara-branca.

Discovery In 1989, R. Grantsau and H. Camargo described a new species of *Amazona* parrot, basing their identification on one dead captive bird, two live examples in captivity and two old, previously misidentified, museum specimens (Grantsau & Camargo 1989, 1990). The designation, as a full and hitherto unknown species, was not universally accepted. Vuilleumier *et al.* in their decennial review of new species felt that there were insufficient data to support species status unequivocally and listed it as a *species inquirenda*. Other authorities such as K. Bosch and H. E. Wolters felt that the birds involved were probably merely the result of individual variation among *A. farinosa* (Bosch 1991). However, in the years after Grantsau and Camargo's original description, and following most of the negative opinions cited above, two further relevant specimens were discovered in European museums. One, held in the Natural History Museum in Tring, England, had been kept as a live specimen in London Zoo from May 1882 to June 1883, when it died and was prepared as a museum skin; to confuse matters still further, this specimen was labelled both as *Chrysotis* (=*Amazona*) *merceneria* (Scaly-naped Amazon, an Andean species) and *Chrysotis ochrocephala* var (Yellow-crowned Amazon, a largely lowland species), but not *Amazona farinosa*, Mealy Amazon. This specimen was noticed by the distinguished Brazilian aviculturalist and parrot authority Nelson Kawall and set aside, but not further worked upon until discovered, by chance, by N. J. Collar in 1992. The second 'new' specimen was discovered in the Berlin Zoological Museum, having lived as a captive in Berlin Zoo from 1910 until it died in 1923. This was labelled as *Amazona farinosa* aberr. *rubricauda Str.*, referring to the eminent German ornithologist Erwin Stresemann (1889–1972). *Rubricauda* means 'red-tailed', one of the distinguishing plumage characteristics of *A. kawalli*. Stresemann himself referred to it as a "curious mutation". This specimen was discovered in 1995 (Collar & Pittman 1996).

The history of the original two live specimens which partly formed the basis of the type description is also interesting. In 1968, Kawall noticed an unusual parrot, resembling *farinosa* but with some distinctive features, in the aviary of José Xavier de Mendonça in Santarem, Pará, Brazil. Kawall brought this bird to São Paulo, where he discovered a second specimen in the aviary of Alcides Vertamatti. Kawall then called a meeting with some parrot breeders and R. Grantsau, resulting in Grantsau and Camargo publishing the type description of the (appropriately-named) *A. kawalli* in 1989. Apparently a photograph exists of a further specimen taken at London Zoo in 1975 (Beolens & Watkins 2003).

▼ White-faced Amazon. Tabajara, Machadinho d'Oeste, Rondônia, Brazil, May 2016 (*Bruno Rennó*).

▼ White-faced Amazon. South of Manaus, Brazil, August 2014 (*Robson Czaban*)

The ultimate fate of this specimen is unknown. Over the course of almost a century, several ornithologists obviously saw *kawalli* without appreciating its distinctness; for example, A. de Miranda Ribeiro reported seeing examples of *farinosa* in the northern Mato Grosso (de Miranda Ribeiro 1920). No less an authority than Helmut Sick photographed a captive bird in a Munducurú village in 1961 which would appear to be *kawalli,* again without realising that a new taxon might be involved (Sick 1961, quoted in Martuscelli & Yamashita 1997).

The controversy over this parrot was finally decisively resolved by Collar and Pittman who demonstrated unequivocally that *kawalli* is a valid species (Collar & Pittman 1996). More recently, significant genetic differences have been demonstrated between the two species (Duarte Caparroz 1995).

Description 35–36 cm. Monotypic. A large, stocky parrot, typical of the genus *Amazona* (large, heavy head, massive bill and square tail), the nape with blackish scaling, overall bright green; red on the outer secondaries; primaries green, blackish towards the tips; bill grey with creamy bare skin at base; iris reddish-orange; tarsi greyish. Differs from *farinosa* in several features, most notably the red at the base of the tail feathers (largely concealed in museum specimens), which also have reduced yellow tips, and yellowish-green instead of red on the leading edge of the folded wing; also a diagnostic area of bare white skin around the bill.

Habitat Lowland tropical rainforest, with an apparent preference for river-edge and permanently flooded forest (Juniper & Parr 1998). There may be a degree of habitat separation from *farinosa.*

Food and feeding Gregarious feeder, in flocks of 2–8 birds. Seen to feed on ten plant species, including leaves of *Hevea brasiliensis,* flowers of *Erythrina* sp., seeds, pulp etc. Not observed feeding near human dwellings (Martuscelli & Yamashita 1997).

Breeding All information comes from the observations of Martuscelli & Yamashita (1997). Breeds in the wet season when the forest is flooded. Seven nests were found, all in cavities, six of them in dead trees, situated 6.5–25 m up. Nests with young in mid-March; in July-August observed in family parties.

Voice The flight call is quite distinct from the deep *chop-chop-chop* of *farinosa* and is, in fact, the best way to distinguish them from that species in the field. Flight call is a single complex note *weeou.*

Movements No data.

Range Endemic. Imperfectly known, due to confusion with *farinosa,* but quite widely distributed in Amazonian Brazil, from the state of Pará to Amazonas and Mato Grosso.

Conservation status Currrently classified by BirdLife International as Near Threatened. It has a wide geographic range, in some parts of which it is quite common (Martuscelli & Yamashita 1997), but severe threats exist for much of this. Apart from habitat destruction, the taking of wild birds, especially nestlings, for the cagebird trade is potentially and probably actually very serious. Traditionally local Indian tribes have kept psittacines as pets; furthermore, trade in wild-taken birds is widespread. On one occasion, Martuscelli and Yamashita saw 150 nestlings of *kawalli* for sale, mostly by Indians; these authors offer very pessimistic predictions as to the conservation status of this species, citing among numerous other factors subsistence hunting for food by both indigenous and other Brazilians (*kawalli* is reported to taste better than macaws *Ara*), habitat destruction by mining activities for nickel and gold, hydro-electric projects, and the lack of actual real protection afforded by 'paper parks'.

Etymology Named after Nelson Kawall, a leading Brazilian aviculturalist and exponent and pioneer of captive breeding He was very active in the programme to save Spix's Macaw *Ara spixii* from extinction. The alternative Portuguese name commemorates E. Garbe who collected the first, albeit unrecognised, specimens in the early 20th century.

SERAM MASKED OWL
Tyto almae

A virtually unknown owl with, apparently, a very restricted distribution.

Type information *Tyto almae* Jønsson *et al.* 2013, Mount Binaiya, above Kanikeh village, Manusela National Park, Seram, Indonesia.

Discovery In 1987 a masked owl, obviously of the genus *Tyto,* was seen by R. Badil and S. Lusli in Manusela National Park on the island of Seram (alternatively Ceram), eastern Indonesia. It was, at the time, assumed to be Lesser Masked Owl *T. sororcula,* the two known races of which occur on the islands of Tanimbar and Buru; although no *Tyto* owl had been recorded from Seram, the record seemed to have remained largely unnoticed. It also remained uncertain whether it pertained to either of the races of *sororcula,* or of a new taxon (Bruce 1999 HBW 5). In February 2012 an international group of ornithologists, organised by the

National History Museum of Denmark, mist-netted a *Tyto* owl in Manusela National Park. Careful genetic and morphological studies clearly indicated that this represented a hitherto undescribed species.

Description 31 cm. Monotypic. Very similar to all *Tyto* barn owls; prominent heart-shaped facial disc pale pinkish-cinnamon, with obvious darker surround; upperparts orangey-buff, head, shoulders and back densely spotted with fuscous; wing feathers including tertials with broad darker bars; rectrices ochraceous-tawny with five fuscous bars; underparts orangey-brown with smaller darker spots; iris very dark brown; bill pale horn; legs fully feathered with orange-brown feathers, feet drab pinkish. *T. sororcula* has more whitish underparts and less clear-cut barring on flight feathers.

Habitat The sole known specimen was taken at 1,350 m in wet, mossy highland forest.

Food and feeding No data.

Breeding Nest and eggs undescribed.

Voice The only vocalisations observed were protest calls from the mist-netted individual; these were a continuous series of drawn-out shrieks (Jønsson *et al.* 2013).

Range Endemic. Known only from Manusela National Park, Seram.

Conservation status Not currently assessed by BirdLife International, but given that the range would appear to be small, that the species has been observed on two occasions only, and that the team that discovered it saw no other individuals in the course of three weeks work in the area, an Endangered or Threatened classification is probably warranted.

Etymology Named after Alma Jønsson, daughter of the senior author, "acknowledging that she had to be without her father while he was out exploring" (Jønsson *et al.* 2013).

▲ Seram Masked Owl, adult female. Mount Binaiya, Manusela National Park, Seram, Maluku province, Indonesia, February 2012 (*Knud Andreas Jønsson*).

SOKOKE SCOPS OWL
Otus ireneae

Probably forms a superspecies with Sandy Scops Owl *O. icterorhynchus.*

Type information *Otus ireneae* Ripley 1966, Sokoke-Arabuko Forest, Kenya.

Alternative names Morden's Scops Owl.

Discovery The type specimen was obtained by A. Williams who made a 10-day collecting trip to the Sokoke-Arabuko Forest in 1965.

Description 15–18 cm. Monotypic. A very small, tufted owl, which occurs in three colour morphs (grey, dark brown and rufous); intermediates also occur. The first two have greyish or dark brown upperparts, the crown streaked blackish, the back mottled and spotted with light and dark, the underparts paler than the back, with small black-tipped white spots, the facial disc brown with paler eyebrows, the primaries brown with whitish bars, conspicuous on the closed wing. The rufous morph is a rich cinnamon-red above and below, the streaking and spotting reduced, paler on the underparts. Iris pale yellow, bill light greenish-yellow with pinkish-grey cere. Juvenile similar to adult.

Habitat *Cynometra-Brachylaena* woodland, rarely in *Brachystegia* woodland; also in thickets. In Kenya at 50–170 m, in Tanzania 200–400 m.

Food and feeding Most active immediately after dusk and before dawn, with a lull around 2300–0300 hrs (Virani 2000). Recorded food items are insects including phasmids and crickets (König *et al.* 2008).

Breeding Nest and eggs undescribed.

▲ Sokoke Scops Owls, bunching together in the presence of an intruder. Sokoke, Kenya, January 2007 (*Tasso Leventis*).

Movements Sedentary. A radio-telemetric study in Kenya showed that pairs foraged in a mean home range of 11 ha, with little overlap with adjacent pairs (Virani 2000).

Voice Song of the male is a series of 5–9 (sometimes more) high whistled notes, *goohk-goohk*, at the rate of about three notes in two seconds, repeated at intervals (König *et al.* 2008).

Range Originally believed to be endemic to the Sokoke-Arabuko Forest in coastal Kenya; however, in 1994–95 significant populations were discovered in the East Usambara Mountains in coastal Tanzania (Evans 1997a, 1997b). A further small population exists in the Marafa Forest north of Sokoke-Arabuko. The Kenyan range is probably about 220 km²; the Tanzanian range, about 97 km². There remains a possibility of further range extensions, though surveys in apparently suitable habitats in both Kenya and Tanzania did not uncover further populations (BLI SF 2013; Cordeiro & Githiru 2000).

Conservation status Currently classified by BirdLife International as Endangered. Factors include the small total range which is in an area of high and increasing human population density. In the Sokoke-Arabuko Forest the bird's population was estimated at about 1,000 pairs in 1995 (Virani 1995). However, later work in 2005 and 2008 suggests a population decline of 22.5% over 16 years (Virani *et al.* 2010). The Tanzanian population occurs at lower densities than the Kenyan. Although much of the Kenyan range seems intact, habitat degradation by removal of timber and charcoal burning is significant. Recent efforts at conservation include community participation aimed at encouraging the local people themselves to become involved in conservation programmes, including income-generating programmes such as bee-keeping, ecotourism and butterfly farming (Anon. 2012). The species is considered to be especially vulnerable to the effects of climate change (Monadjem *et al.* 2013).

Etymology The scientific specific name and the alternative English name honour Mrs Irene Morden, who sponsored the fieldwork which led to the discovery of this species. With her husband William she herself took part in several collecting expeditions to East Africa.

SERENDIB SCOPS OWL
Otus thilohoffmanni

The discovery of a new bird species from Sri Lanka, a country very thoroughly investigated zoologically, came as a great surprise in the ornithological community, the last new bird species from the island having been described in 1872.

Type information *Otus thilohoffmanni* Warakagoda & Rasmussen 2004, Morapitiya-Runakunda Proposed Reserve, Sri Lanka.

Discovery In February 1995 Deepal Warakagoda heard, and recorded, an owl-like call from an area of wet rainforest in the Kitulgala Proposed Reserve in SW Sri Lanka. Over the next six years, at this and in another forest reserve, he heard this mysterious call several times, but never succeeded in glimpsing the vocalist. Other local naturalists were also unfamiliar with the call. Comparison with recordings of other small Asian owls suggested that the closest resemblance was to Reddish Scops Owl *Otus rufescens*, native to the Malay Peninsula, an opinion shared by P. Rasmussen when she heard the recordings. However, in January 2001 Warakagoda was finally able to observe an owl making the call; it was clearly not *rufescens*, nor any of the other small owl species native to Sri Lanka. In February 2001 the new bird was documented by photographs taken in its natural state by Chandima Kahandawala (Warakagoda 2001a, 2001b, 2001c, 2001d). In August 2001 a male was mist-netted and data recorded (this individual was relocated at the same site over two years later). By May 2002, when the bird had been detected in five different forests it was felt that it was safe and ethical to collect a type specimen, which was done in November 2002. This holotype, located in the Sri Lanka Museum in Colombo, remains a unique specimen.

Description 16.5 cm. Monotypic. A small, strikingly rufescent owl, short-tailed, lacking obvious ear-tufts when at rest, but when alarmed erecting small false ear-tufts. Head reddish-brown, supercilium whitish; the back rufous with small blackish chevrons or spots; underparts somewhat paler rufous with blackish triangular spots, the face darker and more deeply rufous; bill ivory-white; legs and feet pinkish-white; iris orange in adult males, yellow in females. Juvenile very similar to adult, but with an incompletely developed facial disc; the eyes are yellow. In skeletal characters differs from other *Otus* species, possibly meriting the erection of a monotypic genus (2013 HBW SV).

► Serendib Scops Owl. Kitulgala Forest, Sri Lanka, March 2006 (*Uditha Hettige*).

Habitat Wet lowland rainforest, 30–1,000m, often in secondary forest with rich undergrowth (König *et al.* 2008).
Food and feeding Few data, but believed to eat mainly insects such as beetles and moths.
Breeding Nest and eggs undescribed; a fully-fledged juvenile observed roosting with an adult in March (Warakagoda 2006).
Voice Not particularly vocal; call is "unobtrusive and easily overlooked". The female gives a short piping tremulous note that rises slightly and falls again in pitch; the male gives a slightly lower-pitched, slightly shorter, less tremulous version.
Movements Presumably sedentary.
Range Endemic. So far detected at only six forest reserves, Kitulgala, Sinharaja, Morapitiya-Runakanda, Kanneliya, Eratna-Gilimale and Peak Wilderness, all located in the wet zone of SW Sri Lanka.
Conservation status Currently listed by BirdLife International as Endangered. Seems to require substantial tracts of undisturbed habitat, the smallest so far being 8.2 km^2. All six locations are in protected areas, which are not, however, totally immune to encroachment. Total population estimated at 200–250 individuals which, given the elusive and retiring nature of the species, might be an underestimate (Warakagoda 2006).
Etymology Named in honour of Thilo W. Hoffmann (1922–2014) "who has for so long done much for nature conservation and ornithology in Sri Lanka" (Warakagoda & Rasmussen 2004). Hoffmann, a multifaceted personality with a wide range of interests, was born in Switzerland, but spent 68 of his 92 years in Sri Lanka, where he was a major force for conservation and wildlife study in his adopted land, becoming, at various times, the President of both the Ceylon Bird Club and the Wildlife and Nature Protection Society (Gunawardana 2014). The common name is taken from the Arabic name for Sri Lanka, first recorded in 361 AD; the origin of the English noun 'serendipity'.

MINDANAO SCOPS OWL
Otus mirus

Although described as a race of Eurasian Scops Owl *O. scops*, or alternatively as a race of Oriental Scops Owl *O. sunia*, the differences in morphology and vocalisations clearly merit full specific status (Marks *et al.* 1999 HBW 5).

Type information *Otus scops mirus* Ripley & Rabor 1968, Hilong-Hilong Peak, Agusan Province, Mindanao Island, Philippines.

▼ Mindanao Scops Owl. Malindang, Mindanao, Philippines, November 2015 (*Robert Hutchinson*).

Discovery The type specimen was collected by D. S. Rabor in April 1963.
Description 19–20 cm. Monotypic. A typical small *Otus* owl, darker and more contrastingly marked than most members of the genus; facial disc with indistinct concentric lighter and darker lines, heavily spotted and blotched above with blackish; paler below with bold sharp black streaks; ear-tufts short; iris brown or yellow; bill greenish-grey; tarsi grey or yellowish-grey.
Habitat Montane rainforest, from 650 m upwards, more common above 1,500 m.
Food and feeding No data; probably small arthropods.
Breeding Nest and eggs undescribed.
Voice A soft double whistled note, *pli-piooh,* recalling a pigeon or dove, uttered in a long series, at intervals of 10–15 seconds (Marks *et al.* 1999 HBW 5).
Movements Probably sedentary.
Range Endemic. Confined to mountains in Mindanao; currently recorded from Mts Hilong-Hilong, Apo, Kitanglad and Lake Sebu (Collar *et al.* 1999).
Conservation status Currently classified by BirdLife International as Near Threatened. Distribution not well known, and probably fairly rare (König *et al.* 2008); much of the suitable habitat is presently somewhat protected by being remote and high altitude, but this may change. Currently only about 10% of the rainforest of the Philippines remains intact (Warburton 2009).
Etymology *Mirus,* Latin, 'astonishing' or 'wonderful'.

NICOBAR SCOPS OWL
Otus alius

A further addition to the extraordinarily complex taxonomy of small insular owls in SE Asia and Wallacea.

▲ Nicobar Scops Owl. Great Nicobar, Nicobar Islands, India, September 2014 (*Jainy Kuriakose*).

Type information *Otus alius* Rasmussen 1998, Campbell Bay, Great Nicobar Island, Bay of Bengal, India.

Discovery The taxonomy of *Otus* owls on islands in the eastern Indian Ocean and Indonesia is very complex and highly confused, with different authors placing various island populations in a variety of species. A specimen taken on Great Nicobar Island in 1966 was listed as a race of Oriental Scops Owl *O. sunia*. A second specimen from the same location in 1977 was tentatively placed in an "expanded" Moluccan Scops Owl (*O. magicus*) taxon; it was then variously classified as a race of *sunia*, or of *magicus*, or as an undescribed '*Otus* sp.' Finally, P. Rasmussen made a detailed examination of the two existing specimens, comparing them to all relevant adjacent taxa, and argued that they represent a hitherto undescribed species (Marks *et al.* 1999 HBW 5; Sangster 1998).

Description 19–20 cm. Monotypic. A small, rather warmly coloured brown *Otus* owl, finely barred above and below, with no streaks on the back and small streaks on the underparts; medium-length ear-tufts, tarsi only partly feathered; eyes yellow, bill dark brownish. Differs from most other geographically adjacent *Otus* species in the lack of streaking on the back and the fine barring above and below. Juvenile plumage unknown.

Habitat The two specimens were taken in coastal forest near sea level.

Food and feeding Stomach contents of the type specimen include a spider and a beetle; of the other specimen, a mangled 10 cm gecko.

Breeding Nest and eggs undescribed. Probably breeds in March or April; the male holotype had enlarged testes on 3 March, the female paratype, taken on 2 April, was approaching breeding condition.

Voice The call of the female was rendered as *ooo-m* or as a "rising, long drawn-out (lasting for 2–2.5 s) single-syllable melancholic moan, repeated after 3–5 s, calling continuously for 30 minutes" (Rasmussen 1998). A long series of piping *weeyu* notes, each lasting less than half a second, repeated every four seconds (Mikkola 2014).

Range Endemic. So far encountered only on Great Nicobar Island; the adjacent Little Nicobar Island has not been thoroughly investigated. Apparently absent from other northerly islands in the group.

Conservation status Currently listed by BirdLife International as Data Deficient. However, bearing in mind that evidence for the existence of the species rests solely on two specimens and a few observations of live birds since, and that the coastal forest on Great Nicobar is under threat from conversion to plantations of coconut, banana and cashew nuts, and that the whole area was severely impacted by the 2004 tsunami, which has resulted in the movement of communities whose livelihood was devastated into new areas, a classification of anywhere from Vulnerable to Endangered would appear justified.

Etymology Latin, *alius*, meaning 'other', reflects that this species is another *Otus* from the Nicobar Islands; it also "encapsulates the family name of Mr Humayun Abdulai, who first collected this species, and contributed a great deal to Indian ornithology" (Rasmussen 1998).

SANGIHE SCOPS OWL
Otus collari

A small, highly restricted *Otus* owl, the separateness of which from adjacent geographic populations was the source of much confusion.

Type information *Otus collari* Lambert & Rasmussen 1998, Sangihe (or Sangir) Island, east Celebes Sea, Indonesia.

Discovery Specimens of a small owl were collected on Sangihe Island (also known as Sangi, Sangir or Great Sanghir) between 1866 and 1887; however, these were believed to be conspecific with Sulawesi Scops Owl *O. manadensis*. Also in 1866 a specimen from the nearby island of Siau was given the specific name of *O. siaoensis*, but subsumed into *manadensis* by later authors. In fact, given the distinctly smaller size of the unique specimen, it might indeed be a valid species (del Hoyo *et al.* 1999). In a 1978 study of Asian members of the genus *Otus* (Marshall 1978), the status of the birds on Siau and Sangihe were left in abeyance in the absence of recordings of their voices, with tentative allocation to the widely distributed Moluccan Scops Owl *O. magicus*, which ranges from Halmahera south to Lombok and Flores in the Lesser Sunda Islands. In 1985 a specimen was taken on Sangihe, whose call, it was believed at the time, was the same as birds on mainland Sulawesi (i.e. *manadensis*). However, in 1996 Lambert tape-recorded an owl on Sangihe which he realised sounded distinct from Sulawesi birds; furthermore, comparison of specimens showed consistent differences between different populations, on which basis Lambert and Rasmussen finally described the species in 1998.

Description 19.5 cm. Monotypic. A small, rather drab *Otus* owl, with moderately-sized ear-tufts; drab-brown above with dark shaft streaks and conspicuous buff spots; underparts paler with fine vermiculations and prominent black shaft-streaks; primaries with dark brown and buff bars, conspicuous on closed wing; iris yellow; tarsi almost completely feathered; bill horn-brown. Juvenile plumage undescribed.

Habitat Quite varied; forest, secondary growth, mixed plantations, etc, sea level to 900 m. Seems to be tolerant of habitat modification, one bird even heard calling in the centre of the town of Tahuna (Riley 2002).

Food and feeding No data on food items; presumably small arthropods.

Breeding No information available.

Voice The call most resembles that of *O. manadensis* in so far as both are whistled; however, that of *collari* is quite distinct, being longer, higher pitched, much sweeter, clearer, more modulated and slurred (Lambert & Rasmussen 1998).

Movements Presumably sedentary.

Range Endemic. Confined entirely to Sangihe Island.

Conservation status Appears to be fairly common in many locations on the island, and in addition seems to adapt to modified habitats including mixed plantations of coconut and nutmeg; much of Sangihe has been thus modified for a century. For these reasons currently classified as Least Concern, albeit with restricted range. Many other species on Sangihe have not been so fortunate; three Critically Endangered endemics – Cerulean Paradise-Flycatcher *Eutrichomyias rowleyi*, Sangihe Shrike-Thrush *Colluroncinchla sanghirensis* and Sangihe White-eye *Zosterops nehrkorni* – only exist in tiny numbers in a very small residual area of habitat, while the endemic race of Red-and-blue Lory *Eos histrio* is already extinct.

Etymology Named in honour of Dr Nigel Collar, presently Leventis Fellow in Conservation Biology, BirdLife International, Cambridge, England, in recognition of his many contributions to and huge influence on bird conservation worldwide.

◄ Sangihe Scops Owl. Near Tamako village, Sangihe, Indonesia, November 2008 (*Philippe Verbelen*).

KARTHALA SCOPS OWL
Otus pauliani

Originally described, rather tentatively, as a race of the Madagascar (Rainforest) Scops Owl *O. rutilus*, although it differs obviously from that species in having very reduced ear-tufts.

Type information *Otus rutilus pauliani* Benson 1960, Mt Karthala, Grande Comore Island, western Indian Ocean.
Alternative names Grande Comore Scops Owl. Local name: 'ndeu'; French: Petit-duc de Karthala.
Discovery In 1958, C. W. Benson, participating in the Centenary Expedition of the British Ornithologists' Union to the Comoro Islands, noticed a feather, which he attributed to either a nightjar or an owl, lining a nest of Humblot's Sunbird *Nectarinia humbloti*. He therefore went out at night to explore forest at 1,800 m on the western slope of Mt Karthala, hearing an unknown call which he attributed to a nightjar. However, when a specimen of the caller was obtained it proved to be a new *Otus* owl. The type specimen was taken at La Convalescence, at 1,700 m, and is now housed in the British Museum (Herremans *et al.* 1991). Benson (1960) felt "inclined to regard this specimen as belonging to a species of its own…however, in deference to the opinion of Professors Berlioz and Stresemann….it is placed as a subspecies of *O. rutilus*". More recent authorities, e.g. König *et al.* (2008), Weick (2006), del Hoyo *et al.* (1999) give the taxon full specific status.
Description 15–20 cm. Monotypic. Much the smallest of the *Otus* owls of the western Indian Ocean, with almost invisible ear-tufts. Two colour morphs (Weick 2006, *contra* BLI SF 2013). Light morph uniformly dark greyish-brown, upperparts with paler spots and fine darker barring; ochre-buff underparts, densely vermiculated and with some darker streaks; facial disc grey-brown with darker concentric lines. Dark morph is rather uniformly dark brown or earth-brown with fine buffish speckles. Iris yellow or brown (possibly age-related), bill greyish-brown, tarsi partially bare.
Habitat Forest and forest-edge, including both primary and degraded forest, 460–1,900 m.
Food and feeding Few data. Stomach contents of the type specimen included beetles; the weak talons suggest that it is mainly insectivorous.
Breeding Nests in cavities, probably lays September to December. Highly territorial (Marks *et al.* 1999 HBW 5).
Voice High whistled *toot* or *choo*, at regular intervals of 1–2 seconds, quickly turning into a faster series of shorter down-slurred *choo* notes repeated at half-second intervals for 10 minutes; female, in duet, a drawn-out *choeiet* (Marks *et al.* 1999 HBW 5).
Movements Probably sedentary.
Range Endemic. Now entirely confined to Mt Karthala, Grande Comore.
Conservation status BirdLife International classification is Critically Endangered. Habitat on Mt Karthala has diminished by 25% since 1983, due to clearing for agriculture, logging, road construction, fires to create cattle pasture, etc. Introduced rats (*Rattus*) and Common Mynas (*Acridotheres tristis*), which are a potential competitor for nest-sites, could be further negative factors. Total population estimated at 1,000 pairs, though a somewhat higher figure is possible (BirdLife International 2013). Creation of a fully protected area of some type, to cover the habitat of this species and those of other Mt Karthala endemics (Humblot's Flycatcher *Humblotia flavirostris*, Grand Comoro White-eye *Zosterops mouroniensis* and Comoro Drongo *Dricrurus fuscipennis*) is urgently needed but has not yet taken place.
Etymology Named in recognition of the contributions of P. Paulian, a French zoologist who studied mammals and birds in the Indian Ocean, including Kerguelen Island, and who discovered the breeding population of what was later named the Amsterdam Albatross *Diomedea amsterdamensis* (q.v.).

▼ Karthala Scops Owl. Grande Comore Island, Comoros, October 2012 (*Pete Morris*).

MOHELI SCOPS OWL
Otus moheliensis

The third new *Otus* owl to come from the Comoro archipelago in recent years (a fourth species, *O. capnodes,* discovered in 1886 on Anjouan Island, was only rediscovered in 1992).

Type information *Otus moheliensis* Lafontaine & Moulaert 1999, at 620 m in the Forêt de Mlédjélé, Mohéli (Mwali) Island, Comoro Islands.
Alternative names French: Petit-duc de Mohéli.

▼ Mohéli Scops Owl. Mohéli Island, Comoros, October 2012 (*Pete Morris*).

Discovery Although Benson (1960) raised the possibility of a scops owl on Mohéli, it was not until February 1995 that a small owl was actually observed by Western ornithologists. In September 1996, during point counts that were being carried out to compare bird populations in different forest types, a new call was heard; the caller was apparently quite common, since at least ten were heard on one night in one location. Siting of mist-nets after dark resulted in the capture of a new *Otus* owl. Subsequent observations in the Forest of Mlédjélé, about 5 km from the original site in late September and October of 1996, located more calling birds; that the callers were indeed the owl was confirmed when a mist-netted specimen called in captivity.
Description 22 cm. Monotypic. A small scops owl with reduced ear-tufts, occurring in two morphs, brown and (apparently very rare) rufous. The former is rufescent-brown, spotted and barred blackish on crown and upperparts; the underparts are more rusty-cinnamon with darker streaks and bars. The rufous morph is bright rufous-cinnamon above and below, with white area on the scapulars and yellowish lines of spots on the primaries; iris yellow; bill horn-coloured; tarsi partially feathered (Weick 2006). Juvenile undescribed.
Habitat Dense humid forest with epiphytes; in lower densities in degraded forest at 450–790 m.
Food and feeding No data; food probably mainly insects and other arthropods.
Breeding Nest and eggs undescribed. Vocalisations in September possibly indicate the breeding season.
Voice Poorly known; male is reported to give a hissing whistle in a series of one to five notes. "*Tyto*-like aspirated hissing screams in long and shorter versions" (Mikkola 2014).
Movements Presumably sedentary.
Range Endemic. Confined to mountainous interior of Mohéli (Mwali) Island, Comoros.
Conservation status BirdLife International classification is Critically Endangered. Although it may occur in fair densities (one individual/5 ha) in prime habitat and half that in degraded forest, the small areas remaining (humid forest in 1998 was only 5% of the island's area, down from 30% in 1968) and the constant destruction of habitat for agriculture are severe continuing threats; introduced species may also be a concern. Total population in 1998 was estimated in the order of 400 individual birds (Lafontaine & Moulaert 1998), with perhaps 260 mature individuals (BLI SF 2013).
Etymology The vernacular and specific scientific names both refer to the island of Mohéli.

MAYOTTE SCOPS OWL
Otus mayottensis

Described by Benson as a race of Madagascar Scops Owl *O. rutilus*; most more recent authorities (e.g. König *et al.* 2008) treat it as a full species.

Type information *Otus rutilus mayottensis* Benson 1960, La Convalescence, Mayotte, Comoro Islands, Indian Ocean.
Alternative names French: Petit-duc de Mayotte.
Discovery The type specimen, presently in the Natural History Museum, Tring, was collected during the British Ornithologists' Union Centenary Expedition to the Comoro Islands in November 1958 at La Convalescence, Mayotte, at an altitude of 400 m. However, there are two specimens from Mayotte in Paris, taken in the late 19th century, labelled '*Scops humbloti*' (Henri Joseph Léon Humblot was sent to the Comoro Islands in 1884 to ensure French sovereignty; he was, apparently, an amateur naturalist who also, incidentally, acquired large tracts of land which he farmed with the aid of slaves – Beolens & Watkins 2003). Normally this name would have priority, but since it does not appear to have been published, Benson's type description takes precedence (Rasmussen *et al.* 2000). Benson described the taxon as a race of Madagascar Scops Owl *O. rutilus,* as do Marks *et al.* (1999 HBW 5). More recent authors (e.g. Rasmussen *et al.* 2000; Weick 2006; König *et al.* 2008) give it full specific status based on morphology and vocalisations, though no DNA evidence has been published.
Description 22–24 cm. Monotypic. Apparently monomorphic. A small owl with reduced ear-tufts, the upperparts greyish-brown with darker vermiculations and streaks, with a fairly prominent nuchal collar; facial disc greyish-brown; underparts brownish with blackish streaks and vermiculations, the centre whitish; primaries and secondaries darker greyish-brown with paler bars; rectrices greyish-brown, indistinctly barred; iris yellow; bill greyish-horn; tarsi totally feathered, toes bare (Weick 2006). Immature (based on one of the Paris specimens mentioned above) has finely patterned underparts and prominently banded tertials and tail (Rasmussen *et al.* 2000).
Habitat Evergreen forest, including degraded wet forest (Lewis 1998).
Food and feeding No data.
Breeding Nest and eggs undescribed.
Voice Similar to, but nevertheless distinct from, that of eastern populations of the Madagascar Scops Owl *O. rutilus*. A series of 3–10 (usually four) hoots, the delivery somewhat slower than *rutilus* but the actual notes slightly longer, rendered as *woohp woohp*. Individuals of *rutilus* did not respond to playback of *mayottensis* songs, though the converse was not tested (Lewis 1998). Dull, monotonous, frog-like grunting (Mikkola 2014).

▲ Mayotte Scops Owl. Pic Combani, Mayotte Island, Comoros, December 2015 (*Ken Behrens*).

Movements Probably sedentary.
Range Endemic. Confined entirely to Mayotte Island, Comoros.
Conservation status Currently classified by BirdLife International as Least Concern, despite the limited geographic range. Described as locally common (Benson 1960). Although geographically part of the Comoro Archipelago, Mayotte is politically part of Metropolitan France, of which it became an overseas department in 2011 and an 'Outermost Region' in 2014, when presumably European Union conservation regulations became applicable. Mayotte is the most densely populated island in the archipelago.
Etymology *Mayottensis*, from Mayotte.

RINJANI SCOPS OWL
Otus jolandae

Another owl species that went unrecognised for more than a century after its original collection.

Type information *Otus jolandae* Sangster *et al.* 2013, Lombok, Lesser Sundas, Indonesia. The holotype was collected in 1896 at an imprecisely specified location in Lombok, probably in northern Lombok.

Alternative names Indonesian: Celepuk Rinjani; local name: Buring or Burung Pok ('Pok-bird').

Discovery In 1896 the colonial civil servant and amateur naturalist Alfred Everett and his associates collected seven specimens of an *Otus* owl on Lombok island which were assigned to the race of Moluccan Scops Owl found on the adjacent islands of Flores, Lembata and Sumbawa, namely *O. magicus albiventris*. In his notes, Everett recorded that the calls heard on Lombok were rather distinctive, but this observation was overlooked by subsequent authors. In 2003 two separate expeditions visited Lombok to study nightjars, and both, independently, recorded owl calls, which could not be ascribed to known species. George Sangster spotlighted the callers on the SE slope of Mt Rinjani and identified them as *Otus* owls; a few days later Ben King, on the SW slope, recorded up to 15 individuals. In 2008 B. Demeulemeester and P. Verbelen, using the recordings of Sangster, photographed the birds; comparison of the photographs with the specimens taken by Everett 120 years earlier (presently housed in the Natural History Museum, Tring, and American Museum of Natural History) showed them to be the same species (Sangster *et al.* 2013; Sykes 2013). On this basis Sangster wrote the type description, also pointing out that the morphometrics of Everette's specimens differed from those of *O. magicus albiventris*.

Description 22 cm. Monotypic. A small owl with obvious ear-tufts, the back warm brown with darker brown streaks and indistinct bars; scapulars whitish with dark markings; the facial disc a mixture of dull cinnamon, dull dark brown and whitish, outlined with whitish; supercilia whitish-grey; primaries dark brown with paler bars; chest rusty-buff to cinnamon with darker streaks, the belly much paler with dark streaks and dark brown speckles; iris golden-yellow; bill brown or fuscous; tarsi fully feathered to base of toes.

Habitat Forest, both continuous and partial, also secondary forest with palm trees; recorded at 25–1,350 m.

Food and feeding No recorded data.

Breeding Nest and eggs undescribed.

Voice The major distinction from adjacent taxa: territorial song is a single whistle without overtones, *pok* or *pork*, quite distinct from the raven-like bark of *O. magicus*; also a high-pitched, ascending *weera* (Eaton *et al.* 2016b).

Movements Probably sedentary and territorial year round.

Range Endemic. Currently found in northern Lombok island, although prior to the destruction of forest on much of Lombok it may have been more widespread in other parts of the island. The status in some suitable habitat in the SW corner of the island has not been determined.

Conservation status Not currently classified by BirdLife International. Observations of Sangster and King suggest that it may be quite common in suitable habitat and that it is not dependent on totally undisturbed forest. Most observations have been made in the 413 km^2 Gunung Rinjani National Park.

Etymology The vernacular name comes from Gunung Rinjani, where recent observations were made and which undoubtedly holds the majority of the current population. The scientific epithet honours Dr Jolanda Luksenberg, wife of G. Sangster, whose work resulted in the description of this species and the nightjar *Caprimulgus meesi* (q.v.).

◄ Rinjani Scops Owls. Lombok, Indonesia, August 2008 (*Philippe Verbelen*).

KOEPCKE'S SCREECH OWL
Megascops koepckeae

A widely distributed but still relatively unknown screech owl of the Peruvian and Bolivian Andes with a convoluted taxonomic history. Forms a clade with *Megascops choliba*, *albogularis*, *clarkii* and *trichopsis* (Tropical, White-throated, Bare-shanked and Whiskered Screech Owls) (Dantas *et al.* 2016).

Type information *Otus choliba koepckei* (sic) Hekstra 1982, Quebrada Yanganuco, Cordillera Blanca, Ancash, Peru.
Alternative names Maria Koepcke's Screech Owl. Spanish: Autillo de Koepcke.
Discovery The first specimen of Koepcke's Screech Owl was collected by A. Morrison in 1939 in Dpto. Ayacucho, Peru, and identified by him as the race *crucigerus* of Tropical Screech Owl *Otus (Megascops) choliba*. In 1969 Maria Koepcke drafted a description of a new taxon, which she named as a new race (*alticola*) of Peruvian Screech Owl *Otus (Megascops) roboratus*. At the same time G. P. Hekstra was working on a monumental publication (issued as a thesis for the Free University of Amsterdam, published in 1982 – Hekstra 1982) on a comprehensive revision of the American *Otus (Megascops)* species. Koepcke was made aware of Hekstra's project by E. Eisenmann, but her tragic death in 1971 prevented any collaboration. Hekstra, basing his evaluation on 23 specimens from Peru and Bolivia, named the new taxon as a new race *koepckei* (sic) of *O. m. choliba* (international nomenclatural rules require the genitive feminine singular ending *ae* appended to the specific honorific). However, the very distinct vocalisations of *koepckeae*, as well as consistent morphological differences argue strongly for full specific status (Fjeldså *et al.* 2012).

▼ Koepcke's Screech Owls, *M. k. hockingi*. Abancay, Peru, September 2010 (*Christian Artuso*).

▼ Koepcke's Screech Owls, *M. k. koepckeae*. Utcubamba Canyon, Peru, October 2012 (*Adam Riley*).

Description 24 cm. Apparently monomorphic, although König *et al.* (1999) speak of "very similar" light and dark morphs, albeit rather variable; a small tufted owl, generally rather dark grey overall, with a darker crown; broadly streaked underparts lacking fine vermiculations; iris yellow, bill greenish-yellow. Feet small and quite delicate. Differs from *roboratus* in its somewhat larger size and lack of a white nape band. Sexes similar. Two subspecies: *M. k. hockingi* differs from the nominate in its more drab grey (rather than chocolate or brown) appearance and its more sparsely marked rear underparts (Fjeldså *et al.* 2012). These authors suggest that differences in song and habitat choice of the two *koepckeae* taxa imply a genetic isolation of long standing, but in the absence of molecular studies have taken the conservative viewpoint of classifying them as races rather than species.

Habitat *M. k. hockingi* lives in temperate Andean woodland, often mist-dependent, at 2,000–4,000 m. The nominate race seems to occur in drier rainshadow intermontane valleys, at 1,400–3,400 m.

Food and feeding Few data. The small feet and claws imply a diet of arthropods, confirmed by stomach contents.

Breeding Nest and eggs undescribed.

Voice Duets (unlike *M. choliba*). Song of the male of the nominate race is a loud, rising and falling series of "hysterical" shrill notes, slowing towards the end; female song is shorter, slower-paced and higher. The calls of the population inhabiting the drier areas of southern Peru are quite distinct, being longer and higher pitched (Fjeldså *et al.* 2012).

Movements Presumably sedentary.

Range Imperfectly known; probably an endemic. Nominate race in western Andes of northern Peru (Lambayeque, Cajamarca and Amazonas, south to Ancash and Lima); *hockingi* in south-central Peru (Apurímac, Ayacucho). The 1939 specimen was from this southern population. Stated by several authors e.g. König *et al.* 1999, possibly to extend to northern Bolivia, but this is denied by Fjeldså *et al.* 2012.

Conservation status Classified by BirdLife International as Least Concern. Although habitat destruction is occurring within its range, in some locations the owl appears locally not uncommon (König *et al.* 1999); range is large, and the species does not approach the threshold for Vulnerable (BLI SF 2014). No separate classification is available for the two races.

Etymology Named in honour of Dr Maria Koepcke (1924–1971), "one of the true pioneers in the 20th century of the ornithology of Peru" (Parkes 1999). With her husband Hans-Wilhelm, she established a biological station in Miraflores, Lima, which became a centre for Peruvian field ornithology (Vuilleumier 1995). She was killed, with all but one of the 92 passengers and crew, when a flight from Lima to Pucallpa in eastern Peru was struck by lightning and crashed in a remote area. The sole, and miraculous, survivor was her 17-year-old daughter Juliane; search efforts were bungled by Peruvian authorities, and it was only after a harrowing eleven-day trek through uninhabited mountain rainforest that Juliane was rescued (Diller 2011). Apart from being honoured with the names *Phaethornis, Pauxi, Cacicus* and *Otus (Megascops) koepckeae*, Dr Koepcke herself was the naming authority for three species: White-chinned Cotinga *Zaratornis stresemanni*, Cactus Canastero *Asthenes cactorum* and Russet-billed Spinetail *Synallaxis zimmeri*.

The race *hockingi* recognises Peter (Pedro) Hocking "in honour of his lifelong efforts to document the Peruvian avifauna" (Fjeldså *et al.* 2012).

CINNAMON SCREECH OWL
Megascops petersoni

A small, virtually unknown owl of wet Andean montane forests; now placed in the Neotropical genus *Megascops*. Regarded by some authorities as a race of Rufescent Screech Owl *M. ingens*, or of Colombian Screech Owl *M. colombianus*, but differs vocally from both; sympatric with *ingens* (König *et al.* 2008). Forms a clade with *M. marshalli, M. hoyi, M. ingens* and *M. colombianus* (Cloud-forest, Hoy's, Rufescent and Colombian Screech Owls) (Dantas *et al.* 2016).

Type information *Otus petersoni* Fitzpatrick & O'Neill 1986, Cordillera del Cóndor, above San José de Lourdes, Dpto. Cajamarca, Peru.

Alternative names Spanish: Autillo de Peterson.

Discovery The type specimen was taken by a joint expedition from Princeton University and Louisiana State University to the Peruvian portion of the Cordillera del Cóndor, a remote mountain range of extreme northern Peru, which also extends across the border into Ecuador. This cordillera is separated from terrain of similar altitude in Ecuador by the valley of the Río Zamora and is remarkably rich in endemics, with no fewer than four recently described species.

Description 21 cm. Monotypic. A relatively small owl with moderately-sized eartufts and generally very cinnamon plumage, the back finely vermiculated with alternate lighter and darker wavy bars; scapulars pale cinnamon; underparts warm cinnamon-buff, the lower breast and belly uniform cinnamon; facial disc warm cinnamon-brown; flight feathers with lighter and darker bars of cinnamon; iris dark brown; bill pale grey-green; tarsi feathered almost to the toes (more than *colombianus*, and less than *ingens*). Considerable variation in plumage, with some specimens much browner, with finer streaking, others much more rufous, the streaking being broader.

Habitat Dense epiphytic forest with heavy undergrowth, at 1,690–2,500 m.

Food and feeding Few data. Specimens from the Ecuador section of the Cordillera del Cóndor had stomach contents of insects, mainly large beetles (Krabbe & Sornoza 1994).

Breeding One nest only so far described, from NE Antioquia in Colombia. The nest was in a natural cavity in a live tree in dense epiphytic forest at an altitude of 1,826 m, located 0.7 m above the ground. The nest was lined with soft spongy wood without added nesting material. An adult was flushed from the cavity on 12 January; one single large, rounded white egg was found on 21 January. Incubation continued until at least 22 February (longer than North American species of *Megascops*). The nest had been predated by 24 February (Freeman & Rojas 2010).

Voice Song is "a simple, long series of separate notes delivered in rapid succession (0.15 sec intervals) sliding slightly up the scale about half a musical note, holding there for about 20 notes, then fading and sliding back to the original pitch. The entire phrase lasts about 4.8–5.8 secs" (Fitzpatrick & O'Neill 1976). Also *wew* calls.

Movements Probably sedentary.

Range Originally thought to be confined to extreme northern Peru and adjacent Ecuador; Fitzpatrick & O'Neill presciently suggested that it might occur in Colombia, based on a 'Bogotá' skin of unknown provenance discovered in the collection of the Academy of Natural Sciences in Philadelphia. More recently a very disjunct population discovered in northern Colombia (NE Antioquia) (Cuervo *et al.* 2008, Freeman & Rojas 2010). In Ecuador, several records well to the north of the eastern known range, in Morona-Santiago and western Napo (Ridgely & Greenfield 2001); in Peru, records from Piura, Cajamarca and Amazonas.

▲ Cinnamon Screech Owl. Arrierito Antioqueno Bird Reserve, Antioquia, Colombia, February 2009 (*Fabrice Schmitt*).

Conservation status Currently classified by BirdLife International as Least Concern. The population may be decreasing in some areas due to deforestation, but much of its habitat is remote. In 1994 described as "remarkably common" in Ecuadorian Cordillera del Cóndor (Krabbe & Sornoza 1994).

Etymology Named in honour of Roger Tory Peterson (1908–1996), arguably the most influential promoter of popular ornithology in the 20th century.

CLOUD-FOREST SCREECH OWL
Megascops marshalli

Another small montane owl species from Andean South America that was first detected by the use of mist-nets. Described as an *Otus*, but presently classified in the exclusively New World genus *Megascops*. It remains one of the least-known owl species in South America.

Type information *Otus marshalli* Weske & Terborgh 1981, Cordillera Vilcabamba, Cuzco, Peru.

Alternative names Spanish: Urcututú de Marshall.

Discovery The type specimen was mist-netted in June 1967 at an altitude of 2,180 m in the Cordillera Vilcabamba, Dpto. Cuzco, Peru.

▲ Cloud-forest Screech Owl, adult. Apa Apa Reserve, La Paz, Bolivia, September 2016 (*Jacob and Tini Wijpkema*).

Description 20–23 cm. Monotypic. A medium-small owl with rather short ear-tufts. Facial discs reddish-brown with a conspicuous dark-brown perimeter and dark at base of bill; upperparts rich chestnut, with blackish barring and mottling, the central back deep reddish-brown with blackish barring; supercilium whitish; scapulars whitish with blackish area near tip, giving a lateral white area barred with blackish; crown reddish-brown, barred black; throat and upper cheeks rufous with dark barring, becoming more whitish with darker markings lower down on belly; primaries and secondaries barred dusky and tawny, rectrices with eight (ten in juvenile) rufous and blackish bars; iris dark brown; bill greyish-yellow; tarsi completely feathered.

Food and feeding Forages in the canopy; no data on food items but probably insectivorous.

Breeding Nest and eggs undescribed. In Peru, based on gonadal condition, probably late June to mid-August (Weske & Terborgh 1981); in Bolivia, breeding probably complete by late August (Herzog *et al.* 2009).

Habitat Wet mossy epiphytic cloud-forest with many bromeliads and a dense understorey, sometimes with *Chusquea* bamboo, pristine or only slightly disturbed, with annual rainfall about 3,000 mm; in Peru 1,900–2,250 m, in Bolivia 1,550–2,580 m (Herzog *et al.* 2009).

Voice In Bolivia, a continuous series of monotonic hoots, beginning quietly, growing in intensity and ending abruptly, lasting from 2.4–9.5 sec; also a trill-like short song. In the Cordillera de Yanachaga, Peru, lower-pitched but faster (Herzog *et al.* 2009).

Movements Probably sedentary.

Range Now known from six scattered localities, three in each of Peru (Dptos Pasco with suitable habitat in Junín; and western Cuzco) and Bolivia, western La Paz (one site) and two sites in Cochabamba. May occur in other unsurveyed areas but apparently truly absent in some drier areas of La Paz, giving a disjunct and patchy distribution stretching north-west to south-east over 1,280 km (Herzog *et al.* 2009).

Conservation status Currently classified by BirdLife International as Near Threatened, based on the small and separated total range (about 13,000 km^2) and its apparent requirement of pristine or near-pristine forest. In some locations, such as around the type locality in Peru and in Cochabamba, Bolivia, it is common (Weske & Terborgh 1981) or relatively common (MacLeod *et al.* 2005). More study needed, but could possibly be downgraded; clearly very sensitive to habitat modification.

Etymology Named in honour of Joe T. Marshall Jr., "in recognition of his long-standing interest in the genus *Otus* and night birds generally and of his contributions to the knowledge of these elusive and fascinating creatures" (Weske & Terborgh 1981).

HOY'S SCREECH OWL
Megascops hoyi

A small screech owl with a limited distribution in north-western Argentina and adjacent Bolivia. Treated by some authors as a race of the allopatric Variable Screech Owl *M. atricapillus*, but distinct both in vocalisations and in DNA (Heidrich *et al.* 1995). Currently placed in the American genus *Megascops*.

Type information *Otus hoyi* König & Straneck 1989, montane forest at La Cornisa, 40 km north of Salta, Argentina.

Alternative names Montane Forest Screech Owl, Yungas Screech Owl. Spanish: Autillo Fresco.

Discovery In September and October 1987 C. König

► Hoy's Screech Owl. Above Yala, Jujuy, Argentina, November 2016 (*Jacob and Tini Wijpkema*).

and R. Straneck observed several screech owls in montane forest in northwestern Argentina. These birds were distinct from Tropical Screech Owl *Megascops choliba* both in habitat choice and in vocalisations. Comparison of a specimen with other *Megascops* taxa from the area established its status as a full species.

Description 24 cm. Monotypic. A small screech owl with short ear-tufts, occurring in three colour morphs, grey, red and brown, the last being the most common. Brown morph has crown and back with darker streaks and vermiculations; a whitish band across the shoulder formed by dark-edged whitish outer webs of scapulars; the primaries and secondaries barred with lighter and darker brown; underparts paler than upperparts, with fine brownish barring and prominent blackish central steaks to feathers; facial disc greyish-brown with fine darker vermiculations bordered by an obvious darker ruff; rectrices densely barred and vermiculated. Grey and red morphs similar but differ in overall background colour. Iris bright deep yellow; bill yellowish-green; tarsi fully feathered.

Habitat Wet montane forest including cloud forest, 1,000–2,600 m, locally to 2,800 m; in Bolivia, also in drier forest at 1,100–1,250 m (Herzog *et al.* 1999).

Food and feeding Prey taken with talons, from upper layers of mature trees, in undergrowth and from the ground. Prey items mostly insects including locusts; one stomach contained small rodent bones.

Breeding Imperfectly known. Nests in cavities in trees including woodpecker holes. Eggs, probably 2–3, probably incubated by female alone, who is fed by the male. Breeding season likely to be September to late October in Argentina; song in August and September, copulation observed in mid-September, recently fledged young in late November (König *et al.* 2008).

Voice Territorial song of male is a long trill of clear staccato notes, about 11 notes per second, lasting 5–20 seconds, beginning softly then increasing in volume to a peak, then rapidly fading away. Female song, sometimes given in duet, is similar but higher in pitch, more tinny and less clear. A second, aggressive or excited song is used when other males intrude on its breeding territory; this is a short staccato sequence of *ou* notes, increasing in volume and ending abruptly. Other calls include a contact call given by both sexes, *chuio* and loud single drawn-out notes rising in pitch (König *et al.* 2008).

Movements No data, but short-range altitudinal movements may occur (König *et al.* 2008).

Range Southern Bolivia (Cochabamba southwards through Chuquisaca) (Krabbe *et al.* 1996) to NW Argentina, southwards through Salta to Tucumán, possibly Catamarca.

Conservation status Currently listed by BirdLife International as Least Concern; however, suspected of being in decline due to continuing habitat destruction. Fortunately much habitat is located on slopes too precipitous for human exploitation.

Etymology Named after G. Hoy, of Salta, Argentina, who collected the type specimen (König & Straneck 1989).

DESERT OWL
Strix hadorami

A taxon resulting from detailed analyses which showed that Hume's Owl *Strix butleri* is two species, one widely distributed but previously unnamed, the other with a more restricted range and almost unknown.

Type information *Strix hadorami* Kirwan *et al.* 2015, Lower Wadi Kelt, Palestinian territory NE of Jerusalem currently under Israeli administration, based on a specimen taken in March 1938.
Alternative names Desert Tawny Owl.
Discovery Hume's Owl *Strix butleri* was considered to be a species of wide but patchy distribution in the Middle East, from Israel, eastern Egypt and Jordan through the Arabian Peninsula. It is one of the least well-known species occurring in the Western Palaearctic. The provenance of the type specimen, taken by a Mr Nash, about whom very little is apparently known, is extremely confused; it probably originated from the coast of present-day Pakistan, although a location in the Arabian Peninsula is not impossible. In 1985, H. Shirihai was examining material at the Natural History Museum, Tring, and noted that this type specimen differed in significant respects from material taken mostly from Israel, as well as conflicting with his own extensive field experience in that country. He proposed to name the Israeli and adjacent populations as a new race of *butleri*, but pressure of other commitments prevented him from so doing. Then, in 2013 Robb *et al.* described a new *Strix* species, the Omani Owl *S. omanensis* (see Invalid Species), based on photographs and recordings, without specimen evidence. Consequently, Kirwan *et al.* (2015) undertook a detailed analysis. Their tentative recommendation, in the absence of specimen material for *omanensis*, was that that species and the type specimen of *butleri* be considered synonymous; but since material from Israel and nearby areas differs significantly from the Omani birds and the *butleri* type specimen, a different species is involved. As the *butleri* type specimen has priority, the populations previously ascribed to *butleri*, by far the most widespread geographically, require a new name, which they designate as *hadorami*. For a detailed account of the morphometric and genetic data which led to this conclusion, see Kirwan *et al.* (2015).

The confusion over the status of these owls appears to have been resolved by a recent paper (Robb *et al.* 2015) published online in August 2015 but not apparently peer-reviewed as yet, which discloses that DNA sequence data from the unique holotype of *S. butleri* and of '*S. omanensis*' are identical; consequently '*omanensis*' is best regarded as the junior synonym of *butleri*, confirming the hypothesis of Kirwan *et al.* [*S. butleri* is now known as Omani Owl, rather than its old name of Hume's Owl. The genetic material for '*omanensis*' was obtained by A. Walsh and

▲ Desert Owl, fledgling (left), male (middle) and female (right). Dead Sea, Israel, May 2016 (*Amir Ben Dov*).

◄ Desert Owl, adult male. Dead Sea, Israel, February 2015 (*Amir Ben Dov*).

M. Robb in March 2015 when they captured a live bird by tape-lure, took a few feathers and blood samples and then released it. Prior to capture the bird gave the characteristic four-note compound hooting. A recent record from NE Iran, confirmed by DNA evidence, extends the range of *butleri* by 1,400 km.]

Description Probably about 30–35 cm. Monotypic. In form a typical *Strix* owl, large-headed and lacking ear-tufts. Upperparts intricately mottled with buff, light sandy-buff and cinnamon; facial discs mainly pale sandy-grey or off-white; underparts pale horn with light rufous barring; vent and undertail-coverts unpatterned near-white; remiges sandy-ochre with five visible bars (primaries) or four (secondaries); rectrices with six dark bars; bare parts not described in life. Differs from Omani Owl by the presence of pale rufous cross-barring over the breast and belly, and paler facial disc.

Habitat Rocky gorges and canyons in arid areas, sometimes in ruins; also in acacias and palm groves. Sea level to 2,800 m.

Food and feeding Nocturnal hunter; prey items (from pellet analysis) include small mammals (rodents and insectivores), some passerine birds, arthropods (scorpions, grasshoppers and beetles) and geckos.

Breeding Relatively little known. One Israeli pair laid five (presumably white) eggs in early May, incubation period 34–39 days, fledging period 30–40 days (Shirihai 1996); a nest in Saudi Arabia had three eggs on 25 February.

Voice Long, soft low hoot followed by two short double notes.

Movements Resident.

Range From eastern Egypt, extreme NE Sudan, east and south Israel, Sinai Peninsula, very patchily across Saudi Arabia, Yemen and SW Oman (Dhofar).

Conservation status Classified (as Hume's Owl) by BirdLife International as Least Concern. Although now realised to be much more common than previously thought, there is evidence of significant decreases in population, at least in Israel, where vehicle mortality is apparently significant (Berkely 1999 HBW 5).

Etymology Named in honour of Hadoram Shirihai, whose observations have significantly broadened our knowledge of 'Hume's Owl'. Shirihai is an ornithologist of extraordinary versatility, having, among many other things, published definitive studies of the birds of Israel, the *Sylvia* warblers and the biology of the Antarctic.

CLOUD-FOREST PYGMY OWL
Glaucidium nubicola

Another new member of the genus *Glaucidium*, long overlooked due to its similarity to the widespread Andean Pygmy Owl *G. jardinii*. Recent studies of the genus have shown it to be greatly more complex taxonomically than previously realised.

Type information *Glaucidium nubicola* Robbins & Stiles 1999, south side of Quebrada San José of the Río Blanco, Carchi province, Ecuador.

Alternative names Spanish: Mochuelo Ecuatoriano.

Discovery In August 1987, during avifaunal inventory work near Mindo, Ecuador, a male pygmy owl was netted. Due to a lack of comparative specimens, no positive identification was made. However, a year later a bird was recorded and collected near Chical, near the border with Colombia. The song recorded was clearly distinct from that of *G. jardinii* while morphologically the specimen undoubtedly belonged to the same taxon as the 1987 example. G. Stiles, on examining the tape recordings of the Chical bird, noted that its voice was most similar to that of the disjunct population of pygmy owls in Costa Rica and Panama, then classified as *G. jardinii costaricanum*. However, molecular studies by Heidrich *et al.* (1995) suggested that *costaricanum* was a subspecies of Northern Pygmy Owl *G. gnoma*. Further specimens of the new taxon, including two taken 60 years previously, all misidentified as *G. jardinii*, were located in North American museums. Careful examination of specimens showed that the new taxon had consistent differences from both *jardinii* and *costaricanum*, and was also clearly differentiated on the basis of molecular DNA studies. In fact, the species was tape-recorded as early as June 1985, at a location in El Oro province in southern Ecuador, but this recording was overlooked until 1990 (Robbins & Ridgely 1990).

Description 16 cm. Monotypic. A small pygmy owl with head and upperparts brown, the crown and sides of the head with fine white flecks; back, mantle and sides of upper breast uniform brown; false 'eye-spots' on back of head dark brown bordered white; scapulars and upperwing-coverts with bold white spots, often tinged rufous; tarsi completely feathered; iris yellow; bill greenish-yellow. Differs from *G. jardinii* in having a shorter tail, less pointed wings and an unmarked mantle; from *G. costaricanum* by the lack of a large white area in the middle of the breast from throat to belly, and in the white, not rufous, borders on the false 'eye-spots' (König *et al.* 2008).

▲ Cloud-forest Pygmy Owl. Above Tandayapa Bird Lodge, Ecuador, November 2010 (*Bob Gress*).

Habitat Very wet primary cloud-forest, always on steep slopes; also in secondary and regenerating forest and forest borders. At one location in Cotopaxi province, Ecuador, occurs in young secondary forest with only ten years of regeneration from pasture (Freile *et al.* 2003). Usually 1,400–2,200 m, but at Buenaventura, El Oro province, Ecuador, as low as 900 m. Usually occurs at lower elevations than *G. jardinii* (Marks *et al.* 1999).
Food and feeding Stomach contents include insects (Orthoptera, Hemiptera) and a small lizard; also possibly small birds (Robbins & Stiles 1999). Partially diurnal.
Breeding Nest and eggs undescribed. Breeding season probably from February to June (Marks *et al.* 1999).
Voice Song is a long series of paired, short whistles, with about 0.2 seconds in between each of the paired notes and 0.32–0.37 seconds between the pairs. More resembles the songs of *G. gnoma* and *G. costaricanum* than that of *G. jardinii.*
Movements Presumably sedentary.
Range Colombia, from Risaralda south through the Andes to northern Ecuador (Carchi, Pichincha and Cotopaxi), with one disjunct record in southern Ecuador (El Oro). May occur at intermediate sites, but suitable remaining habitat is discontinuous.
Conservation status Currently classified by BirdLife International as Vulnerable, based on habitat destruction. The population is estimated to be in the range of 2,500 to 9,998 individuals, rounding out to 1,500 to 7,000 mature individuals. Although much of the relevant habitat is too precipitous for agriculture, logging can and does represent a significant threat. Overall it is suspected that the population is declining; indeed, it has already disappeared from one of its original locations in Colombia due to over-logging (Fierro-Calderón & Montealegre 2010).
Etymology *Nubicola*, Latin, 'cloud-inhabiting', from its preference for cloud-forest.

YUNGAS PYGMY OWL
Glaucidium bolivianum

Probably most closely related to the allopatric Andean Pygmy Owl *G. jardinii.*

Type information *Glaucidium bolivianum* König 1991b, La Cornisa de Jujuy, 40 km north of Salta, Argentina.
Alternative names Spanish: Caburé Yungueño, Mochuelo boliviano.
Discovery Described in 1991 as part of a comprehensive review of the South American pygmy owls conducted by C. König and colleagues, based on extensive studies of vocalisations.
Description 15–16 cm. Monotypic. A typical small pygmy owl, large-headed, lacking ear-tufts, with false eye-markings on the nape. Occurs in three colour morphs, grey, brown and rufous, the brown being the most common, the grey apparently restricted to southern part of range. Grey morph is dusky grey-brown above with pale spots; crown with dense whitish spots; obvious paler eyebrows; facial disc light greyish-brown; upperparts dusky grey-brown with whitish or pale brownish-yellow spots; two 'false eyes' on nape, blackish, bordered with whitish above and pale ochre below; remiges mid-brown with prominent buffy-white spots on the webs, giving a pale line on closed wing; throat and centre of breast and belly whitish, chest dusky-brown, lower chest and belly with broad diffuse brown streaks; tail blackish-brown with five or six pale bars; iris golden-yellow; bill and cere greenish-yellow; tarsi feathered with dirty-yellow toes. Brown morph generally warm dark brown; rufous morph generally rusty orange-brown.
Habitat Wet epiphytic montane cloud-forest with alder (*Alnus*) or *Podocarpus*, mostly 1,400–3,000 m, occasionally down to 900 m.
Food and feeding Feeds in mid or upper levels of forest; prey items include insects, small birds, possibly reptiles. Nocturnal and diurnal, but less active by day than other Andean pygmy owls.

▼ Yungas Pygmy Owl. Manú Biosphere Reserve, Peru, March 2010 (*Matthias Dehling*).

▲ Yungas Pygmy Owl. Leymebamba, Peru, August 2013 (*Fabrice Schmitt*).

Breeding Nest and eggs undescribed; nests in old woodpecker holes. Female alone incubates, fed by male (König *et al.* 1999).

Voice Song starts with 2–3 melodious fluted whistles (often omitted), ending with a tremolo, followed by a series of slightly drawn-out, hollow staccato notes, equally spaced and at a slow tempo; female song similar but higher pitched, sometimes duets with male. Possible contact call is a soft, faint *whoeeo* (König *et al.* 1999).

Movements No data; largely sedentary, but altitudinal movements possible.

Range Eastern slope of Andes, from northern Peru through eastern Bolivia to NW Argentina (Salta, Jujuy and Tucumán).

Conservation status Currently classified as Least Concern; although some populations may have declined due to habitat destruction; widely distributed and in some areas apparently common; some parts of range protected by national parks (e.g. El Rey and Calilegua in Argentina and Cotopata in Bolivia).

Etymology *Bolivianum*, from Bolivia.

SUBTROPICAL PYGMY OWL
Glaucidium parkeri

Another species of small South American owl whose separate specific identity was finally uncovered by painstaking and careful analysis of specimens and tape recordings.

Type information *Glaucidium parkeri* Robbins & Howell 1995, Cerros del Sirá, Dpto. Huánuco, Peru.

Alternative names Spanish: Mochuelo de Parker, Mochuelo de Zamora-Chinchipe.

Discovery In 1969 and 1970, J. Weske and J. Terborgh mist-netted two pygmy owls in the Departamentos of Ayacucho and Huánuco on the eastern slope of the Andes in central Peru. Due to their small size the specimens were identified as Least Pygmy Owl *Glaucidium minutissimum* and remained unremarked in the American Museum of Natural History, New York, for more than 20 years. In the early 1990s, M. Robbins and S. Howell, as part of a study that they were doing on *minutissimum* (Howell & Robbins 1995) had occasion to examine the specimens, concluding that a new species was involved. These authors were sent a recording of a pygmy owl made in 1991 by B. Whitney on Volcán Sumaco, Ecuador; Robbins, on hearing this, realised the song was identical with recordings that he himself had made in the Cordillera de Cutucú in south-central Ecuador and had, at the time, identified as Andean Pygmy Owl *G. jardinii*. The suspicion that all of these represented an undescribed species was confirmed when F. Sornoza collected a specimen in July 1992 in the province of Zamora-Chinchipe in SE Ecuador.

Description 14 cm. Monotypic. A tiny owl lacking ear-tufts. Upperparts dark brown, greyer on crown, conspicuously spotted with white; 'false-eye' markings on nape dark brown, mantle washed olivaceous, scapulars boldly spotted with white; underparts white with the sides of neck and upper chest rufous-brown, and prominent olivaceous-brown lateral streaks on flanks; facial disc greyish-brown, finely speckled white and brown; rectrices blackish with five distinct pale bars; iris yellow; bill yellowish-green; tarsi feathered, toes yellow with darker claws.

Habitat Subtropical evergreen forest at 1,450–1,975 m; all known localities are on outlying ridges (Marks *et al.* 1999); at lower elevations than *jardinii* and higher than other *Glaucidium* species in the *minutissimum* group (Robbins & Howell 1995).

Food and feeding Partially diurnal. Prey probably mainly insects, possibly small invertebrates.

Breeding Nest and eggs undescribed. Other members of the genus are cavity-nesters, frequently using woodpecker holes.

Voice Distinctive and characteristic; quite unlike other South American *Glaucidium* owls. Song is two to four relatively short, low-pitched hoots, with increasing

▲ Subtropical Pygmy Owl. Refugio Los Volcanes, Bolivia, September 2015 (*Richard Hoyer*).

length of pauses between notes, especially between final two notes (Marks *et al.* 1999), somewhat reminiscent of an antpitta.

Movements Presumably sedentary.

Range Imperfectly known and probably discontinuous. Eastern slope of Andes from Volcán Sumaco, Napo, Ecuador, south through southern Ecuador (Cordilleras de Cutucú and Cóndor and Zamora-Chinchipe), central Peru (Huánuco, Ayacucho) and northern Bolivia (König *et al.* 2008). Recent records in the Colombian Andes (Acevedo-Charry *et al.* 2015).

Conservation status Presently unclassified by BirdLife International. Generally regarded as uncommon, but possibly frequently undetected; more fieldwork required.

Etymology Named in honour of the late Theodore A. Parker III. Ted Parker was without doubt the most accomplished field ornithologist that this author has ever encountered, with an aural memory for bird calls and songs that verged on the uncanny. In 1974 he was recruited by George Lowery at Louisiana State University for fieldwork in Peru, an eight-month stint that set him firmly on a career that would see him well on the way to becoming one of the foremost authorities on Neotropical birds. Over the next 20 years he made enormous contributions to our knowledge of the taxonomy, distribution and conservation of the avifauna of South America.

On 3 August 1993, he was killed, along with several other people including Al Gentry, one of the foremost experts on the flora of Ecuador, when their light aircraft flew into a mist-covered mountain in western Ecuador while on a biological survey flight. He is remembered by the specific epithet *parkeri* attached to five full species (*Cercomachra, Glaucidium, Herpsilochmus, Phylloscartes and Scytalopus*) and three subspecies, while the genus *Caryothraustes* has been renamed *Parkerthraustes* (Robbins *et al.* 1997).

AMAZONIAN PYGMY OWL
Glaucidium hardyi

The precise relationships of this species within the *Glaucidium* complex of South America are not completely defined and require further study; its closest relatives may, in fact, be Andean and Yungas Pygmy Owls *G. jardinii* and *G. bolivianum* rather than Least Pygmy Owl *G. minutissimum*.

Type information *Glaucidium hardyi* Vielliard 1989, 20 km south-west of President Medici, Rondônia, Brazil.

Alternative names Hardy's Pygmy Owl. Spanish: Mochuelo amazónica; Portuguese: Caburé-da-amazônia; French: Chevêchette d'Amazonie.

Discovery Described by J. Vielliard in 1989, by virtue of bioacoustic and DNA studies.

► Amazonian Pygmy Owl. French Guiana, September 2011 (*Tanguy Deville*).

Description 14–15 cm. Monotypic. A typical *Glaucidium* pygmy owl, tiny, without ear-tufts and with dummy eye-spots on back of head. Facial disc pale greyish-brown, flecked with fine brownish; crown greyish-brown with small off-white dots and larger scaling; false eye-spots blackish, lined with lighter areas; mantle unspotted rufescent earth-brown; wing coverts with a few inconspicuous white spots; flight feathers barred light and dark; tail dark brown with normally three broken whitish bars, absent in rusty morph; underparts off-white, the sides of the upper breast densely mottled with rufous-brown; flanks and lower underparts boldly streaked reddish-brown; iris bright yellow to golden-yellow; bill yellowish-horn, tinted olive or greenish; cere dirty greenish-yellow; tarsi feathered, toes golden-yellow.
Habitat Primary tropical rainforest, including tall *terra firme* and *várzea*, sea level to 850 m.
Food and feeding Hunts in epiphytic canopy. Food items primarily insects, but also probably small vertebrates.
Breeding Few data. Nests in cavities, especially old woodpecker holes. One nest in central Brazil had almost fledged chicks in mid-September; a second nest in French Guiana in late September, located in an old woodpecker nest 25 m up, had young.
Voice Song is a rapid series of 12–36 notes running into a quavering roll or trill, at a variable tempo of 10–13 notes per second; female voice higher in pitch; sings during day. Voice louder and more ringing than soft trills given by other members of *G. minutissimum* complex (Holt *et al.* 2016).
Movements Presumably sedentary.
Range Widespread in lowland South America, from SE Venezuela (Bolívar), east through the Guianas to northern Brazil (Pará), eastern Peru from Loreto to Madre de Dios, northern and eastern Bolivia, Mato Grosso.
Conservation status Currently classified by BirdLife International as Least Concern; although it may be relatively uncommon, and has lost much habitat to deforestation, it has a very extensive range.
Etymology Named in honour of J. W. Hardy in recognition of his pioneering work in documenting the vocalisations of many birds including the New World owls.

PACIFIC PYGMY OWL
Glaucidium peruanum

A very well known and widely distributed owl, whose separate identity (from Ferruginous Pygmy Owl *G. brasilianum*) was only recently elucidated by C. König and colleagues.

Type information *Glaucidium peruanum* König *et al.* 1991, Ninabamba, near Ayacucho, Dpto. Apurímac, Peru.
Alternative names Peruvian Pygmy Owl. Spanish: Mochuelo Peruano; local: Paca-paca.
Discovery Previously treated as conspecific with Ferruginous Pygmy Owl until careful studies of vocalisations, backed up by DNA evidence, established its full specific status.
Description 15–17 cm. Monotypic (but see below). Exists in three colour morphs. Grey morph: upperparts dark greyish-brown, the mantle and back spotted with white, the forehead with fine white streaking; facial disc indistinct, bordered above by short white eyebrows; 'false eyes' on nape blackish, edged white above and ochre below; flight feathers dark greyish-brown with whitish or yellowish spots on both webs; underparts and throat whitish, the chest densely mottled with greyish-brown, the belly and flanks boldly streaked with greyish-brown; tail dark greyish-brown with six or seven broken whitish bars; iris yellow, bill and cere yellowish-green, tarsi feathered grey-brown, toes yellow. Brown morph similar, generally dark earth-brown; red morph rusty-brown above, the tail rufous, barred with orange-buff or paler russet. Peruvian and Ecuadorian birds have apparently different tail-bars. Birds from SE Ecuador and NE Peru also appear to have different vocalisations, suggesting that further study is warranted.
Habitat Mostly dry or semi-arid habitat, including dry woodland, thorny scrub, farmland with scattered trees and even suburban parks. Sea level to 3,000 m.
Food and feeding Rather few data. Nests in holes, including old woodpecker holes in trees and cacti; also in old nests of horneros (*Furnarius*). Incubation by female alone.
Voice Song of male is a very rapid sequence of short, upslurred staccato notes, 6–7 per second, faster than that of *brasilianum*. Female song similar but higher pitched. When excited, gives a short, high-pitched *chirrp* (König *et al.* 1999).
Movements Apparently sedentary.
Range Pacific South America, from western Ecuador (Manabí) south through western Peru to northern Chile.
Conservation status Currently classified as Least Concern; widely distributed, often in quite heavily modified habitats.
Etymology *Peruanum*, from Peru.

▶ Pacific Pygmy Owl, typical brown morph. Santa Eulalia Road, Lima, Peru, October 2011 (*Dubi Shapiro*).

PERNAMBUCO PYGMY OWL
Glaucidium mooreorum

A species "discovered on the brink of extinction" with a tiny population. Not all authorities accept that it is a valid species, there being mixed opinions as to its separateness from *G. minutissimum*. To complicate matters, König *et al.* (2008) state that *mooreorum* is a synonym for *minutissimum*, but since its distribution is limited to NE Brazil, a new name is required for populations further south in Brazil as well as in eastern Paraguay, which they call *G. sicki*. This treatment is not accepted by the SACC (Proposal 243) and *sicki* is not recognised by the *Handbook of the Birds of the World*, although Mikkola (2014) gives it full specific status.

Type information *Glaucidium mooreorum* da Silva *et al.* 2002a, based on a specimen taken in 1980, Reserva Biológica de Saltinho, Rio Formoso, Pernambuco, Brazil.

Alternative names Portuguese: Caburé-de-Pernambuco.

Discovery There exist only two specimens, both currently in the Universidade Federal de Pernambuco, of pygmy owls from Pernambuco, a location geographically well removed from the ranges of Amazonian Pygmy Owl *Glaucidium hardyi* and Least Pygmy Owl *G. minutissimum*. This caused de Silva *et al.* to examine the material critically as well as comparing songs recorded in 1990 to those of other members of the genus, resulting in their description of it as a new species.

Description 21 cm. Monotypic. A typical *Glaucidium* owl, tiny, with chestnut upperparts, the wing-coverts spotted with cinnamon; flight feathers barred lighter and darker brown; face a mixture of white and brown streaking; rectrices blackish-brown with five incomplete whitish bars; underparts white, the sides of the chest brown, streaked with white, flanks white, coarsely streaked brown; iris yellow; bill greenish-yellow; tarsi and toes orange-yellow. Differs from *G. hardyi* in having a shorter wing-chord and longer tail; from *G. minutissimum* in its longer tail, and from both in its generally paler plumage and in vocalisations.

Habitat Humid lowland forest, sea level to 150 m.

Food and feeding Few data; observed to eat a large cicada; doubtless other insects and possibly small vertebrates.

Breeding Nest and eggs undescribed.

Voice Distinctive, a series of five to seven (usually six) notes, the phrase lasting 1.37–1.52 seconds (da Silva *et al.* 2002). The song of *G. hardyi* has more and shorter notes; that of *G. minutissimum* has fewer and longer notes (da Silva *et al.* 2002).

Range Endemic. Now confined to a tiny area of Pernambuco in east Brazil.

Conservation status Classified by BirdLife International as Critically Endangered, but possibly Extinct. Currently confined to two remaining patches of forest, the Reserva Biológica de Saltinho (4.8 km²) and a 100 ha patch at Usina Trapiche. Absent from several other patches of forest in Alagoas and Pernambuco. Doubtless formerly much more widespread; forest cover in the 'Pernambuco area of endemism' has been reduced from about 40,000 km² to less than 2,000 km² (da Silva *et al.* 2002). Total world population may be as few as 50 individuals (BLI SF 2013). According to Lees *et al.* (2014) there have been no observations since November 2001.

Etymology Named in honour of Dr Gordon Moore and his wife Betty Moore, "who have greatly contributed to biodiversity conservation worldwide". Dr Moore was one of the pioneers in the early days of computer science, being described as one of the "founders of Silicon Valley", and originator of Moore's Law, which predicted an exponential growth of computer power.

ALBERTINE OWLET
Glaucidium albertinum

An almost unknown species, of uncertain distribution, and with virtually no published information on its biology. Placed in the genus *Taenioglaux* by some authors (e.g. König *et al.* 2008).

Type information *Glaucidium albertinum* Prigogine 1983, Musangakye, Zaire (Democratic Republic of Congo).

Alternative names French: Chevêchette du Graben.

Discovery The identity of the small owls from the Albertine Rift area of eastern Zaire, now the Democratic Republic of Congo (DRC), and adjacent Rwanda has been the source of much debate for over 100 years, with various authorities merging Chestnut Owlet *Glaucidium castaneum* with the widely distributed African Barred Owlet *G. capense*, or maintaining their separateness. In 1983 A. Prigogine made a careful analysis of five specimens, taken between 1950 and 1981, one from Rwanda and the remainder from three locations in the DRC, and concluded that a new species distinct from both *capense* and *castaneum* was involved.

Description 21 cm. Monotypic. A very small owl without ear-tufts or very prominent facial disc; upperparts warm maroon-brown, the forehead, crown and nape with creamy spots; back unmarked maroon-brown; outer webs of scapulars yellowish-white, forming a lateral line; flight feathers apart from outer three primaries barred with lighter and darker brown; chin white; throat brown; upper breast brown with yellowish-white bars; belly white with large maroon spots; iris yellow; bill yellowish; legs feathered; toes yellow. Differs from *capense* in having the head spotted, not barred, and in lacking bars on the back.

Habitat Humid open and transitional forest with rich undergrowth, 1,100–1,700 m (König *et al.* 2008), very probably to 2,500 m (BLI SF 2013).

Food and feeding Stomach contents of one specimen contained a beetle and a grasshopper; no other data.

Breeding Nest and eggs undescribed.

Voice Gives short *fyi-fyi* whistles, followed by a long series of *kurr-kurr-kurr* notes (Mikkola 2014). Critical comparison of good recordings of this and other members of the genus would be instructive.

Movements Probably sedentary.

Range Four specimens from three locations in NE DRC; one from Rwanda; one recent sight record from west of Lake Kivu, DRC.

Conservation status Currently listed by BirdLife International as Vulnerable. There is a clear need for more data; from the paucity of records in a well-studied area it may be rare, but given its apparently retiring nature it may be more common than it appears. Forest clearance for agriculture and lumber is widespread; the human population, rapidly increasing naturally and augmented by refugees in an area of perennial political instability, is putting further pressure on the habitat.

Etymology From the Albertine Rift, although the species' range does not approach Lake Albert itself.

LONG-WHISKERED OWLET
Xenoglaux loweryi

The discovery of this extraordinary tiny owl caused a huge flurry of interest at the time in the popular press as well as in ornithological circles – surely one of the most remarkable new Neotropical species in recent years. A close relationship to the pygmy owls (*Glaucidium*), as postulated by O'Neill & Graves (1977), is regarded as unlikely by König *et al.* (2008).

Type information *Xenoglaux loweryi* O'Neill and Graves 1977, 10 km north-east of Abra Patricia, Dpto. San Martín, Peru.

Alternative names Spanish: Mochuelo peludo, Mochuelo de Lowery.

Discovery In August 1976 an expedition from Louisiana State University was camped in near-pristine subtropical forest on the eastern slopes of the Andes, north-west of Rioja in northern Peru. The conditions were physically unpleasant, with continuous pouring rain, and ornithologically unproductive. Despite this, an expedition assistant, Manuel Sánchez (see under Varzea Thrush *Turdus sanchezorum*), managed to put up a number of mist-nets, without success. In view of the conditions it was decided that if no birds were captured by the third day the expedition would move to a more productive and hopefully drier location; but on the morning of the third day, Sr Sánchez returned with what he said was an owl in a bag. The bird taken out of the bag proved to be utterly unlike any other owl ever previously encountered; in the words of O'Neill, "shaped like an *Otus* (screech owl), the size of a *Glaucidium* (pygmy owl) and coloured like a *Lophostrix* (crested owl)."

Description 13–14 cm. Monotypic. Tiny, similar to Elf Owl *Micrathene whitneyi* of Mexico and SW USA, the smallest owl in the world. Warm brown above, densely vermiculated with dark brown to blackish; prominent yellow-white eyebrows; a collar of large whitish spots on lower nape; scapulars with whitish subterminal spots; primaries dull black, the edges of the outer webs with small light spots; rectrices dull brown with lighter and darker mottlings; underparts brown with whitish vermiculations; large long whiskers present around the bill and even longer upswept whiskers at the sides of facial disc, extending beyond the outline of the head; iris amber-orange to orange-brown; bill greenish-grey, tipped yellow; tarsi unfeathered, flesh-pink.

Habitat Humid epiphytic cloud-forest with dense undergrowth, 1,900–2,200 m.

Food and feeding No data; prey probably insects.

Breeding Nest and eggs undescribed. The three specimens taken in August and September by O'Neill and Graves were all adults with slightly enlarged gonads.

Voice Not well known, although several recordings now in existence. Call is a deep, husky *who*, dropping very slightly at the end, repeated at rate of one per three seconds; also a series of three to five similar

► Long-whiskered Owlet. Abra Patricia Lodge, Peru, November 2011 (*Dubi Shapiro*).

whistles followed by a series of faster, slightly higher-pitched notes which could be the song (Weick 2001).

Movements Presumably sedentary.

Range Endemic. Known only from five locations, in the Departamentos of San Martín and Amazonas, Peru.

Conservation status Currently classified by BirdLife International as Endangered. Following the discovery of the species in 1976, two individuals were mist-netted near Bagua, Amazonas, in October 1978. It then 'vanished' for more than 20 years. Subsequently another bird was captured in August 2002 (Balchin 2007), but the first actual observation of a free-flying Long-whiskered Owlet in the wild was not made until February 2007 (Brinkhuizen *et al.* 2012). In 2010 a population was discovered near La Esperanza, Amazonas, with at least five individuals located in one night (American Bird Conservancy 2013). Current BirdLife International population estimates are in the range of 250–1,000 individuals. Recently, information has been provided that affords visiting birders a good chance of seeing the species (Engblom 2010). Some of the known locations, for example at Abra Patricia and La Esperanza, are now fully protected; a reserve created for the Critically Endangered Yellow-tailed Woolly Monkey *Oreonax flavicauda* also protects the owlet. The main threat to other areas is habitat destruction.

Etymology Named in honour of George H. Lowery (1913–1978). Professor Lowery was largely responsible for the pre-eminent status of Louisiana State University in Neotropical ornithology studies, a position maintained to this day. Incidentally, all of the work on *Xenoglaux loweryi*, including the description and the preparation of the colour plate later published with the type description in *The Auk*, was kept secret from Prof. Lowery until, at a museum party, he was presented with the specimens, the manuscript and the plate (Howell & O'Neill 1981).

LITTLE SUMBA HAWK-OWL
Ninox sumbaensis

A small *Ninox* owl, sympatric with the well-known Sumba Hawk-Owl *N. rudolfi*, whose identity was the source of much confusion for a number of years.

◄ Little Sumba Hawk-Owl. Langgaliru National Park, Sumba, Lesser Sundas, Indonesia, July 2010 (*James Eaton*).

Type information *Ninox sumbaensis* Olsen *et al.* 2002, about 45 km west of Waingapu, Sumba, Indonesia.

Alternative names Little Sumba Boobook.

Discovery The presence of an undescribed small owl on the island of Sumba, Indonesia, has been suspected since the 1980s. In view of its small size, and in the absence of specimens or accurate information about its vocalisations, most authorities (e.g. King & Yong 2001) believed it to be an unknown *Otus* (scops) owl, although it was also suggested, rather implausibly, that the reports might have originated from confusion with the sympatric, but much larger, Sumba Hawk-Owl *N. rudolfi* (Marks *et al.* 1999 HBW 5). Finally in December 2001, J. Olsen and S. Trost observed, video-recorded and photographed three pairs of owls in an area east of Waingapu, Sumba. A few days later a specimen, shot by a local hunter, apparently at a location about 4 km from their earlier observations, came into their possession. This single specimen was used for the description of the holotype, with samples of feathers used for genetic analysis; the body was retained by local villagers. The position of the new taxon in *Ninox* rather than *Otus* was unequivocally established by the genetic analysis.

Description 23 cm. Monotypic. A small *Ninox*, its weight perhaps only one-third that of *N. rudolfi* (Olsen 2011). Upperparts light brown, barred with dark brown; face grey with white eyebrows; scapulars edged white with fine black bars; remiges barred with rufous and dark brown; rectrices light rufous-brown, with fine dark brown bars; throat and chest rufous, narrowly barred darker; lower chest and belly white marked with dark chevrons; iris yellow; bill horn-colour, tipped yellow; tarsi yellow with bristles. Differs from *N. rudolfi* by much smaller size, by lack of white spots on crown, and yellow, not brown, eyes (Olsen *et al.* 2002). Juvenile apparently more rufous without chevrons on underparts.

Habitat Known only from primary and secondary forest, at 600–950 m; unlike *N. rudolfi*, does not hunt in open areas outside forest.

Food and feeding No published data.

Breeding Nest and eggs undescribed. Probably breeds late in the year, since local people report young just out of the nest in late November and a dependent fledged juvenile was seen on 24 December (Olsen *et al.* 2009).

Voice Song quite distinct from that of *N. rudolfi*, a single, monosyllabic flute-like *hoot*, uttered every 2.5–3 seconds; female's call a little higher than male's; begging juvenile makes a *churr* (Olsen *et al.* 2009).

Movements Presumably sedentary.

Range Endemic. Confined to Sumba Island, southern Indonesia.

Conservation status Classified by BirdLife International as Near Threatened, with a population placed in the range of 15,000–29,999 total individuals; however, Olsen *et al.* (2002) found it at three locations only. Closed canopy forest covered only 11% of Sumba in 1995 and was much fragmented (Jones *et al.* 1995).

Etymology Both the specific and common names derive from the island of Sumba.

CINNABAR HAWK-OWL
Ninox ios

A striking and distinctive – and still largely unknown – hawk-owl with a very restricted distribution.

Type information *Ninox ios* Rasmussen 1999, Bogani Nani Wartabone National Park, Northern Sulawesi, Indonesia.

Alternative Name Cinnabar Boobook.

Discovery In April 1985 F. G. Rozendaal mist-netted a strikingly bright rufous hawk-owl in Bogani Nani Wartabone National Park, north Sulawesi, Indonesia. It was, at the time, ascribed to a "previously undescribed rufous morph" of Ochre-bellied Hawk-Owl *Ninox ochracea,* a species widely distributed over much of the island of Sulawesi. In 1998 P. Rasmussen examined the specimen in the course of her studies of small owls and realised that, quite apart from its obvious colour differences, it was also distinct from *ochracea* in a number of other features; smaller size, narrower and more pointed wing, more slender tarsi and shorter, more shallow bill which had a pale, not dark, base. On this basis the unique specimen was designated as the holotype of the new species. Since that time a limited number of field observations have been made of live birds (see below).

Description 22 cm. Monotypic. A relatively small *Ninox* owl without ear-tufts; plumage strikingly reddish-brown above; the scapulars with triangular whitish dots; primaries indistinctly barred with chestnut and brown; tail chestnut with narrow darker bars; chest bright rufous, lower chest and belly with indistinct barring; iris yellow, eyelids pink; bill ivory; tarsi short and slender, with fawn feathering.

Habitat Middle elevation forest, 1,120–1,480 m in northern Sulawesi, and at 1,700 m in central Sulawesi

► Cinnabar Hawk-Owl. Gunung Ambang, Sulawesi, Indonesia, March 2009 (*Philippe Verbelen*).

(probably at higher elevations that *N. ochracea*). In Gunung Ambang Nature Reserve occurs in a mosaic of primary forest, cleared areas and cultivation (King 2005), and in selectively logged primary forest (Hutchinson *et al.* 2006).

Food and feeding Little published data; suggested that prey is predominantly flying insects, captured by sallies from exposed branches, sometimes hovering (Hutchinson *et al.* 2006).

Breeding Nest and eggs undescribed.

Voice Song is a two-note *wruck-wruck*, with about 0.5 seconds between notes, quite unlike the song of *N. ochracea* (King 2005). A secondary call is a single, dry, nasal, hissing shriek, becoming louder and harsher then ending abruptly (Hutchinson *et al.* 2006). Populations in southern Sulawesi have a slower-tempo song (Eaton *et al.* 2016b).

Movements Probably sedentary.

Range Endemic. Most records are from the northern arm of Sulawesi, specifically in the east-central part of Bogani Nani Wartabone National Park and Gunung Ambang Nature Reserve, about 30 km to the east. One 1998 record, only positively identified in 2000, at 1,700 m in Lore Lindu National Park, Central Sulawesi, 600 km south of previous locations.

Conservation status Currently classified by BirdLife International as Vulnerable. Seems to be relatively rare, although its known locations are little investigated (Lee & Riley 2001). In the Gunung Ambang location it appears to be able to exist in quite modified or partially degraded habitat. Although both northern sites are nominally protected, habitat destruction, hunting, and the use by local people of mist-nets to catch bats for food are significant problems (Lee & Riley 2001).

Etymology Specific name *ios* from the Greek for rust. It has been suggested (King 2005) that 'Chestnut' would be a more accurate English name than 'Cinnabar'.

CAMIGUIN HAWK-OWL
Ninox leventisi

One of several new *Ninox* owls from the Indonesian and Philippine archipelagos, whose discovery has been the result of careful study of specimens and vocalisations.

Type information *Ninox leventisi* Rasmussen *et al.* 2012, Catarman Mountain, Catarman, Camiguin Sur Island, Camiguin Province, Philippines.

Discovery The 'Philippine Hawk Owl', for the last 60 years treated as one polytypic species, has long been recognised as being very diverse. However, it was not until a very thorough, painstaking study of the group and particularly of its vocalisations was made by P. Rasmussen and her colleagues that the true taxonomy was revealed, involving seven allopatric species and one subspecies. Although specimens have existed in museums for, in some cases, many years, two species and one subspecies were new to science. The holotype of Camiguin Hawk Owl was in fact taken in June 1968 by the distinguished Philippine ornithologist, D. S. Rabor.

Description 25 cm. Monotypic. A medium-sized round-headed owl; crown and nape dark brown, barred buff, mantle warmer brown, scapulars with white patches on outer webs giving a pale patch on the closed wing; tail dark brown, narrowly barred with dull buff; face plain brown; throat with conspicuous white or buff-white patch; underparts brown or tawny, profusely barred with darker brown; iris uniquely grey or very pale yellow-green; bill yellowish, unfeathered parts of tarsi and toes mustard-yellow.

◄ Camiguin Hawk-Owl. Camiguin Sur Island, Philippines, June 2012 (*Bram Demeulemeester*).

Habitat Forest.

Food and feeding No data available.

Breeding Nest and eggs undescribed.

Voice Very low-pitched, typically short strophes repeated after brief pauses, with many rapid irregular barking notes per strophe (Rasmussen *et al.* 2012).

Movements Probably sedentary.

Range Endemic. Restricted to the island of Camiguin Sur, Philippines.

Conservation status Classified by BirdLife International as Endangered; confined to the 238 km² island of Camiguin Sur, of which perhaps 20 km² is virgin habitat. The total population may be fewer than 2,500 mature individuals (BirdLife's Globally Threatened Bird Forum 30/11/2013).

Etymology Named in honour of Anastasio P. Leventis, Vice-President of BirdLife International, in recognition of his many and continuing contributions to bird conservation and ornithological research.

CEBU HAWK-OWL
Ninox rumseyi

A second new *Ninox* from the Philippines, uncovered by the work of P. Rasmussen and her co-workers.

Type information *Ninox rumseyi* Rasmussen *et al.* 2012, Cebu, Philippines, based on a specimen taken in March 1888.

Discovery The description was based on photographs of at least nine live birds, taken by C. Artuso in January 2012, and on the unique specimen, collected by F. S. Bourns and D. C. Worcester in March 1888, Natural History Museum, Tring, ex. Norwich Castle Museum, and labelled *Ninox philippensis spilonota*; the distinctness from other Philippine *Ninox* species was confirmed by detailed analysis of vocalisations.

Description 25 cm. Monotypic. Very similar to other members of the genus in the Philippines; dark brown above, the scapulars with pale edgings; face brown with indistinct whitish eyebrows and white throat; underparts, flight feathers and greater primary coverts brownish, barred buffish; underparts pinkish-buff to rusty-brown, weakly barred with dark brownish; rectrices dark brownish with buff bars; iris lemon- to olive-yellow. More contrastingly patterned above and on the head than Romblon Hawk-Owl *N. spilonota*; most similar to Mindoro Hawk-Owl *N. mindorensis,* but much larger, with more widely spaced bars on rather plain underparts (2013 HBW SV).

Habitat Various types of forest, both interior and edge, occasionally in plantations (Jakosalmen *et al.* 2012).

Food and feeding Very varied: recorded prey includes rats, small birds, colubrid snakes, agamid lizards, skinks, geckos, frogs and toads, large insects including cicadas, mantises, moths, stick-insects and crickets.

Breeding Two nests only found so far, both in natural cavities ca 2–2.5 m up in tall trees. One nest contained two eggs in early May. No further details.

► Cebu Hawk-Owls. Tabunan, Cebu, Philippines, May 2016 (*Ross Gallardy*).

Voice Song highly variable; medium-pitched, irregular strophes delivered rapidly with multiple note types, mainly either gruff staccato clucks and plaintive short downslurs, but various other notes recorded. Sings singly and in duet. Song most similar to that of *N. spilonota* (Romblon Hawk-Owl), but faster, and lacking the long pauses between the notes in a strophe typical of that species (2013 HBW SV).
Movements Probably sedentary.
Range Endemic. Confined to limited areas on Cebu Island, Philippines.
Conservation status Currently assessed by BirdLife International as Endangered (BirdLife's Globally Threatened Bird Forum, 30/11/2013). Cebu has been comprehensively deforested, with only about 17 km^2 (0.3% of the area of the island) in 11 individual patches remaining. Surveys in five of these patches in 2011 revealed 52 owls, extrapolating to a total population of 130–300 pairs. Habitat destruction may be continuing (Jakosalem *et al.* 2012).
Etymology Named in honour of Stephen J. Rumsey (b. 1950), in recognition of his contributions to conservation and of his support for BirdLife International.

TOGIAN HAWK-OWL
Ninox burhani

One of three new hawk-owls recently described from Indonesia.

Type information *Ninox burhani* Indrawan & Somadikarta 2004, Benteng village, Togian Island, Togian archipelago, in Gulf of Tomini near Sulawesi, Indonesia.
Alternative names Togian Boobook.
Discovery The Togian Islands form a small archipelago of about 35 islands located between the northern and eastern arms of Sulawesi. They have until recently been relatively little investigated ornithologically, with apparently rather brief visits in 1871, 1939, 1980 and 1987 (Indrawan *et al.* 2006). Previously, no owls of any species had been recorded from the islands; however, in December 1999 unidentified owl vocalisations were heard and up to three birds were seen. Birds were again seen in February 2001. The type specimen was captured alive using a catapult and small stone. This individual was taken to Jakarta for veterinary care, but, perhaps not surprisingly, died 48 hours later and was prepared as the holotype. Once field observations had ascertained that the species was widespread in the archipelago, it was considered ethical to take a second specimen as the sole paratype, one year later and at a different location.
Description 20 cm. Monotypic. A medium-sized owl lacking ear-tufts. Upperparts dark brown with paler spots and fine bars; scapulars and some coverts with white spots; primaries amber-brown with lighter bars; rectrices rather long, dark greyish-brown with narrow paler bars; face brown with obvious pale supercilia; underparts, throat and chest brown or reddish-brown with paler feather-edgings; belly white with brown streaks; iris orange-yellow, bill olive-grey with a pale creamy tip; ground colour of legs and toes dull yellow or brownish, unfeathered but bristled.

◀ Togian Hawk-Owl. Togian Islands, Central Sulawesi Province, Sulawesi, Indonesia, November 2015 (*James Eaton*).

Habitat In remnants of tropical lowland forest, sea level to 400 m, but also in quite disturbed habitat including mixed gardens, sago swamps, etc.
Food and feeding No data recorded.
Breeding Nest and eggs undescribed.
Voice Song is quite distinct from that of other members of the genus: a throaty, grating *kuk kuk-kukukuk*, the first note often doubled. The alarm note is a throaty grating croak, *rrrrr-wa-waak*. Presumed female song similar to that of male (King 2008); sexes often sing in duet (Eaton *et al.* 2016b).
Movements Presumably sedentary.
Range Endemic. Confined to Togian archipelago. So far encountered on the islands of Batudaka, Togian and Malenge; songs presumed to be of this species heard on the island of Walea Bahi.
Conservation status Currently classified by BirdLife International as Near Threatened; the overall range is small with an estimated population of 2,500–9,999 individuals, equating to 1,667–6,666 mature birds (BLI SF 2013). Part of the species' range is included in Togian Islands National Park, declared in 2004.
Etymology Named after Burhan (many Indonesians have only one name) of Benteng village "in recognition of his knowledge of the birdlife of Togian" (Indrawan & Somadikarta 2004).

PLAIN-TAILED NIGHTHAWK
Nyctiprogne vielliardi

A highly restricted species of small caprimulgid, currently known from three specimens and a handful of field observations. Originally described as a member of the widespread genus *Chordeiles*, but recent convincing arguments suggest that it is more properly placed in the exclusively South American genus *Nyctiprogne* (Whitney *et al.* 2003).

Type information *Chordeiles vielliardi* Lencioni-Neto 1994, Manga, Bahia, Brazil.
Alternative names Bahian Nighthawk, Caatinga Nighthawk. Portuguese: Bacurau-São-Francisco.
Discovery In October 1987, F. Lencioni-Neto obtained two specimens, both males, from a flock of small nightjars observed on the Rio São Francisco in Bahia state, east-central Brazil. Comparison with specimens of all other species occurring in South America clearly established that the specimens represented a new species. The species was not observed again until November 1994, when B. Whitney and his colleagues made the first detailed field observations, along with recordings of vocalisations, at a location in northern Minas Gerais state, some 600 km south of the type locality, but still in the Valley of the Rio São Francisco; these authors also located and identified a further specimen, now in the collection of the Universidade Federal do Rio de Janeiro, taken in April 1994, also in Minas Gerais.
Description 17.5 cm. Monotypic. A small nightjar distinguished by its complete lack of white markings on wings, tail or throat. Upperparts dark chestnut, spotted with chestnut-brown; underparts dark chestnut, spotted and barred with buff; flanks and belly barred with dark chestnut; iris yellowish-brown. Immature more tawny than adult. Description based on two specimens only, both males.
Habitat Most common in riverside habitat bordered by brushy growth and gallery woodland, less common or absent when this habitat has been highly altered.
Food and feeding Foraging appears to be entirely aerial, starting well before dusk; foraging groups dispersed by total nightfall (Whitney *et al.* 2003). Prey items probably small aerial arthropods.
Breeding One nest only described. It was located on the ground, in the depression of a human footprint, on a sandspit. On 9 November it held two pinkish eggs (Whitney *et al.* 2003). No other data.
Voice Song is very similar to that of Band-tailed Nighthawk *N. leucopyga*, the only other member of the genus, and quite distinct from that of Least Nighthawk *Chordeiles pusillus*, supporting reassignment into *Nyctiprogne* (Whitney *et al.* 2003). A three-note call, the first note separated by about 1.5–2 seconds from the remainder, which follow rapidly. When flushed from

▼ Plain-tailed Nighthawk. Bom Jesus da Lapa, Bahia, Brazil, November 2010 (*Ciro Albano*).

roost, call is *bit-bit* (Lencioni-Neto 1994).

Movements Probably sedentary.

Range Endemic. So far only observed in the Valley of the Rio São Francisco, from near Manga in Bahia state (about 10° 30' S) and in the region of Januária and Mocambinho in northern Minas Gerais (about 15° 30' S). Much intervening habitat has not been investigated.

Conservation status Current classification by BirdLife International is Near Threatened. Found in only a limited number of locations and does not appear to adapt well to major habitat modification. However, at some locations appears to exist in good numbers; groups of up to 30 individuals were found roosting in a loose colony, and feeding flocks of 150–200 birds were seen in December 1999, in a 300 m stretch of shoreline (Whitney *et al.* 2003).

Etymology Named in honour of Dr Jacques Vielliard (1944–2010), "in recognition of his contribution to bioacoustics of Brazilian birds".

VAURIE'S NIGHTJAR
Caprimulgus centralasicus

Probably one of the least-known species of bird in the world, represented by one specimen taken 80 years ago.

Type information *Caprimulgus centralasicus* Vaurie 1960, Goma (or Guma) (Pishan), western Sinkiang (Xinjiang), China.

Alternative names Chinese: Zhongya Yeying.

Discovery A specimen was taken in September 1929 by Frank Ludlow in Chinese Turkestan. This was identified by the collector as the nominate race of the Egyptian Nightjar *Caprimulgus aegyptius*. While performing a study of the Palearctic Caprimulgidae, C. Vaurie had occasion to re-examine the unique specimen (housed in the Natural History Museum, Tring) and noted a number of features inconsistent with the original identification, on which basis he described it as a new species (Vaurie 1960). Several authorities (e.g. Cleere 1998) were not convinced that it represented a valid species, but Leader (2009) has marshalled a convincing argument for full specific status.

Description 19 cm. Monotypic. The unique specimen is a female, possibly immature. A small, sandy-buff nightjar with buffy-white throat patches but no nuchal collar; no white patches on primaries, but outer two rectrices with buffish-white tips.

Habitat The type specimen was "shot on sand hills covered with low scrub" at 4,000 ft (1,220 m).

Food and feeding No data.

Breeding Nest and eggs undescribed.

Voice Unknown.

Movements No data; however, the type locality has a very severe winter climate, so it is almost certainly migratory. Given the date of collection of the type specimen, there is no guarantee that it was not on passage from some unknown breeding location further north.

Range Known only from Goma or Guma (now Pishan), Xinjiang, western China.

Conservation status Currently classified as Data Deficient by BirdLife International. However, the habitat around the type locality has undergone major anthropogenic modification since 1929. Reports of several specimens being taken in 1975 at a location 250 km from the type locality (Anon. 1993) are apparently in error, due to confusion with European Nightjar *C. europaeus* (Anon. 1994; P. Leader, *pers comm.* 2014).

Etymology The English name honours Charles Vaurie (1906–1975). Dr Vaurie, who was initially a dentist by profession, was born in France but emigrated to the United States while still a schoolboy. He gave up dentistry when appointed Assistant Curator of Ornithology at the American Museum of Natural History, New York. He published many influential works, including *The Birds of the Palaearctic Fauna.*

MEES'S NIGHTJAR
Caprimulgus meesi

A new nightjar whose identity would doubtless have remained undetected but for a very meticulous analysis of vocalisations.

Type information *Caprimulgus meesi* Sangster & Rozendaal 2004, Nisar, Flores, Indonesia.

Discovery In 2004 Sangster and Rozendaal published a very detailed analysis of the taxonomy and vocal characteristics of the *Caprimulgus macrurus* (Large-tailed Nightjar) complex, which included four then-recognised species, *C. macrurus, C. atripennis, C. celebensis* and *C. manillensis* (Large-tailed, Jerdon's, Sulawesi and Philippine Nightjars). They demonstrated that *macrurus* consisted of three distinct species, the nominate race (with six subspecies) ranging from Pakistan to NE Australia, the former race *andamanicus* confined to the Andaman Islands, and the current taxon, native to Flores and Sumba Islands in the Lesser Sundas archipelago. In fact, specimens had been taken many years previously; the holotype at Nisar, Flores, in 1975, and the paratype on Sumba in 1932. Both had been identified as *C. m. schlegeli*, a race which occurs widely in southern Indonesia, New Guinea and adjacent islands and in NE Australia (Queensland and Northern Territory).

Description 25–29 cm. Monotypic. A typical cryptically-plumaged nightjar, rather greyish in overall tone, with (in the male) a thin white throat-patch, conspicuous white patches on five primaries and on the distal portion of the two outer rectrices. Both known specimens were adult males. Very similar to *C. m. schlegeli* and possibly not distinguishable morphologically.

Habitat Open to dense scrubland with scattered trees, or at the edge of open forest; absent from the edges or interiors of dense forest. Sympatric and possibly syntopic with *C. affinis* on Flores (Sangster & Rozendaal 2004). Sea level to 265 m on Sumba and up to 800 m on Flores.

Food and feeding Few data; the holotype stomach contents were a "conglomerate of small insects and an undamaged cricket" (Sangster & Rozendaal 2004).

Breeding Nest and eggs undescribed; an egg collected in October 1956 on Flores is not definitely attributable to this species. It seems likely that, by analogy with *C. affinis*, the breeding season is August to November with a peak in September and October (Sangster & Rozendaal 2004).

Voice Quite distinct from that of other members of the *C. macrurus* complex; loud and repetitive, a high-pitched note followed by 2–9 lower-pitched waves or downslurred notes (2013 HBW SV); sonograms

▲ Mees's Nightjar. Flores, Lesser Sundas, Indonesia, September 2013 (*James Eaton*).

reproduced in Sangster & Rozendaal 2004. Does not respond to playbacks of the songs of *C. m. schlegeli, C. m. albonotatus, C .m. johnstoni* or *C. m. bimaculatus*, but does react vigorously to playback of its own song.

Movements Probably sedentary.

Range Endemic. Originally described from Flores (several locations in centre and west of the island, and one in the east) and Sumba (mostly in the eastern half). More recently discovered on the islands of Alor and Pantar, a range extension of about 90 km to the east (Trainor *et al.* 2012). The situation in several other intervening islands in the Lesser Sundas has not been investigated.

Conservation status Currently classified by BirdLife International as Least Concern, since it has a substantial geographic range and there is no present evidence of significant population declines.

Etymology Named in honour of Dr Gerlof F. Mees (1926–2013), formerly Curator of the Bird Division of the Rijkmuseum van Natuurlijke Historie in Leiden, the Netherlands, and a foremost authority on various aspects of Asian and Pacific ornithology.

NECHISAR NIGHTJAR
Caprimulgus solala

The description of a new species of bird based solely on one desiccated wing was the cause of some considerable controversy (see below); currently recognised as a "good" species by the IOC and the *Handbook of the Birds of the World*.

Type information *Caprimulgus solala* Safford *et al.* 1995, Nechisar Plains, North Omo Region, Ethiopia.
Discovery During a survey of the Nechisar National Park in September 1990 a decomposing corpse, probably a few days to a week old, was found partly submerged in the soil of a vehicle track. Some other feathers were unfortunately lost, but a complete wing was retrieved. This wing is the sole basis for the description of the species. This has attracted criticism (Forero & Tela 1997) on the grounds that variability of characters, such as the size of the white wing-patch, was not adequately considered; criticism that was vigorously rebutted by the original authors (Safford *et al.* 1997). Both parties agree that DNA comparison with other species would be valuable.
Description Based on one left wing, presumably that of the male. Probably a rather large nightjar with an estimated length of about 25 cm; the wing chord was 18.8 cm. Rather reddish-brown; primary coverts blackish-brown, barred with chestnut-brown; upperwing-coverts with conspicuous rounded buff tips; conspicuous white patches on primaries 7 to 10; tail feathers with white distal corners.
Habitat The unique specimen was found in a completely treeless plain, with short grasses on black lava soil at 1,200 m.
Food and feeding No data.
Breeding No data.
Voice No data.
Range Endemic. The specimen was found on the Nechisar Plains at 6° 00' N and 37° 47' E.
Conservation status Classified by BirdLife International as Vulnerable, but lack of any hard data makes classification difficult. Reports of several large nightjars on the Nechisar Plains in April 2009 (Anon. 2009) require confirmation. Nechisar National Park is threatened by overgrazing by domestic animals, removal of trees for firewood and illegal fishing, though efforts have been made to curtail these activities.
Etymology The scientific name, *solala* (Latin), means one-winged.

PRIGOGINE'S NIGHTJAR
Caprimulgus prigoginei

An enigmatic species, known from one specimen from the Democratic Republic of the Congo (Zaire), but with tantalising suggestions of a much wider distribution.

Type information *Caprimulgus prigoginei* Louette 1990, Malenge, Itombwe, Kivu province, DRC.
Alternative names Itombwe Nightjar. French: Engoulevent de Prigogine.
Discovery In August 1955 a collector working for the distinguished Belgian ornithologist A. Prigogine obtained a female nightjar in the Itombwe Mountains of what was then the Belgian Congo. The specimen was assigned to Bates's Nightjar *C. batesi*, but not without some misgivings from Dr Prigogine himself. In 1989 M. Louette re-examined the specimen and for a number of reasons concluded that it represented a valid species.
Description (female only) 19 cm. Monotypic. A small, rather dark nightjar with a relatively large head, the upperparts and wing-coverts brown with darker spots; the breast brown with tawny and buff spots and bars; belly paler with brown bars; no nuchal collar; small tawny spots on three primaries; outermost rectrices with narrow white tips, remaining rectrices with buffy or tawny tips. By analogy with other species, it is to be expected that in the male the tawny markings on the flight feathers might be replaced by white. Bates's Nightjar is much bigger (29–31 cm).
Habitat The unique specimen was taken at an altitude of 1,280 m, possibly in forest. Observations of what was probably this species, in Lower Guinea, were made in forest with a broken canopy (Dowsett-Lemaire 2009).
Food and feeding No data.
Breeding Nest and eggs unknown.
Voice Not certainly known. Several calls have been heard which might well refer to this species, but unequivocal evidence is lacking. For example, an unidentified "churring" nightjar was heard in forests near Lake Longwe, DRC, in 1952; a rapid "knocking" sound, similar to the call of Swamp Nightjar *C. natalensis* which was recently recorded in the Itombwe Forest could possibly

be *prigoginei* (Cleere 1998). Similar calls have also been heard in northern DRC, in SE Cameroon and Gabon (BLI SF 2013). In all these cases there is presently no hard evidence. Dowsett & Dowsett-Lemaire (2000) also recorded an unknown caprimulgid at 2100 m in dense forest in Cameroon whose call, "a long series of *tchoc-tchoc-tchoc-tchoc*.... notes", was similar to recordings made in the Itombwe Forest; playback of Itombwe birds provoked a reaction in their birds, suggesting that their observations might refer to *prigoginei*. Songs heard in Lower Guinea, believed to be of this species, were described as superficially resembling those of Natal Nightjar *C. natalensis*, but differing in structure of the notes and in timbre; the calls are short and harsh (Dowsett-Lemaire 2009).

Movements Presumably sedentary.

Range Presently known certainly only from the single specimen from Itombwe Forest. However (see above), further fieldwork might well show it to be much more widely distributed.

Conservation status Currently classified by BirdLife International as Endangered, but clearly much more information is needed to make a valid assessment.

Etymology Named after Alexandre Prigogine (1913–1991). Dr Prigogine was born in Tsarist Russia, but in 1921 his family emigrated to Germany, then ultimately to Belgium (a very fortunate acquisition for Belgium; his younger brother Ilya won the Nobel Prize for Chemistry in 1977). A specialist in minerals, he went to work in the then Belgian Congo in 1938. His interest in ornithology probably began about 1946 under the influence of H. Schoutenden, progressing from a hobby to a passion. The specific epithet *prigoginei* is linked in four species (genera *Philodius, Caprimulgus, Chlorocichla* and *Nectarinia*) and Prigogine himself described five new species and numerous races (Louette 1992).

WHITE-FRONTED SWIFT
Cypseloides storeri

A further member of the notoriously difficult genus *Cypseloides*; sometimes regarded as a race of White-chinned Swift *C. cryptus*.

Type information *Cypseloides storeri* Navarro *et al.* 1992, Puerto del Gallo, Tlacotepec, Guerrero, Mexico.

Alternative names White-faced Swift. Spanish: Vencejo Frentiblanco, Vencejo de Cara Blanca.

Discovery In the course of seven years of fieldwork in the Sierra Madre del Sur in Guerrero state, Mexico, a specimen of a medium-sized dark swift was taken at an altitude of 2,500 m in the Sierra de Atoyac in the south of the state. The bird, a male, was found clinging to a towel in camp, late on a foggy night. Careful examination of the specimen led to the conclusion that a new taxon was involved. Subsequent searching through museum collections brought to light three further specimens, from the states of Guerrero and Michoacán, taken by Allan R. Phillips and, uncharacteristically, misidentified by him as Black Swifts *C. niger*. Further specimens have since come to light from Jalisco state (Navarro *et al.* 1993; Ibáñez-Hernández *et al.* 2003).

Description 14 cm. Monotypic. A medium-sized swift with a short square tail, overall sooty brown-black, the chin, lores and forehead greyer; very similar to White-chinned Swift *C. cryptus*, but differs in wing/tail ratios, and in having whitish, not sooty, post-orbital feathers. Suggestions on field identification of *C. storeri* are given by Howell *et al.* (1997) and Sharpe *et al.* (2017).

Habitat Has been seen feeding over various montane habitats, including semi-arid vegetation, pine-oak forest, fields with deep canyons, etc., at altitudes of 1,200–2,500 m; also two documented sightings at 200 m in coastal lowlands (Sharpe *et al.* 2017).

Food and feeding No data. Has been seen in mixed flocks with other *Cypseloides, Chaetura* and *Streptoprocne* swifts.

Breeding Nest and eggs undescribed. One previously misidentified specimen taken in June 1990 at 2,360 m in Jalisco state had enlarged testes. Seems to be frequently associated with canyon waterfalls and may nest behind them.

Voice Call is stated to be distinctive, distinguished from that of Black Swift *C. niger* (which is a series of

▼ White-fronted Swift. Salto Santa Paula, Michoacán, Mexico, June 2016 (*Andrew Spencer*).

low, flat, twittering chirps) by being noticeably huskier, with even spacing, not speeding up into a chatter.

Movements No clear data but may be semi-migratory (Chantler 1998).

Range Endemic. Specimens have come from the Mexican states of Guerrero, Jalisco and Michoacán, although observations, or possible observations, have also been made in Colima and Oaxaca (BirdLife International 2013).

Conservation status Currently classified by BirdLife International as Data Deficient. The difficulty of definitive field identification has made the collection of data and defining clear limits to the range uncertain. It would appear to be uncommon wherever it occurs, most observations involving single birds or very small groups. Sharpe *et al.* (2017) estimate the population to be in the range of 5,000–10,000 birds, while acknowledging that the real figure might well be lower.

Etymology Named in honour of Dr Robert W. Storer (1914–2008), "in recognition of his many contributions to the knowledge of the birds of Guerrero and Michoacán" (Navarro *et al.* 1992).

WHITE-CHESTED SWIFT
Cypseloides lemosi

A member of the somewhat confusing genus *Cypseloides,* which despite being fairly diagnostic and quite widespread remained undiscovered until quite recently.

Type information *Cypseloides lemosi* Eisenmann & Lehmann 1962, Santander, Cauca, Colombia.

Alternative names Giant Swift. Spanish: Vencejo Pechiblanco; Portuguese: Taperuçu-de-peito-branco.

Discovery In October 1951 K. von Sneidern collected a swift in Cauca, Colombia, overall dark in plumage but with a few white chest feathers. This was identified as an abnormal example of Sooty Swift *C. fumigatus.* In October 1957 F. C. Lehmann obtained three specimens, two of which, adults, had broad white pectoral patches; in the third, presumably immature, the white was restricted to three feathers. Similar birds were seen again in 1960, and in 1961 Lehmann was able to take another three specimens. Having discounted the possibility that all these birds were simply partially albinistic examples of one of the other *Cypseloides* species (pointing out some other consistent morphological distinctions), Eisenmann and Lehmann described it as a new species in 1962.

Description 14 cm. Monotypic. Adults are immediately distinguished from all other members of the genus by a conspicuous white chest-patch, rather variable in extent and in the immature reduced to a few white feathers; remainder of plumage sooty-black, somewhat paler below; wings long and pointed; tail with a moderate fork.

Habitat Aerial, over secondary forest, second-growth scrub, pastureland and areas with bare eroded soil; mostly at 1,050–1,300 m, but overall from 250–2,000 m (Roesler *et al.* 2009).

◄ White-chested Swift. Cusco, Peru, August 2012 (*Tim Avery*).

Food and feeding Feeds exclusively aerially. Usually seen in small flocks of 20–25 individuals, but occasionally in greater numbers. Frequently associates with other swifts, such as Chestnut-collared *C. rutilus*, Black *C. niger*, White-chinned *C. cryptus*, Spot-fronted *C. cherriei* and White-collared Swifts *Streptoprocne zonaris*, especially over swarms of flying beetles.
Breeding Nest and eggs undescribed.
Voice A series of *chip-chip-chip* notes, becoming more rapid during social interactions (Howell 2002).
Movements May be partially migratory or nomadic.
Range Initially recorded only from the Cauca region of Colombia, it has now been shown to occur much more widely; in Colombia, in Dptos Cauca, Valle and Amazonas (Downing & Hickman 2002); in Ecuador, in Napo (Downing & Hickman 2002; Ridgely & Greenfield 2001); in Peru, in San Martín, Cuzco, Loreto and Amazonas and in the Cordillera de Cóndor; in western Brazil (Amazonas) (Hansson 2006); and recently in Dpto. La Paz, Bolivia, a range extension to the south of ca 600 km (Roesler *et al.* 2009). It is unclear how many of these records may pertain to migrant birds, since there is presently no unarguable evidence of breeding; Roesler *et al.* (2009) regard breeding in La Paz, Bolivia, as possible.
Conservation status Presently classified by BirdLife International as Least Concern. Although it appears to be nowhere common, it seems to be widely distributed, albeit with an unknown breeding range. Flocks of up to 180 birds have been seen in Napo province, Ecuador (Howell 2002). It has been suggested that it may actually benefit from some clearing of forest.
Etymology Named in honour of Dr Antonio José Lemos-Guzmán (1900–1967), Governor of the Departamento del Cauca and three times Rector of the Universidad del Cauca.

TEPUI SWIFT
Streptoprocne phelpsi

A widespread but only recently recognised species of swift; originally included in the genus *Cypseloides*; for reasons for its transfer to *Streptoprocne,* see Marin & Stiles (1992).

Type information *Cypseloides phelpsi* Collins 1972, Cerro Auyun-tepui, Bolívar, Venezuela, based on a specimen taken in 1938.
Alternative names Phelps's Swift. Spanish: Vencejo de Tepui; Portuguese: Taperuçu-dos-tepuis.
Discovery The Chestnut-collared Swift *Streptoprocne rutilus* is a well-known species of wide geographic distribution; it was originally described (as *Hirundo rutila*) by Vieillot in 1817, based on a specimen from "La Trinité" (Trinidad). C. T. Collins in his 1972 study pointed out that the distinctive population of birds in the Pantepui region of southern Venezuela and adjacent Guyana and Brazil, while believed by many authorities to be identical to Trinidad birds, was in fact very different in several important plumage features and clearly merited full specific status.
Description 16.5 cm. Monotypic. A distinctive swift, only confusible with *rutilus*, which is almost entirely allopatric. Plumage overall blackish, but with a bright chestnut collar, somewhat paler in the female and lacking in juveniles. Differs from *rutilus* by larger size, more deeply forked tail and brighter and more extensive collar.
Habitat Subtropical to upper tropical forest at 200–2,600m; also montane forest grasslands and around cliffs on tepuis (Hilty 2003).
Food and feeding Aerial feeder; no data on diet.
Breeding The one nest so far found is described as a truncated cone of live moss sited on the vertical wall of a small cave among boulders at the edge of a small stream (Chantler 1999 HBW 5). Eggs undescribed.
Voice Often quiet while foraging; in flocks, a squeak followed by a trill and short squeals; also a slow set of reedy or hissing *tic* notes (Hilty 2003).
Movements Few data. Vagrant records outside the known breeding range suggest that it may not be totally sedentary.
Range SW Venezuela (Bolívar, Amazonas), extreme northern Brazil (northern Roraima, northern Amazonas); a few records in NW Guyana (Merume Mountains,

▼ Tepui Swift. Monte Roraima, Roraima, Brazil, January 2016 (*João Quental*).

Roraima). Apparently vagrant records in northern Venezuela (Aragua).

Conservation status Currently classified by BirdLife International as Least Concern, having quite a wide geographic range. Flocks of up to 80 birds seen daily on Cerro Guaiquinima, Bolívar, Venezuela, in March 2000 (Pérez-Emán *et al.* 2003).

Etymology Named after William H. Phelps (1875–1965), a highly successful businessman who might well be described as one of the fathers of Venezuelan ornithology. With his son W. H. Phelps Jr (1902–1988) he founded the Colección Ornitológica Phelps in Caracas, which currently holds some 75,000 specimens.

BARE-LEGGED SWIFTLET
Aerodramus nuditarsus

May form a superspecies with Mayr's Swiftlet *A. orientalis*. Subsumed into Whitehead's Swiftlet *A. whiteheadi* by some authorities (e.g. Beehler *et al.* 1986). However, given full specific status by Chantler (1999 HBW 5).

Type information *Collocalia nuditarsus* Salomonsen 1963, Baroko, Bioto Creek, Papua New Guinea.

Alternative names Schrader Mountain/New Guinea/Naked-legged Swiftlet.

Discovery Described on the basis of material collected by a Danish expedition to New Guinea and Indonesia.

Description 14 cm. Monotypic. A rather dark, slightly fork-tailed swiftlet without an obvious pale rump. Upperparts dark brown with paler fringes on lores and above eye; underparts uniform sooty-grey.

Habitat Aerial feeder, mostly over highlands, but also as low as 30 m altitude.

Food and feeding No data recorded.

Breeding Nest and eggs undescribed.

Voice No data recorded.

Movements Nothing recorded.

Range Endemic to the island of New Guinea, from north of the Gulf of Papua westwards into Irian Jaya.

Conservation status Currently classified by BirdLife International as Least Concern; while the geographic range is fairly restricted, there is no evidence of significant population declines.

Etymology *Nuditarsus*, Latin, 'bare-legged'.

ATIU SWIFTLET
Aerodramus sawtelli

A highly-restricted species of swiftlet, which has variously been considered as a race of Polynesian Swiftlet *A. leucophaeus* and White-rumped Swiftlet *A. spodiopygus*.

Type information *Aerodramus sawtelli* Holyoak 1974, Atiu Island, Cook Islands.

Alternative names Sawtell's Swiftlet, Cook Island Swiftlet. Local name: Kopeka.

◄ Atiu Swiftlets on nest. Anatakitaki Cave, Atiu, Cook Islands, 1990s (*Gerald McCormack*).

Discovery The type specimen, a female with small gonads, was collected in September 1973 by D. T. Holyoak while conducting a survey of Cook Island birds; several islands were visited apparently for the first time, and seven undescribed forms of land birds were collected.

Description 10 cm. Monotypic. A small or medium-sized swiftlet, with a relatively long forked tail, blackish-brown above with pale grey rump; underparts pale grey-brown. Differs from *leucophaeus* in shorter tail, smaller size, bill shape, paler underparts and paler bases to uppertail-coverts.

Habitat Aerial feeder, foraging over both forested and more open areas, but not over coastal coral areas (Fullard *et al.* 2010).

Food and feeding Few data. Faecal analysis (from nesting caves) showed insect remains, especially beetles and dipterid flies (Fullard *et al.* 2010).

Breeding Nests in caves, both in areas where some light penetrates and in total darkness. The nest is a skimpy cup of plant fibres, especially coconut palms, and lichens, consolidated with sticky saliva, located on small ledges high up in caves. Eggs two, white, incubated by both sexes; nests are built from September onwards, last chicks fledged in late April. Sometimes double-brooded. The nest form differs significantly from that of *A. leucophaeus* (Holyoak & Thibault 1978a).

Voice When in the open, a shrill *chreee*; in caves, a series of staccato clicks used in echolocation (McCormack 2005; Pratt *et al.* 1987). Echolocating calls are used strictly to navigate in the total darkness of the nesting caves, not (as in bats) as a technique for locating aerial prey. In the Atiu Swiftlet, in contrast to other Pacific *Aerodramus* species, clicks are apparently always single, not double-pulsed. The clicks have a duration of 2–3 milliseconds each, with a frequency of 6–7 kHz (Fullard *et al.* 2010); when approaching the nest the clicks lengthen (McCormack 2005).

Range Endemic. Confined entirely to Atiu Island, southern Cook Islands.

▲ Atiu Swiftlet. Atiu, Cook Islands, May 2009 (*Gerald McCormack*).

Conservation status Classified by BirdLife International as Vulnerable, based on its limited occurrence (solely on Atiu, an island of under 3,000 ha) and tiny breeding distribution (apparently confined to two limestone caves only, although other breeding sites may be undiscovered). Total breeding population believed to be in the region of 380 pairs in 1999 (Chantler 1999 HBW 5). In one cave, estimates were 148 (Tarburton 1990) and 163–241 birds; in the second cave, 90 nests were counted in 1994 and 106 in 1995 (Fullard *et al.* 2010). While nesting success is low, numbers appear to be stable; principal causes of mortality are predation by land crabs *Birgus latro* and *Cardiosoma longipes*, and starvation. So far Atiu is free of Black Rats (*Rattus rattus*), which on Rarotonga brought an endemic flycatcher to the brink of extinction. Visits by tourists to nesting caves, using artificial light, could potentially have a deleterious effect on breeding, but apparently these are now carefully controlled (Fullard *et al.* 2010).

Etymology Named in honour of Gordon H. Sawtell (1929–2010), Secretary of the Premier's Department of the Government of the Cook Islands, in recognition of his valuable help to D. T. Holyoak in his work.

SCHOUTENDEN'S SWIFT
Schoutendenapus schoutendeni

Probably the least well-known of any species of swift; apparently no confirmed field observations exist and presently known from five specimens only. Currently placed in the genus *Schoutendenapus*, in the subfamily *Apodini*, along with *Apus* and several other genera (Holmgren 1998). Not all authorities have accepted it as a good species (e.g. Hall & Moreau 1992, *contra* Brooke 1971b).

Type information *Apus schoutendeni* Prigogine 1960, Butokolo, Democratic Republic of the Congo (Zaire).

Discovery Known from five specimens collected in eastern Belgian Congo, now DRC, in 1956 and 1959 (Prigogine 1960).

Description 16.5 cm. Monotypic. A small, uniformly blackish-brown swift with a thin, deeply forked tail, bulbous head and tapering outer section of the wings.

Very similar to Scarce Swift (*S. myoptilus*), with which it probably associates, but lacks the paler throat of that species.

Habitat Probably over highland forest; the type locality, Butokolo, has an elevation of 1,470 m.

Food and feeding No data. Probably associates with other species of swift.

Breeding Nest and eggs undescribed. A female specimen collected in February had enlarged oocytes; specimens in October not in breeding condition (Chantler 1999 HBW 5).

Voice Undescribed.

Movements No data to suggest anything other than resident.

Range The known specimens came from eastern DRC, from 28° 11' E to 28°50' E and from 2° 42' S to 5°04' S (i.e. in an area approximately 70 km from east to west and 260 km from north to south) (Chantler & Driessens 1995). Sightings, unconfirmed by specimen evidence, have recently been made in Bwindi Impenetrable Forest, Uganda (ca 1° 00' S, 29° 50' E), and on Mount Tshiaberimu, DRC; if proven these would extend the range significantly to the north and east.

Conservation status Currently classified by BirdLife International as Vulnerable, but in view of the apparent rarity and lack of solid data on numbers a higher classification might be justified. In an area of endemic political instability and rapid population growth, much habitat is vulnerable to degradation.

Etymology Named, very comprehensively, in honour of Henri Schoutenden (1881–1972), a Belgian ornithologist and entomologist who undertook many expeditions to the former Belgian Congo and neighbouring Rwanda and Burundi.

FORBES-WATSON'S SWIFT
Apus berliozi

Very similar to the widespread Pallid Swift *A. pallidus* and originally described as a race of it, with which (along with African Swift *A. barbatus* and Bradfield's Swift *A. bradfieldi*) it forms a superspecies.

Type information *Apus pallidus berliozi* Ripley 1965, Soqotra (Socotra) Island, Yemen.

Alternative names Berlioz's Swift.

Discovery In 1964 A. D. Forbes-Watson was able to visit the then remote island of Socotra (Soqotra), Yemen, during which time he collected a series of swifts which were described as a race of Pallid Swift, *Apus pallidus berliozi*. In 1968 R. K. Brooke described a new swift from Kilifi, coastal Kenya, just north of Mombasa (in fact, an earlier specimen that had been taken in Somalia in 1958 was later discovered). Brooke (1972a) argued convincingly that the new taxon should be treated as a race (*bensoni*) of the Socotra birds, which themselves should have full specific status under the trinomial *A. b. berliozi*.

Description 16 cm. In form a typical *Apus* swift, long-winged, with a deeply-forked tail; overall plain brown, the forehead paler, underparts slightly scalloped; chin and throat whitish, sometimes streaked darker;

▼ Forbes-Watson's Swift (presumed to be *A. b. berliozi*). Left: Raysut, Oman, September 2011. Right: Salalah, Oman, May 2012. *(Hanne and Jens Eriksen)*.

bill black, iris dark brown (nominate). *A. b. bensoni* is darker and browner than nominate, with a darker forehead and reduced scalloping on upperparts (not all authorities accept the validity of this race, e.g. Grieve & Kirwan 2012). For a detailed account of field characteristics, see Porter *et al.* (1994).

Habitat Usually in arid habitats. The nominate race on Socotra is found over much of the island, including over urban areas, dunes, craggy areas, from sea level to 1,200 m. On the wintering grounds in Kenya, seen over coastal lowland forest.

Food and feeding Aerial feeder. Stomach contents include termites, beetles, grasshoppers, ants and bugs (Hemiptera) (Fry *et al.* 1988).

Breeding In Somalia, March to September, possibly to December; nest is a pad, external diameter 110–120 mm, of dried marine weed *Cymodocea*, plant down, fibres etc. glued together with saliva, lined with feathers; located in holes in the roofs of sea caves. Clutch is two white eggs. On Socotra, Forbes-Watson collected 32 specimens in May, mostly in breeding condition. Probably breeds in cracks in cliff-faces (Porter *et al.* 1994).

Voice Call is a "rather flat, screeching *schweee*, not as high-pitched and loud as that of the European Swift *A. apus*". Also a trisyllabic *schweee-weee-eee* (Porter *et al.* 1994).

Movements Partially migratory, but movements not entirely understood. The Socotran population may be migratory (Porter *et al.* 1994; Porter & Suleiman 2013), although Grieve & Kirwan (2012) state that it is sedentary. The race *bensoni* appears to be a migrant or partial migrant, breeding in the mountains of NW Somalia (presence in Ethiopia apparently not confirmed – Redman *et al.* 2011) and wintering in coastal Kenya (Brooke 1972a).

Range *A. b. berliozi*, Socotra Island, over much of the island but especially in the Hajhir Mountains, possibly migratory. *A. b. bensoni*, coastal Somalia and Somaliland (up to 90 km inland), migrant or partial migrant south into coastal Kenya (Chantler & Driessens 1995); also a breeding summer visitor (April to November) to southern Arabia in SW Oman and S Yemen (Grieve & Kirwan 2012, Tibbett 2006).

Conservation status Classified by BirdLife International as Least Concern. Notwithstanding its limited range, the nominate race appears to be common, with flocks of up to 300 birds seen (Porter *et al.* 1994). There is less information about *A. b. bensoni*.

Etymology The vernacular name commemorates A. D. Forbes-Watson (1935–2013), a specialist in the ornithology of Africa who, among other things, described a new species of flycatcher (*Melaeornis annamarulae*) from Liberia. The scientific names honour Prof. J. Berlioz (1891–1975), a French ornithologist who was in charge of the ornithology department of the Paris Museum; and C. W. Benson OBE (1909–1982), who served in the British Colonial Service and who became one of the most influential figures of his day in African ornithology. He is alleged to have tasted all the specimens that he took, claiming that owls have the worst flavour and turacos the best (Beolens & Watkins 2003).

KOEPCKE'S HERMIT
Phaethornis koepckeae

Closely related to Needle-billed Hermit *P. philippii* and Tawny-billed Hermit *P. syrmatophorus*, the straight bill having evolved separately from that of Straight-billed Hermit *P. bourcieri* (Hinkelmann & Schuchmann 1997).

▼ Koepcke's Hermit. Lamas, San Martín, Peru, January 2014 (*Robert Lewis*).

Type information *Phaethornis koepckeae* Weske & Terborgh 1977, Cerros del Sirá, Peru.

Alternative names Spanish: Ermitaño de Koepcke.

Discovery Several expeditions visited the Cerros del Sirá, an isolated mountain range in central Peru, east of the main Andean massif, in the late 1960s. The type specimen was taken in July 1969.

Description 13 cm. Monotypic. A typical, relatively large *Phaethornis* hermit, with prominent facial markings and elongated, white-tipped central rectrices. Bill relatively straight; upperparts olive-green; uppertail-

▲ Koepcke's Hermit. Lamas, San Martín, Peru, January 2014 (*Robert Lewis*).

coverts orangey-red; bright orange below; malar and gular stripes whitish. Female similar with a slightly more decurved bill; immature markedly ochraceous on upperparts.

Habitat Wet tropical forest, usually 270–1,130 m. May be present in good numbers in forest which has been selectively logged (Dauphiné *et al.* 2006).

Food and feeding Nectarivorous, but substantial numbers of arthropods in stomach contents (Remsen *et al.* 1986).

Breeding Nest is typical of the genus, a cup with a cone-shaped base, made of vegetable fibres etc., attached on the underside near the tip of a pendant leaf (Weske & Terborgh 1977). The single nest so far described (in the Cerros del Sirá) had two young in late July.

Voice Song is a ringing, buzzy series of short notes: *bzee-bzee-bzee.* Calls include a ringing rising *tchwee* or *tchwing,* also a descending, accelerating series of high notes (Schulenberg *et al.* 2007).

Movements No data.

Range Endemic. Originally described from the Cerros del Sirá (Dpto. Huánuco) and the Marañon Valley (Amazonas) in northern Peru. Since discovered, found patchily in several intervening areas on the eastern margins of the eastern Andes (Hinkelmann 1988), including sites in San Martín, Ucayali, Cuzco and Madre de Dios (Schulenberg *et al.* 2007).

Conservation status Currently classified by BirdLife International as Near Threatened. Although fairly common in some locations, rapid deforestation is occurring within its elevational range, caused by replacement of forest by oil palm plantations, cattle ranches and by logging. Rarely recorded in second growth, though seems to be able to occur in some areas of selective logging (Dauphiné *et al.* 2006).

Etymology The species is named in honour of Maria Koepcke (1924–1971); see under Koepcke's Screech Owl.

TAPAJOS HERMIT
Phaethornis aethopygus

A taxon with a somewhat confused history, having originally been described as a race of Little Hermit *P. longuemareus,* then as a hybrid of Streak-throated Hermit *P. rupurumii* and Reddish Hermit *P. ruber,* before recent work (Piacentini *et al.* 2009) established it as a valid species.

Type information *Phaethornis aethopygus* Piacentini *et al.* 2009.

Phaethornis longuemareus aethopygus Zimmer 1950, Caxiricatuba, Brazil.

Alternative names Portuguese: Rabo-branco-do-tapajós.

Discovery In 1950 J. T. Zimmer described *P. longuemareus aethopygus,* as a new race of Little Hermit, on the basis of specimens from the right bank of the Rio Tapajós, Brazil, despite the very wide geographic separation from other populations of this species.

Subsequent authors (Hinkelmann 1996) treated it as a hybrid of *P. ruber* and *P. rupurumii*, despite the fact that the latter species is not known to occur anywhere near where the original Zimmer specimens were taken. Finally Piacentini *et al.* (2009), who had access to more specimen material, argued on morphological grounds backed up by field observations, that a new species was involved which became *P. aethopygus* (there is some confusion as to the gender of the word *Phaethornis*, but we follow del Hoyo *et al.* (2013) in giving the specific name a masculine ending; current SACC practice uses the female form *aethopyga*).

Description 9 cm. Monotypic. Typical small hermit with a prominent facial pattern, with pale supercilium and rufescent malar stripe; bill medium-long, slightly decurved; back metallic olive-green; rump and uppertail-coverts bright rufescent; chin white, throat blackish, becoming rufous on chest; belly and undertail-coverts rufous; tail rounded, dusky brownish with reddish shafts, outer webs of most feathers with white bases. Female similar but generally duller, with reduced black on throat.

Habitat Lowland primary forest; also areas disturbed by some logging or fires.

Food and feeding Nectarivorous, no specific data.

Breeding Nest and eggs undescribed. A lekking species; one lek with five males active in early December (Piacentini *et al.* 2009).

Voice Not described; other members of the genus give high-pitched chirps at leks.

Movements No data.

Range Endemic. Brazil, south of the Amazon, in the region of the Tapajós, Xingu and Teles Pires rivers, almost entirely in the state of Pará, marginally into northern Mato Grosso.

▲ Tapajos Hermit. Rio Azul Jungle Lodge, Novo Progresso, Pará, Brazil, July 2016 (*Bruno Rennó*).

Conservation status Classified by BirdLife International as Near Threatened. Conversion of forest for agriculture, which can only become more serious if proposed road developments come to pass, and illegal logging are both major threats. Can tolerate some habitat disturbance, but not outright forest clearance.

Etymology *Aethopygus*, Greek, 'fire-rumped'. The common name is derived from the Rio Tapajós, a tributary of the Amazon.

CHIRIBIQUETE EMERALD
Chlorostilbon olivaresi

An isolated species of very restricted distribution which forms a superspecies with the very widespread Blue-tailed Emerald *C. mellisugus*; probably derived from the dispersal of the race *C. m. gibsoni* (currently found in the Magdalena Valley) during a dry period of the Pliocene or early Pleistocene rather than the geographically closer Andean populations of *mellisugus* (e.g. *C. m. napensis*) (Stiles 1996). May possibly be conspecific with *mellisugus* (Schuchmann 1999 HBW 5).

Type information *Chlorostilbon olivaresi* Stiles 1996, Valle los Menhires, Sierra de Chiribiquete, Caquetá, Colombia.

Alternative names Spanish: Esmeralda Chiribiquete.

Discovery The Sierra de Chiribiquete is an isolated mountain range, consisting of dissected sandstone buttes rising abruptly from the surrounding plains to a maximum height of about 900 m, in Dptos Caquetá and Guaviare in southern Colombia. Because of its topographic uniqueness and pristine nature it was set aside as a national park in 1989, but remained unexplored ornithologically until the 1990s. Two expeditions organised by the Colombian government in 1990 and 1992 were devoted to geological, archaeological and botanical studies, but did bring back one specimen of an unusual hummingbird. This was shown to F. G. Stiles, who suspected that it might represent a new taxon, but the very rough preparation of the specimen made comparisons with other material difficult. In late 1992 a third expedition collected seven specimens and mist-netted and released eight more, allowing a full description to be made (Stiles 1996).

Description 8.5–9 cm. Monotypic. Essentially similar to other members of the genus; male bright green on top and sides; head, chin and throat bright metallic blue, underparts green; thighs white; tail deeply forked and deep blue; bill black above, reddish below. Female green above, with white underparts; post-ocular line white, ear-coverts blackish, tail dusky-greenish, the outer rectrices tipped white. Differs from *mellisugus* in its larger size (body weight typically 35–40% greater), red basal half of the mandible (in males), and in the duller crown of the male and extensive grey bases to the outer rectrices in the female.
Habitat Scrub and savanna-type vegetation on top of mesas; absent from surrounding deep forest. 360–570 m, possibly slightly higher.
Food and feeding Most frequently seen feeding on nectar from *Decagonocarpus cornutus* (Rutaceae). Also takes small arthropods, especially thrips, from *Bonnetia martiana* (Theaceae), and flycatches.
Breeding Nest and eggs undescribed. Gonadal data suggest that the breeding season falls between late November/December and May or July.
Voice Call, a short dry *cht*; song unrecorded.
Movements Probably sedentary.
Range Endemic. Seems to be entirely restricted to the Sierra de Chiribiquete, a range of mountains some 125 km long and 30 km wide, in Dptos Caquetá and Guaviare, Colombia.
Conservation status Classified by BirdLife International as Least Concern; seems to be common or abundant in parts of its limited range, which is at least nominally protected. There seems to be little human pressure on the available habitat.
Etymology Named in honour of Fr Antonio Olivares (1917–1975), formerly Professor of Natural Sciences at the Franciscan Seminary in Bogotá, "in honor of his many pioneering contributions to Colombian ornithology".

OAXACA HUMMINGBIRD
Eupherusa cyanophrys

A very restricted hummingbird, endemic to the state of Oaxaca. Treated as a race of the more widespread Stripe-tailed Hummingbird *E. eximia* by some authorities.

▼ Oaxaca Hummingbird, male. La Soledad, Oaxaca, Mexico, March 2015 (*Nigel Voaden*).

Type information *Eupherusa cyanophrys* Rowley & Orr 1964, 18 km south of Juchatengo, Oaxaca, Mexico.
Alternative names Blue-capped Hummingbird. Spanish: Colibrí Oaxaqueño.
Discovery Prior to 1963 two species of the genus *Eupherusa* were known from Mexico; the widespread Stripe-tailed Hummingbird *E. eximia*, occurring from Mexico to Panama, and the highly restricted White-tailed Hummingbird *E. poliocerca*, from the Mexican states of Guerrero and western Oaxaca (a further species occurs in Costa Rica and Panama). In 1963 J. S. Rowley, in the company of A. R. Phillips, collected a small series of hummingbirds in the Sierra de Miahuatlán, central Oaxaca, well disjunct from the range of *poliocerca*. The males in this series differed significantly from *poliocerca* in plumage and were described by Rowley and Orr as a new species in 1964. Not all authorities treat this taxon as a full species.
Description 10–11 cm, Monotypic. Within its range a distinctive species; body largely iridescent green, wings dark with very obvious chestnut-rufous secondaries, the central tail feathers bright green, the outer rectrices largely white, made obvious by the species' habit of flashing open its tail in flight; crown violet-blue in front, turquoise at the rear. The allopatric White-tailed and Stripe-tailed Hummingbirds lack the distinctive crown. Females may not be identifiable, even in the hand (Howell & Webb 1995).

Habitat Edges of humid evergreen forest with dense understorey, semi-deciduous forest, gallery forest and cloud forest, 700–2,600 m, mostly below 1,800 m (Züchner 1999 HBW 5).
Food and feeding Primarily nectarivorous; has been seen to feed on flowers of *Inga, Kohleria, Lobelia, Malviscus, Manettia, Psittacanthus* and *Hamelia patens*; subordinate to other hummingbirds such as *Amazilia* (Züchner 1999 HBW 5).
Breeding Cup-shaped nest, made of moss, lined with plant down, exterior decorated with lichens, 1.2–6 m up. Clutch two, incubation by female alone; breeding season September to November, May (Züchner 1999 HBW 5).
Voice Song is a slightly liquid, "pebbly" trill, similar to that of White-tailed Hummingbird but jerkier and less hurried (Howell & Webb 1995).
Movements No data, but the fact that it may be present or absent at particular sites suggests some nomadism in response to food availability (Howell 1999).
Range Endemic. Confined entirely to the isolated Sierra de Miahuatlán, southern Oaxaca, Mexico. Can be observed near La Soledad, north of Puerto Angel (Howell 1999; Roberson & Carratello 1997).
Conservation status Currently listed by BirdLife International as Endangered, with a population given in the range of 600–1,700 mature individuals. Habitat destruction became serious from the mid-1960s onwards, when large areas were cut down for maize farming. Further areas of forest were destroyed by a hurricane in October 1997 (BLI SF 2014). It was locally common in some areas in 1997; no more recent data.
Etymology *Cyanophrys*, Greek, from the striking blue crown of the males.

CINNAMON-SIDED HUMMINGBIRD
Amazilia wagneri

Treated as a race of Green-fronted Hummingbird *A. viridifrons* by several authorities; however, Howell (1993) provides strong evidence that full specific status is warranted, since there appears to be some sympatry with *viridifrons*, with little evidence of intermediate forms. This approach is presently under review by the AOU; the current IOC World Checklist treats it as a full species, as does *Handbook of the Birds of the World*.

Type information *Amazilia viridifrons wagneri* Phillips 1966, Oaxaca, Mexico.
Alternative names Spanish: Colibrí flanquicanelo, Amazilia de Wagner.
Discovery Described as a race of *viridifrons* by A. R. Phillips, based on material collected in Oaxaca, Mexico.
Description 10–11.5 cm. Monotypic. Upperparts bronzy-green, crown more blackish-green; pale rufous-chestnut with bronzy-green edges; underparts pure white, bordered by bright cinnamon on upper flanks and sides of throat; base of secondaries with dull rufous patch; bill long, slightly decurved, bright red with a black tip.
Habitat Scrubland, both semi-humid and more arid; deciduous thorn forest, 250–900 m.
Food and feeding Nectar from middle and upper strata of scrubland; also hawks for small insects.
Breeding Nest and eggs undescribed. Birds in breeding condition (based on gonadal data) December to May and August to October.
Voice Call, a dry chattering, also a quiet hard, crackling *zzzzrr'k chiuk* (Howell & Webb 1995).
Movements No data.
Range Endemic. Oaxaca, Mexico, on the Pacific slope, very marginally into western Chiapas.
Conservation status Currently classified by BirdLife International as Least Concern; although the geo-

▼ Cinnamon-sided Hummingbird. Las Palmas, Oaxaca, Mexico, April 2013 (*Andrew Spencer*).

graphic range is rather small, the population appears to be substantial and not believed to approach the threshold of Near Threatened.

Etymology Named in honour of Dr H. O. Wagner (1897–1977), born in Germany but living in retirement in Mexico City, who published extensively in English, Spanish and German on the biology of Mexican hummingbirds.

BOGOTA SUNANGEL
Heliangelus zusii

A mysterious species, known from one single specimen of unknown location.

Type information *Heliangelus zusii* Graves 1993, of unknown location.

Alternative names Spanish: Colibrí de Bogotá.

Discovery In the years around the turn of the 20th century, there was a large (and thoroughly regrettable) trade in bird skins, especially hummingbirds, for the millinery trade. One London auction house alone, between 1904 and 1911, sold 152,000 hummingbirds for this deplorable trade (Doughty 1975). The total number of birds thus sacrificed for female vanity and male cupidity in the years before bird protection laws were enacted in the market countries was doubtless in the millions. Some of these birds became scientific specimens, known under the generic term of 'Bogotá skins'. This does not imply that they were found in Bogotá – in fact, some specimens may have come from as far afield as Venezuela and Ecuador – merely that they were traded via Bogotá.

In 1947 Fr Nicéforo María (a Catholic priest who was an accomplished herpetologist, who is also commemorated in the names of a now extinct race of Yellow-billed Pintail *Anas georgica niceforoi* and Nicéforo's Wren *Thryothorus nicefori*) sent a specimen of a spectacular hummingbird which he had purchased in 1909 to Rodolphe Meyer de Schauensee (see the species account of *Tangara meyerdeschauenseei*) at the Academy of Natural Sciences in Philadelphia, where the specimen is currently housed. This unique specimen caused more than a little controversy. J. T. Zimmer thought that it might be a hybrid between a Long-tailed Sylph *Aglaicercus kingi* and *Heliangelus squamigularis* (another 'species' known from a Bogotá skin). Another opinion suggested a hybrid between *A. kingi* and a *Thalurania*, possibly *furcata* or *colombica* (2013 HBW SV). J. L. Peters, with refreshing honesty, stated in a letter of April 1947 (quoted in Graves 1993) that he was "damned if I know". A. Wetmore, in contrast, believed that the Bogotá specimen represented a valid species. Finally, Hinkelmann *et al.* (1991) resurrected a hybrid origin, between *A. kingii* and *T. furcata*, an intergeneric crossing that is believed to be the origin of the problematic Nehrkorn's Sylph (*Neolesbia nehrkorni*), itself a mysterious 'species' known from four or five specimens of mostly unknown provenance.

In 1993, G. Graves argued convincingly, based on morphological characters, that the 1909 specimen represented a valid species, which he named *Heliangelus zusii*, rejecting the hypotheses that it was *Neolesbia nehrkorni*, a further unknown hybrid or a fraudulently made-up specimen from two different skins. In 2010, Kirchmann *et al.* produced DNA evidence to support Graves's view; their study also indicated that *H. zusii* is a sister to a clade of mid- to high-altitude Andean species currently placed in the genera *Taphrolesbia* (Grey-bellied Comet) and *Aglaiocercus* (sylphs), raising the possibility of generic reassignment of *H. zusii*.

Description 12 cm. Monotypic. Overall dark bluish-black, lower back and rump deep greenish-blue; tail deeply forked, dark purple; gorget and forehead brilliant pale golden-green; whitish post-ocular spot; bill straight, relatively long.

Habitat No data.

Food and feeding No data.

Breeding No data.

Voice No data.

Movement No data.

Range Unknown. It is possible that the specimen could have originated in humid or semi-arid habitat, anywhere from NW Venezuela to northern Peru (Kirchmann *et al.* 2010).

Conservation status Classified by BirdLife International as Data Deficient; however, it would appear to be highly unlikely that the species persists. Habitat destruction in Colombia for coffee and maize production was already well underway at the time of the supposed collection of the unique specimen.

Etymology The common name 'Bogotá' Sunangel itself contains a large element of speculation; the specific name honours Dr Richard L. Zusi (b. 1930), Emeritus curator of birds at the United States National Museum, Washington, DC, "in recognition of his contributions to the systematics of hummingbirds".

ROYAL SUNANGEL
Heliangelus regalis

A spectacular hummingbird; previously thought to be endemic to Peru, but now also recorded in southern Ecuador.

Type information *Heliangelus regalis* Fitzpatrick *et al.* 1979, Cordillera del Cóndor, Peru.
Alternative names Spanish: Colibrí Real.
Discovery In 1975, while conducting a brief survey of the avifauna of the Cordillera del Cóndor, a range of mountains straddling the Peru–Ecuador border, J. W. Fitzpatrick and colleagues obtained a specimen of an immature hummingbird of unknown affinities. In July 1976 a subsequent expedition took a small series of specimens, establishing the identity as a new species of the genus *Heliangelus* (Fitzpatrick *et al.* 1979).
Description 11–12 cm. Male: entire body plumage deep violet-blue, most brightly iridescent on crown; tail long and deeply forked, iridescent violet-blue; bill black, almost straight, with feathering covering base. Female dark green above, rich cinnamon below, with a broad, pale buffy gorget. Male differs from all other members of the genus in lacking a differentiated throat-patch. Race *johnstoni*, a brighter, steely-blue with intense indigo iridescence on crown, throat and upper breast (Graves *et al.* 2011).
Habitat Stunted humid forest and shrubland, 'elfin scrub' consisting of grassland with mossy stunted forest up to 4 m in height, mostly on ridge-tops; possibly maintained from further succession by fire (Heynen 1999 HBW 5); 1,350–2,200 m, but also as low as 550 m (Dauphiné *et al.* 2008).
Food and feeding Nectarivorous; also seems to sally for insects. A favourite food source is the nectar of *Brachyotum quinquenerve* (Melastomataceae) (Seddon *et al.* 1993), although in other locations a wide variety of flowers is visited. Feeds by both hovering and perching (Freile *et al.* 2011).
Breeding Display of male involves a circling flight, the tail alternately fanned and closed (Freile *et al.* 2011). At the type locality, breeding seems to occur in late July

▼ Royal Sunangel, male. Alto Nieva, San Martín, Peru, October 2015 (*David Beadle*).

▼ Royal Sunangel, female. Abra Patricia, Peru, March 2014 (*Jon Hornbuckle*).

(Fitzpatrick *et al.* 1979). One nest in Ecuador with two chicks was "on a sandstone outcrop" (Krabbe & Ahlmann 2009).

Voice A variety of calls under different circumstances; a high, fast *chichup chup*, an emphatic *tchip*, a descending two- or three-note call; in display, a fast chattered series of very high notes and, between disputing males, a high-pitched jumble of notes *jijijiji* etc.

Movements No data.

Range In Peru, Amazonas, San Martín and Cajamarca; in southern Ecuador, Zamora-Chinchipe in the Nangaritza Valley (Krabbe & Ahlman 2009). The race *H. r. johnstoni* was described from the Cordillera Azul, Loreto, Peru (Graves *et al.* 2011).

Conservation status Currently classified as Endangered by BirdLife International. Although it appears to be quite common in some locations (e.g. Hornbuckle 1999), continuing deforestation for cash crops (including marijuana) remains a serious problem. Mining operations and the roadbuilding associated with them, in itself and in opening up areas to further exploitation, is a major concern (Freile *et al.* 2011).

Etymology *Regalis*, Latin, 'royal'; the race *johnstoni* is named in honour of Professor Ned K. Johnston (1932–2003), formerly Curator of Ornithology at the Museum of Vertebrate Zoology, University of California, Berkeley (Remsen & Cicero 2007).

COLOURFUL PUFFLEG
Eriocnemis mirabilis

A spectacular and highly restricted hummingbird, possibly the most endangered in its family. May form a superspecies with Emerald-bellied Puffleg *E. alinae* (Schuchmann *et al.* 2001).

Type information *Eriocnemis mirabilis* Meyer de Schauensee 1967, Charguayaco, Cauca, Colombia.

Alternative names Spanish: Calzadito Admirable, Esmeraldita patiblanca de Munchique, Colibrí de zamarros blancos.

Discovery In April 1967, J. S. Dunning, who was the first person to photograph many species of Neotropical birds using the technique of capturing them and photographing them in an enclosure before release (Dunning 1982), mist-netted an unfamiliar puffleg at Charguayaco, Cauca, SW Colombia. The captured bird was prepared as a specimen by K. von Sneidern and submitted to R. Meyer de Schauensee who described it as a new species in June 1967. Several other specimens caught at the same time were released.

Description 8–9 cm. Monotypic. Male dark shining green with glittering iridescent gorget and frontlet; belly glittering indigo-blue, undertail-coverts coppery-red; leg puffs white with cinnamon tips; tail forked, dark bronzy-green above, golden bronzy-green below; bill fairly short, straight, black, legs pink. Female shining green above, throat whitish with large circular green spots, belly white with conspicuous red spots.

► Colourful Puffleg, adult male defending its territory. Munchique National Park, Cauca, Colombia, February 2010 (*Carl Downing*).

▲ Colourful Puffleg, female. Munchique National Park, Cauca, Colombia, 1998 (*Luis A. Mazariegos*).

Habitat Wet forest and forest borders in subtropical zone, 220–2,800m. Seems to prefer undisturbed cloud-forest, but has been observed at forest edges and in clearings (Heynen 1999 HBW 5).

Food and feeding Nectarivorous. Has been seen to feed from a wide variety of flowers of at least 15 species, from, in order of preference, the Ericaceae, Bromeliaceae, Rubiaceae and Melastomataceae (Ramírez-Burbano *et al.* 2007); feeds in understorey to mid-levels (Mazariegos & Salaman 1999).

Breeding Nest and eggs undescribed.

Voice Female call is a "persistent high-pitched *sip*" (Pearman 1993).

Movements Not certain. It has been suggested that some altitudinal movement may occur (Mazariegos & Salaman 1999); however, capture and recapture of individuals in the same location would indicate otherwise (López-Ordóñez *et al.* 2008).

Range Endemic. Extremely restricted. For two decades after the initial discovery the species was unrecorded. Further sightings occurred in 1987, 1990, 1997 and 1998, all, apparently, within a very short distance of the original type locality, suggesting that the species was "incredibly localised" (Mazariegos & Salaman 1999, Balchin 2007). However, further fieldwork from 2003 onwards uncovered new locations, including one 30 km south of the original type locality. Since suitable habitat survives, it is possible that other populations exist.

Conservation status Currently classified by BirdLife International as Critically Endangered. Until the discovery of the new locations detailed above, estimates of the world population were in the range of 50–249 individuals (BLI SF 2006). This has been changed to 250–999 mature individuals (BLI SF 2014). Main threats are clearing of forest for crops, both legal and illegal, and logging. Prior to the 1980s, the local economy was based on the production of the fruit 'lulo' or naranjilla (*Solanum quitoense*) which required shade and was compatible; however, due to a fungal disease, this became uneconomic, forcing local people to change to logging. Recently, efforts to go back to 'lulo' production and further land acquisition are designed to reverse this trend.

Etymology Latin, *mirabilis*, 'miraculous'.

GORGETED PUFFLEG
Eriocnemis isabellae

A rather atypical puffleg in that it has a distinct iridescent gorget, unlike the majority of the members of this genus. Probably most closely related to Black-breasted Puffleg *E. nigrivestris.*

Type information *Eriocnemis isabellae* Cortés-Diago *et al.* 2007, Serranía del Pinche, Cauca, Colombia.
Alternative names Spanish: Calzadito del Pinche, Zamarrito del Pinche.
Discovery In 2005, during a survey of previously unexplored high altitude habitats in southern Colombia, an unusually coloured male hummingbird, apparently of the genus *Eriocnemis*, was captured, photographed, ringed and released, along with a couple of females. It was initially identified as a Glowing Puffleg *E. vestitus*; however, later comparison with authentic specimens showed that this identification was incorrect. In 2006 further expeditions mist-netted several more males, but no females, giving a holotype and some paratypes described as a new species, *E. isabellae.*
Description 8–9 cm. Monotypic. A relatively small puffleg, with a short bill and moderately forked tail; male blackish with golden olive-green iridescence above; rump glossy green; tail dark steely-blue; gorget glittering violet-blue; breast and belly velvety-black, leg-puffs white, undertail-coverts blue. Female more bronzy-green, grading to bluish on rump; belly velvety-black with light golden-green discs, flanks shining turquoise-green, leg-puffs white (2013 HBW SV).
Habitat Stunted, high-altitude elfin forest, with natural clearings, 2,600–2,900 m, the dominant species being Andean Oak (*Quercus humboldtii*).
Food and feeding Nectarivorous. Has been seen to feed on *Bijaria* and *Cavendishia* (both Ericaceae) and *Cinchona* and *Faramea* (both Rubiaceae).
Breeding Nest and eggs undescribed.
Voice Territorial call a monosyllabic, sharp, frequently-repeated *tuck-tuck*, lower-pitched than some other members of the genus.
Movements No data; altitudinal movements possible.
Range Endemic. Restricted entirely to the Serranía del Pinche, Cauca, Colombia.
Conservation status Currently classified by BirdLife International as Critically Endangered. The current known range is about 10 km^2 or less. The major threats are forest clearance for the cultivation of illegal coca crops; annual damage to the known habitat from this cause is estimated at 8% of the total. The lack of government presence and the general lawlessness associated with illegal drug production does not make the situation any easier. Conservation actions include local initiatives and education. A proposed road into the area has further serious implications.
Etymology Named for Isabella Cortés, daughter of the senior author of the type description.

NEBLINA METALTAIL
Metallura odomae

A high-altitude species that may form a superspecies with several other members of the genus (Heindl & Schuchmann 1998); mitochondrial DNA studies show it to belong to a clade of treeline specialists including Scaled and Black Metaltails *M. aenocauda* and *M. phoebe* Benham *et al.* 2015; sometimes regarded as conspecific with Fire-throated Metaltail *M. eupogon*, but major plumage differences as well as the wide allopatry of the two forms makes this seem unlikely.

Type information *Metallura odomae* Graves 1980, Cerro Chinguela, ca 5 km north-east of Sapalache, Piura, Peru.
Alternative names Spanish: Metalura de Chinguela, Metalura Neblina.
Discovery In October 1977 G. R. Graves, on an expedition organised by the Louisiana State University Museum of Zoology, collected two specimens of an unknown metaltail *Metallura* in the Divisoria de Huancabamba, Piura, Peru. Unfortunately, these specimens were stolen in Peru before detailed comparisons could be made with authentic specimens of other members of the genus. In 1978 a LSUMZ party was able to return to the region and collect a short series of specimens, allowing the description of the new species (Graves 1980).
Description 9 cm. Monotypic. In form a typical metaltail hummingbird, with fairly short, straight bill with the lower mandible undercut, giving a sharply pointed overall shape. Male, upperparts dark shining bronzy-green; throat iridescent reddish-purple; whitish

► Neblina Metaltail, male. Cerro Chinguela, Piura, Peru, June 2014 (*Fernando Angulo Pratolongo*).

post-ocular spot; underparts dull bronzy-green, often with buff scaling; crissum with rufous edgings; retrices bronzy bluish-green above, shining green below, the outermost pair tipped pale grey. Female similar to male but with extensive white or buffy scaling to underparts; throat patch incomplete or scarlet; tail more coppery on upperside.

Habitat High altitudes, 2,600 m to 3,350 m, occasionally up to 3,650 m; mossy elfin forest, forested ravines near treeline, windswept and foggy *páramo*.

Food and feeding Nectarivorous; forages from understorey to canopy. Seen to feed off *Brachyotum* (Melastomataceae), *Berberis* (Berberidaceae) and ericaceous plants. Also hawks for small insects. A very hardy species, seen foraging in snowstorms (Graves 1980).

Breeding One nest only described so far, a mossy cup covered with pale lichens, 2 m up in a small rock-ledge cavity sheltered from the wind; two eggs, incubation by female. This was in late August, but presumed height of breeding season is November–February (Schuchmann 1999 HBW 5).

Voice Calls include a "jerky wiry chatter," also buzzier notes and scratchy chatters (Schulenberg *et al.* 2007).

Movements May show slight altitudinal movement.

Range High mountains of extreme northern Peru (eastern Piura, western Cajamarca) and southern Ecuador (Loja, Zamora-Chinchipe).

Conservation status Currently classified by BirdLife International as Least Concern. In parts of its range apparently quite common; much of the range is remote and inhospitable; however, parts of the range are grazed by cattle and burnt annually. Several areas have at least nominal protected status, including Podocarpus National Park in Ecuador and Tabaconas–Namballe National Sanctuary in Peru. This does not invariably guarantee the absence of damaging activities such as gold-mining and small-scale logging. The Cerro Chinguela in Peru is not apparently protected (BLI SF 2014).

Etymology Named in honour of Babette M. Odom (1911–1984) of Orange, Texas, in recognition of her "long-standing interest in avian natural history" and her support of the LSUMZ field programme in Peru.

▼ Neblina Metaltail, female. Cordillera Las Lagunillas, Zamora-Chinchipe, Ecuador, April 2012 (*Nick Athanas*).

MANGAIA KINGFISHER
Todiramphus ruficollaris

Treated as a race of the more widespread *T. tutus* (Pacific, Polynesian or Chattering Kingfisher) by some authorities (e.g. Fry *et al.* 1992); however, differs substantially in plumage and vocally (Holyoak 1974).

Type information *Halcyon ruficollaris* Holyoak 1974, Mangaia, Cook Islands.

Alternative names Mewing Kingfisher. Maori: Tanga'eo.

Discovery The birds of several of the Cook Islands, including Mangaia (also Mangea or A'ua'u) were not investigated until 1973. In July to September of that year D. T. Holyoak visited a number of islands, during which time he secured specimens of seven new taxa, including a new kingfisher and swift (Atiu Swiftlet, *Aerodromus sawtelli*). The nomenclature of this taxon has been the source of some confusion (Holyoak 1976; Schodde & Holyoak 1977).

Description 22 cm. Monotypic. Male: forehead and crown blue-green; ear-coverts light turquoise; superciliary buff, nape and upper mantle orange-buff; back blue-green, rump and uppertail-coverts deep turquoise; tail deep blue above, blackish below; underparts from chin to crissum entirely white, except for an orange-buff suffusion across upper breast, more pronounced in female (Rowe & Empson 1996b); bill mostly blackish; iris blackish-brown; legs blackish-grey with light yellow soles.

Habitat A forest species, not dependent on water and preferring a continuous forest canopy, including coastal *Barringtonia* forest and indigenous mixed forest, although it will cross open spaces between forest patches. Also mature secondary forest (Rowe & Empson 1996a, 1996b).

Food and feeding Prey is taken at all levels of the forest; food items include worms, caterpillars, grubs, termites, grasshoppers, stick insects, cockroaches, moths, spiders and lizards. Lizards are an important part of the diet during courtship feeding (Rowe & Empson 1996a).

Breeding Breeding season from early October, last fledgings in early February. Nests in cavities excavated in the trunks of dead coconut palms or in decaying limbs of live trees of other species. Frequently numerous holes in a single tree. No nesting material; eggs (2–3, white) laid in bare hole. Incubation 21–23 days by both sexes; young fed by both sexes, fledging period probably about 26 days. In several cases, trios of birds were involved, both with two males (both of whom copulated with the female) and, in one instance, two females and one male. Polygamous males were both seen to defend the nest against intrusive Common Mynas *Acridotheres tristis* (Rowe & Empson 1996b).

Voice Very vocal with a wide repertoire of calls; a strong clear *kek-kek-kek-kek*, a distinctive two-note call (giving rise to the local name for the bird), in aggressive chases a harsh *scraak*, a contact *chucka-chuk*, another close contact call *tui-tui*, mews and croons, etc. (Rowe & Empson 1996b).

Movements No data.

Range Endemic. Entirely confined to the island of Mangaia, 51.8 km^2, southern Cook Islands.

◀ Mangaia Kingfisher, adult. Mangaia, Cook Islands, December 2007 (*Gerald McCormack*).

► Mangaia Kingfisher, adult feeding juvenile. Mangaia, Cook Islands, December 2007 (*Gerald McCormack*).

Conservation status Currently classified by BirdLife International as Vulnerable, with the comment that if decline is suspected, this could be changed to Critically Endangered. In 1992–1993, the population was estimated at 250–450 birds; in 1996–1997, estimates were of 400–700 birds (Baker *et al.* 1998; Kelly & Bottomley 1998). More recent estimates seem to indicate a stable population. Appears to be "surprisingly common" in disturbed areas (BLI SF 2014). Potential threats include clearing of forest, browsing by goats, possible predation by cats and rats, and nest-site competition and disturbance by Common Myna, although kingfishers were common in disturbed areas with a high myna population. A myna control programme has been suggested by the local conservation organisation.

Etymology Latin, 'rufous-collared', referring to the plumage feature which distinguishes this species from *T. tutus*; the English name from the unique location of occurrence.

RUAHA HORNBILL
Tockus ruahae

Originally described as a new race; now currently listed as a full species.

Type information *Tockus erythrorhynchus ruahae* Kemp & Delport 2002, Ruaha National Park, Tanzania.

Alternative names Tanzanian Red-billed Hornbill.

Discovery The Red-billed Hornbill is a very complex group taxonomically, with a variety of different populations scattered across Africa from Senegal to Mozambique. It is perhaps the most thoroughly studied species in its family. In 2002 A. C. Kemp and W. Delport conducted a comprehensive review of the species and described a new race endemic to Tanzania, while at the same time suggesting that all the five races that are clearly separable on the colour of facial plumage and soft parts should be recognised as full species, a viewpoint backed up by DNA evidence, and accepted by the IOC and the *Handbook of the Birds of the World* (2013). A further undescribed taxon may occur in Kenya.

Description 42–48 cm. Monotypic. A relatively small hornbill, plumage entirely black and white; head white, area around eye and central head-streak black; back black with white-spotted shoulders; white patch on middle secondaries; underparts entirely white; iris yellow; bill red with paler base and, in the male, black area at base of lower mandible; legs black with grey soles. Differs from the parapatric *T. (e). erythrorhynchus* (of most of Africa between Mali and Somalia, with the zone of contact in eastern Tanzania) by its yellow, not brown, eye, and black, not pink, circumorbital skin.

Habitat Savanna and semi-arid woodland, mainly in areas with baobabs.

Food and feeding Few data; stomach contents include insect remains including ants, beetle larvae and seeds (Kemp & Delport 2002). *Erythrorhynchus* also takes

▲ Ruaha Hornbill, male. Serengeti National Park, Tanzania, May 2013 (*Markus Lilje*).

► Ruaha Hornbill, female. Serengeti National Park, Tanzania, May 2009 (*Adam Riley*).

small vertebrates such as geckos, bird nestlings and rodents.

Voice Call is similar to that of *erythrorhynchus*, a series of loud accelerating clucking notes.

Breeding Nest and eggs undescribed; presumably similar to those of *erythrorhynchus*, which nests in a natural cavity, the entrance hole of which is reduced in size by the female, using a cement made up of faeces and food remains; clutch 2–7 eggs.

Movements No data indicating movements, although some nomadic wandering might be expected, by analogy with *erythrorhynchus*.

Range Endemic. So far known from the basin between the mountains and lakes of the main Albertine Rift Valley in the east of Tanzania and the Eastern Arc Mountains in the west.

Conservation status Not assessed by BirdLife International, but appears to be locally quite common; several areas of its range are protected by national parks (Ruaha, in central-eastern Tanzania, and Katavi, in the south-west).

Etymology Both the vernacular and scientific names are derived from Ruaha National Park, whose name comes from the Ruaha River.

SIRA BARBET
Capito fitzpatricki

A spectacular barbet from Peru, obviously closely related to the recently discovered Scarlet-banded Barbet (q.v.), and treated as a race of it by some authorities (e.g. Clements: updates and corrections, August 2013), although given full specific status by the IOC World Bird List Version 7.1. Separated geographically from Scarlet-banded by more than 400 km of unsuitable habitat.

Type information *Capito fitzpatricki* Seeholzer *et al.* 2012, 11.45 km WSW of the mouth of Quebrada Shinipo, Cerros del Sirá, Dpto. Ucayali, Peru.

Alternative names Spanish: Cabezón de Sirá, Barbudo del Sirá.

Discovery In late 2008, an investigation of the southern section of the Cerros del Sirá, an isolated area of highlands in the Departments of Ucayali, Junín and Pasco, Peru, was conducted by ornithologists from several institutions in the United States and Peru. Previous very fruitful ornithological work in this range had largely been confined to the northern section of the range, which was separated from the southern section by a lower altitude saddle. In October 2008, M. G. Harvey encountered a unique barbet in a mixed flock at an altitude of 1,225 m. One specimen, a female, was collected. Due to logistical difficulties, further work at this site was not feasible; however, the group established camp at a second site some 18.5 km to the north, in apparently similar habitat. The new species was encountered at this second site; several more specimens were taken, allowing a full description and collection of tissue samples. Apparently local Amerindian people were quite familiar with the new species.

Description 19.5–20 cm. Monotypic. Male: Forehead, crown and nape crimson; upper mantle crimson, tipped pale yellow; white supercilium; lores to ear-coverts black; upperparts, wings and tail black with white feathers on uppertail-coverts; chin, throat and upper chest white; lower chest and flanks crimson; central belly white; lower flanks and thighs black, vent white; iris carmine, naked orbital ring grey-black; bill steel grey with duskier grey on cutting-edge; tarsi slaty-grey. Female similar, but crown less deeply coloured, and outer scapulars red, tipped with yellow. Differs from *C. wallacei* in slightly larger size and quite distinct plumage on lower underparts, *wallacei* lacking the extensive red flanks and black thighs. In addition, the scapular-patch in the female is scarlet in *fitzpatricki*, yellow in *wallacei*. DNA analysis supported the treatment of *wallacei* and *fitzpatricki* as distinct, but closely related, species (Seeholzer *et al.* 2012).

Habitat Tall, lush montane forest with sparse understorey; also lower forest with arboreal epiphytes, with understorey of low trees, shrubs and ferns, 95–1,250 m, possibly higher (2013 HBW SV). Syntopic with Black-spotted Barbet *C. niger auratus.*

Food and feeding Forages in canopy and subcanopy in a slow methodical fashion. Largely frugivorous; stomach contents mostly vegetable, probably fruit, and some insects (Seeholzer *et al.* 2012).

Breeding Nest and eggs undescribed.

◄ Sira Barbet, female. Río Shinipo Valley, Cerros del Sirá, Ucayali, Peru, November 2008 (*Michael G. Harvey*).

Voice Calls include a tityra-like grunting, similar to the corresponding call of *C. wallacei*. Song is a low-pitched purring; roosting birds (in a hole in a dead snag) gave quiet low-pitched groans and grunts (Seeholzer *et al.* 2012).

Movements No evidence of movements.

Range Endemic. Presently known from three locations on the east slope of the southern Cerros del Sirá, Ucayali, Peru; the limits of these are from about 10°25' S to 10°41' S and from about 74°06' W to 74°09' W, an area of about 300 km². Probably does not occur in the northern part of the Cerros del Sirá, where it has not been detected by several expeditions.

Conservation status Classified by BirdLife International as Near Threatened. The range is limited to a narrow altitudinal zone, which at this time is relatively inaccessible and on the boundary of, and within, the Sirá Communal Reserve. Potential threats include mining, logging and oil exploration, all of which are active in the area.

Etymology Named in honour of Dr John W. Fitzpatrick (b. 1951), whose long career at the Field Museum, Chicago, Archbold Biological Station and Cornell Laboratory of Ornithology "has had an immeasurable influence on ornithology and bird conservation" (Seeholzer *et al.* 2012), and who was himself responsible for the description of six new species of birds from Peru (one tody-tyrant, one wren, one tyrannulet, one owl, one antbird and one hummingbird).

SCARLET-BANDED BARBET
Capito wallacei

A spectacular barbet from an isolated mountain range in eastern Peru, one of two such species recently discovered from that country.

Type information *Capito wallacei* O'Neill *et al.* 2000, ca 77 km NW of Contamana, Loreto, Peru.

Alternative names Scarlet-belted Barbet. Spanish: Cabezón de Loreto.

▼ Scarlet-banded Barbet, male. Plataforma, Dpto. San Martín, Peru, July 2014 (*Dubi Shapiro*).

◀ Scarlet-banded Barbet, female. Plataforma, Dpto. San Martín, Peru, July 2014 (*Dubi Shapiro*).

Discovery In 1996, personnel from the Louisiana State University Museum of Natural Science gained access to an unnamed 1,538 m peak by cutting a trail from the east bank of the Río Cushabatay, about 77 km WNW of the town of Contamana, Loreto, Peru. This may be the most isolated and unexplored patch of montane cloud-forest in the whole of South America. Shortly after access was achieved, D. F. Lane, after a strenuous hike, observed a bird which he immediately realised was an undescribed and very spectacular barbet. Four specimens were taken by Lane and Manuel Sánchez, establishing that a new species was indeed involved (Lane 2012). See also xeno-canto.org: XC 62885.

Description 19.5 cm. Monotypic. Male: Crown brilliant scarlet; supercilium white; lores to ear-coverts black; chin, throat and upper breast white; a broad scarlet-red band across central chest; lower chest and belly yellow, becoming paler on vent; wings and tail mostly black; back yellow, rump white; iris blood-red, bill light grey with dark tip; tarsi greenish-grey, soles orange-yellow. Female similar, but with conspicuous yellow patch on scapulars.

Habitat Confined to wet, highly epiphytic cloud-forest, trees usually 10–20 m tall, with a heavy ground cover, with up to 1 m of deep spongy mosses. Apparently only above 1,250 m, below which altitude the nature of the forest changes, becoming drier and less epiphytic.

Food and feeding Occurs in both single-species and mixed flocks, the latter including flycatchers, tyrannulets, warblers and chlorophonias. Feeds mostly 5–15 m up. Food items include fruit and berries and, when in mixed flocks, insects (O'Neill *et al.* 2000).

Breeding Nest and eggs undescribed. Three specimens collected in July retained the bursas of Fabricius, suggesting that they had fledged within the previous six months (O'Neill *et al.* 2000).

Voice Song, quite distinct from that of other *Capito* species, is a long, even-frequency purring, resembling the distant drumming of a medium-sized woodpecker. Also a tityra-like grunting call.

Movements No data, presumably sedentary.

Range Endemic. Currently known only from an unnamed peak labelled on topographic maps as 'Peak 1538', in a range of hills east of the Río Cushabatay system (also east of the Cordillera Azul range) in Loreto, Peru. Apparently occurs only above about 1,250 m; there are substantial areas at and above this altitude, extending some 50 km north of the type locality, which have not yet been investigated (O'Neill *et al.* 2000). Has not so far been found in the Cordillera Azul itself.

Conservation status Classified by BirdLife International as Vulnerable, on the basis of its small (currently) known range. Further fieldwork, if it revealed extensions to the known range, could result in a downgrading of this classification. Presently, the remoteness of the known range is its major protection; human visitation seems to be rare and no habitat modification has occurred above 300 m. Within its known range the species seems quite common (O'Neill *et al.* 2000). The area involved has no formal protection (BLI SF 2014).

Etymology Named in honour of Robert B. Wallace (1918–2002), a businessman and philanthropist "in recognition of his intense interest in, and support of, ornithological exploration by the Louisiana State University Museum of Natural Science" (O'Neill *et al.* 2000).

WHITE-CHESTED TINKERBIRD
Pogoniulus makawai

A highly controversial bird, with radically different opinions among various authorities as to its validity as a species.

Type information *Pogoniulus makawai* Benson & Irwin 1965, 4 miles north of Mayau, Kabompo District, Northern Rhodesia (now Zambia); ca 1,150 m.

Discovery In September 1964, M. P. Stuart Irwin and J. Makawa made a collecting trip under the auspices of C. W. Benson of the Rhodes-Livingstone Museum (currently known as the Livingstone Museum) into the North Western Province of Northern Rhodesia (Zambia). Makawa collected a male tinkerbird in breeding condition. This remains a unique specimen; it is possible that Makawa himself may be the only person knowingly to have seen this bird alive.

Whether *P. makawai* is a good species has been the source of much argument. Goodwin (1965) felt that it probably was, since it was sympatric with Yellow-rumped Tinkerbird *P. bilineatus* (and could not therefore be a race of that species), although he left the window open to the possibility that the Kabompo specimen was merely an aberrant example of *bilineatus*, a viewpoint taken by several authors. Dowsett & Dowsett-Lemaire (1980) pointed out that intensive searches had failed to find the bird, an opinion that they reiterated after 13 further fruitless years (Dowsett & Dowsett-Lemaire 1993). Other ornithologists have agreed with them (e.g. Short & Horne 1985,1988, 2001, 2002).

Against this, Collar & Fishpool (2006) made a detailed examination of all points of view and argued cogently for the specific validity of *makawai*. They pointed out that the distinctive plumage characters of the 1964 specimen, in contrast to *bilineatus*, were too numerous (some 17 features are mentioned and examined) to be accounted for by an abnormal specimen. They also raised the possibility that the 1964 specimen might have been a wanderer from areas not too far distant, in Angola or the Democratic Republic of the Congo (previously Zaire), that have been far less thoroughly investigated than has Zambia. Collar & Fishpool also maintain that even in Zambia, the nature of the *Cryptosepalum* forest is such that an uncommon species might still go undetected.

Description 12 cm. Monotypic. Forehead, crown, mantle and wings black, lacking a pale supraorbital stripe; rump lemon-yellow; rectrices black with yellow edgings; chin and malar stripe black; throat and upper chest cream-white; lower chest, abdomen and flanks pale lemon-yellow with no greenish tinge; iris dark brown; bill mostly black; tarsi and feet whitish-flesh (Benson & Irwin 1965).

Habitat The specimen was taken in *Cryptosepalum* woodland with an understorey of dense thicket.

Food and feeding No information.

Breeding No information; the type specimen, taken in September, had enlarged testes.

Voice No information.

Range The location of the type specimen (see above) is the only information.

Conservation status Currently listed as Data Deficient; if proven to be a good species, it would undoubtedly rate a high category.

Etymology Named after its collector, Jali Makawa (d. 1995). Makawa, who was probably born about 1914 (he himself did not know) was "undoubtedly the most accomplished African bird collector this century". He originally came into the employ of the distinguished authority on African birds, C. W. Benson, as a camp servant, but his "exceptional powers of observation" quickly became apparent, and from then on his career lay in field ornithology, collecting many thousands of specimens, culminating in employment at the Livingstone Museum. He took part in field expeditions to Zambia, Malawi, Kenya, Ethiopia, Botswana, Mozambique and the Comoro Islands. Three subspecies of African birds (a parrot, *Coracopsis vasa makawa*, a bush-shrike *Chlorophoneus olivaceus makawa* and a lark *Spizacorys conirostris makawai*) were named after him, as well as the present tinkerbird; although, as Richard Brooke was to remark, at the time he would have far preferred a case of beer as his just reward!

YELLOW-FOOTED HONEYGUIDE
Melignomon eisentrauti

A widespread but apparently rather rare honeyguide, the circumstances of whose description has caused some controversy.

Type information *Melignomon eisentrauti* Louette 1981, near Grassfield, Liberia.

Alternative names Coe's Honeyguide, Eisentraut's Honeyguide. French: Indicateur d'Eisentraut.

Discovery In March 1980 a honeyguide was mist-netted and collected near Mt Nimba, Liberia. Comparison with a series of Zenker's Honeyguide *Melignomon zenkeri* from the Democratic Republic of the Congo (Zaire) showed it to be distinct, on which basis Louette, in a paper published on 27 March 1981, described it as a new species, *M. eisentrauti*. In fact, in August 1956, W. Serle had collected a specimen in western Cameroon which he ascribed, in a 1959 publication, to an immature plumage of *zenkeri*, and in 1957 M. Eisentraut collected one bird near Mt Cameroon. Both of these proved to be the new species; Louette believed, incorrectly, that Prof. Eisentraut's was the first, for which reason he named it in his honour in his type description.

▲ Yellow-footed Honeyguide. Bobiri Butterfly Sanctuary, Ghana, April 2014 (*Nik Borrow*).

However, in a paper submitted in October 1980 and published in June 1981, P. R. Colston pointed out that the International Union for the Conservation of Nature survey of Mt Nimba had collected 11 specimens of the new taxon between 1965 and 1974. "A copy of this paper (i.e. Colston 1981), proposing a new name, was sent to Dr Louette in October 1980, shortly after it had been submitted for publication. Dr Louette has since seen fit to describe the new honeyguide (i.e. Louette 1981) without informing us of his intention" (Colston 1981). According to Beolens & Watkins (2003), Louette borrowed a specimen, presumably from the IUCN collection, but this is not acknowledged in his 1981 paper. Under international rules of nomenclature, Louette's description and naming has priority.

Description 14.5 cm. Monotypic. Upperparts yellow-olive with a greyer crown; primaries blackish, fringed olive; central rectrices black, the outer largely white with black tips; throat and chest grey; undertail-coverts creamy-brown, thighs yellowish; bill and tarsi yellow. *Zenkeri* is generally darker below, with a dark bill and less white on the rectrices.

Habitat Primary tropical forest, also adjacent secondary forest, up to 750 m (Short & Horne 2002 HBW 7).

Food and feeding Frequents mid-strata and canopy of primary and secondary evergreen lowland forest (Borrow & Demey 2001). Food insects, fruits and their seeds and probably beeswax (Colston 1981); may join mixed flocks (Short & Horne 2002 HBW 7).

Breeding Nest-parasite. Eggs and host unknown, but possibly small woodpeckers, which in Liberia appeared to be breeding at the correct time (mid-March); tinkerbirds, which are frequent hosts of other honeyguide species, by contrast, were not in breeding condition at that time (Louette 1981).

Voice A series of about 13 clear emphatic notes, slightly descending and slowing towards the end *tuu-i tuu-i... tuu tuu* (Rainey *et al.* 2003).

Movements No data, probably sedentary.

Range Discontinuously in West Africa from SE Sierra Leone, NW and NE Liberia, Ivory Coast (Rainey *et al.* 2003; Lachenaud 2006), extreme eastern Nigeria (L. Fishpool & M. Gartshore, pers. comm.), and western Cameroon (Serle 1959; Bowden *et al.* 1995). Further work required in intervening areas.

Conservation status Currently classified by BirdLife International as Data Deficient. Appears to be relatively uncommon in those areas where it is known to occur.

Etymology Named by Louette in honour of Prof. M. Eisentraut (1902–1994), who is perhaps best known as a mammalologist (one bat, two mice and one shrew have the specific name *eisentrauti*; he was also a published poet). Had Colston's prior, but tardily published, description gained priority, his intent was to name it after Dr Malcolm Coe (b. 1930), who made observations of the species in Liberia in 1964.

TAWNY PICULET
Picumnus fulvescens

A new South American piculet whose discovery came not from fieldwork but from the careful examination of existing specimens.

Type information *Picumnus fulvescens* Stager 1961, Garanhuns, SE Pernambuco, Brazil, based on a specimen taken in 1927 by E. Kaempfer.
Alternative names Portuguese: Pica-pau-anão-canela, Pica-pau-anão-de-Pernambuco.
Discovery In 1957 a collection of birds was made in the state of Alagoas, Brazil, by Emilio Dente and acquired by the Los Angeles County Museum. Examination of this collection and comparison with other species of the genus *Picumnus* led to the description of a new species four years later (Stager 1961).
Description 10 cm. Monotypic. A typical Neotropical piculet, small and short-tailed with prominent head markings. Male: top of head black, the forehead feathers with red tips, the remainder with small white spots; upperparts largely rufous-brown or rich fulvous-brown, obscurely streaked; chin and throat whitish or buffy-white; chest, belly and vent rufous or rusty-brown. Female: similar, but red on forecrown lacking. Iris dark brown; bill black with pale grey base; tarsi grey.

▲► Tawny Piculet, female. Murici, Alagoas, Brazil, March 2004 (*Arthur Grosset*).

► Tawny Piculet, male. União dos Palmares, Alagoas, Brazil, November 2010 (*Ciro Albano*).

Habitat Deciduous, semi-deciduous and secondary forest, including degraded secondary scrub; to a lesser extent in *caatinga* (da Silva *et al.* 2011); sea level to 950 m.
Food and feeding Food items insects, especially the larvae and pupae of ants (Ruiz-Esparza *et al.* 2011).
Breeding Nest and eggs undescribed.
Voice A descending series of notes, *driee-driee-driee* or a high *see-see-see-sisi-wi* (van Perlo 2009). Said to drum occasionally (Gorman 2014).
Movements Presumably sedentary.
Range Endemic. Recent fieldwork (Ruiz-Esparza 2011; da Silva 2011) has shown the range to be substantially wider than previously believed; NE Brazil, from coastal and southern Rio Grande do Norte through Pernambuco, Paraíba and Alagoas to Sergipe, west to southern Ceará, marginally in southern Piauí.
Conservation status Currently classified by BirdLife International as Near Threatened, due to the continuing destruction of forest in NE Brazil for a variety of purposes, including logging, conversion to sugarcane plantations and pastureland, and removal of wood as fuel for brick kilns. However, it does seem to show some ability to live in degraded and regenerating secondary forest; further work might justify a downgrading of its status to Least Concern. Occurs in a number of protected locations in several states (BLI SF 2014).
Etymology *Fulvescens,* Latin, from its rusty-brown plumage.

FINE-BARRED PICULET
Picumnus subtilis

A quite widely distributed and, in some locations, fairly common piculet which was overlooked for more than 50 years after the first specimen was taken in Dpto. Puno, Peru, in 1916 by H. Watkins.

Type information *Picumnus subtilis* Stager 1968, based on a specimen taken in 1958, Hacienda Villacarmen, Dpto. Cuzco, Peru.

▼ Fine-barred Piculet, male. Horto Florestal, Rio Branco, Brazil, January 2015 (*Robson Czaban*).

Alternative names Stager's Piculet, Marcapta Piculet. Spanish: Carpintero de Cuzco.
Discovery During a revisional study of the Neotropical piculets of the genus *Picumnus,* examination of specimens from several collections in North America revealed a new species which had hitherto gone undetected due to its resemblance to Plain-breasted Piculet *P. castelnau*, a species which is quite common in Amazonian Peru (Stager 1968).
Description 10 cm. Monotypic. In form a typical piculet, small, dumpy and short-tailed. Male: forehead to nape black, the crown and upper forehead feathers tipped orange-red, nape and sides of crown with small white spots; lores pale buff-white; back, rump and shoulders yellowish-olive, faintly barred with olive-yellow; chin and throat dull white; breast grey, with straw-yellow barring; undertail-coverts straw-yellow; iris brown; bill black with blue-grey base to mandible; tarsi olive-green. Female has crown black with white spots. Juvenile more heavily barred than adults, but ventral barring more diffuse. *P. castelnau* has unbarred whitish underparts, lacks a white-spotted black nape in the male, and in the female has an unspotted black crown.
Habitat Humid tropical forest, lowlands (below 200 m) to 1,100 m. Less associated with *terra firme* than the sympatric Bar-breasted Piculet *P. aurifrons* (Schulenberg *et al.* 2007).
Food and feeding No data.
Breeding Nest and eggs undescribed; season probably April–July.
Voice A series of 4–8 sharp, thin, high-pitched *see-see-see* or *seet-seet-seet* notes, usually falling, but sometimes starting slowly before speeding up. Also repeated, quiet, warbling, squeaky notes. Drum, strong, repeated (2–6)

► Fine-barred Piculet, female. Rio Branco, Brazil, August 2014 (*João Quental*).

bursts of level pitch and volume, rather irregular, rolls often broken, ending in distinct strikes (Gorman 2014).

Movements Presumably sedentary.

Range Mostly Amazonian Peru (north and NE Cuzco, western Madre de Dios, extreme northern Puno (Schulenberg *et al.* 2007); recently in Acre, Brazil (Rêgo *et al.* 2009; Melo *et al.* 2015). Only area of sympatry with *P. castelnau* is in the area of the upper Río Ucayali near its confluence with the Río Urubamba (Stager 1968). Hybrids with *castelnau* have been reported (Winkler & Christie 2002 HBW 7).

Conservation status Classified by BirdLife International as Least Concern; although it has a rather restricted range, in parts of which habitat destruction remains a problem, it is not yet approaching the threshold for Vulnerable status. Fairly easily found at Amazonas Lodge, Peru (Wheatley 1994).

Etymology *Subtilis* refers to the fairly subtle plumage distinctions that allowed this species to hide, undetected, in the collections of several institutions for more than half a century. In fact, a further (and the earliest) specimen, collected in March 1913 in Puno, also by H. Watkins, has turned up since the publication of Stager's type description. It was not identified by the collector but was later labelled *P. castelnau*; it remained in the private collection of Count Josef Seilern, housed in his castle at Lišná, Austro-Hungary (present Czech Republic) until acquired by the Vienna Museum of Natural History in 1986. Its true identity only emerged in 2000 (Schifter 2000).

KAEMPFER'S WOODPECKER
Celeus obrieni

A virtually unknown woodpecker whose rediscovery in the early years of the 21st century, after an 80-year absence, caused a great sensation in the South American birding community.

Type information *Celeus spectabilis obrieni* Short 1973, based on a 1926 specimen from Iruçui (*in errore* for Uruçui), Piauhy (Piauí), Brazil.

Alternative names Caatinga Woodpecker. Portuguese: Pica-pau-do-Parnaíba.

Discovery Some years prior to 1972, C. O'Brien of the American Museum of Natural History showed a puzzling specimen of a female woodpecker, taken by E. Kaempfer in eastern Brazil in 1926, to L. L. Short, an acknowledged expert on the Piciformes. The bird most closely resembled Rufous-headed Woodpecker *Celeus spectabilis*, a rare species found in the western Amazon Basin, from eastern Ecuador to NW Bolivia and extreme western Brazil, but had significant plumage differences from that species; the location of Kaempfer's specimen was over 3,000 km disjunct from the nearest range of *spectabilis*. After O'Brien's retirement in 1972, Short undertook a comparison of the unique specimen with examples of *spectabilis*, at that time unrecorded in Brazil, and described it as a race of that species. In 2006, Tobias *et al.* included

▲ Kaempfer's Woodpecker, male. Caxias, Maranhão, Brazil, June 2012 (*João Quental*).

▲ Kaempfer's Woodpecker, female. Pium, Tocantins, Brazil, February 2016 (*Ciro Albano*).

obrieni in a list of long-lost species that were potential candidates for rediscovery. Amazingly, within weeks of his prediction, the species was found in the state of Tocantins in eastern Brazil, and has since proven to be widespread across several states, if apparently rare. In 2009 Remsen *et al.* elevated the race *obrieni* to full specific status based on morphological and ecological differences from *spectabilis*, although obviously with a close relationship (Benz & Robbins 2011). More recent molecular evidence (de Sousa Azevedo *et al.* 2013) confirms the distinctness of *obrieni* from *spectabilis*, showing that the two forms have been separate since at least the mid-Pleistocene.

Description 26–28 cm. Monotypic. Male. Head including bushy crest rufous-red; lower throat and chest black, underparts creamy-buff; tail black; back plain yellow, scapulars heavily streaked with black, flight feathers chestnut; iris white; bill yellow. Female has less prominent barring on back and abdomen. Juvenile has face and throat extensively blackish, the buffy upper back lacking dark bars and a more reddish-brown crest (Leite *et al.* 2010). *C. spectabilis* differs by having much heavier barring on the back.

Habitat Open gallery forest in *cerrado* woodland, with a strong association with bamboo *Guadua paniculata* (BLI SF 2013). Absent from locations lacking bamboo (Leite *et al.* 2013).

Food and feeding Appears to specialise in ants (*Camponotus* and *Azteca*) found in bamboo canes, especially dry ones. Feeds most frequently at heights of 2–4m, drilling holes in bamboo internodes (Leite *et al.* 2013).

Breeding Nest and eggs undescribed. Breeding season starts in June/July, i.e. at beginning of dry season (Leite *et al.* 2010).

Voice Distinctive laughing, upwardly slurred song, beginning with a shrill squeal or squeak followed by more subdued notes, described as *reeahh-kah-kah-kah-kah* and *swueeah-kluh-kluh-kluh-kluh*. Also brief harsh rasping calls. Female voice different from male (recording of A. do Prado, XC 28191). Both sexes drum, usually on bamboo. Fast, even-paced 2–3 second rolls, usually with a 16–20 second gap between each (Gorman 2014).

Movements Presumably sedentary.

Range Endemic. Recent work has produced numerous records greatly extending the known range, including the states of Tocantins (do Prado 2006; Pinheiro & Dornas 2008); Mato Grosso (Dornas *et al.* 2011), Maranhão (Santos & da Vasconcelas 2007) and in numerous locations in Goiás, including two unrecognised specimens from 1967 and one from 1988 (Hidasi *et al.* 2008; Kirwan *et al.* 2015).

Conservation status Currently classified by BirdLife International as Endangered. Despite the quite extensive total range, it appears nowhere common; the population estimates, in the ranges of 350 to 4,000 individuals, reflect a wide degree of uncertainty. Major threats include habitat loss by conversion to soyabean cultivation, fire and replacement of native habitat with planted *Eucalyptus*.

Etymology The scientific name honours Dr Charles O'Brien (1905–1987), who worked at the American Museum of Natural History in New York for almost 50 years; the English name recognises Emil Kaempfer, a German who made extensive collections in South America between 1926 and 1931, and for whom Kaempfer's Tody-Tyrant *Hemitriccus kaempferi* is also named.

WESTERN STRIOLATED PUFFBIRD
Nystalus obamai

A widely distributed but cryptic species, the distinctiveness of which was first signalled by its characteristic vocalisations.

Type information *Nystalus obamai* Whitney *et al.* 2013, left bank of Rio Madeira, municipality of Porto Velho, Rondônia, Brazil.

Alternative names Spanish: Buco Estriolado del Oeste; Portuguese: Rapazinho-estriado-do-oeste.

Discovery Striolated Puffbirds *sensu lato* have been known to occur from the eastern foothills of the Andes across Amazonian Brazil for a century and a half, under the specific name *N. striolatus* (Meyer de Schauensee 1966); a separate, disjunct or nearly disjunct population in eastern Brazil (southern Amapá, Pará, Mato Grosso and Maranhão) was described as the race *N. s. torridus* (Bond & Meyer de Schauensee 1940). In July 2008 B. Whitney noticed significantly different vocalisations in different Brazilian populations of the 'Striolated Puffbird'. This gave rise to an extensive study of specimens and vocalisations, leading to the realisation that populations west of the Rio Madeira (and, in fact, in extensive areas of Amazonian Ecuador, Peru and Bolivia) were distinct, a conclusion that was confirmed by DNA analyses. This conclusion is accepted by the SACC (Proposal 617), although the suggested split of *N. s. torridus* from nominate *striolatus* was not.

Description 20 cm. Monotypic. In form a typical puffbird; stocky, large-headed with a massive, slightly hook-tipped bill, fairly long-tailed, zygodactyl feet. Plumage essentially identical to that of *N. striolatus*. Crown brownish-black with some rufous margins in front; conspicuous collar, pale ochraceous; back blackish-brown; rectrices dark greyish-brown with about eight yellowish-brown bars; chin plain creamy-white, lower throat and chest brownish-yellow with darker streaks;

▼ Western Striolated Puffbird. Ruta 16 near Siata, Apolobamba, Dpto. La Paz, Bolivia, November 2016 (*Jacob and Tini Wijpkema*).

belly creamy-white, with lateral streaks; iris yellowish-white; bill yellowish-green with darker tip and central culmen; tarsi yellowish-green. Differs from *N. s. striolatus* (its closest geographic neighbour) by having weaker crown spotting and numerous wholly blackish feathers, with no pale terminal tips, on the mantle.

Habitat Humid lowland *terra firme* forest and forest-edge, primary and second-growth, with canopy at least 18 m; absent from *várzea* and dense, closed-canopy forest (Whitney *et al.* 2013). In Ecuador, mostly 800 m to 1,700 m, rarely down to 400 m; (Ridgely & Greenfield 2001); in Peru, up to 1,200 m (Schulenberg *et al.* 2007) (both referring to *N. striolatus* as then recognised); in Bolivia, up to 1,850 m (Whitney *et al.* 2013).

Food and feeding Diet arthropods (Orthoptera, caterpillars etc., including quite large individuals which are beaten into submission against the perch before being swallowed). Forages from perches inside the canopy; also from electric cables along roadsides.

Breeding Nest and eggs undescribed; *N. striolatus* nests in a burrow in an earth-bank, probably excavated by the pair, laying up to four white eggs. In Peru, nests in September.

Voice Diagnostic. May sing in duet. In Peru (described as *striolatus* in Schulenberg *et al.* 2007), a series of slow, quiet whistles, often with pauses between phrases *whi-whi wheee? whi-wooo.* Characteristic stuttering quality of early notes (Whitney *et al.* 2013) distinguishes it from that of *striolatus.*

Movements Presumably sedentary.

Range Western Amazonia, from northern Ecuador (Napo) south through Peru (Loreto, San Martín to Madre de Dios), northern Bolivia (Beni, La Paz) and western Brazil (Amazonas, Acre and Rondônia, to the western bank of the Rio Madeira).

Conservation status Not currently classified by BirdLife International; Whitney *et al.* (2013) state that it is not threatened.

Etymology Named after Barack H. Obama (b. 1961), 44th President of the United States of America, in recognition of his support for moving away from a carbon-based energy policy.

ARARIPE MANAKIN
Antilophia bokermanni

One of the most striking new species to come out of South America in recent years; the discovery of this spectacular and highly endangered piprid caused huge interest in the international ornithological community.

Type information *Antilophia bokermanni* Coelho & Silva 1998, Nascente do Farias, Chapada do Araripe, Ceará, Brazil.

Alternative names Portuguese: Soldadinho-do-Araripe.

Discovery In November 1994 and again in November 1995 G. Coelho heard a vocalisation similar to that of Helmeted Manakin *Antilophia galeata* in a patch of humid forest located at the base of the Chapado do Araripe in Ceará state, NE Brazil, but was unable to see the caller. In December 1996 Coelho and W. Silva recorded the call and were able to make the first sight records of the bird. Two specimens were collected in May 1997 which became the basis for the type description.

Description 15–15.5 cm. Monotypic. Sexually dimorphic. Male is totally unlike any other member of its family. In shape similar to *A. galeata* in that it has a tuft of forward-projecting feathers on the forehead, giving the head a characteristic shape, and (for a manakin) a relatively long tail. Forehead, crown and nape carmine-red, extending in a triangle onto the upper central back; remainder of plumage white, except for remiges and rectrices which are black; iris russet-brown; bill cinnamon-brown or olive-brown; tarsi dusky-brown. Female largely olive all over, the frontal tuft reduced in size. Juvenile resembles female. Female differs from the female of *A. galatea* in its paler belly.

Habitat Evergreen secondary forest, modified forest including both natural and artificial, edges of riparian vegetation; also dense forest away from running water. About 800 m (Linhares *et al.* 2010; Anon. 2012b).

Food and feeding Apparently mostly vegetarian, especially small fruits of species such as *Cordia* spp. (Boraginaceae) and *Cecropia* spp. (Urticaceae); some 21 species identified in its diet (BLI SF 2014). Feeds in lower and middle levels of forest.

Breeding Song activity reaches a peak in September–October (Girão & Souto 2005) with breeding in November–April, although this may be somewhat variable, since females with brood-patches have been captured at the end of July (Mendes de Azevedo Júnior *et al.* 2000). Nest is a cup, suspended between the twigs of a horizontal branch in a variety of species of shrubs (Linhares *et al.* 2010). Breeding was previously thought to be closely associated with watercourses, but recent discoveries of the bird in dense forest on top of the Chapada do Araripe suggest that it is less restricted in its requirements. Eggs two, heavily marked with spots

▲ Araripe Manakin, male. Barbalha, Ceará, Brazil, November 2016 (*Ciro Albano*).

▲ Araripe Manakin, female. Arajara Park, Barbalha, Ceará, Brazil, August 2016 (*Robson Czaban*).

and speckles; incubation by female alone. Plumage of nestlings entirely green (Albano & Girão 2009).

Voice Most vocal in September–October, prior to nesting period. Calls are quite similar to those of *A. galatea*; a musical and warbled *ni-guru,guru-ni* etc. and *wreee pur* (BLI SF 2014).

Movements Appears to be sedentary.

Range Endemic. Until recently believed to be confined to a narrow strip of habitat along the north-west slope of the Chapada do Araripe, less than 1 km wide and less than 40 km long, with a total area of about 28 km^2 (Linhares *et al.* 2010). Recently has also been discovered in an area, albeit very small, of dense forest without streams on the top of the Chapada (plateau) (BLI SF 2014). Since the forested area here is much more extensive than the slopes, this might be highly significant.

Conservation status Currently classified by BirdLife International as Critically Endangered. The total area of suitable habitat, already small, has been further adversely impacted by clearing for agriculture, recreational parks and home building; one area, known to have harboured seven nests, was destroyed by fire in 2004–2005. However, a vigorous conservation campaign is in progress to halt any further environmental degradation and to preserve as much of the remaining habitat as possible (BLI SF 2014). Current population estimates are in the region of 800 individuals.

Etymology Named in recognition of Dr Werner Bokermann (1929–1995) "in honour of his significant contributions to Brazilian zoology" (in fact one mammal and some 13 amphibians are also named after him (Alvarenga 1995)).

CHESTNUT-BELLIED COTINGA
Doliornis remseni

The second species of the genus *Doliornis*, the first having been discovered in the 19th century; distribution may be complicated by confusion with *D. sclateri* (Bay-vented Cotinga).

Type information *Doliornis remseni* Robbins *et al.* 1994, Cerro Mongues, Carchi, Ecuador.

Alternative names Spanish: Cotinga Ventricastaño, Cotinga de Remsen.

Discovery In 1989 a cotinga, which from its general form and shape was clearly of the (then) monotypic genus *Doliornis*, was observed in Podocarpus National Park, in southern Ecuador. This location was well to the north of the known distribution of *D. sclateri*; this observation, and later ones in the same location, were of birds which seemed to differ from *sclateri* in plumage, but the sparsity of the literature then available, especially on the juvenile plumage of *sclateri*, caused some confusion. The possibility of there being an undescribed race of *sclateri*, or even a new species, was considered by the observers, but in the absence of specimen material no conclusion could be drawn (Rasmussen *et al.* 1996). Even more surprising was the discovery of a further population far to the north, in the central Andes of Colombia, which was at the time identified as *sclateri* (Renjifo *et al.* 1994). The new species was finally and unequivocally described in 1994 from a small series of specimens taken in March 1994 in Carchi, northern Ecuadorian Andes (Robbins *et al.* 1994). The Colombian population will almost certainly refer to *remseni*, not *sclateri* (Ridgely & Greenfield 2001).

Description 21 cm. Monotypic. Crown, upper face and lores black; cheeks, forehead, chin and throat grey; upper chest grey, washed fuscous; remainder of underparts deep rufous-chestnut; upperparts blackish-grey, apart from orange-red crest in centre of crown; rectrices blackish-grey; iris dark brown; bill and tarsi black. Female similar but crown more greyish-black. *D.sclateri* has the rufous-chestnut on the underparts more extensive.

Habitat Montane forest and low windswept woodland near the treeline, 2,900–3,650 m, mostly above 3,100 m.

Food and feeding Probably largely frugivorous; recorded food items include fruits of *Escallonia* (Escalloniaceae) and *Miconia* (Melostomataceae).

► Chestnut-bellied Cotinga. Cerro Mongues, Carchi, Ecuador, October 2015 (*Dušan Brinkhuizen*).

Breeding Nest and eggs undescribed; birds in immature plumage collected in March and July, and a bird with enlarged testes in late October (Kirwan & Green 2011).
Voice Undescribed.
Movements No data to indicate anything other than sedentary.
Range Has now been recorded from at least eight scattered locations in the Ecuadorian Andes, from Carchi in the north through Tungurahua (Henry 2008) to Loja and Zamora-Chinchipe in the south; in Colombia in the central Andes (Quindio). Not presently recorded from Peru, but since it has been found very close to the border its occurrence there is likely.
Conservation status Presently classified by BirdLife International as Vulnerable. Population estimates are in the range of 3,500–15,000 individuals; the species occurs in low densities throughout its range and appears generally uncommon. Habitat is being destroyed by the use of fire to produce good grazing, which has lowered the treeline in some locations by several hundred metres. This is especially true in Colombia; greater areas of unspoiled habitat remain in Ecuador.
Etymology Named in honour of Dr J. Van Remsen, Professor and Curator of Birds at the Museum of Natural Science, Baton Rouge, Louisiana, for many years a huge influence on investigative ornithology in South America.

GREY-WINGED COTINGA
Tijuca condita

One of the least-known members of its family, notwithstanding the fact that it occurs only an hour's drive from the second largest city in Brazil.

Type information *Tijuca condita* Snow 1980, based on a specimen taken in 1942, Fazenda da Guinle, Teresopólis, Rio de Janeiro, Brazil.
Alternative names Portuguese: Saudade-de-asa-cinza.
Discovery In October 1942 P. de M. Britto collected a cotinga in the Serra de Orgãos, near Teresópolis in Rio de Janeiro State, Brazil. It was at the time identified as a female of Black-and-gold Cotinga, *Tijuca atra*, a highly sexually dimorphic species which in the female plumage is a dull grey-green. Thirty years later, D. Snow, an acknowledged expert on the Cotingidae, and D. Goodwin re-examined the specimen and noticed some anomalies in the original identification. Subsequently, DNA analysis of a feather of this unique specimen showed unequivocally that it was clearly not *T. atra*, but an undescribed species (Snow 1980; Knox 1980). At the time of its description the species had not been observed since the taking of the original specimen. However, it was rediscovered near the type locality in November 1980 (Scott & Brooke 1993) and has since been seen regularly in several locations (see Range). In fact, no less an authority than H. Sick stated that he had known the bird for a number of years, but

▲ Grey-winged Cotinga. Nova Friburgo, Rio de Janeiro, Brazil, June 2014 (*João Quental*).

had thought that it was the young of *T. atra* "that had not yet learned to call correctly" and "had made no attempt to collect a specimen" (Vuilleumier & Mayr 1987, quoted in Kirwan & Green 2011). The two species are partially sympatric and syntopic.

Description 24 cm. Monotypic. Head and back olive-green, more yellowish on the rump; chin and throat dull grey; bill dark grey above, yellow-olive or greyish-green below. Female similar but duller and somewhat smaller. Differs from female *T. atra* in its smaller size and lack of a pale yellow patch (the shadow of the bright yellow of the male *T. atra*) on the primaries; female *T. atra* has a heavier, brighter yellow bill.

Habitat Bromeliad-rich montane elfin forests, with a canopy height of 5–10 m, and forest patches just below the treeline, 1,340–2,105 m, possibly sometimes lower (Snow 1982; Kirwan & Green 2011).

Food and feeding Primarily frugivorous, especially melastomes, but nine different species have been recorded; also some invertebrate matter, including a caterpillar and a stick insect (Kirwan & Green 2011).

Breeding Nest and eggs undescribed. A female netted in November had a brood-patch (Scott & Brooke 1985) and a bird was seen in mid-November carrying nesting material (Kirwan & Green 2011). The Black-and-gold Cotinga is a lekking species; no leks have been reported yet from the current species but this may be due to lack of observations.

Voice Vocally quite distinct from *T. atra*. Song is a two-note whistle *sooee-wheee*; also a longer slightly undulating *sooooo-ooooo-wheee*. Songs are usually given from concealed perches.

Movements No firm data, but it appears probable that some seasonal vertical movements do occur.

Range Endemic. Currently known from two mountain ranges in Rio de Janeiro State, Serra dos Orgãos and Serra da Tinguá; altitudinal ranges are 1,560–2,105 m in the former and 1,340 m in the latter. Alves *et al.* (2008) used a modelling procedure to predict potential range. Some areas that fell within their predictions did not hold populations of the bird, possibly because the somewhat drier habitat did not support the rich growth of bromeliads apparently necessary. For visiting birdwatchers, it is most easily located on the Pedro-do-Suio trail in Serra dos Orgãos National Park (Honkala & Niiranen 2010).

Conservation status Currently listed by BirdLife International as Vulnerable. The total range is not in excess of 200 km² (of which about 135 km² is protected) and population estimates lie in the range 1,000–2,500 individuals. Three of the sites (Serra dos Orgãos National Park, Tinguá and Araras Biological Reserves) are protected; the population at Nova Caledonia is not, and may be subject to logging. A potential threat is fire, caused accidentally by hikers; part of the habitat at Nova Caledonia was recently burnt (BLI SF 2014).

Etymology *Condita*, Latin, 'stored away' or 'hidden', reflecting the fact that the type specimen lingered for over 30 years in a museum drawer before being detected as a new species (Snow 1980).

CHESTNUT-CAPPED PIHA
Lipaugus weberi

Although the limited range of the species is physically quite easily accessible, the endemic political instability of the area has inhibited ornithological exploration, leading to its only recent discovery.

Type information *Lipaugus weberi* Cuervo *et al.* 2001, La Forsoza, Antioquia, Colombia.

Alternative names Chestnut-capped Cotinga. Spanish: Guardabosques Antioqueño, Piha Antioqueña, Piha Corona Castaña; Local name: El Arrierito ('The Cowboy').

Discovery During survey work to compile inventories of the avifauna of the La Forsoza region in the Cordillera Central of Antioquia, NW Colombia, an unfamiliar cotingid of the genus *Lipaugus* was captured, photographed and released on 31 March 1999. Several plumage features, and its smaller size, were inconsistent with Dusky Piha *L. fuscocinereus*. In May 1999, a second individual was captured and released, causing A. M. Cuervo to consider the possibility of an undescribed species. During August 1999, a rapid evaluation project of an elevational transect on the NE slope of the Cordillera encountered the bird again on numerous occasions – in fact at certain altitudes it was one of the most common species – and mist-netted two birds which became the holotype and a paratype.

Description 20–25 cm. Monotypic. Slate-grey above and below, with the exception of the pale cinnamon vent and the crown, deep chestnut in the adult, duller and more restricted in juveniles; rectrices dark greyish-brown; iris dark brown, orbital ring yellowish; bill blackish; tarsi grey. The male has a peculiar modification of primaries six and seven, with stiffened, comb-like outer webs, absent in the female. Juvenile has broad rufous fringes on the secondaries and outer primaries.

Habitat Restricted to ultra-humid premontane cloud forest, certainly 1,400–1,925m. Prefers pristine forest but will tolerate some selective logging, albeit at lower densities (Kirwan & Green 2011).

Food and feeding Largely frugivorous; diet comprises small to medium-sized fruits of a variety of families. Fruit usually taken by 'hover-gleaning', less frequently from a perched position (Kirwan & Green 2011). Also takes some invertebrates (Cuervo *et al.* 2001).

Breeding Nest and eggs undescribed. By analogy with Dusky Piha *L. fuscocinereus*, it is probably a lekking species. Breeding season not fully known; the female paratype, collected on 27 August, had large ovaries; a juvenile collected in early June had an incompletely ossified skull and was probably a few months old (Cuervo *et al.* 2001).

Voice Quite vocal and conspicuous. Calls, a loud piercing *sreeck*, rising in pitch then abruptly descending, typically repeated at one-second intervals; also a quiet, nasal *gluck-gluck* (Snow 2004 HBW 9). A mechanical wing-whirring is produced by the modified primaries of males when excited.

Movements Presumably sedentary.

Range Endemic. Restricted to a very limited area of Antioquia, Colombia.

► Chestnut-capped Piha. Anori, Columbia, July 2016 (*Fabrice Schmitt*).

Conservation status Recently reclassified by BirdLife International as Critically Endangered, in agreement with the opinion of one of its discoverers (A. Cuervo). Very restricted in habitat, with a small range, calculated to cover between 42 km^2 and 357.5 km^2 (Cuervo *et al.* 2014). Habitat destruction for crops, especially conversion into pastureland and for coffee and plantains has been severe in the area and is ongoing; three-quarters of the original habitat now appears to have been lost, with more than 9% in the decade 2000–2010 alone (Renjifo *et al.* 2014). The species appears to require substantial blocks of primary forest, not less than 30 ha each (Cuervo & Restrepo 2007). Replacement of native forest by exotic species is a further problem. Some areas are protected, including the 4.5 km^2 La Reserva Natural La Forsoza and 13.2 km^2 in the Arrierito Antioqueño Bird Reserve (Sharpe 2015). Further acquisitions and efforts to discourage further conversion of habitat to agricultural use are clearly essential. Although it is known from some 16 sites, it appears to be a lot less common than originally believed (Cuervo 2014), and it is possible that the total population may be as low as 250 birds (Renjifo *et al.* 2014).

Etymology Named in honour of Walter H. Weber of Medellin, Colombia, "for his enormous and ongoing contribution to the Sociedad Antioqueña de Ornitología and for promoting Colombian ornithology and conservation" (Cuervo *et al.* 2001).

BLACK-FACED COTINGA
Conioptilon mcilhennyi

A distinctive cotingid of the western Amazonian lowlands, sufficiently different from the remainder of the family to merit its own monotypic genus.

Type information *Conioptilon mcilhennyi* Lowery & O'Neill 1966, Balta, Río Curanja, Loreto, Peru.

Alternative names Spanish: Cotinga Carinegro; Portuguese: Anambé-de-cara-preta.

Discovery Two female specimens of an unknown cotingid, which could not be placed in any existing genus, were obtained by the 1964–1965 McIlhenny Peruvian Expedition. Among other unusual features was the possession of dense patches of powder-down, a character found in several non-passerine families such as the tinamous, herons and bitterns, some parrots and toucans, but at the time reported only from three passerine species. This prompted further explorations by J. O'Neill into a remote jungle location, where a representative series of specimens was taken. Examination of the available material confirmed that a new genus, as well as a new species, was involved.

Description 23 cm. Monotypic. A typically plump, medium-sized cotinga with a broad, slightly hook-tipped bill. Crown, face and throat black, with a narrow white border from throat to ear-coverts; underparts pale grey; back darker grey, the wings and tail darker still; iris dark reddish-brown; bill dark greyish-brown; tarsi dark olive-grey. Female similar to male but paler underneath

◄ Black-faced Cotinga. Senador Guiomar, Acre, Brazil, January 2015 (*Robson Czaban*).

and substantially smaller. Juvenile plumage not known from specimens; a large nestling seen at a range of 35 m appeared to have no black on the face (Tobias 2003a); a subadult female had much of the plumage tipped or edged with white (Snow 2004 HBW 9).

Habitat Occurs mostly in seasonally flooded swamp forests and other successional vegetation along floodplains, usually close to watercourses (Lloyd 2004); however, also in *terra firme* that has been burnt (Mestre *et al.* 2009). Occurs up to 300 m, rarely up to 450 m; one record at 700 m (Lloyd 2000).

Food and feeding Mostly frugivorous, known to take fruit of figs *Ficus*, *Cecropia* and *Coussapoa*; hovers briefly to pluck items from twigs. Also some insects, including myrmecine ants (Lowery & O'Neill 1966).

Breeding Two nests recorded, both in September in Peru; in Bolivia, lays in late August or September. One nest was located 15 m up in a *Cecropia* in mid-September, the second 35 m up in a huge *Ceiba* tree. Nest-building apparently by female alone, with male in attendance. The nest is very flimsy – "sufficiently small to be invisible" (from 35 m). The progress of the first nest could not be followed; the second held a single nestling on 28–30 September (Tobias 2003).

Voice Vocal and noisy. Song is a squeaky, upwardly inflected *brriiing* or *huuEEE*; contact call, a flat *pew* note. Also a quiet descending *coww* (Kirwan & Green 2011).

Range Originally believed to be endemic to Peru (Madre de Dios, Cuzco and Ucayali), but in fact quite common in some locations in Acre, Brazil (Whittaker & Oren 1999; Mestre *et al.* 2010). More recent records from Pando, northern Bolivia (Tobias 2003).

Conservation status Currently classified by BirdLife International as Least Concern. It is fairly widely distributed, with some areas protected (e.g. Manu National Park). However, some unprotected parts of its range are subject to logging and road development for oil exploration.

Etymology The generic name, *Conioptilon,* from the Greek *konia* 'a fine powder' and *ptilon* 'soft feathers or down', referring to the unusual powder-down of the species; the specific name honours John S. McIlhenny of Baton Rouge, Louisiana, who sponsored the 1964–1965 expedition, and "who has manifested unflagging interest in every aspect of the (Louisiana State University) Museum's program of research" (Lowery & O'Neill 1966).

YUNGAS TYRANNULET
Phyllomias weedeni

A somewhat cryptic species of small flycatcher whose existence would probably have continued to be unsuspected had it not been for the aural abilities of several field ornithologists.

Type information *Phyllomias weedeni* Hertzog *et al.* 2008, Cerro Asunta Pata, 60 km ENE of Charazani, Dpto. La Paz, Bolivia.

Alternative names Spanish: Mosquerito Yungueña, Mosquerito de las Yungas.

Discovery In June 1989 T. A. Parker and M. Gell-Mann observed and recorded some *Phyllomyias* tyrannulets in the Serranía Pilón, Beni, Bolivia, which they tentatively ascribed to Planalto Tyrannulet *P. fasciatus*; however, noting the rather different song (and the wide allopatry from the known range of *fasciatus*) they suspected that an undescribed taxon might be involved. In June 1997 S. Hertzog recorded the songs of three unfamiliar tyrannulets in the Cerro Asunta Pata, Dpto. La Paz, Bolivia. B. Whitney reviewed these recordings and observed that they were reminiscent of, but quite distinct from, the song of *fasciatus*; furthermore, these recordings were identical to those made by Parker and Gell-Mann in Beni eight years earlier. In September 1998 S. Hertzog and J. A. Balderrama returned to the 1997 location where they recorded two birds (which responded to the 1997 recordings) and which were collected as the holotype and a paratype.

Description 11–11.5 cm. Monotypic. A typical *Phyllomyias* tyrannulet, with dark olive-grey crown; upperparts yellow-olive, becoming more yellowish on rump; wings dusky-brown with yellowish or creamy wing-bars on coverts and yellowish edgings on flight feathers; a whitish supercilium including lores and broken eye-ring, bordered by a dark eye-stripe; chin and upper throat whitish; chest greenish-yellow, belly more yellow; iris deep brown; bill black with pinkish base to lower mandible; tarsi black. Sexes similar. Differs from *P. fasciatus* in having a contrasting grey crown and olive back.

Habitat Upper canopy of humid and semi-humid foothill and lower montane forest, 700–1,200 m; also in shade coffee plantations with remnant patches of forest (Hertzog *et al.* 2008).

Food and feeding Limited data. Feeds aerially by short sallies for flying insects in upper canopy. May occur in mixed flocks with other flycatchers and tanagers.

▲ Yungas Tyrannulet. Madidi National Park, Dpto. La Paz, Bolivia, October 2014 (*Richard Greenhalgh*).

Breeding Nest and eggs undescribed. A presumed family group seen in mid-February (Hertzog *et al.* 2008).
Voice Distinctive. Sings antiphonally. Song, a slightly accelerating series of 3–5 whistled notes successively dropping in pitch, the first note longest. In antiphonal duet, one bird gives a very brief burry note and the second follows with three sharp notes, two or three squeaky upslurred whistles, then a sharp but quieter single note. Faster and higher-pitched than corresponding songs of *P. fasciatus* (2013 HBW SV). For a detailed analysis see Hertzog *et al.* 2008.
Movements Probably sedentary.
Range So far found in several locations in Bolivia (Departments of La Paz, Beni and Cochabamba) and in one location in Puno, Peru, very close to the Bolivian border.
Conservation status Currently classified by BirdLife International as Vulnerable. The putative range does not exceed 10,000 km² and in all possibility not all of this is occupied; furthermore, within this area it appears to occur in very low density. Population estimates are in the range of 3,500–15,000 individuals (BLI SF 2014) or 10,000 individuals (Hertzog *et al.* 2008). Destruction of forest for agricultural purposes is the main threat.
Etymology Named "in honor of Alan Weeden, in recognition of his support for conservation work throughout South America" (Hertzog *et al.* 2008).

FOOTHILL ELAENIA
Myiopagis olallai

One of a supremely confusing group of small flycatchers, initially only detected as a new species by careful analysis of vocalisations.

▼ Foothill Elaenia. Podocarpus National Park, Ecuador, July 2015 (*Ross Gallardy*).

Type information *Myiopagis olallai* Coopmans & Krabbe 2000, Río Bambuscaro, 5 km SSE of Zamora, Zamora-Chinchipe, Ecuador.
Alternative names Spanish: Fiofio Submontano.
Discovery In June 1992 P. Coopmans tape-recorded a flycatcher at an elevation of about 1,000 m near Zamora in southern Ecuador. Comparisons of vocalisations of all similar species indicated that a new species was probably involved. Two specimens were taken in 1994 from near Zamora, and in 1996 two more from Volcán Sumaco in Napo, north-central Ecuador. A fifth specimen, taken in Dpto. Ayacucho, Peru and identified as a Forest Elaenia (*M. gaimardii*) was located by N. Krabbe in the American Museum of Natural History, New York. The new species was described in 2000 based on these specimens.
Description 12–12.5 cm. A small greenish-yellow elaenia, very typical of the genus. Crown dark pearly-grey with semi-concealed white coronal patch; lores and ear-coverts mottled grey-whitish; remainder of upperparts olive; wings dusky-brown with three obvious pale sulphur-yellow wing-bars; tertials and (more narrowly) secondaries edged sulphur-yellow; pale supercilium and

darker eyestripe; throat whitish, remainder of underparts yellow with olive flammulations on chest; iris dark brown; bill black with greyish base to lower mandible; tarsi black. Sexes similar. Two new subspecies recently described (Cuervo *et al.* 2014); *M. o. incognita* and *M. o. coopmansi.* Differs from the sympatric (and probably partially syntopic) Forest Elaenia *M. gaimardii* in having a pure white, not yellowish-white, coronal patch, better-defined wing-bars and grey, not grey-brown, crown.

Habitat Humid to wet primary submontane forest and forest-edges, 890–1,500 m.

Food and feeding Limited data. Feeds by making short outwards or upward sallies, hover-gleaning prey from the upper- and undersides of foliage, moss, twigs and branches; forages (*coopmansi*) in mixed flocks in canopy (Cuervo *et al.* 2014). Stomach contents of specimens include a 6 mm black beetle and caterpillars (Coopmans & Krabbe 2000), insect larvae, Hymenoptera and spiders (race *coopmansi*, Cuervo *et al.* 2014).

Breeding Nest and eggs undescribed.

Voice Song is a two-second long harsh trill, rising in pitch distinctly and proceeded by a variable number of introductory notes that also vary in pace and rhythm (Coopmans & Krabbe 2000); also a rapid, descending trilled *t'eerr* (Schulenberg *et al.* 2007). Cuervo *et al.* (2014) give a detailed comparative analysis of vocalisations of the various taxa.

Movements Presumably sedentary.

Range The species has so far been found in disjunct, scattered east-slope Andean locations, but further fieldwork might expand these. In Ecuador, from Sucumbios and Napo south through Pastaza to Zamora-Chinchipe provinces; in Peru, Apurimac and Pasco (BLI SF 2014) and Ayacucho (Coopmans & Krabbe 2000); in Schulenberg *et al.* 2007 shown as occurring in Cuzco. Race *coopmansi,* Antioquia, Colombia; race *incognita,* Venezuelan side of the Sierra de Perijá in Zulia.

Conservation status Classified by BirdLife International as Vulnerable. While it does occur in several protected areas (in Ecuador, Sangay, Podocarpus and Sumaco-Galeras National Parks; in Peru, Yanachaya-Chemillén National Park), outside these the habitat is being destroyed "at an alarming rate", perhaps already resulting in the loss of 50% of the species' range.

Etymology Named "in honor of the late Alfonso Manuel Olalla (d. 1971), in appreciation of his unparalleled contribution to Neotropical ornithology". Sr Olalla also has a species of monkey, Olalla's Titi *Callicebus olallae,* named after him. The subspecies *coopmansi* recognises the huge contribution to Neotropical ornithology of Paul Coopmans, who died at a tragically early age in 2007 (Kirwan & Freile 2008; Krabbe 2008); race *incognita,* Latin, 'unknown', referring to the fact that the subspecies has been described based on one 1940 and two 1951 specimens and is not known in life.

STRANECK'S TYRANNULET
Serpophaga griseicapilla

A species with a confused and confusing history, "a taxonomic rollercoaster" (Lees 2009), described in its present form only in 2008.

Type information *Serpophaga griseicapilla* Straneck 2008, Misión Neuva Pompeya, Chaco, north Argentina.

Alternative names Grey-crowned Tyrannulet, Monte Tyrannulet. Spanish: Piojito de Straneck; Portuguese: Alegrinho trinador.

Discovery In 1959 J. Berlioz described a new tyrannulet, *Serpophaga griseiceps,* on the basis of four specimens taken in Cochabamba, Bolivia. In 1979 M. Traylor determined that *griseiceps* was in fact a juvenile plumage of White-bellied Tyrannulet *S. munda,* an opinion reiterated 10 years later (Remsen & Traylor 1989). However, in 1993 R. Straneck, based on vocal similarities, suggested merging *S. munda* with *S. subcristata* (White-crested Tyrannulet) as one species, while pointing out that a different vocal type existed in Argentina, for which he resurrected the name *griseiceps,* previously demoted to a junior synonym of *S. munda.* In a morphological study, Hertzog & Mazar Barnett (2004) agreed with the synonymy of Bolivian *griseiceps* and *munda,* while pointing out that the differing '*griseiceps*' vocal type discovered by Straneck was also distinctive in its plumage, having a yellow, not white, belly, and in its smaller size. Finally, in 2007 Straneck renamed his birds as *S. griseicapilla,* a name so similar to the original as to be regrettably likely to cause yet further confusion, using as his vehicle a somewhat obscure Argentinian veterinary journal of limited circulation among ornithologists (Jaramillo 2010). *Griseicapilla* thus becomes the senior name for this taxon.

Description 9–10.7 cm. Monotypic. Head and nape grey with white supercilium and darker eyestripe; sometimes concealed white bases to central crown feathers; back drab olive-grey; wings dusky brown with two wing-bars and pale edgings on primaries and secondaries; rectrices dark grey-brown with narrow pale fringes, wider on outer pair of feathers; chin and

▲ Straneck's Tyrannulet. Estancia El Tejano, Paraguay, September 2014 (*Fabrice Schmitt*).

throat whitish-grey; sides of breast grey, belly variably yellowish; iris brown; bill and tarsi black. Sexes similar. Differs from *S. subcristata* in smaller size, shorter tail and lesser amount, or absence of, white crown feathers.

Habitat In breeding season, arid areas with scattered bushes. Outside breeding season in a variety of habitats, including chaco, pampas, *espinal* etc; 30–2,280 m (2013 HBW SV).

Food and feeding Prey items arthropods; forages by moving along branches etc. with little jumps and flutters, 1–6 m or more up (2013 HBW SV).

Breeding Nest and eggs undescribed.

Voice Distinctively different from other members of the genus *Serpophaga*, and in fact closer to the vocalisations of *Inezia* tyrannulets. A high-pitched *twie-tirrrrrrr*; alarm call *tie-tie-tie-tie*.

Movements and Range Migratory, but more data are needed. Breeds in arid regions of western Argentina, from Salta to northern Chubut; possibly also southern Bolivia (Tarija). Outside the breeding season, northern and NE Argentina, with records in Uruguay, Paraguay and southern Brazil (Bencke *et al.* 2001, Farias 2016) (This last was identified as *S. griseiceps*, but the sonogram of the song clearly indicates *griseicapilla*).

Conservation status Currently under review by BirdLife International and consequently not classified. It has a wide range, though more details are needed; appears to be quite common in parts of that range.

Etymology Named after Roberto Straneck of the ornithology division of the Museo Argentino de Ciencias Naturales "Bernardo Rivadavia", Buenos Aires.

MISHANA TYRANNULET
Zimmerius villarejoi

A white-sand forest specialist, probably most closely related to the Red-billed Tyrannulet *Z. cinereicapilla*.

Type information *Zimmerius villarejoi* Alonso & Whitney 2001, Zona Reservada Allpahuayo-Mishana, 25 km WSW of Iquitos, Loreto, Peru.

Alternative names Spanish: Mosquerito de Villarejo, Tiranuelo de Mishana.

Discovery In February 1997 J. A. Alonso, while conducting a faunal inventory along the newly-paved road between Iquitos and Nauta in Dpto. Loreto, Peru, heard and recorded an unknown but distinctive vocalisation

▲ Mishana Tyrannulet. Allpahuayo Mishana National Reserve, Peru, 1999 (*José Álvarez Alonso*).

which was made by a small flycatcher resembling Slender-footed Tyrannulet *Zimmerius gracilipes*. One specimen was taken in March 1997 and four others over the next two years; these, and vocal evidence, allowed the description of a new species in 2001.

Description 10.5 cm. Monotypic. In form a typical *Zimmerius* tyrannulet; small and dainty, with a bulky head, short bill with a rounded culmen and fairly long tail. Plumage entirely green or yellow-green; above greenish-olive, the face paler green; wings dusky, with two narrow wing-bars and wing-stripes formed by edgings and tips to the coverts; inner remiges edged yellow; throat and upper breast greenish-olive, becoming brighter yellow on lower underparts; iris creamy-white; bill quite short and conical, medium brownish above, dull pink below; tarsi bluish-black with yellow soles. Sexes similar. Red-billed Tyrannulet *Z. cinereicapilla* has a contrasting grey crown and red base to lower mandible; the syntopic Slender-footed Tyrannulet *Z. gracilipes* also has a contrasting grey crown, obvious pale supercilia and black lower mandible.

Habitat '*Varillal*', i.e. rather stunted lowland forest on poor white-sand soil, with canopy 12–18 m. Below 200 m.

Food and feeding Forages in canopy or subcanopy, not usually associating with mixed-species flocks for any length of time. Food items fruits, especially berries of mistletoe (Loranthaceae), regurgitating or removing the seeds; also small arthropods captured by short aerial sallies and picked up from leaf margins (Alonso & Whitney 2001).

Breeding Nest and eggs undescribed.

Voice Distinctive. A pair of thin, rising, whistled notes, sometimes single or triple, also a more complex, multisyllabic series of descending notes. Dawn song a simple series of 2–4 evenly spaced whistled notes, each note shorter than the previous one. Alonso & Whitney (2001) give detailed comparisons with the vocalisations of other *Zimmerius* species.

Movements Presumably sedentary.

Range Endemic. Confined to a relatively small area in Dpto. Loreto, Peru, in the drainage of the Río Nanay; also reported from the lower Río Mayo and middle Río Huallaga, San Martín (J. Diaz *in litt.* to BLI SF 2014, not otherwise published). A specimen taken in 1912 near Moyobamba, San Martín, closely matches the Loreto specimens, but lacks any vocal or habitat data. Further field investigation would be desirable.

Conservation status Currently classified by BirdLife International as Vulnerable. The total range is apparently rather restricted; although much of this is protected, removal of trees appears to continue; habitat outside the protected area is more severely impacted. Shany *et al.* (2007) comment that the species seems to be becoming less common.

Etymology Named in honour of the Augustinian priest P. Avencio Villarejo (1910–2000) who explored much of the Peruvian Amazon by canoe in the 1930s and 1940s, and who was much ahead of his time in not regarding the area and its indigenous peoples as merely resources to be ruthlessly exploited.

CHICO'S TYRANNULET
Zimmerius chicomendesi

Although biochemical evidence is presently lacking, almost certainly a close relative of the recently described Mishana Tyrannulet *Z. villarejoi* of Amazonian Peru (Whitney *et al.* 2013a).

Type information *Zimmerius chicomendesi* Whitney *et al.* 2013a, "Transamazonica" Highway, municipality of Humaita, Amazonas, Brazil.
Alternative names Portuguese: Poiaeiro-de-chico-mendesi.
Discovery In August 2009, in southern Amazonas state, Brazil, B. Whitney heard a distinctive but unknown two-note call; playback pulled in a tyrannulet, obviously of the genus *Zimmerius,* but equally obviously separate vocally from other members of the genus, particularly from *Z. villarejoi* which the same author had described in 2001. A good set of recordings was obtained that day. In September 2011 M. Cohn-Haft recorded further calls in another location in Amazonas and obtained a specimen. Later that same year an expedition was mounted which obtained a series of specimens with voice recordings, allowing the description of a new species.

▼ Chico's Tyrannulet. Tabajara, Machadinho d'Oeste, Rondônia, Brazil, July 2013 (*Bruno Rennó*).

Description 10 cm. Monotypic. Morphologically almost identical to Mishana Tyrannulet *Z. villarejoi* (q.v.); a small, greenish flycatcher, with upperparts uniformly olive; the facial region slightly paler greenish; lower belly and ventral regions clear yellow; wings dull blackish with faint olive cast; median and greater wing-coverts fringed yellow; tail same as wing but rectrices with narrow olive-green fringes; iris white; bill brownish-red above, reddish-pink below; tarsi blackish.
Habitat *Campina* woodland; stunted forest, with canopy height 2–6 m with occasional trees of up to 10 m, on sandy soil or rock-like sand, often poorly drained; below 200 m.
Food and feeding Apparently a mistletoe specialist; stomach contents of eight specimens almost entirely seeds and pulp of the widespread species *Oryctanthus alveolatus.* Small quantities of insect parts may have been ingested incidentally with mistletoe berries, although one bird was seen to attempt to capture arthropod prey (Whitney *et al.* 2013a).
Breeding Nest and eggs undescribed. Mated pairs observed in December, though gonads were not in breeding condition then. Other members of the genus make a dome-shaped nest with a side entrance.
Voice Diagnostic. Call is two rapid notes, sometimes one or three; song is rapid series of notes, sometimes with a separate introductory note; also a quiet snarl, probably aggressive.
Movements Presumably sedentary.
Range Endemic. South-central Amazonian Brazil, in southern Amazonas state. Potential range encompasses the Madeira-Aripuana interfluvium, mostly in Amazonas but marginally in northern Rondônia and NW Mato Grosso.
Conservation status Overall presently known range is quite limited in extent. Much of this is fairly remote and not currently under immediate obvious threat. However, extensive forest clearing has occurred along parts of the 'Transamazonica' road (constructed in the 1970s) and might be expected to intensify once this road is paved. Effective habitat protection is essential.
Etymology Named in honour of Francisco 'Chico' Alves Mendes Filho (1944–1988), "a man wise beyond the borders of his time and space", perhaps the most courageous and certainly the most admirable conservationist of the 20th century. One of 18 children, illiterate until early adulthood, he fought for the sustainable use of the Amazonian forest and the rights of its indigenous population; murdered by cattle-ranching interests shortly after his 44th birthday. For a moving tribute, see Whitney *et al.* 2013a.

ANTIOQUIA BRISTLE TYRANT
Pogonotriccus lanyoni

Although described by its discoverer as a *Phylloscartes* tyrannulet, most authorities include the present species in the widespread South American genus *Pogonotriccus*, which may not, in fact, be very closely related to *Phylloscartes*.

Type information *Phylloscartes lanyoni* Graves 1988, based on a 1948 specimen from El Pescado, Antioquia, Colombia.

Alternative names Spanish: Orejerito Antioqueño, Mosquerito Antioqueño.

Discovery During the years 1941 to 1952, the National Museum of Natural History, Washington DC, acquired a huge collection of birds – almost 24,000 specimens in total – made by Melbourne A. Carriker (1879–1965), one of the most assiduous ornithological collectors of his era, specialising particularly in Colombian birds. The analysis of this collection took many years, much of it under the direction of Dr Alexander Wetmore (1886–1978), resulting in the description of several new species. Two specimens of a small flycatcher, taken by Carriker in 1948 at El Pescado, near Puerto Valdivia in the valley of the Río Cauca, Colombia, were acquired by the National Museum in about 1950–1951; but perhaps because Carriker had identified them as Yellow Tyrannulets *Capsiempis flaveola* they were not examined critically for almost 40 years, which resulted in their description by G. R. Graves as a new species. It remains one of the least-known flycatchers in South America. Probably the first ornithologist knowingly to observe the species alive was G. Stiles and companions in June 1990 (Stiles 1990).

Description 11.5 cm. Monotypic. Back and rump bright olive-green; crown greyish-green; wings dark brownish-grey with two bright yellow wing-bars; lores and incomplete eye-ring yellowish-white; face and ear-coverts mottled yellowish; post-auricular crescent indistinct blackish; entire underparts bright yellow, lightly suffused with olive on breast; rectrices olive; iris probably brown; bill blackish with whitish base to lower mandible; tarsi grey. Sexes similar.

Food and feeding Gleans insects from lower surfaces of leaves and twigs (Stiles 1990). Sometimes in mixed-species flocks.

Habitat Semi-deciduous foothill forest, 450–900 m (Stiles *et al.* 1999).

Breeding Nest and eggs undescribed. The male specimen collected by Carriker on 15 May had greatly enlarged testes (Graves 1988). Nesting recorded in March; family group seen in June (2013 HBW SV).

Voice Alarm call, in response to 'pishing', a high, sharp trill, somewhat hard but of low volume *tsip.... tsip....tsirrr* (Stiles 1990); also a short, slightly upslurred *weeet* (BLI SF 2014).

Movements Presumably sedentary.

Range Endemic. Scattered localities in central Colombia; on the east and north slopes of the central Andes in Caldas and Antioquia and on the west slope of the east Andes in Cundinamarca, Santander and Boyacá.

► Antioquia Bristle Tyrant. Cañon del Río Claro, Antioquia, Colombia, April 2015 (*Dušan Brinkhuizen*).

Conservation status Classified by BirdLife International as Endangered. The range is highly fragmented and subject to severe habitat degradation; in parts of its range described as uncommon, although Stiles *et al.* (1999) describe it as common in one location in Boyacá (Monte del Diablo). Population estimates 600–1,700 mature individuals (BLI SF 2014).

Etymology Named in honour of Dr Wesley E Lanyon (b. 1926), formerly President of the American Ornithologists' Union, "in recognition of his research on tyrannid systematics over the last few decades" (Graves 1988).

ALAGOAS TYRANNULET
Phylloscartes ceciliae

The only *Phylloscartes* tyrannulet in its area, with a tiny and highly threatened range.

Type information *Phylloscartes ceciliae* Teixeira 1987, Serra Branca, Murici County, Alagoas, Brazil.

▼ Alagoas Tyrannulet. Jaqueira, Pernambuco, Brazil, May 2016 (*Ciro Albano*).

Alternative names Long-tailed Tyrannulet. Portuguese: Cara-pintada.

Discovery The Museu Nacional of Brazil made several ornithological expeditions in the 1980s to the residual forests of NE Brazil, an area suffering from extreme environmental degradation, during the course of which several new species and subspecies were discovered (Teixeira & Gonzaga 1983a, 1983b, 1985). The type specimen of the current species was collected on 8 May 1984 and several paratypes in May and November of the same year.

Description 12 cm. Monotypic. A small, long-tailed tyrannulet, typical of the genus but with especially conspicuous wing-bars. Generally olive above with obvious white supercilium and ear-coverts, the latter with an ashy-black ocular stripe, extending to the outer edge of the ear-coverts in a crescent, with further stripes lower down and behind eye; lores also ashy-black; wings dusky brown with two yellowish-white bars on the coverts; generally whitish below; tail, frequently cocked, long, blackish, with yellowish-green borders to rectrices; below, whitish on chin and throat, sides of breast with greenish wash; remainder of underparts whitish with a light lemon-yellow wash; iris chestnut, bill black, tarsi dark bluish-grey. Sexes similar.

Habitat Highland evergreen forest, 400–550 m.

Food and feeding Few data. Frequently seen in mixed flocks with various tanagers (Roda *et al.* 2003), flycatchers, antbirds and furnariids. Scans surfaces of leaves and branches for small insects (Teixeira 1987).

Breeding Nest and eggs undescribed; probably breeds September to February.

Voice A peeping sequence *djü, djü,* sometimes sharper and faster *ürürüt, ürürüt,* and a quiet, isolated *thüp.*

Movements Apparently sedentary.

Range Endemic. Now restricted to isolated patches of remnant forest (the localities currently known) in NE Alagoas and eastern Pernambuco states, Brazil.

Conservation status Currently classified by BirdLife International as Endangered; by other authors (e.g. Silveira *et al.* 2003) as Critically Endangered. The known

sites are isolated and fragmented; as long ago as 1987 Teixeira stated that the Atlantic forests of NE Brazil were "in the final stages of destruction", a process that has continued apace since, with habitat near this type locality of Murici reduced from 70 km² in the 1970s to 30 km² in 1999. Two sites in Peruambuco (Pedra Talhada and Reserva Privada Frei Caneca); efforts to protect the type locality so far unsuccessful. Estimated population 250–1,000 mature individuals, or 350–1,500 individuals in total (BLI SF 2014).
Etymology Named in memory of Cecilia Torres (1952–1985), wife of the discoverer.

RESTINGA TYRANNULET
Phylloscartes kronei

A species very similar to the much better known and far more widespread Mottle-cheeked Tyrannulet *P. ventralis*, whose separate identity was again first suspected because of distinctive vocalisations.

Type information *Phylloscartes kronei* Willis & Oniki 1992, Jardim Europa, Ilha Comprida, São Paulo, Brazil.
Alternative names Portuguese: Maria-da-restinga.
Discovery In July 1983 E. O. Willis and Y. Oniki were investigating scrubby restinga woodland at the northern end of Ilha Comprida, near Curitiba in São Paulo State, Brazil, when they heard a distinctive vocalisation from a small *Phylloscartes* flycatcher. In plumage it most resembled Olive-green Tyrannulet *P. virescens*, a completely allopatric species occurring north of the Amazon, rather than the sympatric *P. ventralis*. Later, two calling birds were collected which established the specific separateness of the taxon. In fact, a specimen, currently housed in the Museum of Zoology of the University of São Paulo, was collected on the mainland in the Ribeira Valley in 1898 by R. Krone but was not recognised until Willis & Oniki's investigation 90 years later.
Description 12 cm. Monotypic. A typical small *Phylloscartes* tyrannulet, rather long-billed and long-tailed; generally greenish-olive above; eye-ring and narrow supercilium yellowish; dusky eye stripe and lores, extending to a dusky post-auricular crescent; yellowish ear-coverts; face yellowish, finely mottled; wings darker dusky with two prominent yellowish-white wing-bars; underparts generally yellow, paler on throat and brighter abdomen; iris brown, bill black with paler base to lower mandible, tarsi black. Sexes similar. The sympatric, but not apparently (at least mostly) syntopic *P. ventralis* is noticeably less yellow on the face.
Habitat Woodland edges and second growth on sandy soil (*restinga*); apparently prefers swampy areas with pools of standing water (Remold & Ramos Neto 1995). Mostly at or close to sea level, but inland, records up to 450 m (Mazar Barnett *et al.* 2004) and 820 m (Meyer 2016).
Food and feeding Insectivorous and frugivorous. Insects are caught by sally-flights and by gleaning plant surfaces; food items include arthropods, especially Coleoptera and Hymenoptera, with fruits of *Clusia criuva* and *Ternstroemia brasiliensis* (Gussoni 2010; Gussoni & Santos 2011) and *Myrsine coriacea* (Meyer 2016); often in mixed flocks with tanagers and parulid warblers.

▼ Restinga Tyrannulet. Ilha Comprida, Sao Paulo, Brazil, September 2008 (*Carlos Gussoni*).

◄ Restinga Tyrannulet, juveniles in nest. Reserva Bicudinho-do-brejo, Guaratuba, Paraná, Brazil, October 2012 (*Carlos Gussoni*).

Breeding Most data come from the work of C. Gussoni. Breeding season is September to January. The nest, built by the female, is a closed ovoid with a side-entrance, built of dry leaves with an outer covering of green moss, glued by spider-webs, the nest-chamber lined with soft seeds. Nests are situated in vegetation, from 0.72–3 m above the ground (see photograph). Clutch is 2–3 unmarked white eggs, incubated by female only, for 12 days. Nestlings are fed and attended by both parents (Gussoni 2011, 2014; Remold & Ramos Nieto 1995; Willis & Oniki 1992).

Voice Song is a rapid twittering, more prolonged than that of *P. ventralis*; call quite distinct, a frequent loud disyllabic *chuwup* or *psirip* (Clay *et al.* 1998).

Movements Apparently sedentary.

Range Endemic. Coastal SE Brazil, in the states of São Paulo, Paraná, Santa Catarina south to extreme NE Rio Grande do Sul (Clay *et al.* 1998; Bencke *et al.* 2000), including the islands of Ilha Comprida and Ilha do Cardoso; inland in the floodplain of the Rio Ribeira.

Conservation status Currently classified by BirdLife International as Vulnerable. The main threat is habitat destruction, mostly for beachfront dwellings, the city of São Paulo being a short drive away from some of the prime areas; the construction of a bridge linking Ilha Comprida with the mainland and extensive clearance for roads has exacerbated this situation. Several areas of occurrence do enjoy a measure of protection (Fitzpatrick 2004 HBW 9), for example the Reserva Bicudinho-do-brejo (which also protects Marsh Antwren, q.v.), and the Parque Natural Municipal da Lagoa de Parado. In suitable habitat the population may be quite dense; one 21 ha area averaged 1.04 pairs per ha (Gussoni 2014). Current population estimates are in the range of 3,500–15,000 individuals (BLI SF 2014).

Etymology Named after Ricardo Krone, who was born in Germany, set up a pharmacy in Iguape and became the premier zoologist of the Ribeira Valley at the end of the 19th century; he collected the first specimen in 1898 (albeit unrecognised for almost 90 years). In fairness to Sr Krone, it must be pointed out that the specimen was misidentified (as *P. ventralis*) by no less an authority than Hans von Berlepsch, one of the most distinguished German ornithologists of the late 19th century (Willis & Oniki 1992). Sr Krone was also a noted speleologist who published a book on the caves of the Ribeira Valley and described a cave-dwelling catfish *Pimalodella kronei* (Beolens *et al.* 2014).

BAHIA TYRANNULET
Phylloscartes beckeri

A very restricted and still largely unknown tyrannulet, found in two Brazilian states only. Taxonomically probably closest to Mottle-cheeked Tyrannulet *P. ventralis* and Restinga Tyrannulet *P. kronei.*

Type information *Phylloscartes beckeri* Gonzaga & Pacheco 1995, 7 km SE of Boa Vista, Bahia, Brazil.

Alternative names Portuguese: Borboletinha-baiana.

Discovery During a short survey in 1992 and 1993 of the tiny remnants of the once extensive montane forest that formerly covered substantial parts of the state of Bahia, L. P. Gonzaga and J. F. Pacheco discovered two new species, a tyrannulet and a spinetail (q.v.). As with several other members of the (rather unsatisfactory)

► Bahia Tyrannulet. Camacan, Bahia, Brazil, September 2016 (*João Quental*).

genus *Phylloscartes* (Graves 1988), vocalisations were crucial in identification.

Description 12 cm. Monotypic. Very similar to other *Phylloscartes* tyrannulets (Clay *et al.* 1998). Bright olive-green above, greyish centre to crown; lores and eyestripe dusky with supraloral and eye-ring buffy; ear-coverts yellowish, bordered obscurely with dusky; wings dusky-olive with two obvious pale yellow wing-bars; rectrices dusky-olive; throat and breast off-white with pale yellow flecking; belly and vent yellowish, more olive at sides of breast; iris dark brown, bill brown, pearly-white at base of lower mandible, tarsi pale grey. Sexes similar. Differs from the widespread Mottle-cheeked Tyrannulet (*P. ventralis*) in its buffy, rather than yellow, facial markings and from Alagoas Tyrannulet (*P. ceciliae*) in the more yellow underparts.

Habitat Montane primary and mature secondary forest with large trees, 750–1,200 m.

Food and feeding Snatches and hover-gleans from leaf surfaces. Feeds mostly in the canopy, often in mixed-species flocks. Identified stomach contents includes beetles, flies, bugs and Lepidoptera.

Breeding Nest and eggs undescribed. Probably breeds in September–February.

Voice Song is variable, rather soft and twittery, including short trills; quieter and more varied than those of *P. ventralis* and *P. ceciliae*. Call a short, soft *tik*.

Movements Probably sedentary.

Range Endemic. Currently known from nine locations, seven in Bahia state and two in northern Minas Gerais.

Conservation status Currently classified by BirdLife International as Endangered. It was, presumably, far more widespread before the very extensive destruction of forest in Bahia and adjacent states for agriculture and logging. Now occurs in small surviving forest remnants. Some habitat is protected, e.g. Serra Bonita Private Reserve and Amargosa and Chapada da Diamantina National Park, all in Bahia state; however, even in the last location, illegal logging and charcoal production apparently occurs. Population estimates in the range of 3,500–15,000 individuals (BLI SF 2014, Parrini *et al.* 1999).

Etymology Named in honour of the entomologist Professor Johann Becker, in recognition of his efforts to preserve the remnants of the Atlantic forests of NE Brazil.

CINNAMON-FACED TYRANNULET
Phylloscartes parkeri

A quite widely distributed but still remarkably little-known tyrannulet from lower elevations of the eastern Andes.

Type information *Phylloscartes parkeri* Fitzpatrick & Stotz 1997, near Hacienda Amazonia, Madre de Dios, Peru.

Alternative names Spanish: Orejerito de Parker.

Discovery In 1899 the German collector O. Garlepp obtained a small yellowish flycatcher from Dpto. Cuzco, Peru. C. E. Hellmayr in 1927 presciently suggested that an "undescribed form" was involved (Hellmayr 1927), but nothing further was heard of the bird for almost another half century. In 1972 J. Terborgh and J. Weske

▲ Cinnamon-faced Tyrannulet. San Pedro, Manu Road, Cusco, Peru, October 2009 (*Daniel Lane*).

mist-netted and photographed a small flycatcher in Dpto. Apurímac, Peru. They suspected that this bird was of the same species as the Garlepp specimen, but no further examples were encountered. However, in 1981 J. W. Fitzpatrick collected several examples in the lower Cosñipata Valley of SE Peru and in 1982 D. F. Stotz obtained a further specimen in Dpto. Beni, Bolivia. The holotype, from which the type description was made, was collected in 1983 in Dpto. Madre de Dios, Peru. The species has since proved to be of quite wide occurrence in foothill forests in SE Peru (Fitzpatrick & Stotz 1997).

Description 12 cm. Monotypic. A rather distinctive *Phylloscartes* tyrannulet, with characteristic cinnamon supercilium and facial markings. Crown medium-grey; remainder of upperparts olive; wings blackish with two bright yellow wing-bars; rectrices olive; entire underparts bright yellow with indistinct olive streaking; iris brown, bill black, tarsi grey. Juvenile has paler rufous lores and more mottled breast than adult. Sexes similar. The only other *Phylloscartes* flycatcher with cinnamon facial markings is the widely allopatric Rufous-lored Tyrannulet *P. flaviventris*, from which it differs in its cinnamon-rufous lores, greyish-green, not dark brownish, crown, olive streaking on the underparts and less clearly-defined yellow wing-bars.

Habitat It appears to require mature montane evergreen forest, 700–1,200 m and does not occur in second-growth forest.

Food and feeding Usually occurs in mixed-species flocks, almost always foraging in upper levels of the canopy, by hover-gleaning and snatching prey from leaf surfaces. Stomach contents of Peruvian specimens entirely arthropod including beetles, homopteran and hemipteran bugs, ants, wasps, spiders, etc.

Breeding Nest and eggs undescribed. Based on gonad condition of specimens and presence of juveniles, breeding season probably peaks in August to November (Fitzpatrick & Stotz 1997).

Voice Song is "a high, springy, laughing chatter that quavers, rising slightly *chi chi tchrrEEEEeeeEEEEEew*; call a high *tchew*" (Schulenberg *et al.* 2007).

Movements Presumably sedentary.

Range Lower altitudes of eastern slope of Andes, from northern Peru, from Huánuco south through Junín, Cuzco, western Madre de Dios, Puno and in northern Bolivia (Beni and La Paz) (Parker *et al.* 1991; Walker *et al.* 2006; Fitzpatrick & Stotz 1997).

Conservation status Currently classified by BirdLife International as Least Concern. The species has a quite extensive geographical range which includes several protected areas including Manu National Park, and in some locations appears "locally fairly common" (Schulenberg *et al.* 2007). However, it does appear to lack an ability to adapt to disturbed or secondary forest and the population is probably declining (BLI SF 2014).

Etymology Named in honour of the late Theodore (Ted) Parker (see under Subtropical Pygmy Owl *Glaucidium parkeri*).

ACRE TODY-TYRANT
Hemitriccus cohnhafti

A highly restricted tody-tyrant, most closely related to the allopatric Yungas Tody-Tyrant *H. spodiops* of Bolivia; assigned to the subgenus *Snethlagea*.

Type information *Hemitriccus cohnhafti* Zimmer *et al.* 2013, Estrada da Pedreira, 10 km east of Assis Brasil, Acre, Brazil.

Alternative names Portuguese: Maria-do-acre; Spanish: Titirijí de Acre.

Discovery In September 2009, while conducting field surveys to the east of the town of Assis Brasil in Acre state, western Brazil, K. Zimmer and A. Whittaker found an unfamiliar tody-tyrant which they suspected to be a new species. Recordings and observations were made; however, it was not certain at that time whether the birds might represent an outlying population of Yungas Tody-Tyrant, found in middle elevations in Bolivia. Subsequent work in early 2010 and in June 2012 resulted in the acquisition of several specimens and further recordings; these and DNA phylogenetic analysis allowed the description of the Acre population as a new species (Zimmer *et al.* 2013).

Description 11 cm. Monotypic. In form a typical *Hemitriccus*, being a rather large-headed flycatcher with a medium-length tail. Head and nape greenish-olive; back to rump green; throat and breast olive-green with creamy-yellowish streaks; mid/lower belly to vent sulphur-yellow; primaries and secondaries blackish with yellow edgings to outer webs; wing-coverts blackish with yellowish-tawny tips, giving two conspicuous wing-bars; rectrices blackish with dark greenish inner webs; iris variable, cream to orange; bill black above, blackish below with creamy base; tarsi grey. Sexes similar. Differs from Yungas Tody-Tyrant by more obvious buffy lores and more distinct wing-bars, as well as vocalisations.

Habitat Second-growth and forest-edge vegetation on poor sandy soils, with stunted forest 5–12 m in height; so far found only in vicinity of rivers.

Food and feeding Food appears to be arthropods, caught by gleaning from under-surfaces of leaves; does not apparently forage in mixed-species flocks.

Breeding Nest and eggs undescribed.

Voice Song is a trill, typical of the genus, lasting only 0.2 seconds, comprising some eight notes; the corresponding trill of *H. spodiops* is much longer, with more than 30 notes and is higher pitched (see Zimmer *et al.* (2013) for comparative sonograms). Sometimes given in sequences of several bursts. Also several other calls, *keek*, *keep* etc.

Movements Presumably sedentary.

Range Currently known from a small area of Acre state, western Brazil, including a new site 161 km to the north-east of the type locality (de Melo *et al.* 2015),

▲ Acre Tody-Tyrant. Ramal do Noca, Rio Branco, Acre, Brazil, August 2015 (*Bruno Rennó*).

and recently in adjacent Madre de Dios, Peru (Harvey *et al.* 2014). Given that these locations are very close to Bolivia, separated only by the 100 m wide Acre River, occurrence in that country is also very probable.

Conservation status Not currently assessed by BirdLife International. The known distribution seems to be rather patchy and very small. Although the species occupies some second-growth habitat, major clearing for ranching (as has occurred extensively nearby in recent years) would obviously constitute a threat; probably best treated as Data Deficient (Zimmer *et al.* 2013).

Etymology Named in honour of Dr Mario Cohn-Haft, "in recognition of his numerous and ongoing contributions to our understanding of the marvellous avifauna of Amazonian Brazil" (Zimmer *et al.* 2013).

CINNAMON-BREASTED TODY-TYRANT
Hemitriccus cinnamomeipectus

A very restricted and still largely unknown tody-tyrant from a few remote areas of northern Peru and southern Ecuador; curiously, probably most closely related to Kaempfer's Tody-Tyrant *H. kaempferi*, which is allopatric by 4,000 km in SE Brazil (Fitzpatrick & O'Neill 1979).

Type information *Hemitriccus cinnamomeipectus* Fitzpatrick & O'Neill 1979, above San José de Lourdes, Cordillera del Cóndor, Cajamarca, Peru.

Alternative names Spanish: Titirijí pechicanelo, Tirano Tody pechicanelo.

Discovery During the course of several ornithological explorations by staff of the Louisiana State University Museum of Natural Science and the Field Museum of Natural History, Chicago, of some little-known isolated ridges in the Peruvian Andes, a series of eight specimens of a new tody-tyrant was taken in three different locations; on this basis a new species was described.

▲ Cinnamon-breasted Tody-Tyrant. Alto Nieva, San Martín, Peru, October 2014 (*Daniel Lane*).

Description 10 cm. Monotypic. In form a typical *Hemitriccus* tody-tyrant; very small, rather large-headed with a spatulate bill. Upperparts dark olive, browner on crown and with a buff-orange nuchal collar; face, lores, cheek, ear-coverts, throat and breast bright cinnamon; remiges and rectrices dusky, with broad pale yellow edgings on innermost secondaries; belly contrastingly pale yellow; iris light reddish-brown; bill greyish with flesh-coloured base to the lower mandible; tarsi greyish or pinkish-grey. Sexes similar; juvenile has brownish wash on upperparts, throat and breast pale sandy.

Habitat Thick, stunted mossy cloud-forest (Clock 2004 HBW 9). At the northern end of its range, in Ecuador, in stunted sclerophyllous scrub, about 3–5 m high with dense understorey; 1,600–2,200 m.

Food and feeding Forages alone, in pairs or occasionally in mixed-species flocks, mostly in the understorey, 2–3 m up, sometimes into mid-levels at 4 m (Ágreda *et al.* 2005). Stomach contents include unidentified insect parts.

Breeding Nest and eggs undescribed. In northern Peru, breeding season (based on gonad condition and moult) probably coming to an end by late June. No other data.

Voice A short rattle, *dddddrr-rrt* and a loud high-pitched *wheek* repeated 3–4 times (Ágreda *et al.* 2005). Song consists of a single descending *prrrrr* (Begazo *et al.* 2001).

Movements Presumably sedentary.

Range Eastern Andean slopes, from southern Ecuador (Morona-Santiago and Zamora-Chinchipe – Ágreda *et al.* 2005, Krabbe & Sornoza 1994) and three locations in northern Peru (Cordillera del Cóndor, Cajamarca, northern Cordillera de Colán, Amazonas and Abra Patricia, San Martín). These locations are only about 300 km from end to end, and this species appears to be very discontinuous even within this range.

Conservation status Currently classified by BirdLife International as Vulnerable. The total range is small and apparently fragmented. In Peru, deforestation in the Cordillera de Colán is proceeding rapidly. In Ecuador, in the Cordillera del Cóndor, much forest has been destroyed by gold- and silica-mining activities.

Etymology *Cinnamomeipectus*, Latin, 'cinnamon-chested'.

LULU'S TODY-FLYCATCHER
Poecilotriccus luluae

A very attractive and endangered little tyrannid, which was known for many years prior to its formal description. Forms a superspecies with Rufous-crowned Tody-Flycatcher *P. ruficeps*.

Type information *Poecilotriccus luluae* Johnson & Jones 2001, 5.6 km south of Corosha, Dpto. Amazonas, Peru.
Alternative names Lulu's Tody-Tyrant, Johnson's Tody-Tyrant. Spanish: Titirijí de Lulu.
Discovery In August 1970 N. K. Johnson collected a mated pair of tody-flycatchers in the NE Andes of Peru. Although obviously distinct from other *Poecilotriccus* species, more specifically Rufous-crowned Tody-Flycatcher *P. ruficeps*, Dr Johnson was unwilling to describe a new species on the basis of inadequate material. Although further investigations were planned, these were rendered unnecessary, since several expeditions from Louisiana State University, from 1974 to 1983, collected a series of examples. Further work involved the collection of more specimens and gathering vocal recordings. With exemplary professional courtesy, LSU staff gave Johnson & Jones full access to all material and deferred to them the authorship. In the meantime the species was observed by a number of other ornithologists (e.g. Davies *et al.* 1997; Hornbuckle 1999).
Description 10 cm. Monotypic. In form a typical *Poecilotriccus* flycatcher; small, dumpy-bodied, short-tailed and large-headed. Head bright mahogany-red with white chin; lower nape grey, back olive, wings black with two narrow yellow wing-bars and yellow-white tertial edgings; throat with rufous band, and a narrow lower whitish band; breast-band yellowish olive-green, lower chest, belly and vent orange-yellow to

▼ Lulu's Tody-Flycatcher. Abra Patricia, Peru, April 2015 (*Miguel Lezama*).

yellow-ochre; rectrices olive-brown; iris colour variable, brown, red-brown, purple-brown, café or maroon; bill black; tarsi grey, pinkish- or bluish-grey or grey-flesh. Sexes identical.

Habitat Humid montane forest, 1,850–2,900 m; most often in or near bamboo thickets, but also second growth and forest edge (Schulenberg *et al.* 2007).

Food and feeding Forages by sally-gleans, mostly from underside of leaf surfaces. Food comprises various insects, including Hymenoptera. Forages in pairs, below 10 m.

Breeding Nest and eggs undescribed.

Movements Presumably sedentary.

Range Endemic. Currently known from six locations in NE Peru (Amazonas to San Martín) (Johnson 2002; Schulenberg *et al.* 2007; Hornbuckle 1999; BLI SF 2014).

Conservation status Classified by BirdLife International as Endangered. The main threat is forest clearance for logging and agriculture; this is especially severe in Cordillera de Colán, one of the main strongholds of the species.

Etymology The species was named in honour of the late Lulu May Van Hagen (1912–1998) in recognition of her generous and dedicated support of research in avian genetics (Johnson & Jones 2001).

RUFOUS TWISTWING
Cnipodectes superrufus

A distinctive flycatcher from western Amazonia which is a sister species to the much more widely distributed Brownish Twistwing *C. subbrunneus.*

▼ Rufous Twistwing, male. North bank of the Río Tahuamanu at Oceania, near Iberia, Peru, October 2004 (*Daniel J. Lebbin*).

Type information *Cnipodectes superrufus* Lane *et al.* 2007, Pakitza, Madre de Dios, Peru.

Alternative names Spanish: Mosquero Rufo, Alitorcido Rufo. Portuguese: Flautim-Rufo.

Discovery In 1984 a rufous-plumaged flycatcher was observed in Tambopata, Peru, by an experienced observer without at the time being identified. Retrospectively this would become the first record of Rufous Twistwing (Tobias *et al.* 2008). In February 1990 a large rufous flycatcher was mist-netted and collected in Manu National Park, Dpto. Madre de Dios, Peru. Provisionally it was left unidentified, but was later identified as a Rufous Casiornis *Casiornis rufus*, a rare austral migrant in Peru. This specimen was not seriously examined until 2002, when D. F. Lane realised that it was an unknown taxon, obviously of the (formerly) monotypic genus *Cnipodectes*. In 2003 F. L. Lambert videotaped an individual near Cocha Cashu Biological Station in Manu and, by an extraordinary coincidence, T. Valqui reported the species from Dpto. Cuzco, making the first tape recordings. In fact, a further specimen had been taken in 1997 in Dpto. Cuzco but had been labelled *Attila bolivianus* (Dull-capped Attila). Further fieldwork has now identified the species in several other Peruvian locations as well as in adjacent Brazil and Bolivia.

Description 18.4–24 cm. Monotypic. A medium-large flycatcher with a broad, flat bill, large rictal bristles and curiously modified outer primary feathers. Overall rufous, slightly crested in male, the lores, chin and throat paler; back to rump deep rusty-brown; rectrices chestnut; wings dark brown; underparts generally chestnut-rufous, becoming paler on belly and flanks; iris red or maroon in adult, possibly paler or grey in juvenile; bill dusky-brown above, light fleshy-pink below; tarsi grey or grey-pink. Sexes similar, but female has no crest and is noticeably smaller.

Habitat Lowland forest; seems to be associated with stands of mature *Guadua* bamboo, preferring areas with discontinuous tree canopy (Tobias *et al.* 2008), but also occurs in river-edge thickets; 250–550 m.

Food and feeding Feeds low in understorey, not above 3 m; prey is arthropods.

Breeding Nest and eggs undescribed. By analogy with *C. subbrunneus*, and consistent with observations showing that males may not react aggressively to playback, probably has a polygamous mating system. A paratype collected at the end of February had enlarged testes (10 × 5 mm).

Voice Makes a mechanical non-vocal noise using the modified primary feathers; this is probably the "accelerating and decelerating buzzing noise" observed by T. Valqui, the bird at the time flying about 4 m up with slower and deeper wingbeats than normal. Several different true vocalisations noted; a series of squeaky, rising *quec* or *cueet* calls; a stereotyped series of loud, descending whistled notes, sometimes preceded by sharp ticking sounds (Lane *el al.* 2007); loud, non-stereotyped scolding notes; and a loud whistled *hyew* call (Tobías *et al.* 2008).

Movements Presumably sedentary.

Range Imperfectly known, since much suitable or apparently suitable habitat has not been surveyed. SE Peru (Dptos Cuzco and Madre de Dios); western Brazil (Acre) (de Melo *et al.* 2015); and extreme NW Bolivia (Dpto. Pando).

Conservation status Data lacking; Tobías *et al.* (2008) suggest that a provisional classification of Vulnerable would be appropriate. Total potential range is about 89,000 km^2, encompassing 13 locations where the species has been encountered; however, it does not appear to be abundant in most of these and in Acre, Brazil, is apparently rare (Mestre *et al.* 2010). Although bamboo is often the first colonist of disturbed areas, leading to increased opportunities for bamboo specialists such as Inca Wren *Thryothorus eisenmanni* (Brewer 2001), Rufous Twistwing may require more mature stands, which are themselves vulnerable to human harvest.

Etymology *Superrufus* (Latin) refers to the striking rufous coloration and the larger size of the species.

ORANGE-EYED FLATBILL
Tolmomyias traylori

A widely distributed but relatively uncommon flycatcher, whose ultimate discovery was once again the result of the uncanny aural virtuosity of the late Ted Parker.

Type information *Tolmomyias traylori* Schulenberg & Parker 1997, north bank of Amazon River, 5 km ESE of Orán, Dpto. Loreto, Peru.

◄ ▼ Orange-eyed Flatbill. Left: Río Morona, Loreto, Peru, July 2001. Right: Quebrada Shimigay, Río Napo, Loreto, Peru, December 2015. (*Daniel Lane*).

Alternative names Orange-eyed Flycatcher. Spanish: Picoplano ojinaranja, Picoplano ojiamarillo.

Discovery In June 1983 T. Parker, who was working in lowland forest habitat near the Río Napo in eastern Peru, observed two flycatchers feeding young just out of the nest. He realised at this time that this was a species which he had never seen, or more especially heard, before. In February 1984 J. S. Dunning mist-netted and photographed an unusual tyrannid downstream from Iquitos, Peru; Parker was not able to recognise the species from Dunning's photographs, but realised that it was the same as the birds which he had seen eight months previously (Stap 1990). A subsequent search through collections in various institutions located no fewer than seven specimens, taken over a wide geographic range, from Dpto. Putumayo in southern Colombia to Dptos Amazonas and Loreto in Peru; the oldest of these, taken in 1887 by J. Hauxwell (for whom Hauxwell's Thrush *Turdus hauxwelli* and Plain-throated Antwren *Myrmotherula hauxwelli* are named), had sojourned in the collection of the Natural History Museum, Tring, unrecognised for over a century. Hauxwell, incidentally, employed Amerindians with blowpipes and darts to collect specimens, which is less damaging than bird-shot. These historic specimens became paratypes. The holotype was collected by A. Capparella in July 1984 in Dpto. Loreto, Peru. The type description was, for Parker, published posthumously.

Description 13.5 cm. Monotypic. A typical *Tolmomyias* flycatcher, quite small, large-headed and dumpy with a broad, flat bill. Olive-green above, the crown contrastingly dusky-grey; supraloral area and ear-coverts buff; forehead cinnamon; wings dusky-brown with two yellow wing-bars and yellow edgings to flight feathers; dusky rectrices with narrow edgings; throat buffy-white with chest ochraceous-buff; belly bright yellow; iris pale yellow-orange; bill black, lower mandible pale purple; tarsi dull purple. Sexes similar.

Habitat Midstorey of relatively mature *várzea* forest, sometimes on river islands, below 400 m.

Food and feeding Few data. Insectivorous. Forages by aerial sallies directed at leaves (Schulenberg & Parker 1997). Sometimes in mixed-species flocks.

Breeding Nest and eggs undescribed. Two adults attending juveniles in June in northern Peru.

Voice The distinctive call is described as a "two-parted and buzzy *wheeeeezzz-birrt* or *psi-trrrrrrr* given at rather long intervals and sometimes with a few other buzzy notes appended" (Ridgely & Greenfield 2001) or as "a rising-falling raspy whistle followed by a low chatter *ZREEE'chirr'rr*" (Schulenberg & Kirwan 2012a).

Movements Presumably sedentary.

Range Western Amazonia, from SE Colombia (Dpto. Putumayo; possibly extreme southern Guainía), eastern Ecuador, mostly along the Río Napo (Sucumbios and Napo), also SE Pastaza; northern Peru (Dptos Amazonas and Loreto). Currently no Brazilian records, although these are quite probable.

Conservation status Currently classified by BirdLife International as Least Concern. Although it does not appear common in any part of its distribution, it has a wide geographical range.

Etymology Named in honour of Melvin A. Traylor (1915–2008) "in recognition of his outstanding contributions to ornithology" (Schulenberg & Parker 1997). Traylor, who held the rank of Major in the United States Marine Corps, was severely wounded in the assault on Tarawa in 1943, losing an eye, but became one of the leading forces in African, Asian and South American ornithology in the post-war years. For much of his professional career he was associated with the Field Museum of Chicago, becoming assistant curator of birds in 1956. Curiously, he had examined some of the misidentified specimens of *T. traylori* prior to its discovery, noting a couple that "didn't quite fit" (Schulenberg & Kirwan 2012a).

WHITE-THROATED PEWEE
Contopus albogularis

One of the least-known species of the genus *Contopus*.

Type information *Myiochanes albogularis* Berlioz 1962, Maripasoula, Maroni Basin, French Guiana.

Alternative names French: Moucherolle à bavette blanche; Portuguese: Piuí-queixado.

Discovery Described by J. Berlioz on the basis of material collected near Maripasoula, on the borders of French Guiana and Suriname.

Description 13 cm. Monotypic. In form a typical *Contopus* flycatcher, notably smaller and darker than most members of the genus; generally uniform dark sooty-grey, the wings and tail somewhat darker; obscure narrow whitish eye-ring; underparts sooty-grey with a contrasting white throat; iris dark brown; bill black above, yellow below; tarsi black.

Habitat Edges of lowland humid forest, 400–700 m.

Food and feeding Captures prey by aerial sallies from a high exposed perch, returning to the same perch. Diet mainly flying insects.

► White-throated Pewee. Mont Galbao, French Guiana, August 2008 (*Thierry Nogaro*).

Breeding Nest and eggs undescribed.
Voice Call is a slightly squeaky, sharp *hwick,hwick*. Dawn song is a repeated *free-bit, wheeyr, wheeyr* (Ridgely & Tudor 2009).
Movements Presumably sedentary.
Range Central and eastern Suriname (Mees 1974), southern French Guiana, extreme north-east Brazil (Amapá).
Conservation status Classified by BirdLife International as Least Concern. Although generally not common, it has a wide geographic range, considerable portions of which are protected (Brownsberg Nature Park, Suriname; Parc National Amazonien de Guyane; Parque Nacional Montanhas do Tumucumaque, Brazil).
Etymology *Albogularis,* Latin, 'white-throated'.

SALINAS MONJITA
Xolmis salinarum

A large showy flycatcher inhabiting one of the harshest, most inhospitable areas of South America. Sometimes placed in genus *Neoxolmis.*

Type information *Xolmis rubetra salinarum* Nores & Yzurieta 1979, 'Monte de las Barrancas', Salinas Grandes to the north-west of Córdoba, Argentina.
Alternative names Salinas Tyrant. Spanish: Monjita Salinera.
Discovery Originally described in 1979 by M. Nores & D. Yzurieta as a race of the much more widely distributed Rusty-backed Monjita *X. rubetra*, based on a specimen taken in September 1977. However, most more recent authorities (e.g. Ridgely & Tudor 1994; Olrog 1984; Farnsworth & Langham 2004 HBW 9) classify it as a valid species, a treatment in agreement with the conclusions of the SACC (Proposal 350).
Description 16.5 cm. Monotypic. Crown rufous; back rufous-brown; rump greyish-white; wings black with large white patch involving greater and median coverts; tertials black with white edgings; rectrices black with white edgings on outer three; conspicuous white supercilium; ear-coverts whitish with dark crescent behind; underparts white; iris dark brown; bill and tarsi black. Female similar but with streaked neck. Differs from *X. rubetra* in smaller size, lack of buffy flanks and chest streaking, and in having unstreaked ear-coverts and more white on the wing-coverts (this white patch appears to be used in a wing-raising display which may act as an isolating mechanism from *rubetra*).
Habitat Restricted to the borders of open saltflats, often close to saline watercourses or standing water; specifically in areas with a *Salicornia* mat covering the salt-impregnated soil or in isolated groups of thorn bush (*Atriplex, Heterostachys, Prosopis reptans* and *Cyclolepis genistoides*); less commonly at the edge of chaco woodlands bordering the salt flats. Occurs at 100–200 m.
Food and feeding No data on diet but doubtless insectivorous. Feeds on ground and in low bushes, usually in pairs, but will flock during the austral winter.

◄ Salinas Monjita, male. Salinas Grandes, Córdoba province, Argentina, December 2006 (*Nick Athanas*).

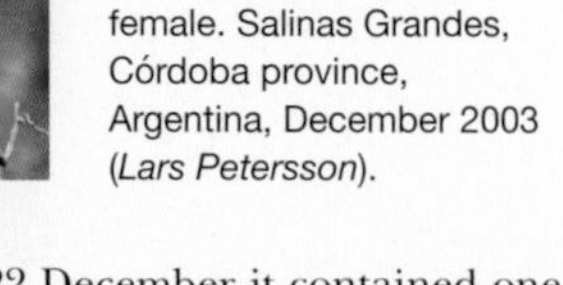

◄▼ Salinas Monjita, female. Salinas Grandes, Córdoba province, Argentina, December 2003 (*Lars Petersson*).

Breeding One active nest only so far described (Cobos & Miatello 2001). Nest was a cup 13 cm wide, 8 cm high, with an internal diameter of 5 cm and a depth of 5 cm, located in the centre of a bush at a height of 40 cm above the ground. On 22 December it contained one egg and one chick; egg colour creamy-salmon with small brown specks and a few brown marks, concentrated at the blunt end. Three other empty nests were of similar construction and location, 40–50 cm up.
Voice No data. Appears to be a rather silent species.
Movements Relatively sedentary; after the breeding season may be found in small flocks of up to 15 birds, more rarely up to 65, but these appear to be roving rather than migratory.
Range Endemic. Confined to arid regions of west-central Argentina (eastern La Rioja, extreme southern Catamarca, SW Santiago del Estero and NW Córdoba; recently in NW San Luis).
Conservation status Classified as Near Threatened, mainly because of its small range. Given the inhospitable nature of its habitat, unlikely to be much impacted by human activity; small-scale salt mining in Córdoba is not considered a threat.
Etymology *Salinarum*, Latinised from Spanish 'salina', a saltflat.

ACRE ANTSHRIKE
Thamnophilus divisorius

A spectacular sexually dimorphic antshrike with a limited range in Brazil and adjacent Peru.

Type information *Thamnophilus divisorius* Whitney *et al.* 2004, Morro Queimado, Serra da Jaquirana, municipality of Mâncio Lima, Acre, Brazil.
Alternative names Portuguese: Choca-do-Acre; Spanish: Batará de Acre.

Discovery In July 1996, during a rapid ecological evaluation of the Parque Nacional da Serra do Divisor, on the western borders of Brazil with Peru, a group of ornithologists spent 18 days at six sites within the park. Notwithstanding the limited field time available, several species of birds previously unknown or poorly known in Brazil were found, along with a hitherto unknown species of antshrike (Whitney *et al.* 2004).

▲ Acre Antshrike, male. Serra do Divisor National Park, Acre, Brazil, November 2016 (*Ricardo Plácido*).

◄ Acre Antshrike, female. Serra do Divisor National Park, Acre, Brazil, November 2016 (*Ricardo Plácido*).

Description 21–23 cm. Monotypic. Highly sexually dimorphic. Male is uniformly dark blue-grey, slightly darker on the crown, wings and tail; underwing-coverts mottled greyish; may have small white spots on some wing-coverts and outer rectrices; iris chestnut-brown; heavy hooked bill black; tarsi dull bluish-grey, soles yellow. Female bluish-grey on face and upperparts, wings and tail somewhat darker; underparts brownish-orange with some greyish or creamy-white mottling on chin and throat; bill dark grey with pale horn base.

Habitat Stunted ridgetop woodlands with a broken canopy and dense understorey; trees mostly of the Fabaceae family, with an understorey of ferns and terrestrial bromeliads, on thin, sandy soils. Holotype taken at 500 m; 400–600m in Peru (Schulenberg & Kirwan 2011).

Food and feeding Few data; seen to take a caterpillar and a 3 cm orthopteran (Whitney *et al.* 2004). Forages low down, mostly by gleaning plant surfaces, sometimes by upward sallies.

Breeding Nest and eggs undescribed.

Voice Usually first detected by vocalisations. The loud song is described as "an accelerating, descending series of rich notes with a distinctive, higher pitched, bisyllabic terminal bark: *kuk kuk-kuk-kuk-ku-ku-ku-ku'wah'AH*" (Schulenberg *et al.* 2007). Both sexes sing. The song of females is slightly lower in frequency than the song of males (Whitney *et al.* 2004). Calls include a "cawed *aw* and a longer, descending *awwr*" (Schulenberg *et al.* 2007), as well as an apparent food solicitation call from a juvenile, described as "a quiet chatter" (Whitney *et al.* 2004, Schulenberg & Kirwan 2011).

Movements Presumably sedentary.

Range Restricted to stunted forest on tops of ridges in the Serra do Divisor, Acre, Brazil, and adjacent similar habitat in the Sierra del Divisor, Ucayali, Peru.

Conservation status Despite its limited geographic range, rated by BirdLife International as Least Concern. Although some habitat degradation has taken place, the Brazilian portion of the range is protected in Parque Nacional da Serra do Divisor, while the presently known Peruvian distribution is within the Zona Reservada Sierra del Divisor.

Etymology *Divisorius*, Latinised form of the Portuguese 'divisor', or dividing (range).

RONDONIA BUSHBIRD
Clytoctantes atrogularis

◀ Rondonia Bushbird, male. Machadinho d'Oeste, Rondônia, Brazil, September 2013 (*André Grassi Corrêa*).

▼ Rondonia Bushbird, female. Machadinho d'Oeste, Rondônia, Brazil, March 2014 (*Bruno Rennó*).

The second member of the peculiar genus *Clytoctantes*; its distribution is very imperfectly known, making conservation strategies difficult.

Type information *Clytoctantes atrogularis* Lanyon *et al.* 1990, Cachoeira Nazaré, Rondônia, Brazil.

Alternative names Black-throated Bushbird. Portuguese: Choca-de-garganta-preta.

Discovery Over a period of several months in 1986 S. Lanyon, D. Stotz and D. Willard from the Field Museum of Natural History, Chicago, conducted extensive ornithological investigations along the Rio Jiparaná in Rondônia state, Brazil. On 22 October, a female of what, from its bizarre recurved bill, was obviously a bushbird was mist-netted; from its plumage it was immediately clear that it was totally distinct from the only other member of this genus, *C. alixii*, which is highly allopatric, occurring in northern Colombia and western Venezuela. Two black-plumaged males were seen at the same time but no further specimen could be obtained. Although Lanyon *et al.* expressed some reservations about describing a new species from one female specimen, it was sufficiently distinctive to allow this to happen.

Description 17 cm. Monotypic. Male entirely black with massive upturned black bill. Female chestnut-brown on head and ear-coverts, becoming brown on back and grey on uppertail-coverts; underparts light chestnut, becoming grey on crissum; chin and throat conspicuously black; rectrices black; bill massive, laterally compressed with uniquely upturned lower mandible and cutting-edge, black; iris dark brown; tarsi black.

Habitat Initially found in lowland humid *terra firme* forest with dense vine tangles; also observed (in Amazonas state) in *campinarana,* a unique vegetation type on white-sand soil with dense, low canopy and no emergents or lianas (Guilherme & Sousa Santos 2013).

Food and feeding Few data. Stomach contents of one specimen consisted of arthropod fragments (Guilherme & Sousa Santos 2013). Has been seen to open up dead stems in search of arthropods (Whitney 2005). Usually feeds fairly low down.

Breeding Nest and eggs undescribed. The two female specimens had ovaries 5 × 2 mm with minute ova (22 October, Rondônia) (Lanyon *et al.* 1990) and ovaries granulated 3 × 3 mm (24 August, Rondônia) (Guilherme & Sousa Santos 2013).

Voice Male song or call, a very loud, trilled whistle *tree-tree-tree* (Lanyon *et al.* 1990). What may be an alarm call is described as a "fairly loud, multisyllabic tremulous whistle becoming slightly protracted and fainter at the end" (Whitney 2005).

Movements Presumably sedentary.

Range Endemic. Has been recorded in widely scattered locations in Amazonian Brazil; Rondônia (three locations); Amazonas (two locations); Mato Grosso (two locations) (Whitney 2005, Whittaker 2009, Guilherme & Sousa Santos 2013).

Conservation status Originally classified as Critically Endangered (the habitat at the type locality has since been destroyed). Downgraded to Vulnerable (BLI SF 2014) after discovery in several new locations. It appears to be uncommon in all locations.

Etymology *Atrogularis*, referring to the diagnostic black throat-patch that distinguishes the female from the female of *C. alixii*.

BROWN-BACKED ANTWREN
Epinecrophylla fjeldsaai

A member of the 'stipple-throated' assemblage of antbirds, now placed in a new genus *Epinecrophylla* (Isler *et al.* 2006) which may not in fact be closely related to the other traditional *Myrmotherula* group; *fjeldsaai* forms a superspecies with the more widely distributed Stipple-throated Antwren *E. haematonota* and Foothill Antwren *E. spodionota*.

Type information *Myrmotherula fjeldsaai* Krabbe *et al.* 1999, near Río Tiputini, SSW of Pompeya, Provincia de Napo, Ecuador.

Alternative names Yasuni Antwren. Spanish: Hormiguerito Yasuní.

Discovery It has been appreciated for some time that Ornate Antwren *E. ornata* and White-eyed Antwren *E. leucophthalma* show distinct variations in the colour of their backs, but since the different colour forms are geographically separated they have been classified as races of the relevant species. However, in the case of Stipple-throated Antwren *E. haematonota* this appears not to be the case. Brown-backed birds (*fjeldsaai*) are apparently sympatric with red-backed ones (*haematonota*) in Loreto, Peru, without interbreeding (one apparently intermediate specimen, from an unknown location, and now presumably lost, was mentioned by Hellmayr, 1910). A brown-backed specimen taken by R. Olalla in 1963 was labelled as *haematonota,* but later identified as *leucophthalma* (Ortiz-Crespo *et al.* 1990). However, in 1992 P. Greenfield re-examined this specimen; in 1994 N. Krabbe collected a brown-backed bird in Napo, Ecuador, and a search through the collections of several institutions located a further six specimens. In addition, four photographs exist. All of these birds appear consistent in plumage characters and consistently different from specimens of *haematonota,* allowing for the description of the brown-throated birds as a new species.

Description 10–11 cm. Monotypic. In form very similar to other species of the genus; small, short-tailed, plump-bodied with a relatively large bill. Male: crown olive-brown, back dark yellowish-brown; wings and tail darker rufous-brown, with conspicuous white spots on median and lesser coverts and buff tips on the greater coverts; sides of face, breast and upper belly light grey; remainder of underparts, apart from the white-spotted black throat, pale yellowish-brown; iris greyish-brown;

◄ Brown-backed Antwren, male. Napo Wildlife Center, Orellana, Ecuador, November 2011 (*Dušan Brinkhuizen*).

bill blackish with grey-blue cutting edge; tarsi grey-blue. Female: sides of head ochraceous; chin and throat white with dark streaks; underparts buffy-brown.

Habitat Lowland evergreen forest, mostly *terra firma*, to a lesser extent *várzea*, especially areas with an understorey of palms; 150–300 m.

Food and feeding Insectivorous; prey items include Orthoptera, Coleoptera and spiders. Food taken mostly from curled-up dead leaves, usually 1–8 m up. Usually in mixed flocks with other antwrens, antbirds, greenlets etc.

Breeding Nest and eggs undescribed. The male holotype (16 July) had testes 3 × 1 mm.

Voice The loud song is a trill of abrupt sibilant notes, first acending then dropping in pitch, similar to that of *E. haematonota*. Differs from other *Epinecrophylla* species in the pace of individual notes.

Movements Presumably sedentary.

Range Eastern Ecuador (Napo and Pastaza) and western Peru (Loreto).

Conservation status Classified by BirdLife International as Least Concern. The geographical range is quite large and a substantial part of it is protected by Yasuní National Park in Ecuador; there appears to be no evidence of significant recent population declines.

Etymology The newly erected generic name *Epinecrophylla* is derived from three Greek roots signifying 'on the dead leaf', referring to the feeding technique of this and related species; the specific name honours Professor Jon Fjeldså (b. 1942) of the Zoological Museum of Copenhagen, in recognition of his huge contributions to the ornithology of Africa and South America and his great influence on conservation in both continents.

RIO DE JANEIRO ANTWREN
Myrmotherula fluminensis

A virtually unknown bird, whose taxonomic status is a subject of much debate and whose very existence is uncertain.

Type information *Myrmotherula fluminensis* Gonzaga 1988, 4 km south-east of Santo Aleixo, Majé, Rio de Janeiro, Brazil.

Alternative names Portuguese: Choquinha fluminense.

Discovery In July 1982 L. P. Gonzaga was banding birds in a disturbed habitat in the Serra dos Orgãos, Rio de Janeiro state, Brazil, when he mist-netted an antwren that he was not able to identify. Suspecting that a new taxon was involved he collected it; this specimen, currently housed in the Museu Paraense Emilio Goeldi in Belém, Brazil, is unique.

Description 9.5 cm. Monotypic. The specimen is a male. Overall slate-grey, paler on lower underparts; black bib from chin to chest; white dots on shoulders; two white wing-bars formed by edgings to greater and lesser coverts. Most resembles Ihering's Antwren *M. iheringi* but differs in being slightly larger, with a more slender bill and longer, more graduated, tail. DNA analysis to investigate the suggestion that it is a hybrid between Unicoloured and White-fringed Antwrens *M. unicolor* and *M. axillaris* was inconclusive (Collar in BLI SF 2012).

Habitat Subtropical or tropical moist lowland forest; 20 m.

Food and feeding Stated to forage low down in thick vegetation and vine tangles, often in mixed flocks.

Breeding No Data.

Voice No data.

Movements No data.

Range Endemic. Confined to a tiny area in the Serra das Orgãos.

Conservation status Critically Endangered, if not already extinct. Prior to the discovery, most of the lowland forest in the putative range was destroyed. If it does persist, it will be in the Serra do Mar Ecological Reserve, where observations of several birds were made in March 1997 (Curson & Lowen 1997); this requires further confirmation.

Etymology *Fluminensis*, Latin, referring to a river (i.e. Rio de Janeiro).

ALAGOAS ANTWREN
Myrmotherula snowi

A highly restricted and endangered antbird, another potential victim of the massive destruction of the Atlantic forests of Brazil. Forms a superspecies with *M. unicolor.*

Type information *Myrmotherula unicolor snowi* Teixeira & Gonzaga 1985, Pedra Branca, Alagoas, Brazil.
Alternative names Snow's Antwren. Portuguese: Choquinha-de-alagoas.
Discovery In 1979 the Ornithology Department of the National Museum, Rio de Janeiro, while investigating forest habitat in Alagoas, NE Brazil, collected several specimens of a small grey antwren, which were at the time classified as a new subspecies of Unicoloured Antwren *M. unicolor.* The new taxon was highly disjunct from the nominate race, there being a gap of over 1,300 km between the closest points of the ranges of the two forms. In a study published in 1997, Whitney & Pacheco recommended that the Alagoas population be elevated to full species rank, based on the distinctive plumage of the female, differences in measurements and probable vocal differences. This viewpoint has been accepted by most authors (e.g. Ridgley & Tudor 2009; Zimmer & Isler 2003 HBW 8).
Description 10–15 cm. Monotypic. Small, short-tailed antwren with a fairly substantial bill. Male very similar to *M. unicolor,* largely pale grey overall without wing-bars; throat feathers black, tipped grey. Female rufous-brown overall, with pale throat and chin and darker wing-coverts; differs from female *unicolor* in more rufescent plumage. Both sexes differ from *unicolor* in having a longer, wider and deeper bill, shorter tail and longer wings.
Habitat Wet forests commonly clouded in mist, usually near small creeks where the undergrowth is rich in lianas, 400–550 m (Roda *et al.* 2009).
Food and feeding Insectivorous. Stomach contents include spiders, beetles, ants and cockroaches; seen to take caterpillars and Orthoptera (Teixeira & Gonzaga 1985; Whitney & Pacheco 1997). Sometimes in mixed flocks, especially with White-flanked Antwren *M. axillaris,* furnariids and woodcreepers.
Breeding Nest and eggs undescribed. Female with egg in oviduct in early February; adults attending juveniles in May (Teixeira & Pacheco 1985).
Voice Song is "a series of 3–6 downslurred, clear-whistled syllables…delivered in about 2–4 seconds" (Whitney & Pacheco 1997). Also a *kleek* contact-call and a two- or three-syllable descending vocalisation (possibly an alarm call).
Movements Presumably sedentary.
Range Endemic. Originally thought to be confined to fragmentary forest patches in Alagoas state (Whitney & Pacheco 1995); more recently discovered in several locations in Pernambuco state, extending its range northwards by about 500 km (Roda *et al.* 2003, Roda *et al.* 2009).

◀ Alagoas Antwren, male. Murici, Alagoas, Brazil, January 2011 (*Ciro Albano*).

Conservation status Currently classified by BirdLife International as Critically Endangered. It is confined to small, fragmented patches of forest, now widely separately by intervening areas cleared for sugarcane and other crops. Further forest clearance continues. World population estimated at 30–200 mature individuals (BLI SF 2014). The survival of the species clearly requires rigorous protection, from clearance and charcoal gathering, of what little habitat remains. Recently described as "racing to global extinction" (Lees 2015).

Etymology Named in honour of David W. Snow (1924–2009), one of the most distinguished British ornithologists of his time. Educated in Oxford, his career included stints at the Edward Grey Institute, the New York Zoological Society's Field Station in Trinidad and the Charles Darwin Research Station in the Galápagos (of which he was Director) before returning to the UK to become Director of the British Trust for Ornithology. In addition to the Alagoas Antwren, the cotingid genus *Snowornis* is named for him.

▲ Alagoas Antwren, female. Murici Ecological Station, Alagoas, Brazil, June 2007 (*Ciro Albano*).

CAATINGA ANTWREN
Herpsilochmus sellowi

A widely distributed species, known from a large number of specimens, whose separate identity was established by vocalisation studies.

Type information *Herpsilochmus sellowi* Whitney *et al.* 2000, 2 km E of Boa Nova, Bahia, Brazil.

Alternative names Portuguese: Chorozinho-da-caatinga.

▼ Caatinga Antwren, male. Aquiraz, Ceará, Brazil, June 2011 (*Ciro Albano*).

Discovery Traditionally the Pileated Antwren *Herpsilochmus pileatus*, first described in 1823 from Bahia state, was believed to have a range covering most of eastern Brazil (e.g. Ridgely & Tudor 1989); indeed, earlier authors sometimes merged this with Black-capped Antwren *H. atricapillus*. However, a careful study of vocalisations by B. Whitney and his colleagues showed that the true *pileatus* is confined to a relatively small area of

▼ Caatinga Antwren, female. Caetité, Bahia, Brazil, July 2009 (*João Quental*).

coastal Brazil (thereby meriting the new common name of Bahia Antwren), while a larger area of interior Brazil is inhabited by a vocally distinct population, which additionally shows consistent differences in plumage and measurements. This was named as a new species, *H. sellowi*, with the common name of Caatinga Antwren, in recognition of its habitat. The new taxon is parapatric with *H. pileatus*, while being sympatric, and in places syntopic, with *H. atricapillus* (Whitney *et al.* 2000).

Description 10.5–11.5 cm. Monotypic. Male: crown and nape black; lores white, supercilium greyish-white, back grey; wings black with two conspicuous white wing-bars formed by edgings on coverts and spots on shoulders; tail graduated, black with white tips and white sides; underparts white, greyish on breast, flanks and crissum; iris brown; bill black, grey below; tarsi light grey, soles yellow. Female similar, but crown and forehead with dull buffy edgings, supercilium less well defined. Differs from *H. pileatus* by narrower bill and absence of a black loral spot; from *H. atricapillus* by smaller size and shorter tail.

Habitat *Caatinga* woodland and scrub of various types; especially in semi-deciduous woodland rich in vines and terrestrial bromeliads. Mostly 300–900 m, locally up to 1,100 m and down to near sea level in northern parts of range.

Food and feeding Forages in middle and upper strata of trees and bushes, usually in pairs, often in mixed flocks. Insectivorous; stomach contents include Orthoptera, Hemiptera and Coleoptera.

Breeding One nest so far described. It was a loose, insubstantial unlined cup of fungal hyphae, grass blades, tendrils, leaf fragments and spider webs, built by both sexes, 3.6 m up in a *Byrsonima* tree. Construction was underway on 2 September; seven days later it contained two eggs, light beige with brown spots concentrated at the blunt end (da Silva *et al.* 2008).

Voice Loud song is diagnostic, a rapid series of up to 40 notes, at a uniformly rapid pace without change in duration of individual notes.

Movements Presumably sedentary.

Range Endemic. Interior Brazil, patchily from Rio Grande do Norte, southern Ceará and central Maranhão south through western Bahia to NW Minas Gerais. Recent records from Alagoas and Pernambuco (Pereira *et al.* 2014). A disjunct population in Pará (Serra do Cachimbo). However, Santos *et al.* describe this population as *Herpsilochmus* aff *sellowi* (i.e. akin to *sellowi*) and state that it is "more related to *H. sellowi*" without giving further specifics (Santos *et al.* 2011).

Conservation status Presently classified by BirdLife International as Least Concern. It has a wide distribution, though apparently rather discontinuous. In some locations described as very common (BLI SF 2014). However, habitat destruction within its range is widespread.

Etymology Named after the German, Friedrich Sellow (1789–1831), who came to Brazil to collect natural history specimens in 1814. Here he met Prince Alexander Philipp Maximilian II of Wied-Neuwied (of Wied's Flycatcher fame), whom he accompanied in the field for several months. Over the next few years he made important collections, of over 5,000 birds, 250 mammals and 100,000 invertebrates. He was drowned in the Rio Doce in Minas Gerais in 1831, at the age of 42 (Stresemann 1948).

ASH-THROATED ANTWREN
Herpsilochmus parkeri

A very restricted antwren, part of the *Herpsilochmus pileatus* clade which includes *H. sellowi, H atricapillus* and *H. motacilloides.*

Type information *Herpsilochmus parkeri* Davis & O'Neill 1986, ca 15 km NW of Jirillo (or Jerillo), Dpto. San Martín, Peru.

Alternative names Parker's Antwren. Spanish: Tiluchí de Garganta Ceniza, Tiluchí de Parker.

Discovery In the autumn of 1983, an expedition from the Louisiana State University Museum of Zoology (LSUMZ), while working on the eastern foothills of the Peruvian Andes in the Department of San Martín, encountered an unfamiliar *Herpsilochmus* antwren and were able to collect several specimens.

Description 11.5–12.5 cm. Monotypic. Male: crown jet-black; supercilium greyish-white; lores and eyestripe jet-black; ear-coverts pale greyish-white; back grey; wings black with two conspicuous white bars on the coverts and white on outer webs of remiges; tail relatively long, black with increasing amounts of white on the rectrices from the centre to the outer feathers; chin and breast pale grey; sides of breast and flanks light grey; centre of abdomen white; undertail-coverts greyish-white; iris brown; bill black above, medium-grey below; tarsi and feet bluish-grey. Female has an orange-cinnamon forehead; crown between sepia and black; supercilium yellowish-ochre; eyestripe between sepia and black; chin, throat and breast yellow-ochre, the throat paler; abdomen white, flanks pale grey.

► Ash-throated Antwren, male. Moyobamba, Peru, June 2015 (*Ross Gallardy*).

Habitat Semi-stunted montane woodland with trees up to 12 m, with dense understorey and abundant epiphytes; also in taller trees with closed canopy of 20–30 m. 1,250–1,450 m (Zimmer & Isler 2008 HBW 8).
Food and feeding Forages in mid-levels to canopy, often in pairs or family groups. Often in mixed-species flocks with woodcreepers, furnariids etc. Stomach contents insects, especially beetles, also bugs, spiders etc. Not known to follow ant-swarms.
Breeding Nest and eggs undescribed. Probably breeds in the May–September dry season.
Voice Male song is an accelerating and slightly descending chippered trill with several well-spaced introductory notes, frequently echoed by the female whose song is similar, but softer, shorter and perhaps higher pitched (Ridgely & Greenfield 2001). Call, a rich, short *tchew*, often doubled or tripled (Schulenberg *et al.* 2007).
Movements Presumably sedentary.
Range Endemic. Confined to Dpto. San Martín, Peru; initially one area only, around the type locality northwest of Jirillo, more recently one individual observed near Abra Patricia. Some apparently suitable habitat near the type locality remains to be explored.
Conservation status Classified by BirdLife International as Endangered. The total range is very small and subject to deforestation; population estimates lie in the range of 350–1,500 individuals (BLI SF 2014). Although described as "relatively common" at the type locality, for the survival of the species legal protection of habitat is required (Begazo *et al.* 2001).
Etymology Named in honour of the late Ted Parker (see under Cloud-forest Pygmy Owl *Glaucidium parkeri*).

ANCIENT ANTWREN
Herpsilochmus gentryi

A quite widespread and in places not uncommon species which, nevertheless, went undetected until its vocalisations were studied. Forms a superspecies with *H. stictocephalus*.

Type information *Herpsilochmus gentryi* Whitney & Álvarez 1998, P. J. Lores, Loreto, Peru.
Alternative names Spanish: Tiluchí Antiguo.
Discovery In the early 1990s J. Álvarez A. and B. M. Whitney spent much effort in recording avian vocalisations in Dpto. Loreto, northern Peru. In September 1994 Whitney listened to a recording made by Álvarez of an unfamiliar song; realising that it most resembled that of Todd's Antwren *Herpsilochmus stictocephalus*, a highly disjunct species whose nearest geographic location was in southern Guyana, more than 1,900 km to the north-east, Whitney and Álvarez mounted a brief expedition to the Tigre River region of Loreto. This resulted in new tape recordings being made of several individuals and the collection of two specimens. Subsequent work revealed that this species has a wider distribution in Peru, and is also found in eastern Ecuador (Pastaza) (Whitney & Álvarez 1998; Shany *et al.* 2007).
Description 10–11 cm. Monotypic. Male: crown and nape black, supercilium whitish, eye and loral stripe black; upperparts dark grey with some black markings; scapulars black, edged white; wings black with conspicuous white bars formed by edgings to greater and

◄ Ancient Antwren, male. Mishana National Reserve, Allpahuayo, Peru, 1999 (*José Álvarez Alonso*).

lesser coverts; rectrices graduated, black with white tips; underparts with olive-tinged flanks. Female similar, but crown with small orange spots and underparts darker; iris dark brown; bill blackish above with greyish cutting-edges, lower mandible greyish; legs bluish-grey, the soles yellowish-orange

Habitat Several types of humid tropical lowland forest growing on nutrient-poor soils, including *irapayal*, dominated by trees over 40 m high, and *varillal* (more stunted trees on white-sand soils). In Ecuador, tall *terra firme* forest on dry ridge-tops (Zimmer & Isler 2008 HBW 8).

Food and feeding Insectivorous, including adult and larval Lepidoptera, probably also spiders. Feeds in canopy and upper levels, 15–40 m up, usually in mixed-species flocks.

Breeding Nest and eggs undescribed. Examination of gonads and moult of specimens suggests that the breeding season is between January and March (Whitney & Álvarez 1998).

Voice Both sexes sing. Male song is a series of up to 15 notes, lasting a little longer than two seconds, increasing and decreasing in pitch and intensity; female song similar but shorter. Resembles the song of Todd's Antwren, but has more notes with less overall change in frequency. Other calls include a *chup*, a longer *tink* and an abrupt rattle.

Movements Presumably sedentary.

Range Endemic. Extreme eastern Ecuador (Pastaza) and in the drainages of the rivers Marañon, Tigre, Corrientes, Pacacoro and Pastaza in NE Peru (Loreto), extending to the outskirts of the city of Iquitos.

Conservation status Currently classified by BirdLife International as Near Threatened. The main threat is habitat loss, especially in the environs of Iquitos; much of the remainder of the range is not presently under great pressure.

Etymology Named in honour of Alwyn Gentry, "one of the most gifted and productive field botanists of all time" (Whitney & Álvarez 1998), who spent much time investigating the white-sand plant communities of western Peru. Dr Gentry died in the same light aircraft crash as Ted Parker, in western Ecuador in August 1993.

ARIPUANA ANTWREN
Herpsilochmus stotzi

Sister species to Predicted Antwren *H. praedictus*, from which it is separated by the Rio Madeira.

Type information *Herpsilochmus stotzi* Whitney *et al.* 2013, left bank of the Rio Aripuanã, municipality of Manicoré, Amazonas, Brazil.

Alternative names Portuguese: Chorozinho-do-aripuanã.

Discovery In June and November 1986, D. F. Stotz and his colleagues obtained a small series of specimens of a new *Herpsilochmus* antwren in a remote area in northern Rondônia, Brazil. At the time these were believed to be a disjunct population of Black-capped Antwren *H. atricapillus*, whose nearest distribution was in Mato Grosso state, some 550 km distant (Stotz *et al.* 1997). However, in a study of the Bahia Antwren *H. pileatus* complex, the

▲ Aripuana Antwren, male. Right bank of the Machado River, Rondônia, Brazil, February 2014 (*Fabio Schunck*).

▲ Aripuana Antwren, female. Transamazônica Highway (BR 230), Amazonas, Brazil, June 2012 (*Fabio Schunck*).

Stotz specimens were re-examined and in 2000 (Whitney *et al.*) it was suggested that a species distinct from *atricapillus* was involved; further study, involving vocalisations and DNA evidence, resulted in a formal description as a new species, *H. stotzi* (Whitney *et al.* 2013). This conclusion is supported by the SACC (Proposal 585).

Description 11.3 cm. Monotypic. In plumage a typical *Herpsilochmus* antwren, sexually dimorphic. Male has a black crown with whitish streaks, grey-white supercilium and black eyestripe; back grey; wings blackish with conspicuous whitish spotting on shoulders and greater and lesser coverts; throat whitish-grey, remainder of underparts grey, paler in centre of breast; tail black, the outermost rectrices white with dark base, other rectrices black with white tips, narrowest on central feathers; iris dark brown; bill black above, grey below; tarsi dark grey. Female similar in plumage pattern, but clear grey of male replaced by buffy-tinged grey, crown markings more pronounced. Male very similar to, and possibly indistinguishable from, Black-capped Antwren *H. pileatus*; told from Predicted Antwren *H. praedictus* by plainer grey mantle. Females from other members of the genus by paler, creamy-white throat contrasting with orange forehead.

Habitat *Campinarana* forest, i.e. old-growth forest with canopy height of 12–20 m, growing on nutrient-poor and poorly drained soils; often with understorey composed of the palm *Lepidocaryum tenne*. Mostly below 200 m (Whitney *et al.* 2013).

Food and feeding Forages in canopy, usually in mixed-species flocks. Stomach contents of eight specimens comprised insects; in two cases, spiders as well.

Breeding Nest and eggs undescribed; nests in July and August, with newly fledged birds being fed by adults.

Voice Distinctive; a rapid series of notes, up to 20 per second, on one pitch, slowing towards the end, but generally slower than that of Predicted Antwren *H. praedictus* (sonograms in Whitney *et al.* 2013).

Movements Presumably sedentary.

Range Endemic. Western Amazonian Brazil, in the Madeira-Aripuanã interfluve, in southern Amazonas, NE Rondônia and NW Mato Grosso.

Conservation status Not currently classified by BirdLife International, but Whitney *et al.* (2013) found it common and do not regard it as facing any immediate threat, though roadbuilding is a potential longer-term problem.

Etymology Named in honour of Douglas F. Stotz, who discovered the bird in 1986. Dr Stotz has had a long Association with the Field Museum of Natural History in Chicago and is currently Senior Conservation Ecologist.

PREDICTED ANTWREN
Herpsilochmus praedictus

One of two new species of *Herpsilochmus* antwrens discovered from western Amazonian Brazil; probably a sister species to the other, the Aripuana Antwren *H. stotzi.*

Type information *Herpsilochmus praedictus* Cohn-Haft & Bravo, 2013, 30 km west of Humáita, Amazonas, Brazil.

Alternative names Purus Antwren, Madeira-Purus Antwren. Portuguese: Chorozinho-esperado.

Discovery In 1988 M. Cohn-Haft observed a black-and-white antbird, obviously of the genus *Herpsilochmus*, in *terra firme* forest in the Urucu basin, Amazonas, Brazil; however, in the absence of specimens or recordings, nothing further could be done. Among known species it most closely resembled Spot-backed Antwren *H. dorsimaculatus*, a widely allopatric species essentially confined to the north of the Amazon River. Had it been that species, Dr Cohn-Haft's observations would have shown a major range extension. The absence of a *Herpsilochmus* in that area was unexpected, leading to predictions that such birds had to be present, but had simply been undetected. In 1999 Dr Cohn-Haft was conducting survey work near Humáita, Amazonas, when similar birds were again encountered, on both sides of the Rio Madeira, itself a formidable ecological barrier. Fortunately, M. & P. Isler had provided recordings of numerous antwren species, subsequently published (Whitney *et al.* 2000). It was immediately obvious that two new vocal types were involved, on opposite banks of the Rio Madeira, and that both differed from previously described species. Consequently specimens were taken, allowing the description of two new species; this one and Aripuana Antwren *H. stotzi* (q.v.). The SACC unanimously recommended acceptance of both as full valid species (Proposals 585, 586).

Description 10 cm. Monotypic. Sexually dimorphic. Very similar to several other *Hersilochmus* species; male essentially black, white and grey, with a black crown and eyestripe, white supercilium, black wings with conspicuous broad white edgings to coverts and tertials; back medium-grey, washed ochraceous, with darker markings; rectrices graduated, blackish, with conspicuous white tips, especially on outermost feathers which are largely white; throat whitish, remainder of underparts unmarked grey; iris dark brown; bill black above, pale grey below; tarsi slate-grey. Female has a dusky rusty forehead, the crown with white streaks, and an unmarked grey back. Differs from *H. stotzi* (its closest relative) in female plumage in having a more uniform and rusty forehead and longer and paler crown streaks; male very similar, the present species tending to have more black on the back.

Habitat Several types of forest; low-stature forest on sandy soils adjacent to savannas (but not the savannas themselves); also tall *terra firme* (upland) forest on weathered clay soils and seasonally flooded forest along small blackwater rivers; below 500 m (Cohn-Haft & Bravo 2013).

Food and feeding Forages in pairs, often in mixed-species flocks, usually in upper levels of canopy, 5–30 m up. Food items include insects; no detailed data.

◄ Predicted Antwren. Borba, Amazonas, Brazil, July 2016 (*João Quental*).

Breeding Nest and eggs undescribed.
Voice Both songs and calls resemble those of other members of the genus but are nevertheless diagnostic. Loud song, a uniformly paced softly purring, rolling trill lasting about two seconds, given by both sexes. Call, a dull *pwip*, differing from the *tchwee* of *H. stotzi*.
Movements Presumably sedentary.
Range Endemic. Presently known only from the state of Amazonas, Brazil, stretching in the east to the Rio Madeira and Rio Solimões; west probably to the Rio Juruá, but imperfectly known.
Conservation status Currently not classified by BirdLife International; Cohn-Haft & Bravo (2013) do not believe it to be at risk, given the substantial range, its apparent tolerance for several habitat types and the recent creation of several national parks, albeit not yet enforced.
Etymology The common and scientific names refer to the fact that the occurrence of the species was predicted before the actual discovery.

SINCORA ANTWREN
Formicivora grantsaui

A very restricted species, most closely related to the antwren group containing Rusty-backed Antwren (*F. rufa*), Southern White-fringed Antwren (*F. grisea*) and (oddly) Marsh Antwren *F. (Stymphalornis) acutirostris* (q.v.) than to other members of the genus *Formicivora*.

Type information *Formicivora grantsaui* Gonzaga *et al.* 2007, Cumbuca Valley, 3.5 km north-east of Mucugê, Serra da Sincorá, Bahia, Brazil.
Alternative names Portuguese: Papa-formiga-do-sincorá.
Discovery In January 1997, in the north-east of the Serra do Sincorá range in Bahia state, Brazil, D. Buzzetti and A. Carvalhaes observed and recorded a *Formicivora* antwren resembling Rusty-backed Antwren *F. rufa*, but whose voice appeared to be different from that species. Two years later A. Carvalhaes collected two male specimens. Comparison with antwren specimens of *rufa* showed significant plumage differences. The matter was finally settled in November 2002 when L. Gonzaga, who had previously done extensive work on other members of the genus, returned with A. Carvalhaes and made extensive recordings of vocalisations and collected further specimens of the bird in question as well as adjacent populations (not syntopic) of *rufa*, establishing the validity as a new species. In fact, a specimen was taken as early as 1965 by R. Grantsau; he was, however, unwilling to describe it as a new taxon without additional material (Grantsau 1967). Also, in January 1997, B. Whitney and K. Zimmer observed a pair of adult *Formicivora* antwrens feeding young which they suspected, on the basis of plumage and vocalisations, to be an undescribed species (Gonzaga *et al.* 2007).
Description 12–13 cm. Monotypic. Sexually dimorphic. Male: upperparts deep brown; face, cheeks, throat, chest and belly black, separated from the brown

► Sincora Antwren, female left, male right. Lencois, Bahia, Brazil, January 2016 (*Ciro Albano*).

upperparts by a white supercilium which extends all the way to the upper flanks; lower belly and vent grey; upper-wing-coverts black with two white wing-bars and white spots on shoulders; remiges brownish-grey; tail long and frequently cocked, with rectrices graduated, grey and black with conspicuous white tips; iris dark brown, bare orbital skin greyish-black; bill all black in male, in the female with grey base to mandible; tarsi dark plumbeous-grey, soles yellow. Female lacks the black face, mask and underparts and white borders; instead it has white underparts with broad black lateral streaks.

Habitat Scrubby vegetation around sandstone rock outcrops (known locally as '*campo rupestre*'), 850–1,100 m. *F. rufa* is sympatric but prefers flatter areas on laterite soils.

Food and feeding Few data; forages in pairs or family parties. Presumably insectivorous.

Breeding Nest and eggs undescribed. Adults feeding juveniles at the end of January (Gonzaga *et al.* 2007).

Voice Both sexes sing in a duet. The loud song is similar to other members of the genus, but slower; a steady repetition of very short notes with a descending frequency. Distinctive alarm call consists of two or more notes. Gonzaga *et al.* (2007) provide a very detailed analysis of vocalisations.

Movements Presumably sedentary.

Range Endemic. Confined to the Serra do Sincorá in the Espinhaço Range, Diamantina region, Bahia, Brazil, where it is known from four localities; might well occur in other locations within this area.

Conservation status Classified by BirdLife International as Near Threatened. The total geographic range is quite limited, not exceeding 200 km² in extent. Recent protection and rehabilitation of some areas from the effects of mining should have a positive effect; part of the habitat is now protected by the Chapada Diamantina National Park, but accidental fires such as have occurred recently are a concern.

Etymology Named in honour of Rolf Grantsau, born in Kiel, Germany, in 1928, who has made major contributions to our knowledge of the ornithology of Brazil.

PARANA ANTWREN
Formicivora acutirostris

Originally placed as the monotypic example in the newly erected genus *Stymphalornis*, on the basis of the anatomy of the syrinx, but now considered to belong in *Formicivora*. Until the discovery of the Marsh Antwren *F. paludicola* (q.v.) in eastern São Paulo it was believed to be the only example of its family to live in reedbeds.

▼ Parana Antwren, male (left) and female (right). Itapoá, Santa Catarina, Brazil, May 2011 (*Ciro Albano*).

Type information *Stymphalornis acutirostris* Bornschein *et al.* 1995, Balneário, Ipacaray, Matinhos, Paraná, Brazil.
Alternative names Marsh Antwren. Portuguese: Bicudinho-do-brejo.
Discovery During researches into the breeding biology of Unicolored Blackbird *Agelasticus cyanopus* in July 1995, a team from the Museu de Histórica Natural 'Capão da Imbuia' and the Museu Nacional-UFRJ were most surprised to come upon a hitherto undescribed antbird, in a habitat unique in this family. A female holotype and male paratype were taken, allowing the description of the new species.
Description 14 cm. Monotypic. Sexually dimorphic. A long-tailed and long-billed antwren. Male largely chestnut-olive above with blackish-grey on the face and underparts; wing-coverts and shoulders blackish with two rows of bold white spots on the coverts and white spotting on the shoulders; flight feathers edged with chestnut-olive; rectrices graduated, black, the outer ones narrowly tipped with whitish or grey. Female has the throat, chest and upper belly whitish, boldly streaked with black. Iris dark chestnut, bill black, brownish cutting edges, tarsi ashy-blue (Rienert & Bornschein 1996).
Habitat Littoral marshes with *Scirpus* (Cyperaceae), *Typha* (Typhaceae) and Pteridophyta grasses and bushes, riverine marshes and other lowland wetlands, often subject to fluctuating water levels. Sea level or near thereto.
Food and feeding Forages in the lower levels (up to 60 cm) of marsh vegetation. Prey items comprise arthropods.
Breeding Nest is unique in the Thamnophilidae in response to the unique habitat of this species (and *F. paludicola*, q.v.). Nest is built between August and February; it is a cup of dry fibres, straw, silk and frequently leaf fragments, attached to (often vertical) marsh vegetation in a manner not found in other members of the family. Eggs two, white with varying amounts of irregular brown spotting (Reinert *et al.* 2012).
Voice Loud song consists of doublets of abrupt notes, the first note more intense and lower pitched, repeated in a slightly uneven and variably paced series. Calls are various squealing notes (Zimmer & Isler 2003 HBW 8).
Movements Presumably sedentary.
Range Endemic. Coastal marshes in southern Paraná and Santa Catarina, Brazil, contained within an area with a total north–south extent of some 80 km, much narrower from east to west.
Conservation status Currently classified by BirdLife International as Endangered. Confined to a limited series of narrow coastal or near-coastal marshes, a habitat typically under threat from a variety of causes including landfilling, sand extraction, disturbance from boat traffic, fires, etc. Invasion by exotic species of vegetation is also a serious potential problem. Population estimate in 2007, 17,700 individuals (Reinert *et al.* 2007). Appears to be remarkably long-lived for a small bird with a body-mass of no more than 10 g, with individuals of proven ages of 16 and 14 years and many examples exceeding 8 years, though older females may "lose interest" in breeding (Bornschein *et al.* 2015).
Etymology The now defunct generic name *Stymphalornis* is derived from the mythical man-eating Greek Stymphalian birds (*Ornithes Stymphalides*) which lived in the thick vegetation around Lake Stymphalis in Arkadia and which were destroyed by Herakles as his sixth labour. *Acutirostris*, Latin, 'sharp-billed'.

MARSH ANTWREN
Formicivora paludicola

A sister species to Parana Antwren *F. acutirostris* (q.v.) with an even more restricted distribution. The IOC and others treat this form as a subspecies of Parana Antwren.

Type information *Formicivora paludicola* Buzzetti *et al.* 2013, Córrego Taboão doParateí, São Paulo state, Brazil.
Alternative names Portuguese: Bicudinho-do-brejo-paulista.
Discovery In October 2004 D. Buzzetti observed and tape-recorded a female and a subadult male of an antwren which at first glance closely resembled the then recently described Parana Antwren *F. acutirostris*. However, the location, in São Paulo state, was more than 300 km from the nearest portion of the range of that species, and the habitat, a non-coastal marsh at an altitude of 640 m, was also quite distinct from that used by *acutirostris*. Suspecting that a new species might be involved, Buzzetti returned the following day and collected an adult male and a female. In February 2005 L. F. Silveira discovered a new population of the same bird in an area due shortly to be flooded by the construction of a dam; this population numbered about 100 birds. A plan was developed (see Conservation status) to preserve these birds by transferring them to a safe location; in the early stages of this work M. R. Bornschein, who was the senior author of the type description of *F. acutirostris*, collected further

◄ ▼ Marsh Antwren, male (top) and female (bottom). Biritiba Mirim, São Paulo, Brazil, September 2010 (*Elis Simpson*).

specimens, confirming first impressions that the São Paulo birds represented a new species. In fact, a specimen taken in 1945 by H. Camargo and misidentified as a Rufous-backed Antwren *F. rufa* may well have been the first example of *F. paludicola;* the specimen was so damaged during collection that the collector did not bother to preserve what would have been the type specimen of a new species (Buzzetti *et al.* 2013).

Description 14 cm. Monotypic. Sexually dimorphic. Male: crown, back, mantle and uppertail-coverts dark greyish-brown; forehead grey; lores black; supercilium white; ear-coverts, throat, underparts and vent black; rectrices black with white tips on outer two feathers; lesser wing-coverts white, median and greater coverts black, tipped white; primaries and secondaries dark greyish-brown; iris brown; bill black; tarsi dark bluish-grey. Female: crown, back, mantle and uppertail-coverts dark greyish-brown; forehead and lores grey; supercilium white; ear-coverts, throat and underparts white with black spots; lesser wing-coverts white, median and greater coverts black with white tips; primaries and secondaries very dark brown; soft parts as in male. Males differ from *F. acutirostris* in having black, rather than grey, cheeks, ear-coverts, throat, belly and thighs, and dark greyish-brown, as opposed to dark brown, crown, upperparts and uppertail-coverts. Females have brown upperparts, instead of dark-greyish brown in *acutirostris.* Both sexes have a shorter exposed culmen than in *acutirostris.* Molecular work also distinguishes the two forms, which probably diverged from each other in the range of 250,000–640,000 years ago (Buzzetti *et al.* 2013).

Habitat Marshland dominated by cattails (*Typha*), bulrushes (*Schoenoplectus*) and other Cyperaceae and Poaceae; 600–760 m.

Food and feeding Feeds in pairs or family groups in the lower strata of marsh vegetation; diet includes mosquitoes, mantises and caterpillars.

Breeding Nest and eggs undescribed. Males appear to start breeding in subadult plumage. A pair collected in early October had brood-patches, indicating reproductive activity at that date, and that both sexes incubate. Copulation observed in October, and juveniles recorded from January to March. Del-Rio and Silveira (2016) observed brood-patches on captured birds between October and February; four nests (undescribed) were found from October to late November.

Voice Buzzetti *et al.* (2013) give a very detailed analysis of vocalisations. Loud song is a sequence of rapidly but evenly repeated two-note phrases. Alarm call is a two-note phrase. Contact call is a single note. Also recorded are distress, display, foraging and aggressive calls.

Movements Sedentary; did not colonise newly created suitable habitat even though this was only 1 km from an existing population.

Range Endemic. Known from 15 locations in five municipalities, all within the state of São Paulo. Extensive searches in more than 50 other marshes did not find any further populations.

Conservation status Not currently classified by BirdLife International. Its distribution is geographically very small, probably only about 1,420 km², and it occurs only in limited areas of suitable habitat within that area. Furthermore, these limited areas, located only 50 km from the largest city in South America, are under severe pressure from a variety of causes, including encroachment by housing, fish farming, drainage for agriculture and fire; in addition exotic invasive plants, especially the grass *Urochloa arrecta*, creates unfavourable habitat. Buzzetti *et al.* (2013), estimate the population as being in the range of 250–300 individuals; within the state of São Paulo it is classified as Critically Endangered (Silveira 2010). A significant portion of the population was in the area of the Paraitinga dam and was certainly headed for destruction as the impoundment filled; as a drastic recourse, 72 birds were captured and relocated to several sites. Most of these transplanted individuals were still present three years later and some had bred in their new sites. Del-Rio & Silveira estimate that its former range was in the region of 300km². Buzzetti *et al.* (2013) point out that immediate effective action is essential if Marsh Antwren is to be saved from extinction.

Etymology *Paludicola*, Latin, a dweller in fens or marshes.

ORANGE-BELLIED ANTWREN
Terenura sicki

This species was described from a female specimen which does indeed have an orange belly; confusingly, the male is entirely black and white, lacking any orange. Forms a superspecies with the allopatric Streak-capped Antwren *T. maculata*.

Type information *Terenura sicki* Teixeira & Gonzaga 1983a, Serra Branca, Murici, Alagoas, Brazil.

Alternative names Alagoas Antwren. Portuguese: Zidede-do-nordeste.

Discovery In early 1979 an expedition from the Museu Nacional of Brazil conducted ornithological investigations in what remained of the Atlantic forests of NE Brazil, the main objective being to locate the endangered Alagoas Curassow *Mitu mitu* (this species has since become extinct in the wild; it exists in

◄ Orange-bellied Antwren, male. Jaqueira, Pernambuco, Brazil, October 2016 (*Ciro Albano*).

▼ Orange-bellied Antwren, female. Pedra D'Anta Private Reserve, Lagoa dos Gatos, Pernambuco, Brazil, March 2013 (*Stephen Jones*).

captivity with a population of about 130 birds, some of which are hybrids). This expedition was highly productive in unexpected ways, since it led to the discovery of several species not previously known from the state of Alagoas, as well as the description of two new species, Orange-bellied Antwren and Alagoas Foliage-gleaner *Philydor novaesi* (q.v.). The type specimen of the former was collected in February 1979.

Description 9.5–10.5 cm. Monotypic. Sexually dimorphic. Male: crown and nape streaked black and white; back and rump black with a few white streaks; scapular patch white (concealed); wings black with white covert-edgings forming two wing-bars; tail black; underparts white with blackish streaks on sides of chest. Female has streaked black-and-white head and nape; back and rump rufous-brown; underparts orange; iris brown; bill brown above, lead-grey below; tarsi leaden-grey. The female in the type description was in fact immature plumage (Teixeira *et al.* 1988).

Habitat Upper and mid-upper levels of semi-humid upland evergreen forest at 200–700 m, with one record in lowland forest at 76 m (BLI SF 2014, Albano 2009).

Food and feeding Few data. Feeds in pairs or family groups, sometimes in mixed flocks with a variety of tanagers, gnatwrens, furnariids etc. (Roda *et al.* 2003), mostly in upper levels of forest 7–20 m up, occasionally down to 5 m. Food items arthropods; beetles, cockroaches, spiders etc.

Breeding Breeding season probably November to February. Only one nest known; a small cup made of moss and fungal filaments, situated in a horizontal fork of a branch about 10–12 m up (Zimmer & Isler 2003 HBW 8). Eggs undescribed.

Voice Song is a loud dry rattle; call a sharp *chip*; another call is a longer, downslurred note.

Movements Presumably sedentary.

Range Endemic. Confined to the states of Alagoas and (recently discovered) Pernambuco (Albano 2009). So far it has been found in eight locations.

Conservation status Currently classified as Endangered. Much of the known existing habitat has been (and continues to be) destroyed by conversion to sugar-cane plantations, pastureland and by (frequently illegal) logging. The remaining forest at Murici, after much reduction in area, is now apparently protected (Anon. 2001).

Etymology Named in honour of Helmut Sick (1910–1991). Dr Sick was one (and possibly the most distinguished) of a long line of German scientists who made lasting contributions to the ornithology of Brazil, running from Prince Alexander Maximilian II (1782–1867), through Friedrich Sellow (1789–1831), Emilie Snethlage (1868–1929) and Rolf Grantsau (b. 1928). He was marooned in Brazil on a collecting trip by the outbreak of war in 1939; when Brazil joined the Allies in August 1942 he was interned as an enemy alien but remained in Brazil after the end of hostilities and later became a Brazilian citizen. One of his special achievements was the rediscovery, in 1978, of a population of the (then) Critically Endangered Lear's Macaw *Anodorhynchus leari*. He published a huge volume of work, but is perhaps best known for *Ornitologia Brasiliera-Um Introduçao* published in English as *Birds in Brazil – A Natural History* (Vuilleumier 1995).

PARKER'S ANTBIRD
Cercomacroides parkeri

Described as a member of the large genus *Cercomacra* (and still considered to be so by some authorities); however, the clade including the present species has a number of consistent differences, in morphology, vocalisations, nest architecture and molecular phylogeny, necessitating the erection of a new genus *Cercomacroides* Tello *et al.* 2014; the other members of the clade are *C. nigrescens*, *C. laeta*, *C. tyrannina* and *C. serva* (Blackish, Willis's, Dusky and Black Antbirds).

Type information *Cercomacra parkeri* Graves 1997, La Bodega, north side of Río Negritu, Dpto. Antioquia, Colombia, based on a specimen taken by M. A. Carriker Jr on 16 June 1951.

Alternative names Spanish: Hormiguero de Parker.

Discovery Dusky Antbird *Cercomacra tyrannina* is a polytypic species whose range, with a total of four subspecies, extends from northern Oaxaca in Mexico through to the northern bank of the Amazon in eastern Brazil. In 1997 G. Graves published a careful analysis of museum specimens from the foothills of the Cordilleras Occidental and Central, Colombia, and came to the conclusion that in this region two species are involved, *C. tyrannina* (including the nominate subspecies and *C. t. rufiventris*) and an entirely distinct population which he named Parker's Antbird *C. parkeri*.

Description 13.5–14.5 cm. Monotypic. Sexually dimorphic. Male is extremely similar to *C. tyrannina*, slate-grey above, paler grey below, palest on the chin; wings darker grey with white interscapular patch and two white wing-bars formed by tips of the greater and median coverts, with small white spots on the shoulders;

► Parker's Antbird, male. Cerro Montezuma, Risaralda, Colombia, December 2011 (*Daniel Uribe*).

▼ Parker's Antbird, female. Reinita Cielo Azul, Colombia, January 2010 (*Nigel Voaden*).

rectrices darker grey with narrow white tips; iris rich chocolate-brown; bill blackish; tarsi plumbeous. Differs from *tyrannina* by more olivaceous flanks. Female is more distinctive: crown, upperparts and wings olive-grey with a tawny tinge; wings with cinnamon edgings on coverts; sides of head greyish with paler supercilium; throat and underparts tawny-buff, flanks tinged olive. Distinguished from *tyrannina* by brownish-grey rather than rufous-brown sides to face. In both sexes the tail is longer than in nominate *tyrannina*.

Habitat Borders and undergrowth of wet montane evergreen forest and tall second-growth woodland, including small forest fragments as well as continuous woodland, at 1,100–1,950 m. Tends to occur at higher altitudes than *tyrannina*, with apparently no overlap; this may not always have been the case, however (Graves 1997).

Food and feeding Few data; stomach contents comprise insects, especially beetles, and spiders. Feeds in pairs or family groups, in undergrowth near the ground, sometimes to mid-levels.

Breeding Nest and eggs undescribed. The type specimen taken in June was in "laying" condition; fledglings seen in April and July.

Voice The best field distinction from *C. tyrannina*. Song is a forceful *pur, pee-pee-pee-pi-pi-pi-pr* (Ridgely & Tudor 2009). Both males and females sing, the female song being higher pitched and shorter. Call is a rapid sputter of notes (Zimmer & Isler 2003 HBW 8).

Movements Presumably sedentary.

Range Endemic. Andes of west-central Colombia (west slope of West Andes, northern and eastern slopes of Central Andes and western slope of East Andes); specimens from Dptos Antioquia, Valle del Cauca, Caldas, Bolívar and Risaralda.
Conservation status Currently classified by BirdLife International as Least Concern; the overall range is quite extensive, though it does appear to be declining in numbers due to habitat destruction.
Etymology Named in honour of the late Ted Parker, in Graves' words (1997) "a savant – the most talented field ornithologist of our generation".

MANU ANTBIRD
Cercomacra manu

Shares a clade with six other species; the Rio de Janeiro, Grey, Mato Grosso, Bananal, Rio Branco and Jet Antbirds (*C. brasiliana, C. cinerascens, C. melanaria, C. ferdinandi, C. carbonaria* and *C. nigricans* (Tello *et al.* 2014)).

Type information *Cercomacra manu* Fitzpatrick & Willard 1990, Shintuya, Madre de Dios, Peru.
Alternative names Spanish: Hormiguero del Manu; Portuguese: Chororó-de-manu.
Discovery In 1975 J. Fitzpatrick and D. Willard several times mist-netted an unusual antbird near the Manu River in Manu National Park, Peru. These were tentatively identified as Jet Antbirds *C. nigricans*, a species found patchily from Panama and Colombia to Venezuela and Ecuador, and believed to occur in Amazonia solely on the basis of one specimen taken in 1964 at Balta, Dpto. Ucayali, Peru (O'Neill 1969). Because the taking of specimens was prohibited in Manu National Park, the identity of the mist-netted birds was not confirmed. Subsequently, several specimens were collected outside the Park, allowing the description of a new species of *Cercomacra* antbird; the 1964 specimen also belonged to this new species (Fitzpatrick & Willard 1990).
Description 14–15 cm. Monotypic. Sexually dimorphic. Male, largely blackish all over, slightly greyer on the flanks; outer scapulars white; two wing-bars formed by narrow white edgings to the coverts; small white spots on shoulders; rectrices graduated, black, with white tips. Differs from *C. nigricans* by narrower white tips on rectrices and less jet-black appearance. Female: crown, upperparts and wings olive-brown, with two white wing-bars and small white shoulder-spots; sides of face and underparts grey, the throat obscurely streaked white, the lower flanks brownish-olive; iris pale sandy-brown; bill black above, mottled silvery-grey below; tarsi pale grey.

▲ Manu Antbird, male. Alta Floresta, Mato Grosso, Brazil, August 2016 (*João Quental*).

► Manu Antbird, female. Los Amigos Biological Station (CICRA), Madre de Dios, Peru, July 2003 (*Daniel J. Lebbin*).

Habitat A bamboo specialist (Kratter 1998). Usually found in relatively pure stands of *Guadua* bamboo, tending to be absent if bamboo is much mixed with other vegetation; it does, however, occur in second-growth woodland adjacent to bamboo (Parker & Remsen 1987) and in tall riverine forest (Fitzpatrick & Willard 1990). From 400 m (possibly lower) to 1,200 m.

Food and feeding Forages from 2–15 m up, usually in pairs or in family groups, sometimes briefly joining mixed flocks. Hops actively from perch to perch, sometimes sally-gleans. Stomach contents include caterpillars, Orthoptera and other small arthropods (Fitzpatrick & Willard 1990).

Breeding One nest so far described, in September in Madre de Dios, Peru; in Acre, Brazil, male carrying food either to nest or to juveniles in December. Nest is a pensile pouch, suspended by its rim, woven into 3 mm bamboo branchlets near where they are forked; 3.5 m up in dense bamboo. Male apparently incubating on 11 September. Eggs undescribed (Kratter 1998).

Voice Pairs duet. Distinct from that of the sympatric *C. cinerascens*. Male song is a deliberate harsh *hert-CHUCK-hert-CHUCK* etc. the second syllable heavily accented; a second call given by both sexes is a rapid, slightly descending whinny of 9–13 guttural staccato notes. Also a rapidly repeated *Chut-up CHUT-up* etc. given in syncopation by both sexes (Fitzpatrick & Willard 1990).

Movements Presumably sedentary.

Range Widespread but apparently rather disjunct; more work remains to be done in intervening areas. Peru: south-western Ucayali, northern Cuzco south across Madre de Dios, probably also in northern Puno; northern Bolivia (Pando and La Paz); western Brazil (Acre), also Mato Grosso, Tocantins, Maranhão and Pará (Beadle *et al.* 2003; Kirwan *et al.* 2015a; Whitney 1997).

Conservation status Currently classified by BirdLife International as Least Concern. The overall geographic range is quite large, if apparently (according to present data) discontinuous. Some areas (e.g. Manu National Park) are effectively protected. It may, possibly temporarily, benefit from some human activity as bamboo is frequently the first colonist of disturbed areas (Beadle *et al.* 2003).

Etymology Both the common and the scientific names are derived from Manu National Park, which gets its name from the Manu River.

MANICORE WARBLING ANTBIRD
Hypocnemus rondoni

'Warbling Antbird' *sensu lato* is widespread, ranging from the Guianas to Bolivia; it has been shown to be taxonomically complex, with perhaps six valid species involved (Isler *et al.* 2007). The description of the current species is merely a part of this. Genetically closest to Rondonia Warbling Antbird *H. ochrogyna* (Whitney *et al.* 2013).

Type information *Hypocnemus rondoni* Whitney *et al.* 2013, left bank of Rio Roosevelt in the Municipality of Colniza, Mato Grosso, Brazil.

Alternative names Portuguese: Cantador-de-rondon. Note: Donegan (SACC 2007, Proposal 299) suggests Ant-Warbler as the generic common name for the whole genus.

Discovery During survey work in the vicinity of Manicore, Amazonas, Brazil in 2000, B. Whitney and M. Cohn-Haft recognised that the vocalisations of the local population of Warbling Antbird *H. cantator sensu lato (H. striata)* were distinct from those of other populations. However, material was limited and no specimens could be taken; hence in the major study published in 2007 (Isler *et al.*), no clear conclusions could be drawn, and the form was referred to as "*taxon novem*" (a new form), possibly a subspecies of *H. striata*. Subsequently, further work by Whitney *et al.* (2013) established the vocal distinctiveness of the new form, and this and the taking of a series of specimens allowed the description of a new species, confirmed by biochemical data. These conclusions were unanimously accepted by the SACC (Proposal 588).

Description 11.3 cm. Monotypic. Very similar to other members of the Warbling Antbird complex; sexually dimorphic. Male: head, nape, face mantle and chest black and off-white, with a conspicuous black crown-stripe; whitish supercilium, blackish eyestripe and white-streaked black ear-coverts, throat and chest; lower belly whitish in centre, orange-buff on flanks; back, wings and shoulders olive-brown, except coverts which are blackish with conspicuous white or buffy tips; rectrices dark yellowish-brown; iris brown; bill black above, grey below; tarsi yellowish-grey. Female similar in pattern, but whitish of head replaced by buff, the tips to the lesser coverts buffish and the greater coverts buffish-brown, not blackish.

Habitat *Terra firme* forest, especially where openings have been made by tree-falls etc., where denser understorey occurs; below 500 m.

Food and feeding Forages in understorey, only joining mixed-species flocks when these pass through its territory. Identified stomach contents comprise insects (Coleoptera and Orthoptera).

Breeding Nest and eggs undescribed; nest of Guianian Warbling Antbird *H. cantator* is a pensile pouch, placed quite low down in vegetation; eggs two, very variable in coloration in the various taxa of the *cantator* complex (Zimmer & Isler 2003 HBW 8).

Voice Distinctive in comparison to other species of the *cantator* complex (Isler *et al.* 2007). Song is usually four, sometimes three or five, notes, the first short and screechy, the remainder in successively higher frequencies, tempo more rapid than other '*cantator*' taxa and never ending with raspy notes. Most similar to the song of Spix's Warbling Antbird *H. striata* to the north, but differs significantly from songs of Rondonia Warbling Antbird *H. ochrogyna* and Peruvian Warbling Antbird *H. peruviana* (Whitney *et al.* 2013).

Movements Presumably sedentary.

Range Endemic. Western Amazonian Brazil; southern Amazonas, NE Rondônia and NW Mato Grosso.

Conservation status Not currently classified by BirdLife International. Whitney *et al.* (2013) do not believe it to be presently threatened, but point out that the overall range is very restricted, that much of the habitat is unprotected, and that future hydro-electric schemes may drastically alter the area.

Etymology Named in honour of Candido Mariano da Silva Rondon (1865–1958), a seminal figure in the history of Brazil. Commissioned into the Brazilian Army in 1888, he was involved in the coup d'état which overthrew Pedro II, the last Emperor of Brazil, and established the Brazilian Republic. In the early 20th century he explored huge areas of western Brazil, discovering a number of major rivers. Ahead of his time, he advocated strongly for the rights of the aboriginal peoples of Brazil (his mother was Amerindian), efforts which led to the establishment of a national park for indigenous peoples on the Xingu River. The Brazilian state of Rondônia is named in his honour.

▼ Manicore Warbling Antbird. Left: male, right bank of the Machado River, Rondônia, Brazil, February 2014. Right: female, left bank of the Roosevelt River, Colniza, Mato Grosso, Brazil, August 2011. (*Fabio Schunck*).

ALLPAHUAYO ANTBIRD
Percnostola arenarum

Forms a sister species with Black-headed Antbird *P. rufifrons*; may not be closely related to other *Percnostola* antbirds (Isler *et al.* 2007).

Type information *Percnostola arenarum* Isler *et al.* 2001, Mishana, Río Nanay, Loreto, Peru.

Alternative names Spanish: Hormiguero del Allpahuayo.

Discovery The white-sand habitat of Dpto. Loreto, Peru, has in recent years yielded several new species (Whitney & Álvarez 1998; Álvarez & Whitney 2001). A study of *Percnostola* antibirds in this region revealed a population of birds with plumage and vocalisations distinct from Black-headed Antbird, a widely distributed species ranging from Colombia and Venezuela through the Guianas to NE Brazil. The type specimen was collected in October 1998; however, a poorly preserved specimen taken in 1865 and currently housed in the Natural History Museum, Tring, previously assigned to *P. rufifrons*, was shown on careful examination to be *arenarum* (Isler *et al.* 2001).

Description 14–15 cm. Monotypic, but see below. Sexually dimorphic. Male: generally slate-grey overall, with a darker grey crown, black throat, darker grey shoulders with two white wing-bars and white spots on shoulders. Female: crown, upperparts and ear-coverts dark grey, wing coverts blackish with two light cinnamon-rufous wing-bars and spots on shoulders; underparts largely reddish yellow-brown with centre of throat and centre of belly white; iris grey; bill black with bluish-grey lower mandible; tarsi bluish-grey. Male differs from male *P. rufifrons* in having a dark grey, less blackish, crown, absence of crest, grey irides and shorter tail; female differs in having a clear white central belly patch.

Habitat Understorey of lowland evergreen forest known locally as *varillal*, a dense stunted forest with canopy height of 10 m, not seasonally flooded, on pure white sandy soil; up to 150 m (Zimmer & Isler 2003).

▲► Allpahuayo Antbird, female (top) and male (bottom). Tierra Blanca, Peru, June 2013 (*Fabrice Schmitt*).

Food and feeding Singly, in pairs or family groups, not usually in mixed-species flocks, low down or on the ground. Prey items comprise insects and probably spiders.
Breeding Nest and eggs undescribed.
Movements Presumably sedentary.
Voice Loud song is a rapidly delivered decelerating series of flat-sounding whistles on the same pitch; other calls include a short, upslurred querulous whistle and a short rattle (Zimmer & Isler 2003).
Range Endemic. Confined entirely to Dpto. Loreto, Peru, where it is restricted to white-sand plains in the region of Ríos Tigre, Nanay and Marañón. A second population has been found in *varillal* forest near the Río Morona in western Loreto (Shany *et al.* 2007); these authors suggest that this might represent a different subspecies. Readily found in the Allpanuayo–Mishana Reserve.
Conservation status Currently classified as Vulnerable. Notwithstanding the reserve status of most of its highly specialised habitat, numerous illegal and damaging activities continue, including illegal harvesting of wood, clearing of land for homesteads, etc.
Etymology *Arenarum*, from the Latin 'arena' (sand), reflecting the species' dependence on specific white-sand forest.

STRESEMANN'S BRISTLEFRONT
Merulaxis stresemanni

Probably forms a superspecies with Slaty Bristlefront *M. ater*, but vocally and morphologically distinct. Probably (along with the near-mythical Kinglet Calyptura *Calyptura cristata* and Alagoas Foliage-gleaner *Philydor novaesi*) the most endangered species in South America today.

Type information *Merulaxis stresemanni* Sick 1960, based on a specimen taken in the 1830s near Salvador, Bahia, Brazil.

Alternative names Portuguese: Entufado-baiano.
Discovery Described by H. Sick in a comprehensive study of Brazilian Rhinocryptidae published in 1960, based on two specimens only, the first a male taken in Bahia state by F. Kaehne sometime between 1831 and 1838, the second a female, also collected in Bahia, in 1945. Sick noted the significant differences between these specimens and *M. ater*, on which he based his type description (Sick 1960). Since that time a third specimen, a female, has surfaced in the Senckenberg

▼ Stresemann's Bristlefront, male. Macarani, Bahia, Brazil, September 2016 (*Ciro Albano*).

Museum, Frankfurt am Main, Germany; it probably dates from the first half of the 19th century, but as to location the label merely states 'Brasilien'.

Description 19.5 cm. Monotypic. Sexually dimorphic. Male uniformly slaty-black, with a crest of stiff, narrow upward-pointing feathers on the forehead; tail long and graduated; iris dark brown; bill black; tarsi dark brown. Female: crown, sides of head and upper back slaty, lower back, wing-coverts and rump very dark brown; tail black; underparts bright rusty-red, washed brown at sides; frontal bristles as male. A bird in apparently female plumage was observed singing in 2008, suggesting that full adult male plumage is not acquired until the second year (Whitehouse & Ribon 2008). Distinguished from *M. ater* by significantly larger size, lack of brownish wash on flanks and heavier bill. Female has richer chestnut underparts than those of *M. ater*.

Habitat Limited recent records suggest that it inhabits forest somewhat drier than adjacent humid valley forest, below 200 m.

Food and feeding Feeds on or near ground, or on fallen tree trunks. Seen to prey on insects.

Breeding One nest so far described, found by D. Pioli and G. Malacco in the Mata do Passarinho Reserve of the Fundaçao Biodiversitas. It was situated at the end of a tunnel, about 2 m long and about 1 m up, in an earth bank, partly covered by hanging vegetation, in October 2012. No details of eggs, clutch size etc., so far published (Anon. 2013e).

Voice Very different from *M. ater*. Characteristic falling series of musical whistled notes, slightly ascending at the end, of 10–12 seconds duration. Also semi-musical *tink* call.

Movements Presumably sedentary.

Range Endemic. Original two specimens from Bahia state, Brazil, where it was rediscovered in 1995. It now no longer appears to persist at this last location. However, it has since been found in one location in Minas Gerais state on the border with Bahia.

Conservation status Currently classified by BirdLife International as Critically Endangered. It presumably was never common; despite the fact that extensive collecting occurred in Atlantic Brazil in the 19th century, before the massive habitat destruction that now prevails, only three specimens are known. For more than 100 years it was not encountered, until the single female noted above was collected in 1945; it no longer occurs at that location. In 2004, a tiny population was discovered on the borders of Bahia and Minas Gerais. The total range of the species appears to be 35 km² or less; the habitat is very vulnerable to further degradation by clearance, fires etc. The current world population may be as low as ten individuals "stranded in an island of forest" (BirdLife International website, May 2016). Several substantial land purchases have been made recently to protect this and other endangered species of birds and animals.

▲ Stresemann's Bristlefront, female. Macarani, Bahia, Brazil, December 2016 (*Ciro Albano*).

Etymology Named in honour of Erwin Stresemann (1889–1972), a distinguished German ornithologist whose name has been attached to the vernacular or scientific names of a number of species or subspecies, from parrots and lories to orioles and owls.

DIADEMED TAPACULO
Scytalopus schulenbergi

A relatively distinctive species (for a *Scytalopus* tapaculo), probably closest to the Puna Tapaculo *S. simonsi*; although somewhat similar in plumage to Silver-fronted Tapaculo *S. argentifrons* of Costa Rica and Panama, probably not closely related (Krabbe & Schulenberg 2003 HBW 8).

Type information *Scytalopus schulenbergi* Whitney 1994, 4 km west of Chuspipata on main road between La Paz and Coroico, Province of Nor Yungas, Dpto. La Paz, Bolivia.

Alternative names Spanish: Churrín Diademado, Tapaculo Diademado.

Discovery In February 1992, while conducting an avifaunal survey in humid temperate forest on the eastern Andean slope near La Paz, Bolivia, B. Whitney observed and tape-recorded an unknown tapaculo.

▲ Diademed Tapaculo, female. Cotapata National Park, Dpto. La Paz, Bolivia, November 2014 (*Samantha Klein*).

Along with O. Rocha of the Museo Nacional de Historia Natural, La Paz, he returned in March 1993 and obtained a series of the new taxon, along with further extensive tape recordings. This was sufficient to allow the unequivocal description of a new species.

Description 10 cm. Monotypic. A small dark tapaculo with, in the male, a distinctive white forecrown. Upperparts and tail largely dark grey, back washed with dull brown; rump dull orange-rufous with dusky bars; forecrown and supercilium silvery-white, contrasting with narrow black mask on lower forehead; upperparts grey, lightest on throat, the lower belly with silvery sheen; flanks and vent usually dull orange-rufous. Female similar to male but silvery facial markings less prominent and more brown above; iris dark brown, upper mandible black, paler at base of lower mandible; tarsi dark yellow-brown. Differs from other tapaculos in the general area in its facial markings and less well-defined dark markings on the lower flanks.

Habitat Undergrowth of humid forest, especially with *Chusquea* bamboo, up to the treeline (2,750–3,400 m). Replaced at higher altitudes by Puna Tapaculo *S. simonsi*, at lower by Trilling Tapaculo *S. parvirostris*.

Food and feeding Forages on ground or in lower levels of dense vegetation. Stomach contents insects (Whitney 1994).

Breeding One nest described, in shape typical of the genus, a spherical ball of moss and lichens with a few twigs, with a side-entrance, located just under the ground in a slope covered with moss and leaf-litter. Breeding season presumed to be September–January; in Peru, nestlings in mid-October (Krabbe & Schulenberg 2003 HBW 8).

Voice Male song loud and distinctive, a trill of downstroke notes lasting 7–15 seconds, beginning with a few spaced notes that accelerate with falling pitch and increasing volume until becoming level (Krabbe & Schulenberg 2003 HBW 8). Whitney (1994) gives a detailed comparative analysis of vocalisations of this and related species.

Movements Presumably sedentary.

Range Andes from SE Peru (Cuzco, Puno) to central Bolivia (La Paz, Cochabamba).

Conservation status Currently classified by BirdLife International as Least Concern. Although the range is restricted, geographically and altitudinally, in many areas it appears to be quite common and the population appears to be fairly stable (BLI SF 2014).

Etymology Named in honour of Dr Thomas S. Schulenberg, for many years one of the foremost field and taxonomic ornithologists of the Andean region of South America, instrumental in the discovery and naming of numerous new species.

BOA NOVA TAPACULO
Scytalopus gonzagai

An endangered montane species of tapaculo, separated from the more widely distributed Mouse-coloured Tapaculo *Scytalopus speluncae* by careful analysis of morphometrics, plumage colour and vocalisations, backed up by genetic analysis.

Type information *Scytalopus gonzagai* Mauricio *et al.* 2014, Serro do Rio Preto, municipality of Iguaí, Bahia, Brazil.

Alternative names Bahia Mouse-coloured Tapaculo, Gonzaga's Tapaculo. Portuguese: Macuquinho-preto-baiano.

Discovery Mouse-coloured Tapaculo is a quite well-known species with a wide distribution in SE Brazil, marginally into Argentina. Isolated populations of tapaculos, initially ascribed to this species, were found in two mountain ranges (Boa Nova and Serra das Lontras) in Bahia state, Brazil, in the early 1960s and in 1999. Both locations were allopatric from the range of *speluncae*. Careful study of specimens from these populations, as well as detailed analysis of vocalisations, indicated that a new species was involved. This has been unanimously accepted by the SACC (Proposal 643).

▲ Boa Nova Tapaculo. Boa Nova, Bahia, Brazil, April 2008 (*Ciro Albano*).

Description 12 cm. Monotypic. A typical *Scytalopus* tapaculo in form, being small, plump and short-tailed. Above and below largely blackish-grey, the chin rather paler; flanks brown with blackish-grey barring; crissum brown with blackish bars; bill black; tarsi and toes black with yellowish claws. Differs from *speluncae* in having barred flanks, a relatively shorter tail and more robust bill. Biochemical sequencing of tissue samples by Mata *et al.* (2009).

Habitat Undergrowth of humid, montane evergreen forest, mostly undisturbed primary forest with canopy 15–25 m, but also a mixture of primary and disturbed forests; rarely wholly in disturbed forest; 660–1,140 m.

Food and feeding Feeding behaviour typical of genus; hops on ground over rocks and fallen logs. No data on diet.

Breeding Nest and eggs undescribed.

Voice The song differs from that of all populations of *speluncae* in being slower in pace; call-note *kreew* also differs. Both sexes sing. Mauricio *et al.* (2014) give a very detailed comparative analysis of the vocalisations of two populations of *speluncae* and of the new species.

Movements Presumably sedentary.

Range Endemic. Confined to two isolated montane massifs in SE Bahia state, eastern Brazil; five localities in the municipalities of Boa Nova and Iguaí and, disjunctly, in the municipality of Arataca. These locations are separated by unsuitable lowland habitat.

Conservation status Not yet classified by BirdLife International, but the authors of the type description recommend a status of Endangered. The overall range is very small (5,885 ha) and the two main centres of occurrence are isolated; a preliminary population estimate is 2,883 individuals. A little more than half of the total range is protected in two national parks (Serra das Lontras and Boa Nova). The remaining area is under great pressure from clandestine timber extraction and clearing for agriculture.

Etymology Named in honour of Luiz Antonio Pedreira Gonzaga, a Brazilian ornithologist who inaugurated the ornithological exploration of the forests of Bahia, and who was himself responsible for naming two new species (a flycatcher and a spinetail). It has been pointed out (J. V. Remsen, in SACC Proposal 643) that the proposed original English name of Bahia Mouse-coloured Tapaculo is a guaranteed source of confusion, since there already is both a Bahia Tapaculo (*Eleoscytalopus psychopompus*) and a Mouse-coloured Tapaculo (*Scytalopus speluncae*); a suitable name avoiding this is Boa Nova Tapaculo.

MARSH TAPACULO
Scytalopus iraiensis

A highly restricted and endangered tapaculo, most closely related to the more widely distributed Mouse-coloured Tapaculo *S. speluncae*. Given the very conservative morphology of the genus, relationships are most reliably determined by genetic investigation (Mata *et al.* 2009).

Type information *Scytalopus iraiensis* Bornschein *et al.* 1998, right bank of Rio Irai, Quatro Barras, Paraná, Brazil.

Alternative names Tall-grass Wetland Tapaculo, Wetland Tapaculo. Portuguese: Macuquinho-da-várzea.

Discovery In September 1991 M. R. Bornschein and M. Pichorim were investigating the avifauna of humid habitat along the Rio Irai in Paraná State, Brazil, when they were surprised to see three individuals of Sickle-winged Nightjar *Eleothreptus anomalus*, a little-known and endangered species. Dr Bornschein and B. L Reinert returned several times to investigate the area further when they observed small dark birds flying heavily just above the fields; on obtaining a specimen this was obviously a *Scytalopus* tapaculo. During further work up to 1998, five specimens were taken and tape recordings were made; careful morphological examination and vocal analysis showed clearly that the birds were quite distinct from Mouse-coloured Tapaculo *S. speluncae* and were thus described as a new species *S. iraiensis*.

Description 10.5 cm. Monotypic. A dark grey, rather long-tailed tapaculo, darker above, dark ashy below; iris dark brown, bill blackish, tarsi dull flesh-colour.

Habitat Unlike others in the genus, uniquely tied to marshy habitats. Favours tall, dense, seasonally flooded grassland dominated by a fairly uniform cover of spike-rush (*Eleocharis*), other Cyperaceae and grasses. Locally, also patchy tussocks reaching 20–40 cm above water; 750–950 m (Krabbe & Schulenberg 2003 HBW 8). Kiemann & Vieira (2013) identify four types of habitat vegetation and four ecological zones.

Food and feeding Perch-gleans from low vegetation; climbs up and down stems. Prey items comprise arthropods (cockroaches, bugs, beetles and mosquitoes) (Bornschein *et al.* 1998).

Breeding Nest and eggs undescribed. A December specimen had partially developed gonads; May–June specimens were not in breeding condition.

Voice Song consists of a series of churred notes, increasing in pitch, lasting about 30 seconds; Bornschein *et al.* (1998) give a very detailed analysis of vocalisations.

Movements Presumably sedentary.

Range Endemic. Quite widespread but very disjunct; originally thought to be endemic to a small area of Paraná state, but since discovered in Rio Grande do Sul and several locations in Minas Gerais (Bornschein *et al.* 2001, Vasconcelos *et al.* 2008). More recently, Mlíkovský (2009) discovered nine adult and two probable juvenile specimens in the Institute of Zoology of the Polish Academy of Sciences, Warsaw, collected in 1922 by the Polish zoologist T. Chrostowski (1878–1923) and misidentified as *S. speluncae*. These came from five locations in Paraná, giving a westward range extension

◄ Marsh Tapaculo. São José dos Pinhais, Paraná, Brazil, November 2012 (*Sergio Gregorio*).

of about 100 km from the locations discovered by Dr Bornschein. Whether these populations are still extant is not known.

Conservation status Currently classified by BirdLife International as Endangered, with an estimated population in the range of 350–1,500 individuals; however, Kiemann & Vieira (2013) examined occurrence records and give their estimates of population size as 31,584 ±7,140 mature individuals. Although the total known range is quite extensive, it inhabits highly fragmented and isolated patches of habitat. Presently known from about 20 locations. Main threats are habitat destruction by dam construction (indeed, the type locality is now flooded), drainage for agricultural uses, sand extraction and burning.

Etymology *Iraiensis*, from the Rio Irai in Paraná state.

DIAMANTINA TAPACULO
Scytalopus diamantinensis

Most closely related to Rock Tapaculo *S. petrophilus* of SE Bahia state. The taxonomy of *Scytalopus* species in Brazil is still the subject of some debate and differing opinions (see Proposal 329 to the SACC). Treated as a valid species in HBW SV (2013).

Type information *Scytalopus diamantinensis* Bornschein *et al.* 2007, Capão do Vale, municipality of Ibicoara, Bahia, Brazil.

Alternative names Portuguese: Tapaculo-da-chapada-diamantina.

Discovery As part of a study to investigate the complex and confusing taxonomy of the Brazilian tapaculos, M. Bornschein and his colleagues mounted an expedition to the Chapada Diamantina in Bahia state during August 2006. During this time specimens were taken and tape recordings of different types of vocalisations made. Careful analysis of this material, genetic and morphological, and comparison with equivalent data from other Brazilian members of the genus led to the conclusion that a new endemic species was involved.

Description 10–11cm. Monotypic. Predominantly grey with a blackish tail; crown, nape and back blackish-grey, rump blackish-grey with brown and black barring; underparts medium-grey, the chin and throat paler; centre of belly whitish; flanks and lower belly cinnamon-brown with blackish bars; iris dark-brown; bill black with paler base; tarsi yellow-brown. Female is browner above. Identical in plumage to Planalto Tapaculo *S. pachecoi* as adults; however, immature males of *diamantinensis* are separable from *pachecoi* by the barring pattern on the upperwing-coverts.

Habitat Forest, both mature and secondary, including early successional stages after logging on terrain varying from almost flat to steeply sloping, often with bracken (*Pteridium aquilinum*), at 800–1600 m.

Food and feeding Feeds on the ground and in lower levels of vegetation, up to 2 m above the ground.

► Diamantina Tapaculo. Ibicoara, Bahia, Brazil, November 2016 (*Ciro Albano*).

Breeding Nest and eggs undescribed.

Voice Song is not consistently distinguishable from that of *S. pachecoi*; usually faster-paced (with some overlap) and lower pitched (with slight overlap with *pachecoi* and wide overlap with *Scytalopus* sp. nov). *Diamantinensis* does, however, have a diagnostic call not found in other tapaculos, a single *tcheep*. Bornschein *et al.* give a detailed comparative analysis of vocalisations.

Movements Presumably sedentary.

Range Endemic. Currently known exclusively from the Chapada Diamantina, Bahia; the nine known locations are all situated in the eastern part of this range, from about 12° 00' S to 13° 41' S and from 41° 13' W to 41° 53' W, i.e. within a rectangle of approximate dimensions 185 km north to south and 75 km east to west (Bornschein *et al.* 2007).

Conservation status Currently classified by BirdLife International as Near Threatened. The total range is fairly small and habitat disturbance, by clearing for agriculture, woodcutting and fires, is widespread. Part of the species' range is protected by the Chapada Diamantina National Park and the Malimbu-Iraquara State Environmental Protection Area; does also appear to have tolerance for some level of disturbance.

Etymology Both the vernacular and scientific names are derived from the Chapada Diamantina.

ROCK TAPACULO
Scytalopus petrophilus

Probably most closely related to Brasilia Tapaculo *S. novacapitalis*, itself only described by H. Sick in 1958. The taxonomy of the Mouse-coloured Tapaculo *S. speluncae* group remains extremely confused. Recent work (Mata *et al.* 2009) shows that *speluncae* and the present species are not even sisters (Whitney *et al.* 2010).

Type information *Scytalopus petrophilus* Whitney *et al.* 2010, Serra da Piedad, Caeté, Minas Gerais, Brazil.

▼ Rock Tapaculo. Serra da Mantiqueira, Passa Vinte, Minas Gerais, Brazil, October 2012 (*Bruno Rennó*).

Alternative names Espinhaço Tapaculo. Portuguese: Tapaculo serrano.

Discovery A population of rather pale grey *Scytalopus* tapaculos has been known to occur in upland areas of the Espinhaço Range of Minas Gerais, Brazil, since 1989. However, considerable confusion (and different interpretations of the same evidence) prevailed for some years, with the new form sometimes attributed to the Brasilia Tapaculo *S. novacapitalis*. In June 2010 B. Whitney and colleagues published a detailed analysis of the plumage and vocalisations of this population, describing a new species endemic to a limited area of Minas Gerais state.

Description 10.5 cm. Monotypic. A typical *Scytalopus* tapaculo in form; plumage pale grey, darker on the crown, lighter on the throat; belly whitish, sometimes with light grey tinge; flanks and lower underparts yellowish-brown with black barring; iris dark brown; bill black, tipped grey; tarsi brownish-cream.

Habitat Quite varied; open rocky areas with low trees, shrubs and grasses; elfin forests and cloud-forests in canyons; partially deforested and burnt areas; and in second-growth semi-deciduous woodland. Mostly 900–2,100 m.

Food and feeding Forages mostly on the ground, alone or in pairs. Stomach contents of five specimens was insects.

Breeding Nest and eggs undescribed. Birds in breeding condition September to early December.

Voice Song very similar to those of Planalto Tapaculo *S. pachecoi* and Diamantina Tapaculo *S. diamantinensis*; a long repetition of one note, often lasting one minute or more. Contact call is diagnostic, a short downslurred *pzeeu*.

Movements Presumably sedentary.

Range Endemic. Minas Gerais, Brazil, mostly in the Espinhaço Range; more recently in the north of the state (Oliveira Mafia 2015); south into São Paulo (de Souza 2013). Total extent of records, at least 700 km north to south and 200 km east to west. In southern Minas Gerais overlaps with *S. speluncae.*

Conservation status Presently not classified by BirdLife International. Appears to be rather patchily distributed within its limited range, parts of which are subject to planting of biologically-inert *Eucalyptus* plantations.

Etymology *Petrophilus*, from Greek 'rock-loving', from one of its major habitats.

PLANALTO TAPACULO
Scytalopus pachecoi

The late discovery of this species was caused by confusion with the more widespread Mouse-coloured Tapaculo *S. speluncae*, with which it is sympatric. It is possible, however, that its closest relationships may not be with that species, but with other *Scytalopus* populations found in the Brazilian interior (Mauricio 2005).

Type information *Scytalopus pachecoi* Mauricio 2005, Cerro das Alunas, Capão do Leão, Rio Grande do Sul, Brazil.

Alternative names Spanish: Churrín de Pachero; Portuguese: Tapaculo ferrerinho.

Discovery The taxonomy of tapaculos of the genus *Scytalopus* has been the source of endless debate (for a detailed account of the general confusion and fluctuating opinions surrounding this genus, see Mauricio 2005). This author conducted a detailed study of the plumage and other morphological features of the tapaculos of SE Brazil, backed up by careful analysis of vocalisations, and realised that an unrecognised species was involved. However, all authorities agree that the final word on tapaculo systematics has yet to be written.

Description 12 cm. Monotypic. A typical *Scytalopus* tapaculo; upperparts, chest and upper belly rather uniformly pale grey, slightly paler on the throat, wings and tail slightly darker; flanks and thighs buff or rufous, conspicuously barred with blackish; iris dark brown, bill black, tarsi brownish. Much lighter on the underparts than all populations of *S. speluncae.*

Habitat Primary and secondary forest, from lowlands to upland *Araucaria* forest, probably favouring small stream valleys and bamboo tangles; from near sea level to 1,500 m (2013 HBW SV).

Food and feeding Forages in forest understorey, especially within areas of terrestrial bromeliads and bamboo tangles; prey is arthropods (2013 HBW SV).

Breeding Nest and eggs undescribed.

Voice For a detailed analysis, see Mauricio (2005). The song is characteristic in having an accelerated trill-like ending (also shared with the widely allopatric Brasilia Tapaculo *S. novacapitalis* and a possibly further undescribed form), and a unique monosyllabic semimetallic contact call.

Movements Probably largely sedentary, although local dispersal between forest patches likely (2013 HBW SV).

Range Three disjunct populations. Serra do Sudeste, southern Rio Grande do Sul, Brazil; in the Planalto above the Serra Geral escarpment in NE Rio Grande do Sul and adjacent SE Santa Catarina; and in northern and eastern Misiones, Argentina, and in adjacent Rio Grande do Sul. In fact there are numerous specimens, some dating back to 1919, from Argentina and Brazil which were misidentified as *S. speluncae* (e.g. Bertoni 1919; Giai 1951). Probably also occurs in western Paraná and Santa Catarina. Reports from Dpto. Caazapa, Paraguay (identified as *speluncae*) (Brooks *et al.* 1995) require further investigation.

▼ Planalto Tapaculo. Urupema, Santa Catarina, Brazil, January 2012 (*João Quental*).

Conservation status Currently classified by BirdLife International as Least Concern. Although the range is quite restricted and the population may be declining, the thresholds for Near Threatened status do not appear to be approached.

Etymology Named in honour of José Fernando Pacheco, in view of his major contributions to the ornithology of Brazil in general and the taxonomy of *Scytalopus* tapaculos in particular. The English name recognises the distribution in the Planalto Meridional; the Portuguese is the diminutive derived from 'ferreiro', a worker in iron, from the birds' metallic call-note.

CHOCO TAPACULO
Scytalopus chocoensis

Formerly included within Nariño Tapaculo *S. vicinior* but clearly differentiated by vocalisations.

Type information *Scytalopus chocoensis* Krabbe & Schulenberg 1997, El Placer, Esmeraldas, Ecuador.

Alternative names Spanish: Churrín del Chocó, Tapaculo del Chocó.

Discovery Nariño Tapaculo *S. vicinior* was described as a race of Pale-throated Tapaculo *S. panamensis* by Zimmer (1939). During the 1980s, fieldwork in Ecuador demonstrated that two taxa were involved, differentiated vocally and elevationally, although extremely similar in plumage. Since the specific name *vicinior* referred to the higher altitude population, a name was required for the low altitude species, which was provided by the 1997 monograph of Krabbe and Schulenberg.

Description 11 cm. Monotypic. A fairly small tapaculo with relatively heavy bill and dark-barred brown flanks. Male: upperparts dark grey, crown and mantle with indistinct dark feather tips; remainder of upperparts dark grey, washed with dark brown; rump brown, rump and uppertail-coverts barred dusky; underparts grey, paler on throat; flanks, lower belly and undertail-coverts dark reddish-brown, barred blackish. Female similar, but brownish wash to upperparts more extensive; iris dark brown; bill black or blackish; tarsi blackish-brown.

Habitat Dense undergrowth of wet, mainly primary forest, occasionally forest borders, at 250–1,250 m; in Panama, 1,340–1,465 m. Over most of its range replaced at higher altitudes by *S. vicinior*, without apparent overlap (2013 HBW SV).

Food and feeding Feeds on the ground. Stomach contents of specimens comprised unidentified insects, in one instance beetles.

Breeding One nest described from Darién, Panama, at an altitude of 840 m. The nest was a spherical ball of fine dark rootlets and some moss, of outside and inside diameter about 12 cm and 8.5 cm, respectively, with a side-entrance hole about 4 cm in diameter. The nest was situated on the ground, concealed by overhanging dead palm leaves. On 15 August it held two young; fed

◄ Choco Tapaculo, male. La Union Road, Esmeraldas province, Ecuador, May 2013 (*Nick Athanas*).

by both parents (Christian 2001). Females with active gonads collected in February.

Voice Song is a series of sharp, high-pitched and well-enunciated notes introduced by a stutter *p-d-d-d-pi-pi-pi-pi...* lasting some 15–25 seconds. Scold-note is sharp, high-pitched 3–8 notes *chiu-chiu-chiu* (Ridgely & Greenfield 2001).

Movements Presumably sedentary.

Range NW Ecuador (eastern Esmeraldas and south-western Imbabura), western Colombia north to Darién (Cerro Pirre), Panama.

Conservation status Currently classified by BirdLife International as Least Concern. Although the species' range is geographically restricted, populations appear to be stable. Occurs in several protected areas in Ecuador and Colombia.

Etymology *Chocoensis,* derived from the Chocó biographical region, itself derived from the Department of Chocó, western Colombia.

MAGDALENA TAPACULO
Scytalopus rodriguezi

A vocally distinct species, perhaps most closely associated with *S. robbinsi* of south-eastern Ecuador and *S. stilesi* of northern Colombia (2013 HBW SV); possibly more than one species is involved.

Type information *Scytalopus rodriguezi* Krabbe *et al.* 2005, Finca Merenberg Natural Reserve, San Agustín municipality, Dpto. Huila, Colombia.

Alternative names Upper Magdalena Tapaculo (nominate race). Spanish: Churrín de Alto Magdalena, Tapaculo de Alto Magdalena.

Discovery As long ago as 1980, R. Ridgely and S. Gaulin, while investigating the avifauna of the Finca Merenberg in the Cordillera Central, Dpto. Huila, Colombia, regularly heard tapaculo vocalisations that they were unable to ascribe to any definite species (Ridgely & Gaulin 1980). In 1986, B. Whitney also tape-recorded a *Scytalopus* that neither he nor N. Krabbe was able to identify. Unfortunately, the endemic unrest in that part of Colombia made the area too dangerous to visit for a number of years. However, by 2002 the situation had stabilised sufficiently to allow further field-work. A. Cuervo was able to make recordings of the *confusus* subspecies of White-browed Tapaculo *S. atratus,* ruling out that species as the source of the mysterious calls. Finally, in February 2003, N. Krabbe and his colleagues were able to visit Finca Merenberg, where a number of individuals in song were recorded and three male specimens were taken. On this basis a new species was described (Krabbe *et al.* 2005). In 2013, Donegan *et al.* described a population geographically very disjunct from the Finca Merenberg birds, in the Serranía de los Yariguíes, Santander, some 580 km to the north. Although they pointed out that, using criteria such as those established by Isler *et al.* (1998), "there is a reasonable case to rank [this population] as a species",

► Magdalena Tapaculo. Reinita de Cielo Azul Reserve, San Vicente de Chucurí, Dpto. Santander, Colombia, June 2012 (*Dubi Shapiro*).

they took a conservative viewpoint in describing it as a race of *rodriguezi*, namely *S. r. yariguiorum*.

Description 11.5 cm. Monotypic. Essentially similar to a number of montane tapaculos of NW South America; plumage largely darkish grey, slightly paler on the throat, with lower flanks and rump cinnamon, barred blackish; iris dark brown; bill black; tarsi dusky-brown. *S. r. yariguiorum* has a darker mantle, smaller overall mass and a shorter tail (Donegan *et al.* 2013).

Habitat Dense understorey of tall, humid forests at 1,900–3,000 m (Donegan *et al.* 2010).

Food and feeding Feeds low down or on the ground; prey items include insects, especially beetles.

Breeding Nest and eggs undescribed. Does not breed January–February.

Voice Song is one of the simplest of any tapaculo, a single note, repeated 4–5 times per second, most commonly in bouts of 2–5 or more phrases, the phrases themselves lasting 2–60 seconds. Krabbe *et al.* (2005) provide a detailed analysis of vocalisations. The song of *S. r. yariguiorum* is distinct, being lower-pitched (Donegan *et al.* 2013).

Movements Presumably sedentary.

Range Endemic. The nominate race is currently known with certainty from two locations in Dpto. Huila, Colombia. *S. r. yariguiorum* has so far been found in the Serranía de los Yariguíes, Dpto. Santander.

Conservation status The nominate race is currently classified by BirdLife International as Endangered. Apparently suitable habitat remaining is estimated at only 169 km^2, and deforestation continues, even within the Finca Merenberg Natural Reserve. Efforts to create a protected area in the Serranía de los Minas, the species' stronghold, are underway. Total population is currently placed in the range of 1,500–7,000 mature individuals. *S. r. yariguiorum* is also restricted geographically, and is considered to be in danger of extinction (Anon. 2017). Part of its range is protected by the Reserva Natural de las Aves Reinita Cielo Azul, of the ProAves Foundation, Colombia.

Etymology Named in honour of José Vicente Rodríguez Marecha, "in recognition of his dedication and contribution to ornithology and conservation in Colombia" (Krabbe *et al.* 2005).

STILES'S TAPACULO
Scytalopus stilesi

Another tapaculo almost identical in plumage to several other species in the region, but with distinctive vocalisations. It is suggested that its closest relatives are *S. robbinsi* and, more distantly, *S. rodriguezi*.

▼ Stiles's Tapaculo. Los Yarumos Ecopark, Manizales, Caldas province, Colombia, June 2009 (*Daniel Uribe*).

Type information *Scytalopus stilesi* Cuervo *et al.* 2005, Finca Canales, Vereda Cajamarca, municipality of Amalfi, Dpto. Antioquia, Colombia.

Alternative names Spanish: Churrín de Stiles, Tapaculo de Stiles.

Discovery Over a period of some ten years, a number of ornithologists observed and tape-recorded a *Scytalopus* tapaculo in mid-elevation humid forests in the Cordillera Central of Colombia. After examining recordings of these vocalisations, N. Krabbe suspected that an undescribed species might be involved. More recent fieldwork in Dpto. Antioquia by A. M. Cuervo resulted in the collection of eight specimens and the making of further recordings. Analysis of these, backed up by genetic studies on tissue samples, confirmed that a new species was involved (Cuervo *et al.* 2005).

Description 11 cm. Monotypic. Extremely similar to several other tapaculo species of the region; largely dark grey above and below, darkest on wings, tail and back, the throat paler; rump brown with indistinct darker bars; flanks and lower belly bright brown with blackish scalloping; iris dark brown; bill black with horn-coloured base to lower mandible; tarsi greyish-brown. Female paler grey, especially on the throat; more brown on wings; tarsi more pinkish. Larger and less blackish than Blackish Tapaculo *S. latrans*; smaller than local race (*confusus*) of White-crowned Tapaculo *S. atratus*, and lacking white crown; smaller and longer-billed than Spillmann's Tapaculo *S. spillmanni*.

Habitat Cloud-forest with many palms and ferns, at 1,420–2,130 m. Often separated from other sympatric or near-sympatric species by elevation and habitat.

Food and feeding Feeds on or near ground in dense understorey; prey items comprise arthropods.

Breeding Nest and eggs undescribed. Birds with enlarged testes in January and February; observed feeding fledglings in June and July.

Voice Diagnostic. Song is a series of monotonous *churrs*, from 3–5 seconds, sometimes up to 15 seconds in duration; faster and lower pitched than those of El Oro Tapaculo *S. robbinsi*. Females may sing in duet with males. Calls two or three syllables, *cu-wi?*, *cu-cui-wi?* Cuervo *et al.* (2005) give detailed analysis of vocalisations with comparisons of those of adjacent species.

Movements Presumably sedentary.

Range Endemic. Northern half of the Cordillera Central of Colombia, in Dptos Antioquia, Caldas and Risaralda.

Conservation status Classified by BirdLife International as Least Concern. Although much habitat destruction has occurred within its range, the species has a wide geographic distribution and appears to be quite common in some locations.

Etymology Named in honour of F. Gary Stiles. Dr Stiles has made major contributions to Neotropical ornithology, initially in Central America where he was the co-author, with Alexander Skutch, of the *Guide to the Birds of Costa Rica* (1989); more recently he has worked in Colombia and northern South America.

TATAMA TAPACULO
Scytalopus alvarezlopezi

A species very similar in plumage to several other Colombian *Scytalopus* tapaculos, the distinctions being rather subtle; vocally very distinctive. Forms part of a clade that includes El Oro Tapaculo *S. robbinsi*, and the Colombian species, Magdalena Tapaculo *S. rodriguezi* and Stiles's Tapaculo *S. stilesi*.

Type information *Scytalopus alvarezlopezi* Stiles *et al.* 2017, 8 km NE of Geguadas, Alto de Pisones, municipality of Mistrató, Risaralda, Colombia.

Alternative names Alto Pisones Tapaculo. Spanish: Tapaculo de Tatamá.

Discovery In June 1992 F. G. Stiles, in NW Dpto. Risaralda, Colombia, repeatedly heard churring trills, which he initially ascribed to a frog. Shortly afterwards he observed a tapaculo which he realised was the source of the calls. Subsequently he was able to make recordings and obtain an adult male specimen. Comparison of this specimen and its vocalisations with reference material suggested that a new species might be involved. In 2004 C. D. Cadena performed a biochemical examination, using a toepad of the Stiles specimen, which indicated that the bird was probably an undescribed species, most closely related to *S. robbinsi* of Ecuador and *S. rodriguezi* and *S. stilesi* of Colombia. Dr Stiles nevertheless decided to wait for more material before describing a new species. Finally, in April 2015 (the species appears to be extraordinarily shy and retiring, even by the extreme standards of *Scytalopus* tapaculos), a second male paratype was mist-netted, allowing the description of the species.

Description 11.5 cm. Monotypic. A typical *Scytalopus* tapaculo in form, medium-sized, black above with the rump slightly tinged dark brown; below dark greyish-black, the rear flanks, extreme lower abdomen and crissum broadly and indistinctly barred black and dark rufous; primaries and tail dark brownish-black; iris dark brown, bill black with pale flesh-white gape; tarsi and feet dark brown. Female apparently very similar (Stiles *et al.* 2017).

Habitat Dense understorey vegetation in cloud forest

▼ Tatama Tapaculo. Cerro Montezuma, Risaralda, Colombia, April 2015 (*Julian Heavyside*).

ravines and slopes between 1,300 m and 1,750m on Pacific slopes; also locally on east-facing slopes at 2,000–2,200 m, where the ridge-line does not exceed 2,200–2,300 m. Does not appear to occur in second growth. In the Cerro Montezuma, sandwiched altitudinally between Choco Tapaculo *S. chocoensis*, which occurs up to about 1,300 m, and Narino Tapaculo *S. vicinior*, which takes over from 1,750 m to 2,100 m.

Food and feeding Appears to be very terrestrial, walking and hopping along the ground, briefly fluttering up to 50 cm to capture arthropod prey from foliage. Stomach contents contained small beetles and a spider.

Breeding Nest and eggs undescribed. The two male specimens, taken on 2 April and 6 June, had moderately enlarged testes.

Voice Characteristic and distinct, both from its closest relatives and from other nearby tapaculos. Song is a "seemingly endless, machine-like series of short phrases of about 7–9 nearly identical notes" (Stiles *et al.* 2017, who give a very detailed analysis of vocalisations, with comparative sonograms of this and other closely related species).

Movements Presumably sedentary.

Range Endemic. Based on vocalisations, the Pacific slope of the Western Andes (locally on some eastern-facing slopes) of Colombia, in western Antioquia, Chocó, Risaralda and Valle (Stiles *et al.* 2017).

Conservation status A restricted-range species which does not appear to tolerate significant habitat modification. Nevertheless, Stiles *et al.* (2017) do not suggest any classification higher than, at the most, Vulnerable, since in certain areas it appears to be common and its habitat, at least in the Tatamá region, is fairly continuous and for the most part not currently threatened.

Etymology Named in honour of Humberto Álvarez-López, the "dean of Colombian ornithology", in recognition of his "many contributions to the knowledge and study of this country's birds over nearly half a century".

EL ORO TAPACULO
Scytalopus robbinsi

An endangered tapaculo, almost identical in plumage with the more widespread Choco Tapaculo *S. chocoensis*.

Type information *Scytalopus robbinsi* Krabbe & Schulenberg 1997, 9.5 km west of Piñas, El Oro, Ecuador.

Alternative names Ecuadorian Tapaculo. Spanish: Churrín de El Oro, Tapaculo de El Oro.

Discovery In 1997 N. Krabbe and T. Schulenberg published a monumental review of the species limits and natural history of the notoriously complex group of *Scytalopus* tapaculos of Ecuador. In this study they named three new species (El Oro Tapaculo *S. robbinsi*, Chusquea Tapaculo *S. parkeri* and Chocó Tapaculo *S. hocoensis*), as well as proposing full specific status for no fewer than 14 taxa considered by earlier authors (e.g. Zimmer 1939, Peters 1951) as races of other species. The present species differs from *S. chocoensis* in all vocalisations, as well as genetically.

Description 11 cm. Monotypic. Relatively small; male mostly dark mouse-grey above, the feathers slightly

▼ El Oro Tapaculo, male. Buenaventura Reserve, El Oro province, Ecuador, December 2013 (*Agustin Carrasco*).

▼ El Oro Tapaculo, male. Buenaventura Reserve, El Oro province, Ecuador, January 2015 (*Claudia Hermes*).

tipped blackish; lower back and rump dark brown; upperparts slightly paler grey than upperparts; rectrices blackish; lower sides, flanks, lower belly and undertail-coverts cinnamon-brown with blackish bars; iris dark brown; bill blackish; tarsi brown or dark brown. Female has brown nape and mantle, lower belly cinnamon-brown with blackish bars.

Habitat Undergrowth of wet forest, in foothills and lower subtropical zone, at 700–1,250 m. Altitudinally separated from Blackish Tapaculo *S. latrans*, which occurs in same mountain range.

Food and feeding Forages on or near ground. Stomach contents of specimens comprised insects including beetles.

Breeding Nest and eggs undescribed. Song activity peaks in September and February.

Voice Song of male (and possibly female) a minute-long series of double notes, the second part lower than the first. Female calls include single notes and series; Krabbe & Schulenberg (1997) give a detailed series of sonograms.

Movements Probably sedentary.

Range Endemic. Southern Ecuador, on Pacific slopes in provinces of El Oro and Azuay.

Conservation status Currently listed by BirdLife International as Endangered. Total geographic range is very small; habitat destruction, especially in Azuay province, is severe and continuing. Present in the Fundación Jocotoco Buenaventura Reserve near Piñas but apparently less common there than in previous years (BLI SF 2014). The future of the species is clearly dependent upon persuading landowners to retain suitable habitat.

Etymology Named in honour of Mark B. Robbins (b. 1954), in acknowledgement of "his substantial contribution to Neotropical ornithology" (Krabbe & Schulenberg 1997).

PERIJA TAPACULO
Scytalopus perijanus

A further overlooked species of *Scytalopus* tapaculo, unearthed using evidence from vocalisations and nucleic acids.

Type information *Scytalopus perijanus* Avendaño *et al.* 2015, above Vereda El Cinco, municipality of Manaure, Dpto. Cesar, western slope of Serranía de Perijá, Colombia.

Alternative names Spanish: Tapaculo de Perijá.

Discovery The history of *Scytalopus* tapaculos in the Serranía de Perijá, a spur of mountains noted for a high level of endemism, forming the northern boundary between Colombia and Venezuela, is extremely confused. As long ago as 1941–42, M. A. Carriker (q.v. under Pale-billed Antpitta *Grallaria carrikeri*) collected a long series of tapaculos from six locations on the Colombian slope of the Serranía de Perijá; these were mistakenly ascribed to the *nigricans* race of White-crowned Tapaculo *S. atratus* (Carriker 1954). Between 1951 and 1978 an even longer series was taken on the Venezuelan slope, which was variously identified as Caracas Tapaculo *S. caracae* (Ginés *et al.* 1953), Merida Tapaculo *S. meridanus* (Hilty 2003) or Brown-rumped Tapaculo *S. latebricola* (Ridgely & Tudor 1994). More recently it was hypothesised that the Venezuelan population could represent an undescribed species (Lentino *et al.* 2004) or a subspecies related to *S. meridanus* or to Mattoral Tapaculo *S. griseicollis*. Obviously some clarification was highly desirable.

Finally, in September 2006 two specimens were taken in cloud-forest on the Colombian slope, and more revealingly a series of 16 specimens, backed up by sound recordings, in the same general area which was not far from the location of the original Carriker specimens. Molecular phylogenetic analysis showed the these to be highly divergent from all other *Scytalopus* species, a conclusion backed up by vocal, morphological and ecological data, on which bases the new species was described.

▼ Perija Tapaculo. Serranía de Perijá, NW Venezuela/N Colombia, July 2008 (*Andres Cuervo*).

Description 18.5 cm. Monotypic. In form and plumage a typical *Scytalopus* tapaculo, rather small with upperparts dark neutral grey, a brown nuchal patch, more prominent in the female; chin, throat, breast and centre of belly medium to light neutral grey, darker on sides of breast; lower belly and flanks clay- or tawny-coloured; iris dark brown; bill dull horn with paler tip; tarsi brown.

Habitat Several different forest types; lower montane to upper montane humid forest, including elfin forest, *subpáramo* and *páramo*. 1,370–3,230 m.

Food and feeding Forages low down or on the ground; stomach contents of specimens comprised insect remains.

Breeding One nest only so far described. Nest was a globular structure made of mosses, grasses and rootlets in a subterranean cavity with a short tunnel entrance, similar to those of other members of the genus. On 13 July it contained two nestlings.

Voice Avendaño *et al.* (2015) give a very detailed analysis of vocalisations, with comparisons of those of other relevant tapaculos. Song is distinctive, a series of *churrs* lasting up to three seconds.

Movements Presumably sedentary.

Range Recorded at 19 sites on both slopes of the Serranía de Perijá, nine in Venezuela (Zulia) and ten in Colombia (Dptos La Guajira and Cesar), from about 9° 56' N to 10° 50' N and about 72° 44' W and 73° 03' W.

Conservation status Not currently classified by BirdLife International, but the discoverers suggest a status of Endangered as appropriate. There has been much destruction of habitat on both sides of the border, for crops both legal and illegal. None of the Colombian range is protected, but 300,000 ha on the Venezuelan side is nominally covered by the Sierra de Perijá National Park. However, general lawlessness in the area may make this less effective than it might be.

Etymology The English, Spanish and scientific names all reflect the range of the species.

CHUSQUEA TAPACULO
Scytalopus parkeri

Forms a superspecies with Spillmann's Tapaculo *S. spillmanni*.

Type information *Scytalopus parkeri* Krabbe & Schulenberg 1997, 20 km SSW of San Lucas, Loja, Ecuador.

Alternative names Spanish: Churrín de Chusquea, Tapaculo de Chusquea.

Discovery Fieldwork by N. Krabbe in SE Ecuador in the 1990s demonstrated that the tapaculo population there was clearly distinct from Spillmann's Tapaculo

◄ Chusquea Tapaculo. Tapichalaca, south of Loja, Ecuador, June 2013 (*Eduardo Carrión L.*).

S. spillmanni in both voice and habitat, despite being extremely similar in plumage. Its specific status was later confirmed by genetic studies (Arctander & Fjeldså 1994); the type specimen, whose vocalisations were recorded, was taken by Dr Krabbe in 1991.

Description 11.5 cm. Monotypic. In plumage very close or identical to *S. spillmanni*, which is however completely allopatric. Largely dark mouse-grey above, with faint brown wash on nape, lower back, rump and edges of remiges; below, pale mouse-grey, the central lower belly with silvery feather-tips; lower belly bright ochraceous-buff; lower sides, flanks and crissum tawny or cinnamon-brown with blackish barring; iris dark brown; bill blackish; tarsi light brown. Sexes similar.

Habitat Dense stands of *Chusquea* bamboo and adjacent forest undergrowth, 2,750–3,350 m in Ecuador, to 2,900 m in northern Peru (2013 HBW SV).

Food and feeding Forages on or near the ground; stomach contents of specimens insects including beetles and in a few instances plant material.

Breeding Breeding season appears to be protracted, possibly all year except during dry periods, with specimens in breeding condition in March, June, September and November, and juveniles collected in December and February. Three nests so far described. One, found in October 2004 in the Tapichalaca Biological Reserve, Zamora-Chinchipe, Ecuador, was a typical *Scytalopus* mossy ball, about 15 cm in diameter, with a side-entrance, located in a crevice in a low bank. On 7 October, the nest contained two chicks, covered in long grey down, fed by both adults with Lepidoptera, phasmids, beetles and Orthoptera (Greeney & Romborough 2005, Greeney 2008).

Voice The best distinction from other tapaculos. Male dawn song is a trill, initially descending, lower-pitched and slower-paced than that of *S. spillmanni*. Day song a series of relatively low-pitched trills, similar to that of *spillmanni* but with repeated breaks. Also various shorter calls (Ridgeley & Greenfield 2001), including a buzzy *bzeww* (Schulenberg *et al.* 2007). Sexes duet, the female component being higher-pitched (Krabbe & Schulenberg 1997, who give several sonograms).

Movements Presumably sedentary.

Range Ecuador (SW Morona Santiago, western Zamora-Chinchipe, eastern Loja, also disjunctly in south-central Morona Santiago); highest ridges of the Cordillera del Cóndor and northern Peru (eastern Piura and NW Cajamarca).

Conservation status Currently classified as Least Concern. Although the geographic range is rather small, there is no evidence of major population declines. Appears to be fairly common in some locations; several parts of the range are protected e.g. Podocarpus National Park and the Tapichalaca Reserve of the Fundación Jocotoco, Ecuador.

Etymology Named in honour of the late Ted Parker (see under Subtropical Pygmy Owl *Glaucidium parkeri*), who was the first ornithologist to record and collect the species. Appropriately, the vehicle for the description of this and two other *Scytalopus* species, *Ornithological Monographs* No. 48, was subtitled *Studies in Neotropical Ornithology Honoring Ted Parker.*

JUNIN TAPACULO
Scytalopus gettyae

An all-dark tapaculo, with its closest relatives (based on genetic studies) the sympatric Tschudi's *S. acutirostris* and Rufous-vented *S. femoralis* Tapaculos, and the allopatric Blackish *S. latrans* and Long-tailed *S. micropterus* Tapaculos (Cadena & Cuervo, in prep).

Type information *Scytalopus gettyae* Hosner *et al.* 2013, Huaytapallana Cordillera, between Calabaza and Toldopampa, Dpto. Junín, Peru.

Alternative names Spanish: Tapaculo de Junín.

Discovery While conducting a survey of birds and their pathogens along an elevational gradient on the eastern slope of the Andes in Dpto. Junín, Peru, in September and October 2008, P. Hosner collected four specimens of an all-dark tapaculo with an unfamiliar song. Careful analysis of the vocalisations, plumage and morphometrics of these specimens, along with genetic studies, indicated that a new species was involved.

Description 11–12 cm. Monotypic. An all-dark tapaculo, with plumage uniformly fuscous-brown above and below; iris dark brown; bill black; tarsi dusky-brown. Juvenile plumage undescribed, but a recently published photograph (see overleaf) shows extensive pale scalloping on underparts and shoulders, with an obvious whitish wing-bar formed by edgings to the primary and secondary greater coverts. Very similar in plumage to several other species; smaller than the syntopic Large-footed Tapaculo *S. macropus*, diagnostically darker than Trilling and Tschudi's Tapaculos *S. parvirostris* and *S. acutirostris*, and lacks the brown and black flank barring of Rufous-vented Tapaculo *S. femoralis*.

Habitat Has so far been detected in dense, low, secondary vegetation with shrubs, ferns and *Chusquea* bamboo

▲ Junin Tapaculo, juvenile. Apalla-Andamarca Road, Dpto. Junín, Peru, September 2013 (*Nick Athanas*).

thickets, under forest canopy adjacent to primary forest, at 2,400–3,200 m. Uncertain as to whether it occurs in inaccessible primary forest in the area.

Food and feeding Feeds low down (below 2 m) or on the ground; stomach contents of all specimens were insects.

Breeding Nest and eggs undescribed. All three adult specimens (1–7 October) had enlarged breeding-condition testes.

Voice Song distinctive; repeated, energetic arpeggio of phrases consisting of a single loud high-frequency note, followed by 2–5 lower and weaker notes that become louder and higher; also a distinctive metallic, sharp descending call-note (Hosner *et al.* 2013).

Movements Presumably sedentary.

Range Endemic. Presently known only from a very restricted area of Dpto. Junín; two locations only, about 5 km apart, in one drainage area on the eastern slope of the Andes. Hosner *et al.* (2013) suggest that there are areas nearby that would be worthy of exploration; careful examination of specimens in current collections would also be desirable.

Conservation status Currently unclassified by BirdLife International and should be listed as Data Deficient. However, from the very limited area in which it has so far been found, a provisional classification of Endangered might be justified. Much of the habitat around the type locality remains in good condition (Hosner *et al.* 2013).

Etymology Named for Caroline Marie Getty (b. 1957) "in honor of her long-term dedication to nature preservation" (Hosner *et al.* 2013). Ms Getty works for the National Fish and Wildlife Foundation.

BAHIA TAPACULO
Eleoscytalopus psychopompus

The taxonomic status of this form is not entirely clear; closely related to White-breasted Tapaculo *E. indigoticus*. Following molecular work by Mauricio *et al.* (2008), *psychopompus* and *indigoticus* occupy the newly erected genus *Eleoscytalopus*, possibly more closely allied than other tapaculos to the bristlefronts (*Merulaxis*).

▼ Bahia Tapaculo. Igrapiuna, Bahia, Brazil, April 2008 (*Ciro Albano*).

Type information *Scytalopus psychopompus* Teixeira & Carnevalli 1989, near Valenca, Bahia, Brazil.

Alternative names Chestnut-sided Tapaculo. Portuguese: Macuquinho-Baiano.

Discovery Discovered as part of the ongoing investigations by staff of the Ornithology Department of the Museu Nacional de Rio de Janeiro into forest habitats in NE Brazil, which has resulted in the description of several new species.

Description 11 cm. Monotypic. Upperparts dark bluish-grey, becoming reddish-brown on the lower back and rump; small whitish spot on lores; iris dark brown; bill blackish above, pale grey to yellow below; tarsi yellowish-brown. Distinguished from *indigoticus* by unbarred flanks, blue-grey, not barred-cinnamon, thighs and longer bill.

Habitat Mostly mature wet lowland forest; also degraded pioneer vegetation along waterways. In small swamps around rivers or swampy parts of the river itself; at 15–20 m (BLI SF 2014).

Food and feeding No data.
Breeding Nest and eggs undescribed. From gonadal condition of specimens, in breeding condition in October (Teixeira & Carnevalli 1989).
Voice Call a sequence of 27–28 short notes, about 9 per second, similar to that of *E. indigoticus*; also a regular, frog-like slightly ascending call, *frrrrrrrooww*, lasting about three seconds. Alarm call is a low frequency short note (BLI SF 2014).
Movements Presumably sedentary.
Range Endemic. Imperfectly known due to its very secretive nature. Confined to small scattered locations in the state of Bahia.
Conservation status Currently classified by BirdLife International as Critically Endangered, although with further work this may be downlisted, especially since it may occur in modified habitats. No longer present in the locations where the original type specimens were obtained, but has recently been located in six localities in four further municipalities. Occurs in two protected areas (Una Biological Reserve and the Reserva Ecológica da Michelin), where the wardening against illegal hunting and forest clearing is very effective. The Una reserve contains some 64–68 pairs; rare and patchy in other locations. More recent surveys have found it in six different locations, including the two above. Upper population range about 400 individuals.
Etymology Greek *psychopompos*, literally 'soul-guide', a mythological spirit responsible for escorting the souls of the recently deceased to the afterlife; in several cultures associated with birds, for example Whip-poor-will *Caprimulgus vociferus*.

ELUSIVE ANTPITTA
Grallaria eludens

A little-known species, formerly placed in the genus *Thamnocharis* with Ochre-striped Antpitta *G. dignissima*, with which it forms a superspecies.

Type information *Grallaria eludens* Lowery & O'Neill 1969, Balta, Dpto. Loreto, Peru.
Alternative names Spanish: Tororoi de Ucayali; Portuguese: Tovacuçu-xodo. Local name: Du xau.
Discovery In 1966, J. P. O'Neill spent eight months collecting specimens of birds near Balta, Dpto. Loreto, in Amazonian Peru, a location which had already yielded several species new to science. In the summer of 1966, he obtained a specimen, taken by one of the local Indians, of a new antpitta. In 1967 he returned to the location where, with the help of the Indians (who knew the species well, having a local name for it), a series of specimens was obtained, allowing the description of a new species with detailed anatomical data to be made in 1969.
Description 19 cm. Monotypic. In form a typical *Grallaria* antpitta, almost tail-less, with large powerful legs and thick bill. Upperparts including lores and ear-coverts olivaceous-brown; throat white; remainder of underparts heavily streaked black, on a buff base on the breast and upper belly, white base elsewhere; iris brown; bill dusky or horn-coloured, pinkish-flesh below; tarsi blue-grey.
Habitat Humid dense lowland *terra firme* forest with dense understorey, at 120–500 m.
Food and feeding Few data. Feeds on or near the ground in dense vegetation. Diet presumably principally invertebrates. Probably does not follow mixed flocks of other species.
Breeding Nest and eggs undescribed.
Voice Song consists of two whistles, very similar to that of *G. dignissima* but with a longer introductory note and the second note beginning with an abrupt rise (Whittaker & Oren 1999). Alarm call *churr*.
Movements Presumably sedentary.
Range Imperfectly defined. Known from several scattered locations in Amazonian Peru (eastern Loreto and Ucayali, NE Madre de Dios) and western Brazil (Amazonas and Acre) (Whittaker & Oren 1999; Willis 1987).
Conservation status Currently listed by BirdLife International as Least Concern. The geographic range appears to be quite large. Although there has been significant habitat destruction (and the species is probably intolerant of much habitat modification), with probable population declines, these do not seem as yet sufficient to trigger a higher classification. The only part of its range currently with formal protection is Manu National Park, Peru.
Etymology Both the common and scientific names recognise the extremely secretive nature of the species. Apparently well deserved; one May, A. Whittaker spent five hours trying to observe a singing bird in dense bamboo-dominated understorey in the Rio Juruá area of western Brazil without catching so much as a glimpse of the singer (Whittaker & Oren 1999).

CUNDINAMARCA ANTPITTA
Grallaria kaestneri

A rather small antpitta, probably most closely related to Santa Marta Antpitta *G. bangsi.*

Type information *Grallaria kaestneri* Stiles 1992, 3 km ENE of Monterredondo, Municipio de Guayabetal, Dpto. Cundinamarca, Colombia.
Alternative names Kaestner's Antpitta; Spanish: Tororoi de Cundinamarca.
Discovery In October 1989, P. Kaestner, who was at the time the United States Consul in Bogotá, Colombia, explored a recently opened road leading to the town of El Calvario, Dpto. Meta, Colombia, a location about 50 km south-east of the centre of Bogotá. At an altitude of 2,250 m he heard an unfamiliar antbird song. By playback he was able to lure the bird into sight; it was obviously an antpitta but he was not able to ascribe it to any known species. Suspecting that an undescribed species might be involved, he returned to the site in May 1990 with F. G. Stiles. During this visit recordings were made and a specimen was taken, from which it was obvious that it was a new species. Dr Stiles returned to the site in January 1991, when two further specimens were collected, along with data on population densities, elevational range and behaviour. Unfortunately further work was precluded for several years by civil unrest and guerrilla activity.
Description 15–16 cm. Monotypic. Generally dull olive-brown above; throat whitish; breast greyish-olive with narrow white shaft streaks; iris dark brown; bill black above, grey below; tarsi plumbeous or light purplish-grey.
Habitat Lower levels of humid montane forest and mature secondary forest with dense dark understorey. Can apparently tolerate some degree of habitat disturbance.
Food and feeding Feeds low down on forest floor or very close to it, flipping over leaf-litter, or digs briefly in soft ground. Prey items include beetles, cockroaches, spiders, katydids and earthworms.
Breeding Nest and eggs undescribed. Probably breeds in second half of year, coinciding with rainy season (Krabbe & Schulenberg 2003 HBW 8).
Voice Song consists of three sharp, clear, whistled notes *wirt, wiirt, weert!,* sometimes omitting the last note; also an aggressive call of two piercing notes, given by both sexes.
Movements Presumably sedentary.
Range Endemic. Amazonian slope of eastern Andes in Dpto. Cundinamarca and Dpto. Meta, Colombia.
Conservation status Currently listed by BirdLife International as Endangered; Cortés *et al.* (undated) suggest that Critically Endangered might be more appropriate. Known from only three locations which have suffered significant environmental destruction in recent years; in one of its former strongholds (Farallones de Medina) it no longer occurs. The major threat is logging; clear-cutting is very damaging, though the species may tolerate some selective logging.
Etymology Named after Peter G. Kaestner who first detected the species. Mr Kaestner began his career as a Peace Corps volunteer before entering the U.S. Foreign Service in 1980, being posted to India, Papua New Guinea, Colombia, Malaysia, Namibia, Guatemala and Brazil. He is listed in the *Guinness Book of World Records* as being the first person to see a member of all of the 150 families of birds then recognised (Beolens & Watkins 2003).

◄ Cundinamarca Antpitta. Monterredondo, Colombia, January 2017 (*Ross Gallardy*).

JOCOTOCO ANTPITTA
Grallaria ridgelyi

Possibly the most spectacular of all the antpittas; the discovery of the Jocotoco Antpitta caused a great sensation in the birdwatching community. Probably most closely related to Chestnut-naped Antpitta *G. nuchalis*, a more widespread species that ranges from Colombia to Peru.

Type information *Grallaria ridgelyi* Krabbe *et al.* 1999, Quebrada Honda, Zamora-Chinchipe, Ecuador.
Alternative names Spanish: Tororoi Jocotoco.
Discovery In late 1997 R. Ridgely and several colleagues put together a trip, with a main emphasis on obtaining good recordings of vocalisations, to an area of southern Ecuador not far from Podocarpus National Park. On 20 November, while investigating a mule-trail running down a steep valley locally called Quebrada Honda, they heard a soft, measured hooting call, unfamiliar to all of them; but the bird fell silent, leaving them intrigued but puzzled. However, a couple of miles further down the trail, the same call, this time very close, burst out, allowing a recording to be made. Playback induced first one then a second bird, clearly of a totally unknown species, to emerge from the dense understorey, and both obligingly (and atypically for an antpitta) remained in plain view for 20 minutes. The following day some poor-quality photographs were obtained. In December F. Sornoza obtained the type specimen at the same location, and in January 1998 a full-scale expedition was mounted which captured and photographed the first live specimen (Ridgely 2012; Krabbe *et al.* 1999). In fact, the call was quite well known to local people, who believed it to be that of an owl (the bird does indeed sometimes sing at night – Ridgely 2012).
Description 20–22 cm. Monotypic. Uniquely plumaged. A rather large *Grallaria*, large-headed, plump and almost tailless with a black cap and ear-coverts; upperparts, wings and tail brownish-olive; large white patch in front of eye, separated from white throat by a narrow black band; breast and belly greyish-white; sides of breast and undertail-coverts brownish-olive; iris dark red; bill black; tarsi bluish-grey. Immature similar, but face-patch less well-defined, crown chestnut with fine black streaks; base of lower mandible fleshy orange (Greeney & Juiña 2010).
Habitat Lower levels of dense wet montane evergreen forest, with bryophytes and bamboo, at 2,300–2,680 m. Apparently seems to require the existence of a stream, even a small one, but with permanent flow.
Food and feeding Forages low down or on the ground. Stomach contents of specimens comprise invertebrates (insects, including beetles and ants, millipedes and earthworms). Becomes easily habituated to feeding stations at which worms are provided, making for easy observation by visitors. May be a commensal feeder, in the wild following large mammals such as bears and tapirs (Greeney 2012a).
Breeding So far only one nest described. This was a deep, bulky cup made of dead leaves, mostly dicotyledonous or bromeliad, flush against a dead tree trunk in a large clump of epiphytes, 3.6 m above the ground. On 8 November the nest held one large chick which fledged about 15 November. It is not known whether Jocotoco Antpittas routinely lay one egg (two is more typical of the genus – Greeney *et al.* 2008), or whether in this case one egg was lost; however, observations of a single dependent chick with attending adults have been made several times (Greeney & Gelis 2005a). The nest was only 30 m from a worm feeding station, which both adults patronised; it is suggested that a reliable source of food may result in two broods per year (Greeney & Juiña 2010).
Voice Call is diagnostic; male song is a series of low hooting notes, at intervals of 1–2 seconds, sometimes prolonged for a minute or more (Ridgely & Greenfield 2001). Other calls, given by both sexes, are a doubled *hoo-coo* and a more guttural *hoo-krr*. Call of juvenile is a slightly drawn-out single *woooo* note (Greeney & Gelis 2005).
Movements Presumably sedentary.
Range Southern Ecuador in Zamora-Chinchipe. Initially only in an area now in the Tapichalaca Biological Reserve; now also detected in parts of Podocarpus National Park, including a range extension to the Cordillera de Tzunantza, some 30 km north-east of the type locality (Heinz *et al.* 2005). Recently found in Hito Jesus, Cajamarca, Peru, immediately adjacent to the Ecuadorian border (O'Neill 2006).
Conservation status Currently classified by BirdLife International as Endangered. The main threat is habitat destruction for agricultural purposes, mainly cattle grazing (indeed, the Tapichalaca Reserve was only created by Fundación Jocotoco in the nick of time to save a substantial portion of habitat). An area on Cerro Toledo, 10 km north-east of the type locality where the species was found in 1998, was by 2002 largely cleared, with no birds present (Heinz *et al.* 2005). Substantial parts of the species' total range are protected in the Tapichalaca Reserve and in Podocarpus National Park (Sornoza 2000), although within the latter illicit logging and mining apparently occur (BLI SF 2014).
Etymology Named in honour of its discoverer, Robert

▲ Jocotoco Antpitta, adult. Tapichalaca Reserve, Zamora, Chinchipe, Ecuador, December 2008 (*Nick Athanas*).

S. Ridgely. Dr Ridgely has been one of the most influential figures in the ornithology of Central and South America for many years, with a major contribution to conservation. He is the author of field guides to Panama and Ecuador, as well as the monumental *Birds of South America*. Formerly at the Philadelphia Academy of Natural Sciences, he is now with the American Bird Conservancy and is honorary president of World Land Trust US. Among his numerous awards are the Eisenmann Medal (2001), the Chandler Robbins Award (2006), the Ralph Schreiber Conservation Award (2011) and the Allen Award (2013).

PALE-BILLED ANTPITTA
Grallaria carrikeri

A spectacular antpitta from Peru, often regarded as a sister species of Chestnut-naped Antpitta *G. nuchalis*, but possibly most closely related to Jocotoco Antpitta *G. ridgelyi*.

Type information *Grallaria carrikeri* Schulenberg & Williams 1982, Cordillera de Colán, SE of La Peca, Dpto. Amazonas, Peru.
Alternative names Spanish: Tororoi de Carriker.
Discovery During faunal surveys by personnel from the Louisiana State University Museum of Zoology in Dpto. Amazonas, northern Peru, in 1978 a new and distinctively plumaged antpitta was collected. In 1979 a number of further specimens were obtained. However, a specimen in juvenile plumage, originally misidentified as *G. nuchalis*, had been taken not far from the type locality as early as 1976.

Description 19 cm. Monotypic. Face, lores, ear-coverts and throat black; crown blackish-brown, nape and back olivaceous-brown; wings and tail dark chestnut; chest and belly slate-grey, scalloped with blackish, the centre of the belly paler; flanks and crissum buff-olive; iris crimson or reddish-brown; bill ivory, tarsi blue-grey.
Habitat Humid montane forests, with trees of up to 30 m with broken canopy and heavy epiphytic and dense understorey, with much *Chusquea* bamboo, 2,350–2,900 m (Schulenberg & Kirwan 2012b).
Food and feeding Feeds on or near ground. Stomach contents of specimens comprised arthropods, especially caterpillars and beetles. Nestlings fed earthworms as well as unidentified prey (Wiedenfeld 1982).
Breeding One nest only described so far. It was a shallow cup, external dimensions 19 × 20 cm, internal 12 × 14 cm, 3 m up in the sloping trunk of a partially fallen

► Pale-billed Antpitta. Above San Lorenzo, Pomacochas, Peru, April 2015 (*Miguel Lezama*).

tree, made of sticks and wet decaying leaves, lined with rootlets. On 14 October the nest held two chicks of estimated age seven days; fed and brooded by both adults (Wiedenfeld 1982).

Voice Song is a series of six notes, irregularly spaced, given in about three seconds, at intervals of about 6–10 seconds; that of *G. nuchalis* is longer and consists of an accelerating series of nearly 20 notes. Call of *carrikeri* is a series of high metallic notes (Schulenberg *et al.* 2007).

Movements Presumably sedentary.

Range Endemic. Eastern slope of northern Peruvian Andes, from Amazonas (Cordillera de Colán) south to La Libertad. Southern limits of range imperfectly known.

Conservation status Currently classified by BirdLife International as Near Threatened, uplisted from Least Concern because of its small range. Within this range there is significant pressure from human immigration, aided by recent road developments, with clearing of forest for cattle pasture (Begazo *et al.* 2001).

Etymology Named in honour of Melbourne A. Carriker (1879–1965), one of the most influential field ornithologists of his day, who did much of the foundation work on the ornithology of Central America and northern South America. He was also an expert on the then little-studied field of feather-lice (Mallophaga).

CHESTNUT ANTPITTA
Grallaria blakei

A cryptic species that forms a superspecies with the more widely distributed Rufous Antpitta *G. rufula*.

Type information *Grallaria blakei* Graves 1987, eastern slope of the Cordillera de Carpish, Dpto. Huánuco, Peru.

Alternative names Spanish: Tororoi Castano.

Discovery In the 1960s specimens of a small uniform antpitta were collected in the Cordillera de Carpish of Peru, but were incorrectly identified as *G. rufula*, although it was noted at the time that they were larger

◀ Chestnut Antpitta. Apalla Junin, Peru, September 2013 (*Nick Athanas*).

and darker than the local race (*obscura*) of *rufula*, with the suggestion that a new species might be involved (Tallman 1974). In a study of the genus published in 1987, G. Graves pointed out that the Carpish birds consisted of two clearly differentiated forms, separable easily by plumage and apparently altitudinally separated in habitat preferences, on which basis he described a new species.

Description 15 cm. Monotypic (but see below). Plumage essentially chestnut-brown of different shades; brighter burnt sienna on face, throat and breast; flanks, thighs and undertail plain brown; feathers of centre of belly with indistinct grey bars in birds from Dpto. Huánuco, less obvious in birds from Dpto. Amazonas and lacking in birds from Dpto. Pasco; iris brown; bill black; tarsi slate-grey. Plumage differences in the southern (Pasco) population, along with apparent differences in vocalisations, suggest that further work is required. Differs from *G. rufula* in its richer-coloured plumage, lack of an eye-ring and in vocalisations.

Habitat On or near the ground in dense undergrowth with bamboo (*Chusquea*) stands in humid montane forest, 1,700–2,500 m, locally to 3,100 m (Schulenberg *et al.* 2007).

Food and feeding Feeds on or near ground. No data on diet.

Breeding Nest and eggs undescribed. Probably nests from December to April or May; specimens with retained juvenile feathers in early August and late November (Graves 1987).

Voice The song of Chestnut Antpitta in most of its range (from Amazonas south to Junín) is described as "a rapid, monotone, slightly accelerating series of chiming notes *chew'chu'u'u'u'u'u'u'u'u'u'u*" (Schulenberg *et al.* 2007); and as "3.2–4.6 s long, given at 4–11 s intervals, a ringing, evenly pitched series of 38–52 notes at 2 kHz given at an even pace of 11–12/s" (Krabbe & Schulenberg 2003 HBW 8); however, the apparent song of Chestnut Antpitta in central Peru in Pasco is different, "a single chiming note *clew*" (Schulenberg *et al.* 2007; Schulenberg & Kirwan 2012c). Other calls, a brief *weeoo* (Barnes *et al.* 1997).

Movements Presumably sedentary.

Range Endemic. Disjunct. From Cordillera de Colán in northern Peru (Amazonas); Huánuco, San Martín and La Libertad; and in Pasco. Skeletal material from much further south, in Ayacucho, may refer to *blakei*, but requires confirmation (Graves 1987).

Conservation status Currently listed by BirdLife International as Near Threatened. The overall range is small; although significant parts of it are protected by remoteness and lack of roads, habitat destruction is widespread in the remainder.

Etymology Named after Emmet R. Blake (1908–1997), Emeritus Curator of Birds at the Field Museum, Chicago, where he spent much of a long and fruitful career. Among other works he was the author of the uncompleted *Manual of the Neotropical Birds* and the first field guide to the birds of Mexico.

URRAO ANTPITTA
Grallaria fenwickorum

A species whose discovery and description has generated considerably more than its fair share of controversy (see below). Most closely related to Brown-banded Antpitta *G. milleri*, another Colombian endemic.

Type information *Grallaria fenwickorum* Barrera & Bartels 2010, Colibrí del Sol Bird Reserve, Vereda El Chusqual, Páramo del Sol, Urrao, Antioquia, Colombia. [Syn. *Grallaria urraoensis* Carantón-Ayala & Certuche-Cubillos 2010]

Alternative names Fenwick's Antpitta, Antioquia Antpitta. Spanish: Tororoi de Urrao, Tororoi de Fenwick.

Discovery During regular monitoring activities conducted in forests near the Páramo de Frontino, Antioquia, Colombia, in September 2007, D. Carantón captured an unfamiliar antpitta that was measured and released. It could not be photographed but appeared not to resemble any *Grallaria* species known from the area. A second individual was captured and released, after examination and photography in early February 2008, not far from the first location. Later that same month a specimen was accidentally obtained; a bird had apparently landed on a furled mist-net, entangling a foot, and subsequently died. This was prepared as a study skin. Comparison of this with museum material confirmed its phenotypic distinctiveness. A second specimen was deliberately taken in March 2008. On this basis, Carantón and his assistant K. Certuche wrote a paper describing the new species as *Grallaria urraoensis,* Urrao Antpitta, submitted on 19 January 2010 and accepted on 6 May 2010 for the journal *Ornitología Colombiana.*

In the meantime, L. Barrera and A. Bartels captured a bird in January 2010 and took measurements and a number of feathers before banding and releasing the live bird. On this basis they published a description

▲ Urrao Antpitta. Reserva Natural de las Aves Colibrí del Sol, Urrao, Antioquia, Colombia, January 2010 (*Nigel Voaden*).

as a new species in *Conservación Colombiana* which was issued on 18 May 2010, in advance of the publication of Carantón and Certuche's paper. In this paper they acknowledged the involvement of Carantón but failed to mention Certuche. It is noteworthy that Carantón and Certuche had submitted their paper to *The Condor* in late 2009, but it was not accepted for publication since a reviewer had suggested that proper permits had not been obtained for the collection of the 2008 specimens.

The circumstances surrounding the publication of this new species generated much controversy, and continue to do so (Regalado 2011). Diametrically opposite viewpoints are expressed in editorials in *Conservación Colombiana* (Comité Editorial 2010) and *Ornitología Colombiana* (Cadena & Stiles 2010). The first description has priority, so the name *fenwickorum* is used by HBW; however, the IOC uses *urraoensis,* which is also the preference of the SACC (Proposal 479). The common name Antioquia Antpitta has been suggested as a compromise between the two parties.

Description 16–17 cm. Monotypic. A medium-sized, rather plain antpitta. Upperparts olive-brown, darker on tail; underparts and flanks grey with paler grey tips; sides of chest tinged olive-grey; central belly whitish-grey, tinged yellowish; iris very dark brown; bill bluish-grey, darker at base; tarsi bluish-grey. Juvenile quite different; upperparts black with broad rufous-brown tips, lower underparts creamy-buff; bill bright orange, tarsi dark pink (Barrera & Bartels 2010).

Habitat Very wet montane forest with 2,000 mm of rain annually; found in both primary and secondary forest, with dense understorey of epiphytes and bamboo (*Chusquea*), at 2,500–3,200 m, but possibly as low as 2,000 m, especially in areas with tree-fall or landslides (Carantón & Certuche 2010).

Food and feeding Forages on or near the ground. Stomach contents of the two known specimens were insect parts, with fragments of beetles the only identifiable remains.

Breeding Nest and eggs undescribed. The two male specimens (February and March) both had enlarged testes; vocal activity peaks between February and May; and an adult with an old brood-patch was captured and released in June. A fledgling observed in mid-June was fed earthworms by both parents (Carantón & Certuche 2010).

Voice Song is three notes, progressively increasing in pitch and length *tu-tuui-tuuet*; call is a single sharp note, higher pitched than the song.

Movements Apparently sedentary.

Range Endemic. Currently known from an extremely small area in the western Andes of Colombia in Antioquia state; known only from the Colibrí del Sol Bird Reserve and immediately adjacent areas.

Conservation status Critically Endangered, with a very small population in a very small range. Actual known population is in the region of 24 territories; by extrapolation into apparently suitable habitat, estimated at 57–156 territories (Barrera & Bartels 2010). The total extent of occurrence estimated by BirdLife International as ca 80 km^2, with a population in the range of 70–400 individuals (BLI SF 2014). Main threat is habitat destruction for pasture, and potentially for mining; this latter was until recently discouraged by civil disorder, but with increasing political stability pressure from this source will increase.

Etymology The name *fenwickorum* (genitive plural) honours the Fenwick family of Virginia, USA, whose personal resources were instrumental in funding the purchase of the 731 ha Colibrí del Sol Reserve, in which the greater part of the known world range of *G. fenwickorum* is protected. In actual fact, the main impetus for the creation of the reserve was the protection of the Critically Endangered Dusky Starfrontlet *Coeligena (bonapartei) orina*, which was described from the area in 1957 and not seen again for 50 years; the discovery of a second Critically Endangered species in the same location was purely serendipitous. The alternative name *urraoensis* is derived from the place name Urrao, of indigenous origin.

ALTA FLORESTA ANTPITTA
Hylopezus whittakeri

A cryptic species, part of the taxonomically complex Spotted Antpitta *H. macularius* group which probably comprises three or four genetically separate species.

Type information *Hylopezus whittakeri* Carneiro *et al.* 2012, Belterra, Floresta Nacional do Tapajós, Sucupira base, km 117 on Highway BR163, Pará, Brazil.

Alternative names Portuguese: Torori-de-alta-floresta.

Discovery In 2009, A. Whittaker reported the occurrence of an "undescribed taxon in the *Hylopezus macularius* complex", observed during an avifaunal inventory of the Pousada Rio Roosevelt, a species-rich lowland area in Amazonas state, Brazil. In 2012 Carneiro *et al.* published a very detailed analysis of the complex, with thorough analysis of morphological and vocal characteristics along with comparative DNA work of the various populations of the group; their conclusion (that the group comprises four valid species of which one, the subject species, was undescribed), was recommended for acceptance by the SACC (Proposal 622). In fact, tape recordings of the new species exist from as early as 1999 and 2002, while what was to become the holotype had been collected in 2002, and two paratypes in 1974 and 1986 (Carneiro *et al.* 2012).

Description 14 cm. Monotypic. A typical almost tail-less long-legged antpitta. Crown dark grey with fine blackish streaks; area around eye conspicuous buffy-ochre, edged below blackish, throat white, malar stripe blackish; upperparts olive-brown, streaked yellow-ochre on scapulars and upper mantle; upperwing-coverts olive-brown, tipped cinnamon-buff, underparts white, heavily streaked black on chest, flanks buffy; iris dark brown; bill greyish-black with base of lower mandible pink. Very similar to Spotted Antpitta *H. macularius* but

▼ Alta Floresta Antpitta. Cristalino Lodge, Alta Floresta, Mato Grosso, Brazil, June 2012 (*Andrew Whittaker/Birding Brazil Tours*).

slightly larger, with shorter tarsus; separable from *H. m. paraensis* probably only by voice and genetics.
Habitat Lower levels of humid lowland forest with dense understorey; prefers wet or flooded areas of *terra firme* forest, but in south of range also in drier transitional forest; from near sea level to 500 m.
Food and feeding Forages low down in dense vegetation; no data on diet.
Breeding Nest and eggs undescribed. The type specimen, collected on 23 July 2002, had testes measuring 4 × 2 mm.
Voice Diagnostic. The loud song consists of 4–6 (usually 5) whistled, hooting notes; differs from other members of the *macularius* complex in length and spacing of notes. Interestingly, as long ago as 1989 T. Parker recognised that the song of the population in Alta Floresta was distinctive (Zimmer, in SACC Proposal 622).
Movements Probably sedentary.
Range Endemic. Occupies area between rivers Xingu and Madeira, Amazonas, Brazil; not yet recorded from Bolivia, though one specimen from Rondônia, about 150 km from border (Carneiro *et al.* 2012).
Conservation status Not presently assessed by BirdLife International; however, it appears to be very sensitive to habitat modification or fragmentation. The smallest undisturbed patch of forest in which individuals were found was 19 ha; absent from several patches smaller than this (Carneiro *et al.* 2012).
Etymology Named in honour of Andrew Whittaker (b. 1960), a British-born ornithologist, author and tour leader "whose contributions to Amazonian ornithology over the last 20 years resulted in the description and rediscovery of several species" (Carneiro *et al.* 2012).

OCHRE-FRONTED ANTPITTA
Grallaricula ochraceifrons

An almost unknown species; probably forms a superspecies with Peruvian Antpitta *G. peruviana*.

Type information *Grallaricula ochraceifrons* Graves *et al.* 1983, 10 km road distance NE of Abra Patricia, Dpto. San Martín, Peru.
Alternative names Spanish: Ponchito Frentiocre.

▼ Ochre-fronted Antpitta, male. Abra Patricia, NE Peru, March 2014 (*Jon Hornbuckle*).

Discovery In August 1976 G. Graves and colleagues mist-netted a strikingly patterned *Grallericula* antpitta near Abra Patricia, Peru, in the Eastern Cordillera of the Andes. Two further specimens were collected in the Cordillera de Colán in Dpto. Amazonas in 1978. The species has been observed since, with several more specimens taken in Dpto. San Martín in 2002 (Hornbuckle 1999; BLI SF 2014).
Description 10.5 cm. Monotypic. Sexually dimorphic, unlike most *Grallericula* antpittas. Upperparts olive-brown; throat white with black malar stripe; underparts white with black scalloping on the chest, the sides of the breast tinged buff; flanks streaked olive-brown. Male has an ochraceous-buff forecrown, lores and eye-ring; female has olive-brown lores and forehead; generally darker on underparts; iris dark brown; bill black with pink base to lower mandible; tarsi pinkish-grey.
Habitat Dense undergrowth in wet, epiphytic cloud-forest, both stunted and tall, 1800–2500 m.
Food and feeding Forages low down in dense vegetation; no data on diet.
Breeding Nest and eggs undescribed; specimens taken in July and August were not in breeding condition.
Voice A sharp, slightly descending whistle *whiew?;* also four notes, the last strongest, longest and rising *few-few-few-wheea* (Spencer, XC 45298, January 2010).
Movements Presumably sedentary.
Range Endemic. Presently known from two locations only, in Dptos San Martín and Amazonas, Peru, south of the Rio Marañón.

Conservation status Currently listed by BirdLife International as Endangered. Known from only two sites, and not present in all apparently suitable habitat. Clearing of forest in its limited range has been severe and is continuing, even in the nominally protected Alto Mayo Forest. The newly designated Abra Patricia Alto Nieva Private Conservation Area aims to protect this species and Long-whiskered Owlet *Xenoglaux loweryi* (BLI SF 2014).

Etymology *Ochraceifrons*, Latin, 'ochre-fronted'.

LONG-TAILED CINCLODES
Cinclodes pabsti

Probably most closely related to Bar-winged Cinclodes *Cinclodes rufus* and Cordoba Cinclodes *C. comechingonus,* with which it may form a superspecies. However, in a phylogenetic study published in 2004, Chesser suggested that *pabsti* is an isolate from the other species of the genus, which fall into a Patagonian and central Andean highland clade and a north-central Andean and Pacific clade.

Type information *Cinclodes pabsti* Sick 1969, between Tainhas and Taimbézinho, Rio Grande do Sul, Brazil.

Alternative names Pabst's Cinclodes. Portuguese: Terezinha Pedreirinho.

Discovery In early 1966 H. Sick undertook an ornithological exploration of the southernmost corner of Brazil, an area which he had not previously visited. This was at a time of exceptionally wet weather in Brazil, with numerous landslides, but the state of Rio Grande do Sul was largely spared this. The main objective was to gather data on Black-necked Swan *Cygnus melanocoryphus* and Chilean Flamingo *Phoenicopterus chilensis*; however, passerines were also studied and a specimen of a new cinclodes was obtained in February 1966. This bird was sufficiently well known to local people to have several local names.

Description 21–22 cm. Monotypic (but see remarks under *C. espinhacensis*). A notably large, long-tailed, long-legged cinclodes, almost reminiscent of a Chalk-browed Mockingbird *Mimus saturninus.* Upperparts greyish-brown, with conspicuous white supercilia, dark brown lores and ear-coverts; two broad cinnamon-buff wing-bars and wing-band; tail dark brown with buff tips to outer rectrices; throat conspicuously white, remainder of underparts yellowish buffy-brown; iris brown; bill black; tarsi grey or black.

Habitat Grassland, pasture and agricultural land,

▼ Long-tailed Cinclodes. Aparados da Serra, Brazil, April 2011 (*Fabrice Schmitt*).

often near water; frequently near houses, sometimes in urban areas.

Food and feeding Terrestrial; runs rapidly on open ground, sometimes perching on rocks or fence posts. Diet arthropods.

Breeding Nest is an open platform of grass, feathers and string, sited in rock crevices or in tunnels excavated in banks; sometimes in cavities in buildings. Breeding season November in Santa Catarina state, September–November in Rio Grande do Sul (Sick 1993).

Voice Song is a prolonged trill *tsiewrrrrrrr*, increasing in volume and then decreasing; call is a descending *tseeoo*.

Movements Apparently sedentary. Has shorter, rounder wings than the migratory Bar-winged Cinclodes *C. rufus*.

Range Endemic. SE Brazil; SE Santa Catarina, NE Rio Grande do Sul. A population in Minas Gerais, 1,000 km to the north, is the subject of taxonomic debate (see under *C. espinhacensis*).

Conservation status Currently classified by BirdLife International as Near Threatened. Although it can co-exist very well with some traditional agricultural practices, it is suspected of declining in numbers due to the planting of open rangeland with pine trees and orchards, with, in some areas, up to 60% of available land so planted. Occurs in several actual or proposed national parks.

Etymology Named in honour of Dr Guido Frederico João Pabst (1914–1980), a Brazilian botanist famed for his work on orchids (the genus *Pabstia* is named for him) (Beolens & Watkins 2003). He sponsored Dr Sick's work in southern Brazil.

OLROG'S CINCLODES
Cinclodes olrogi

Treated as a race of Grey-flanked Cinclodes *C. oustaleti* by some authorities, or alternatively of Bar-winged Cinclodes *C. fuscus* (Nores 1986), but quite distinct in plumage. Ironically, Olrog himself regarded it as a race of *oustaleti*; recent molecular data show that it is a sister species to *oustaleti* (itself a 'split' from *C. fuscus*). Vocal differences suggest maintaining each as a separate species (Schulenberg 2010).

Type information *Cinclodes olrogi* Nores & Yzurieta 1979, Pampa de Achala, Sierras Grandes de Córdoba, Argentina.

Alternative names Spanish: Remolinera de Olrog, Remolinera Chocolate.

Discovery The type description is based on an adult male specimen taken in early April 1977 at an altitude of 2,200 m on the Pampa de Achala, Córdoba.

Description 17 cm. Monotypic. A typical cinclodes, rather small. Crown blackish-brown, separated from dark lores and ear-coverts by a white supercilium; upperparts dark grey-brown, faintly tinged chestnut; primaries dark brown with conspicuous white bases; tail dull grey-brown, outer rectrices tipped dull rufous; throat and upper chest whitish, becoming dull brown on lower chest and belly, the chest with indistinct dark scalloping; iris dark brown, bill and tarsi dark.

Habitat Open rocky grassland, often with cliffs and canyons; especially near streams. 1,600–2,800 m, but

◀ Olrog's Cinclodes. Near Quebrada de la Condorito Nature Reserve, Cordoba state, Argentina, November 2010 (*Graham Ekins*).

lower in austral winter.

Food and feeding Feeds actively on ground, swiftly running and pecking; frequently beside streams and in wet margins. Food items arthropods.

Breeding Excavates burrows in soft banks, up to 0.5 m in length; also in crevices in rocks and cavities in structures. Nest is a pad of hair and fibres; eggs, two. Breeding season is November–December.

Voice Distinct from other *Cinclodes* species but no data (Schulenberg 2010).

Movements Some altitudinal movement; in winter down to 900 m.

Range Endemic. Central Argentina (western Córdoba, in Sierras Grandes and Sierra de Comechingones) and NE San Luis (Sierra de San Luis). Total range within an overall area measuring about 350 km north to south and 140 km east to west.

Conservation status Currently classified by BirdLife International as Least Concern. Although the overall range is quite small and the population may have declined, it does not appear to be approaching the threshold for Near Threatened. The rocky nature of the habitat gives it some protection.

Etymology Named in honour of Claës Christian Olrog (1912–1985). Dr Olrog was born and educated in Sweden, but moved to Argentina in 1948, where he became one of the country's foremost ornithologists (and mammalogists). Notably, unlike many other immigrant scientists, he learned flawless unaccented Spanish (or, at least, Argentinian-accented Spanish). He was one of the pioneers of bird-ringing in Argentina. He was author of the first guide to Argentinian birds (*Las Aves Argentinas: una Guía de Campo,* 1959), as well as a posthumous book on the mammals of southern South America.

PERIJA THISTLETAIL
Schizoeca perijana

Treated as conspecific with several other isolated mountain-top thistletails under the name *fuliginosa* by some authorities (e.g. Vaurie 1971); however, Remsen (2003 HBW 8) maintains them as separate full species.

Type information *Schizoeca perijana* Phelps Jr 1977, camp 'Frontera 2', Sierra de Perijá, Zulia, Venezuela.

Alternative names Spanish: Piscuiz de Perijá.

Discovery Between 1941 and 1953 M. A. Carriker collected some 23,000 bird specimens in Colombia for

▼ Perija Thistletail. Serrania del Perija, Colombia, September 2015 (*Luis Eduardo Urueña*).

the Smithsonian Institution, Washington DC. Carriker provided full field notes on his collection; however, they remained unstudied for many years. In 1977 W. H. Phelps Jr named the new species, based on material collected in 1974 in the Sierra de Perijá, a range forming the boundary between northern Colombia and Venezuela, on the Venezuelan side of the border; in the meantime, S. Hilty and W. Brown studied the Carriker material (Carriker himself died in 1965) and discovered many significant specimens showing noteworthy range extensions, as well as a series of nine specimens of *A. (S.) perijana*, taken in July 1942 and unrecognised for 40 years (Hilty & Brown 1983).

Description 19–22 cm. Monotypic. In form a typical thistletail, small-bodied and fine-billed, with a long, graduated ragged tail. Upperparts greyish-brown; faint greyish supercilium; wing-coverts edged chestnut, giving a chestnut-brown patch on the closed wing; cinnamon-buff patch on chin; iris reddish-brown; bill black, with greyish base to lower mandible; tarsi grey.

Habitat *Páramo* grassland, or else in the forest and forest-grassland border-region, with dense undergrowth; 2,900–3,400 m.

Food and feeding Usually seen in pairs, foraging on arthropods and berries.

Breeding Nest and eggs undescribed. Adults in breeding condition in July (Fjeldså & Krabbe 1990).

Voice Song is similar to other thistletails; three or four slightly hoarse notes followed by a short, more tinkling sequence (Cuervo, XC 22194).

Movements Presumably sedentary.

Range Imperfectly known, since access to the area has been hazardous for many years due to the activities of Colombian guerrillas, apparently on both sides of the border. Has a rather narrow altitudinal range, all within the Sierra (or Serranía) de Perijá in Zulia, Venezuela, and Guajira, Colombia.

Conservation status Currently classified by BirdLife International as Endangered. The species is known from a limited number of locations in a small overall range. The main threat appears to be habitat destruction, much of it for illegal cultivation of narcotics. Part of the Venezuelan range is nominally protected by the Sierra de Perijá National Park, though due to the general lawlessness of the area there is no active management (BLI SF 2014). Habitat on the Venezuelan side does appear to be more intact (Restall *et al.* 2006). Recent land acquisitions on the Colombian side have created the Chamicero del Perijá Bird Reserve, while civil disorder in the area has diminished (Anon. 2015a). Current estimated population is in the range of 150–700 mature individuals (BLI SF 2014).

Etymology Both the vernacular and scientific names are derived from the Perijá Range.

VILCABAMBA THISTLETAIL
Asthenes vilcabambae

Forms a superspecies with several other allopatric high-altitude thistletails and sometimes treated as conspecific with them; conversely, Hosner *et al.* (2015b) present evidence for the full specific status of *A. f. ayacuchensis*, a treatment that we have followed.

◀ Vilcabamba Thistletail. Vilcabamba la Vieja, Vilcabamba, Peru, February 2015 (*Miguel Lezama*).

Type information *Schizoeca fuliginosa vilcabambae* Vaurie *et al.* 1972, Cordillera de Vilcabamba, Cuzco, Peru.
Alternative names Spanish: Piscuiz de Vilcabamba.
Discovery Described by Vaurie *et al.* in 1972, based on a study of material collected by J. S. Weske and J. W. Terborgh and by J. P. O'Neill, from 1967 onwards, originally as a race of White-chinned Thistletail *S. fuliginosa*; however, Braun & Parker (1985) argued for full specific status.
Description 18–19 cm. Monotypic. In form a typical *Schizoeca* thistletail, small, fine-billed with a very long, highly graduated and deeply-forked tail consisting of fine, pointed feathers with diffuse barbs. Overall rather dull brown above without an obvious eye-ring but with an inconspicuous pale supercilium; underparts more grey; chin light ochraceous; lower chest and belly with indeterminate paler scaling; iris brown to greyish-brown, bill blackish to dark horn with paler grey base to lower mandible; tarsi grey.
Habitat High-altitude *páramo* grassland, elfin forest and forest-edge with dense undergrowth. Mostly 2,830–3,600 m, rarely down to 2,500 m.
Food and feeding Few data. Forages singly or in pairs; prey items presumably arthropods.
Breeding Nest and eggs undescribed.
Voice Song begins with a series of slower, slightly ascending notes, which gradually become a rapid trill, ascending towards the middle then descending to an abrupt end (Hosner *et al.* 2015b).
Range Endemic; central Peru, in the Cordillera de Vilcabamba.
Conservation status Despite the restricted geographic range, classified by BirdLife International as Least Concern, as the population appears to be stable.
Etymology The species is named for its geographic location.

AYACUCHO THISTLETAIL
Asthenes ayacuchensis

Described by Vaurie *et al.* as a race of the Vilcabamba Thistletail *A. vilcabambae* (q.v.). However, Hosner *et al.* (2015b) argue convincingly for full specific status, based on vocalisations and molecular date.

Type information *Schizoeca fuliginosa ayacuchensis* Vaurie *et al.* 1972, Puncu, 30 km NE of Tambo, Ayacucho, Peru.
Alternative names Spanish: Piscuiz de Ayacucho.
Discovery In 1972 Vaurie *et al.* described the Vilcabamba Thistletail, *Asthenes (Schizoeca) vilcabambae,* as two geographically separated races, the nominate and the subspecies *ayacuchensis,* the latter from Dpto. Ayacucho, Peru. In a short communication published in 2015, Hosner *et al.* (2015b), presented vocal and biochemical evidence for full specific status of the latter. This has been accepted by the IOC but not, as yet, by the SACC.
Description 18–19 cm. Monotypic. In form a typical *Schizoeca* thistletail, very similar to Vilcabamba Thistletail *Asthenes vilcabambae* but lacking the pale scaling on the lower chest and belly; face blacker, gular patch more extensive and conspicuous.
Habitat A variety of stunted forest understorey and forest-edge habitats, including forest-edges adjacent to agricultural clearings, often with *Chusquea* bamboo.
Food and feeding No recorded data.
Breeding Nest and eggs undescribed.
Voice Song is a rapid high-pitched trill, lasting about two seconds, at a steady frequency; rather quiet and low in amplitude compared to other members of the genus (Hosner *et al.* 2015b).
Range Currently known from only three small valleys on the eastern Andean slope of Ayacucho, although further work might extend this. Confined to a narrow altitudinal belt with a total area of probably less than 500 km^2.
Conservation status Given the small known range, which is under considerable threat from burning and grazing, Hosner *et al.* (2015b) express some conservation concerns.
Etymology The specific name refers to the sole area of occurrence, Dpto. Ayacucho.

CIPO CANASTERO
Asthenes luizae

A restricted-range species whose relationships remain to be precisely defined; vocally most similar to Rusty-vented Canastero *A. dorbignyi*, but in plumage most resembles Patagonian Canastero *A. patagonica*.

Type information *Asthenes luizae* Vielliard 1990, Alto da Boa Vista, Serra do Cipo, Minas Gerais, Brazil.

Alternative names Portuguese: Lenheiro-da-serra-do-cipo, João-cipo.

Discovery In August 1988 M. Pearman and J. Hurrell observed a furnariid in the Brazilian state of Minas Gerais, which they immediately recognised as being a canastero (*Asthenes*). Since the location of their observation (the Serra do Cipo) was almost 2,000 km from the known range of any other member of this genus, the importance of their observation was immediately apparent to them. In fact, unbeknown to them, two other British ornithologists, S. Cook and B. Forrester, had independently seen the same species in the same location just a few weeks previously. Pearman carried out extensive fieldwork in 1988 and 1989. He was then informed that J. Vielliard had collected specimens in 1987, but had not published any information. Consequently Pearman published his observations in a paper submitted in January 1990; in this paper he made detailed observations of plumage, behaviour and vocalisations, but refrained from describing the new bird beyond "an undescribed Canastero (*Asthenes*)" (Pearman 1990). In the meantime J. Vielliard submitted a paper (in October 1989, published in 1990) describing, in rather sketchy detail, the new species, citing a specimen taken in December 1985 by F. Lencioni (Vielliard 1990). This description is based on two specimens, the holotype in the Zoological Museum, São Paulo University, and a paratype held privately. Consequently, Ferreira de Vasconcelos *et al.* (2008) redescribed the species in more detail, using the original holotype and ten additional specimens.

Description 17 cm. Monotypic. In form a typical *Asthenes* canastero, small, terrestrial, with a graduated and rather long tail. Upperparts greyish-brown, the crown warm brown; supercilium narrow and whitish; chin and upper throat white, finely streaked black; underparts grey, washed dusky-olive on centre of belly, deeper colour on rear flanks and more chestnut on undertail-coverts; tail dusky-brown, outer feathers rufous-chestnut. Iris blackish to brownish, upper mandible dark grey with black tip, lower mandible with paler grey base; tarsi dull pinkish or grey.

Food and feeding Mostly terrestrial, moving rapidly over rocky ground while gleaning in crevices; diet arthropods.

Breeding Nest is an untidy mess of twigs with a side entrance, leading to a chamber lined with cactus-down, feathers and fibres, invariably located in the shrub *Vellozia nivea*, about 12–28 cm above the ground. Eggs two, plain white. Nesting season September–November. Nest parasitism by Shiny Cowbirds *Molothrus bonariensis* probably very common; two active nests found were both parasitised (Gomes & Rodrigues 2010). For a photograph of an active nest, see Remsen (2003 HBW 8).

Voice Song is a musical series of 10–15 notes, the first eight or so loud and sharp, the last three contrastingly lower. Call note a high-pitched metallic *jlit*. Pearman (1990) gives a detailed analysis of vocalisations.

◀ Cipo Canastero. Santana do Riacho, Minas Gerais, Brazil, May 2013 (*Ciro Albano*).

Range Endemic. Originally found in a very small area – 10 km² – in the Serra do Cipo, Minas Gerais, Brazil. However, a number of other populations have been found in numerous other localities (all in Minas Gerais), from north-east of Belo Horizonte, disjunctly to the south-east of Montes Claros, a discontinuous distribution of about 350 km north to south (Cordeiro *et al.* 1998; Ferreira de Vasconcelos 2002).
Conservation status Currently classified by BirdLife International as Near Threatened. To a considerable degree it is protected by the unsuitability of its rocky habitat for agricultural purposes. Some areas are subject to fires set to clear areas for pasture, but this does not apply to the bulk of the range. Parts of the range have now been colonised by Shiny Cowbirds as a commensal to stock-raising and human settlement; in those areas nesting losses are potentially very damaging as the canastero is naïve to the parasite; however, in much of the range cowbirds do not as yet occur. Considerable parts of the range are protected by several parks (Ferreira de Vasconcelos 2002; Ferreira de Vasconcelos *et al.* 2008). Current population in the range of 50,000–100,000 individuals (BLI SF 2014).
Etymology Named after Luiza, the wife of F. Lencioni, the collector of the first specimen "for her constant help in his ornithological searches" (Vielliard 1990).

BAHIA SPINETAIL
Synallaxis whitneyi

A form with a complex and highly disputed taxonomic history; the current state of confusion is best summarised in part 7 of "A Classification of the Bird Species of South America", published by the SACC (26 May 2014). "*Synallaxis whitneyi/cinerea* was formerly considered a junior synonym of *S. ruficapilla.* Pacheco & Gonzaga (1995) showed that this population merits species rank, which they named *S. whitneyi.* Whitney & Pacheco (2001) then showed that *whitneyi* was a synonym of *cinerea.* More recently, however, Stopiglia & Raposo (2006) proposed that *whitneyi* is indeed the correct name. SACC proposal passed to change back to *whitneyi*...Bauernfeld *et al.* (2014), however, concluded that *cinerea* is the correct name."

To add further to the confusion, an analysis by Stopiglia *et al.*, published in 2013, suggests that only *ruficapilla* and *infuscata* can be considered valid species, and that *whitneyi* is a synonym of the former. Ridgely & Tudor (2009) treat *whitneyi* as a valid species, a course of action which for the purpose of this book we have followed, while realising that the last word on the subject has yet to be written. In most reference books (e.g. HBW 8, 2003) Bahia Spinetail is used as the common name of *S. cinerea*; if the specific status of *whitneyi* is validated an unambiguous vernacular name will be needed.

Type information *Synallaxis whitneyi* Pacheco & Gonzaga 1995, 7 km south-east of Boa Vista, Bahia, Brazil.
Alternative names Portuguese: João-baiano.
Discovery In July 1992 J. F. Pacheco and L. P. Gonzaga were investigating the remnants of the vanishing patches of Atlantic forest east of Boa Vista, Bahia, Brazil, an area where they had recently discovered a new species of flycatcher, the Bahia Tyrannulet (q.v.). Here they encountered a spinetail whose song resembled that of Plain or Pinto's Spinetail *S. infuscata,* a species whose limited distribution in NE Brazil was some 800 km away from their site; however, in plumage it bore some similarities to Rufous-capped Spinetail *S. ruficapilla*, again widely disjunct to the south-west. Capture of specimens revealed that they had only eight rectrices – as in *ruficapilla* – as opposed to the ten of *infuscata.* On this basis they described the Bahia birds as a new species, *S. whitneyi.*

▼ Bahia Spinetail. Boa Nova, Bahia, Brazil, December 2016 (*Ciro Albano*).

Description 16 cm. Monotypic. A typical *Synallaxis* spinetail in form, with long graduated, spiky-tipped rectrices and contrasting cap and wings. Forehead, crown and nape amber; mantle, rump and flanks dark brownish-olive; lores and ear-coverts blackish-grey, forming a mask through the eyes; throat dark grey; centre of belly glaucous; upperwing-coverts dark amber; iris reddish-brown; bill dark grey; tarsi olivaceous.

Habitat Undergrowth of humid forest, especially dense tangles of vines, ferns and bamboo near the forest edge, 750–1,000 m. Not, however, restricted to bamboo (cf. *S. ruficapilla*) (Pacheco & Gonzaga 1995).

Food and feeding Forages from ground level to 5 m up, usually below 2 m. Stomach contents arthropods: ants, bugs, beetles, Hymenoptera, cockroaches etc.

Breeding Nest and eggs undescribed.

Voice Pacheco & Gonzaga (1995) give a detailed analysis of vocalisations. The song is a doublet of two linked syllables, often repeated; it differs in detail from the corresponding vocalisation of *S. ruficapilla,* and greatly from that of *S. infuscata.* Females of *whitneyi* also sing, at a higher frequency than males. Other calls include a scold, a loud low *tschrrr,* common to all three taxa.

Range Endemic. No more than ten pairs have been found in the remnant forests of the Serra da Ouricana, eastern Bahia. It has subsequently been found in three discrete areas of Chapada da Diamantina National Park in central Bahia (Parrini *et al.* 1999). Recently, however, it has also been discovered in a number of additional localities: Mata Escura; Fazenda Limoeiro (Ribon *et al.* 2002); Serra Bonita, near Camacan; southern Chapada Diamantina (Parrini *et al.* 1999); a forest belt 5 km wide running along the coast between Itubera and Camamu (P. C. Lima *in litt.* 2003), and at Serra das Lontras in southern Bahia (Silveira *et al.* 2005) (BLI SF 2014).

Conservation status Currently classified by BirdLife International as Vulnerable. The main threat is habitat destruction for agriculture; some areas of the Chapada da Diamantina National Park appear to be subject to illegal logging (BLI SF 2014). Population estimates are in the range of 600–1,700 mature individuals.

Etymology Named in honour of Bret M. Whitney (b. 1955), an American ornithologist who specialises in, among many other things, sound recordings and vocalisations, a skill which lead to the discovery of the Cryptic Warbler *Cryptosylvicola randrianasoloi* (q.v.).

APURIMAC SPINETAIL
Synallaxis courseni

Very closely related to the widespread Azara's Spinetail *S. azarae* and regarded by some authorities as a superspecies with that and with Sooty-fronted Spinetail *S. frontalis,* both of which are allopatric.

Type information *Synallaxis courseni* Blake 1971, Bosque Ampay, Dpto. Apurímac, Peru.

Alternative names Blake's Spinetail, Coursen's Spinetail. Spanish: Pijuí de Apurímac.

◄ Apurimac Spinetail. Abancay, Dpto. Apurímac, Peru, February 2014 (*Carlos Calle*).

Discovery Based on an adult male specimen taken by Peter Hocking "during his extensive travels (in Peru) as a missionary" and donated to the Field Museum of Natural History, Chicago (Blake 1971).
Description 19–20 cm. Monotypic. In form a typical *Synallaxis* spinetail, rather long-tailed. Forehead, face, back and most of underparts dark grey; rear of crown and nape dark rufous; wings mostly bright orange-rufous except for dark brownish tips to remiges; 10 rectrices, dark sooty-brown; throat-patch sooty-black with pale grey feather margins; iris chestnut; bill black with blue-grey base to lower mandible; tarsi greenish-grey.
Habitat Dense undergrowth in humid *Podocarpus* forest, including areas of second-growth following clearing or landslides; areas of *Chusquea* bamboo; fragments of residual cloud-forest surrounded by cleared agricultural patches (Lloyd 2009).
Food and feeding Forages in lower levels of dense understorey, occasionally up to 4.5 m, and on the ground. Food items arthropods, no further details.
Breeding Nest and eggs undescribed. Adults in breeding condition in December; juveniles seen in March.
Voice Highly vocal. Song is indistinguishable from that of *S. azarae* (to playback of whose songs it responds strongly); a sharp nasal-like *keet-kweet*, often repeated endlessly. Other calls, a low chatter and a low-pitched squeaky trill (Lloyd 2009).
Movements Presumably sedentary.
Range Endemic. Originally thought to be confined to a limited area around Nevada Ampay, north of Abancay, Dpto. Apurímac, Peru; more recently found in several other locations, including between Apurímac and Ayacucho and in the Vilcabamba Mountains.
Conservation status Currently classified by BirdLife International as Vulnerable, due in the main to its restricted geographic range. Recent range extensions may change this. The original location is protected by the Ampay National Sanctuary but areas outside this are subject to habitat clearance for agriculture. Total population estimated in the Sanctuary is about 1,000 individuals; outside, no estimates (BLI SF 2014).
Etymology Named in honour of C. Blair Coursen, in recognition of his "interest in neotropical birds and timely generosity" (Blake 1971).

ORINOCO SPINETAIL
Synallaxis beverlyae

The relationships of this species have not been totally clarified. In plumage closest to Pale-breasted Spinetail *S. albescens* (with which it is syntopic), but vocally more resembles Dark-breasted, Chicli and Cinereous-breasted Spinetails (*S. albigularis, S. spixi* and *S. hypospodia*).

Type information *Synallaxis beverlyae* Hilty & Ascanio 2009, from an unnamed small island in the Río Orinoco, about 22 km south of Puerto Ayacucho, Venezuela.
Alternative names Rio Orinoco Spinetail; Spanish: Pijuí del Orinoco.
Discovery In January 1998 S. Hilty was travelling by boat up the Río Orinoco, south of Puerto Ayacucho, which at that point separates Colombia from Venezuela, when he heard unfamiliar vocalisations coming from a small river island; one was from a flycatcher (*Stigmatura* sp.), the other of a spinetail. Tape recordings were made of both; these, and subsequent observations, confirmed that the spinetail song was from an undescribed *Synallaxis* species. In May 1999 a series of specimens was taken on the island by staff of the Colección Ornitología Phelps in Caracas, followed by more in January 2001. Subsequently, in July 2005 and August 2006, D. Ascanio discovered the new species far downstream, at two locations in the state of Delta Amacuro.

▼ Orinoco Spinetail. Isla Chivera, Delta Amacuro state, Venezuela, December 2016 (*David Ascanio*).

▲ Orinoco Spinetail. Isla Chivera, Delta Amacuro state, Venezuela, May 2015 (*David Ascanio*).

The easternmost of these was some 800 km in a direct line from the site of the original discovery.

Description 13–16 cm. Monotypic. In overall appearance a typical *Synallaxis* spinetail; a fine-billed, small furnariid with a long and graduated tail of spiky-tipped feathers. Crown orange-rufous; face and forehead grey, with whitish supercilium; upperparts mostly light greyish-brown with a contrasting cinnamon-rufous patch on shoulders; tail slightly more rufescent than back; chin and throat whitish, becoming more grey on chest; undertail-coverts drab beige; iris light grey to pale brownish-yellow; bill horn-grey, paler at base of lower mandible; tarsi horn-grey.

Habitat Confined to river islands or banks immediately adjacent; in successional vegetation, with stunted and taller trees. Areas subject to seasonal flooding and often rather changing and impermanent, especially the downstream locations.

Voice Distinctive from that of the syntopic *S. albigularis*; neither species responds to playback of the other species' song. Song is a series of 6–9 notes, occasionally fewer or more in response to playback, in series of notes with pauses in between. Calls include a harsh rattle.

Movements Presumably sedentary.

Range Imperfectly known. Currently occurs on small riverine islands, in at least six known locations, in Amazonas and the borders of Delta Amacuro and Monagas states, Venezuela (Hilty & Ascanio 2010). Also in a very limited area on the Colombian bank south of Puerto Ayacucho (Donegan *et al.* 2010). The precise border between Venezuelan states, and between Venezuela and Colombia, is sometimes difficult to locate, due to the changes in the river channels which form the boundaries.

Conservation status Currently listed as Near Threatened. The known range, although widespread, is small; further fieldwork along the Orinoco is desirable. Some habitat destruction by conversion for corn and black bean cultivation and for tourist purposes has occurred (D. Ascanio, pers. comm.; BLI SF 2014).

Etymology Named in honour of Beverly, wife of the senior author of the type description, acknowledging "her unwavering support....during her husband's long absences and her invaluable field assistance in Latin America" (Hilty & Ascanio 2009).

BOLIVIAN SPINETAIL
Cranioleuca henricae

A sister species of Stripe-crowned Spinetail *C. pyrrhophia;* may form a superspecies with a number of other species.

Type information *Cranioleuca henricae* Maijer & Fjeldså 1997, 3 km north of Inquisivi, La Paz, Bolivia.

Alternative names Inquisivi Spinetail. Spanish: Curutié Boliviano.

Discovery In December 1993 the senior author, while conducting an ornithological survey in the montane valleys of the Inquisivi area, Dpto. La Paz, Bolivia, heard several spinetail songs in an area of dry forest. These were quite reminiscent of the song of Stripe-crowned Spinetail *C. pyrrhophia*; however, when the singers were finally seen, they were very clearly not of that species, having a solid rufous cap rather than the distinctively black-and-buff striped cap of *pyrrhophia.* In January 1994 the new form was found to be common, and two specimens were taken, followed by a third in January 1995. In October 1996 the species was found in northern Dpto. La Paz, some 200 km to the north-west, indicating a wider distribution than first suspected.

Description 14–15 cm. Monotypic. Crown rufous; white supercilium; back brownish-olive, shading to tawny on rump and rufescent on uppertail-coverts; wings rufous-chestnut; tail graduated with exposed barbs at the tips, rufous-chestnut; throat white, remaining underparts pale greyish-olive, darker on undertail-coverts; iris warm brown; bill pinkish with culmen and tip of lower mandible sooty; tarsi olive-yellow.
Habitat Understorey of dry, seasonally deciduous forest in rainshadow valleys, 1,800–3,000 m, possibly up to 3,300 m; at one location about 20 km south-east of the type locality, abundant at 1,300 m (Seeholzer *et al.* 2015). Has been found in exotic *Cupressus* groves but absent from orchards and *Eucalyptus* plantations (BL I SF 2014).
Food and feeding Forages in pairs, occasionally joins mixed-species flocks, from understorey to canopy. Food items arthropods.
Breeding Nest and eggs undescribed. A possible juvenile in January (Maijer & Fjeldså 1997).
Voice Song is a bouncy, accelerating and decelerating series of sharp notes, sometimes ending in a short, trilled *trrrrt* lasting 1.5–2 seconds, sometimes up to four seconds; also a longer, more irregular song, 3.5–14 seconds, going up and down in pitch but tending to descend overall, often as a duet. Contribution of female weaker and higher in pitch (Seeholzer *et al.* 2015). Call a rich churring *t-t-t* or *tittttt,* repeated (Remsen 2003 HBW 8).
Movements Presumably sedentary.
Range Endemic. Bolivia, in Dptos La Paz and Cochabamba (Herzog *et al.* 1999b).
Conservation status Currently classified by BirdLife International as Endangered. Much of the suitable habitat has been severely modified or destroyed by clearing, cutting for firewood, plantation of non-native *Eucalyptus* etc. Habitat destruction and overgrazing has also led to an increase in destructive landslides. Current population is estimated in the range of 2,500–10,000 individuals, but is declining (BLI SF 2014). At a newly reported location at an altitude of 1,300m, it was unexpectedly abundant, with about 60 individuals vocalising on one 3 km stretch of road (Seeholzer *et al.* 2015). The species does appear to have some toleration of habitat modification (Mayer 1999).
Etymology Named in honour of the mother of the senior author, Henrica G. van der Werff, "who sent her unwilling 14-year old son to the Dutch Youth League for Nature Study, where his interest in birds was aroused" (Maijer & Fjeldså 1997).

▲ Bolivian Spinetail. Inquisivi, Bolivia, October 2011 (*Fabrice Schmitt*).

DELTA AMACURO SOFTTAIL
Thripophaga amacurensis

Probably most closely related to Orinoco and Striated Softtails *T. cherriei* and *T. macroura,* both of which are very widely allopatric.

Type information *Thripophaga amacurensis* Hilty *et al.* 2013, along the Caño Acoima, Delta Amacuro, Venezuela.
Alternative names Amacuro Softtail. Spanish: Colesuave del Amacuro.
Discovery In October 2004 S. Hilty and D. Ascanio, working from M/V *Clipper Adventurer* in the southern parts of the delta of the Orinoco River, made a pre-dawn excursion by Zodiac up the Caño Acoima, one of the numerous channels within the delta; during this, an unfamiliar duetting vocalisation was heard. This was recorded; shortly after dawn the pair of singers was observed by all three authors of the type description, who saw an obvious similarity to Orinoco and Striated Softtails, while also noting plumage differences. During the course of the morning several other pairs of birds were located along about 1 km of streamside vegetation. In late December 2004 a short series of specimens was taken by staff of the Colección Ornitología Phelps in Caracas, allowing comparisons with voucher specimens of *T. cherriei* (itself an almost unknown species which has been seen only by a

◄ Delta Amacuro Softtail. El Toro, Delta Amacuro state, Venezuela, December 2016 (*David Ascanio*).

handful of ornithologists since its discovery in 1902); these established the distinctiveness of the new taxon.

Description 17 cm. Monotypic. In form much like the other *Thripophaga* softtails; a medium-sized furnariid with a long, graduated tail and fairly heavy pointed bill with a slightly decurved upper mandible. Head and upper back dull olive-brown with prominent yellow-buff streaks; scapulars and lower mantle warmer brown; tail cinnamon; closed wing warm chestnut-brown; chin cinnamon-buff, throat rufous-chestnut, chest brownish-olive with pale buffy streaks, becoming less well-defined on belly; vent ochraceous-tawny; iris rusty-brown; bill steel-grey with dusky base to upper mandible, paler flesh-pink base to lower mandible; tarsi dull yellowish-green. Sexes similar, except that female has slightly paler throat-patch.

Habitat Seasonally flooded forest, with trees 20–25 m in height, with fairly open understorey, usually near small streams and rivers. Type specimen taken at 35 m above sea level.

Food and feeding Forages at 4–18 m up, usually above 10 m; singly or in loose pairs. Rummages in vine-tangles and leaf debris caught up in vines and forks. No data on diet.

Breeding Nest and eggs undescribed. Other members of the genus build ball-shaped nests with entrances on the underside (Remsen 2003 HBW 8).

Voice Song is distinctive, usually given in duet; a long chattery rattle from one bird, joined by the other for a variable length of time, after which there is a gradual reduction in volume and tempo. The duet ends typically with only the initial bird still singing (Hilty *et al.* 2013).

Movements Presumably sedentary.

Range Endemic. Currently recorded from only four sites in the southern portion of the Orinoco Delta; total range estimated at 32–48 km^2, although further fieldwork might result in the expansion of this (Hilty *et al.* 2013).

Conservation status Not currently classified by BirdLife International, but the very small known range would suggest concern. At present the low human population has not had significant effects on the habitat; however, in higher and drier areas, including one where the species was found, clearance for cattle ranching could be an issue. A possible future concern could be oil-sands exploitation in areas adjacent to the Orinoco Delta.

Etymology Both the vernacular and scientific names come from the place name Amacuro, itself of Amerindian (Warao) origin, meaning roughly 'a quilt of rivers'.

PINK-LEGGED GRAVETEIRO
Acrobatornis fonsecai

One of the most remarkable discoveries in Brazilian ornithology for many years, requiring erection of a monotypic genus. The Pink-legged Graveteiro has many features in anatomy, plumage and behaviour that mark it out as very different from all other furnariids. Its relationships are not entirely clear; the plumage, voice and nest architecture suggests a connection to the greytails (*Xenerpestes*), two species very highly disjunct in Panama, Colombia and Ecuador, but relationships to Orange-fronted Plushcrown (*Metopothrix aurantiaca*) or the treerunners (*Margarornis*), or possibly the *Asthenes* canasteros or *Cranioleuca* spinetails, have all been postulated (Remsen 2003 HBW 8).

► Pink-legged Graveteiro. Camacan, Bahia, Brazil, December 2012 (*Ciro Albano*).

Type information *Acrobatornis fonsecai* Pacheco *et al.* 1996, Serra das Lontras above Itatingui, municipality of Arataca, Bahia, Brazil.

Alternative names Portuguese: Acrobata.

Discovery In January 1988, while searching for potential habitat for the Critically Endangered Stresemann's Bristlefront *Merulaxis stresemanni*, B. M. Whitney investigated a slope cloaked in undisturbed Atlantic forest in the Serra das Lontras in Bahia state, Brazil. Heavy rain precluded observations; in November 1994 Whitney suggested to J. S. Pacheco and P. S. M. da Fonseca that they examine the area further. On 17 November, while observing a mixed flock foraging in shade trees in a cocoa plantation, a strange pair of birds, one grey and one brown, was seen and studied for several minutes. Very obviously these were of a species not known in Brazil, and possibly not known at all. Consequently, the area was revisited in January 1995 by a larger group including the authors of the type description, and four specimens, two of each colour-type, were taken. A further visit in October 1995 resulted in the discovery of a large number of nests, the definition of the range and extensive recordings of vocalisations, allowing the description of a new genus and species.

Description 13–14 cm. Monotypic. Adults are basically grey and black. Crown black, forehead with grey spotting; supercilium grey; back grey, faintly scalloped with black; lower back and uppertail-coverts paler grey; wings black and grey; throat and breast grey, faintly streaked paler; lower belly slightly tinged olive; tail quite long, graduated, with spiny and forked appearance, grey with the distal area blackish; iris pale grey; bill dark brown above, pinkish below; tarsi strikingly bright pink. Sexes similar. Juvenile, uniquely in a furnariid, is quite distinct; generally tawny or reddish-yellow, with duskier wings, tail and crown.

Habitat Now apparently totally confined to cocoa plantations, which are traditionally grown under existing shade trees, often after much thinning; up to 550 m. Presumably, prior to the introduction of cocoa, it inhabited moist lowland forest which has now been destroyed. Absent from undisturbed montane forest.

Food and feeding Notable in its feeding behaviour. Usually found in mixed-species flocks including tanagers, flycatchers, warblers, vireos etc. Very acrobatic, frequently clinging upside-down in the manner of a Xenops, gleaning from live foliage, the bark of dead limbs etc. Food items comprise arthropods: beetles, midges, ants, bugs, insect larvae, insect eggs and spiders.

Breeding Nests are untidy, bulky, obvious constructions, sited with little regard for concealment, located mostly in tall mature leguminous trees, at considerable heights; in shape globular or ovoid, made of coarse twigs, with an entrance tunnel leading to a simple moss-lined chamber measuring about 6 × 6 cm. Very frequently more than one nest (in fact up to five) are located in a tree, but it appears that only one is active at one time; the additional nests, which may be smaller than the active nests, are 'dummies' and not used, but possibly act as decoys against predators, and possibly as a source of nesting material. Immature birds have been observed to work alongside the adults in nest-construction; it is not known whether these helpers were the young of previous broods. Breeds in austral spring, especially September–October. Eggs undescribed. Clutch size probably two or three; young fed by both sexes (Pacheco *et al.* 1996).

Voice Song is a simple series of very short, piercing syllables, starting slowly then speeding up, lasting about 4–8 seconds, though longer in response to playback. Pairs duet; the female contributions are irregularly paced bursts of sharp chips. Foraging and flight calls are short sharp syllables delivered at irregular intervals.

Movements Presumably sedentary.

Range Endemic. A limited area, inland, in SE Bahia and NE Minas Gerais, Brazil.

Conservation status Currently listed by BirdLife International as Vulnerable. Essentially none of the original (postulated) habitat remains; almost the entire range of the species is coincident with areas where cocoa is grown. In this area of Brazil this has been practiced since the middle of the 18th century. The traditional method of cocoa production involves growing cocoa trees under the cover of much larger, scattered shade-trees; Pink-legged Graveteiro has obviously adapted to this well, living in highly anthropogenic habitats (otherwise it doubtless would have become extinct 250 years ago). Unfortunately, this traditional method of cocoa production is endangered, by a combination of low prices and an infestation of the accidentally introduced fungal disease *Crinipellis perniciosa*, leaving landowners sometimes with no alternative but to cut down the large trees for cash and convert the land to cattle pasture. As Pacheco *et al.* (1996) point out, under these circumstances large-scale land purchase may be essential for the survival of the species (BLI SF 2014; Anon. 2013b).

Etymology The new generic name *Acrobatornis* comes from the Greek 'acrobatic bird', as does the suggested Portuguese vernacular name. The specific name honours P. S. M. da Fonseca, "a multi-talented friend of many years, not only because he was the first to gasp in wonder at the living bird, but also in recognition of his unending encouragement and deep generosity" (Pacheco *et al.* 1996). The English vernacular name is derived from the Portuguese 'gravetos' (twigs and sticks), with reference to its characteristic and conspicuous nests.

ALAGOAS FOLIAGE-GLEANER
Philydor novaesi

Forms a superspecies with Black-capped Foliage-gleaner *P. atricapillus*, a species with an extensive but disjunct distribution in SE Brazil.

Type information *Philydor novaesi* Teixera & Gonzaga 1983b, Serra Branca, municipality of Murici, Alagoas, Brazil.

Alternative names Portuguese: Limpa-folha-do-noreste.

Discovery In early 1979 D. M. Teixera and L. P. Gonzaga of the Federal University of Rio de Janeiro were conducting a survey in some of the residual fragments of the Atlantic Forest in Alagoas state, Brazil, to ascertain the status of the endangered Alagoas Curassow *Mitu mitu* (this species has since become extinct in the wild, being now represented by some 130 individuals, not all pure-bred, in captivity). Incidentally during this study, in February 1979, two specimens of an unknown furnariid were mist-netted out of a mixed flock of passerines moving low-down through the forest. Comparison with numerous specimens of other *Philydor* species established the validity of the specimens as a new species.

Description 18 cm. Monotypic. A medium-sized furnariid, essentially rufous-brown all over, the rump more rufous and the uppertail-coverts brighter rufous; wings darker rufous-brown; tail rounded, bright rufous; underparts paler rufous; cap dark brown, faintly spotted on forehead; conspicuous pale buff eye-

◀ Alagoas Foliage-gleaner. Jaqueira, Pernambuco, Brazil, October 2008 (*Ciro Albano*).

ring and supercilium; iris brown; bill blackish-brown above, ivory below; tarsi greenish-horn. Differs from *P. atricapillus* in less warm tone and lack of blackish crown and lower facial stripe.
Habitat Tropical lowland evergreen forest and humid, hilly second-growth forest, 400–550 m (Remsen 2003 HBW 8).
Food and feeding Usually in pairs, in lower and mid-storey, often in mixed flocks. Recorded prey is arthropods, including beetles, ants, Orthoptera and insect larvae.
Breeding Nest and eggs undescribed. One immature seen in March.
Voice Song is a slightly descending series of whistles, *uu-uu-uu*; alarm call, *thurr.*
Movements Presumably sedentary.
Range Endemic. Confined to very limited areas of Alagoas and adjacent Pernambuco states, Brazil.
Conservation status Critically Endangered; very probably extinct. More than 95% of the forest in the 'Pernambuco Centre of Endemism' has been lost to cattle ranching, sugarcane production and other purposes. After the species' discovery in 1979, four further specimens were taken between 1983 and 1986. Some 6,000 ha of the original site was protected as the Murici Ecological Station in 2001; however, forest degradation has continued, with illegal logging, charcoal extraction, fires and hunting; reserve guards have been threatened (Whittaker 2001). There were no records between 1992 and 1998; subsequently there were numbers of observations, usually of single birds or pairs, from 1998 onwards. In 2003 a separate population was discovered at Frei Caneca Reserve, Pernambuco (Mazar Barnett *et al.* 2005), with undocumented reports from another nearby location. Despite extensive searching, the last documented record, at Frei Caneca, was in September 2011, with a further sight record in April 2012. No further observations have been reported since then (Lees *et al.* 2014).
Etymology Named in honour of the distinguished Brazilian ornithologist Dr Fernando da Costa Novaes (1927–2004).

CRYPTIC TREEHUNTER
Cichlocolaptes mazarbarnetti

A species with close affinities to Alagoas Foliage-gleaner *Philydor novaesi* (q.v.), itself a recently described species; arguments are made that both species more properly belong in the genus *Cichlocolaptes*, previously monotypic.

Type information *Cichlocolaptes mazarbarnetti* Mazar Barnett & Buzzetti 2014, based on a specimen taken in 1986 at Serra Branca, Alagoas, Brazil.
Alternative names Portuguese: Gritador-do-noreste.
Discovery During fieldwork in October 2002, J. Mazar Barnett and D. Buzzetti observed an ovenbird at the Murici Ecological Station in Alagoas state, Brazil, that in plumage resembled Alagoas Foliage-gleaner *Philydor novaesi*, but differed in general morphology and voice, being far more reminiscent of Pale-browed Treehunter *Cichlocolaptes leucophrys*, a species found much further south in Brazil. Later the same workers observed Alagoas Foliage-gleaner in Pernambuco state and realised that the Murici birds differed markedly from these. This led to a re-examination of two specimens taken at Murici in 1986 and identified at that time as *novaesi*; the substantial differences between these and specimens of genuine *novaesi* led to the description of the new species, as well as the opinion that both it and *novaesi* properly belong in *Cichlocolaptes*. Claramunt (2014) performed a detailed morphometric analysis and concluded that *mazarbarnetti* is a valid species.
Description 20–22 cm. Monotypic. Crown and forehead jet-black; nape, back and rump cinnamon-brown; tail pale orange-rufous, darker distally in centre; throat, sides of head, and supercilium pinkish-buff; ear-coverts pinkish-buff streaked dusky; breast and belly cinnamon; flanks brown; wings rich earth-brown; iris brown; bill black above, paler below; tarsi (of dried specimen) greyish-olive. Differs from *novaesi* in larger size, heavier bill, uniformly blackish crown, dark patches on sides of neck and absence of buffy eye-ring.
Habitat Dense, humid forests on hilly terrain, with trees of up to 25 m with dense growth of epiphytic mosses, bromeliads and orchids.
Food and feeding Usually feeds in mixed-species flocks, foraging in mid-levels and subcanopy 8–20 m up, often rummaging around in bromeliads. No data on diet.
Breeding Nest and eggs undescribed.
Voice Mazar Barnett & Buzzetti (2014) give a detailed analysis of vocalisations and comparisons with those of Alagoas and Black-capped Foliage-gleaners *Philydor novaesi* and *P. atricapillus* and Pale-browed Treehunter *Cichlocolaptes leucophrus*. Song is a fast, dry rattle followed by a series of 4–8 loud raspy notes. Alarm call is 1–3 notes, two descending and ascending notes followed by a raspy note.

Movements Presumably sedentary.

Range Endemic. Presently known from two sites only; the type locality at Murici in Alagoas state and at Frei Caneca in Pernambuco state, some 60 km to the south, although not all suitable habitat has been thoroughly searched.

Conservation status Not currently assessed by BirdLife International; however, Mazar Barnett & Buzzetti suggest that Critically Endangered would be appropriate. The species is presently known from two locations only, with an area of less than 100 km², with probably no more than 3,000 ha of suitable surviving habitat. Habitat destruction has not ceased; the species appears to be reliant upon abundant bromeliad growth, which does not occur in young secondary forest. Logging in apparently optimum habitat continues. "Sadly our expectations for the long-term survival of this species are not high" (Mazar Barnett & Buzzetti 2014). Recent reports (Anon. 2015b) suggest that the total population may not exceed ten birds, probably optimistic in view of the fact that the last definite record was in 2007 (Pereira, Dantas *et al.* 2014); evaluated as Extinct in the latest Brazilian Red List.

Etymology Named, by the junior author of the type description, for the senior, Juan Mazar Barnett. In modern practice it is not customary for the author of a type description to name a species after himself. Sadly, in this case, Sr Mazar Barnett, Argentinian by birth, died at the tragically early age of 37 and the type description was published posthumously. He was one of the most prolific authors in modern Latin American ornithology (as a brief glance at the bibliography this book will attest), publishing more than 50 papers in refereed journals (including several involving the description of new species). His premature death robbed South American ornithology of one of its most promising scientists (Naka 2013; Naka 2014). The Portuguese name 'Gritador' means one who screams, in reference both to the loud vocalisations of the species and to a Brazilian legend of a hunter who accidentally shot his brother and whose soul can be heard in the forest, screaming in pain, as he searches for him.

INAMBARI WOODCREEPER
Lepidocolaptes fatimalimae

A semi-cryptic species, very similar to Rondônia Woodcreeper *L. fuscicapillus* (itself a newly proposed split), but well differentiated vocally.

Type information *Lepidocolaptes fatimalimae* Rodrigues *et al.* 2013, Municipality of Guajará, Amazonas state, Brazil.

Alternative names Portuguese: Arapaçu-do-inambari. Spanish: Trepatroncos del Inambari.

Discovery In a study published in 2013, Rodrigues *et al.* made a careful examination of the Lineated Woodcreeper *L. albolineatus* complex, a group with a confusing taxonomic history. Previously, *albolineatus* was treated as a polytypic species with five races (e.g. Marantz *et al.* 2003). Rodrigues's mitochondrial DNA study showed the existence of five groups within the *albolineatus* complex, all of which were already named; however, one population from south of the Solimões-Amazon and west of the Rio Madeira, to which the name *fuscicapillus* had been incorrectly applied, was vocally, morphologically and genetically distinct and was named as a new species, *L. fatimalimae*. The SACC recommended acceptance of this treatment (Proposal 620).

◄ Inambari Woodcreeper. Estrada da Pedreira, 10 km east of Assis, Brazil, September 2009 (*Andrew Whittaker/Birding Brazil Tours*).

Description 17–19 cm. Monotypic. A relatively small, fine-billed woodcreeper. Upperparts chestnut-brown, richest on the tail; chin buff, chest brown with prominent longitudinal streaks, heaviest on chest; iris brown; bill horn-grey; legs greyish. Differs from *fuscicapillus* in having an unspotted head which is the same colour as the back, absent or inconspicuous post-ocular stripe and chest stripes tipped brown rather than black.
Habitat Humid lowland and foothill *terra firme* forest, sometimes in *várzea*.
Food and feeding Forages in canopy of primary and disturbed forest. Mostly in pairs and usually in mixed-species flocks. Prey items comprise arthropods.
Breeding Nest and eggs undescribed. Presumably in tree cavities like its closest relatives.
Voice A rapid series of short notes, at the rate of about ten notes per second, the sequence lasting two or three seconds. Rodrigues *et al.* (2013) give a detailed analysis and comparison with other members of the *albolineatus* complex.
Movements Presumably sedentary.
Range Western Amazonia, Brazil. Acre and Amazonas, west of the Rio Madeira and south of the Rio Solimões; eastern Ecuador; Peru, from Loreto and San Martín south through Huánuco to Cuzco and Madre de Dios; Bolivia from Beni and La Paz to Santa Cruz.
Conservation status Not currently classified by BirdLife International, but widespread in many remote areas that are not presently excessively impacted by human activity; also appears to tolerate some slight habitat modification, so presumably should be Least Concern.
Etymology Named in honour of Sra Fatima Lima, in recognition of her many years' service as curator of the collections in the Museu Paraense Emilio Goeldi, Belém, Brazil.

CAMPBELL'S FAIRY-WREN
Chenorhamphus campbelli

An isolated and almost unknown species with a confusing taxonomic history.

Type information *Malurus campbelli* Schodde & Weatherly 1983, Mt Bosavi, Papua New Guinea.
Discovery In February 1980, and again in February and November 1981, R. W. Campbell, working under the auspices of the Australian bird-banding scheme on Mt Bosavi in western Papua New Guinea ("one of the most remote and inaccessible localities in New Guinea") mist-netted and photographed five examples of an unusual fairy-wren. These were measured, described and photographed, but no specimens were taken. On this basis Schodde & Weatherly (1983) described a new species, *Malurus campbelli,* separated from the more widely distributed Broad-billed Fairy-wren *M. grayi* because of its smaller size, narrower bill, pure black, not dusky-blue crown and other plumage characters. The elusive nature of the bird as well as the inaccessibility of its distribution made further study difficult, but several specimens in different plumages were taken by Campbell and R. D. MacKay in November 1982 (Schodde 1984). However, in a study of 31 of the 37 known specimens of *M. grayi* (including all the '*campbelli*' specimens), LeCroy & Diamond concluded that the apparent distinctions in plumage of '*campbelli*' fell within the range of *grayi,* and instead classified it as a race of that species, a conclusion in line with Beehler *et al.* (1986) and Vuilleumier *et al.* (1992). Notwithstanding all of this, in 2011 Driskell *et al.* published a study of mitochondrial and molecular DNA, showing that *campbelli* was highly divergent from *grayi.* They also showed that, along with two other monotypic New Guinea malurids (Orange-crowned Fairy-wren *Clytomyias insignis* and Wallace's Fairy-wren *Sipodotus wallacii*), *grayi* and *campbelli* form a clade distinct from the more widely distributed Australasian genus *Malurus,* placing them in a genus *Chenorhamphus.*
Description 11–13 cm. Monotypic. A typical fairy-wren in form, small, waif-like, with a large head, long fine legs and a lengthy, graduated tail. Sexually dimorphic, although not to the degree of Australian *Malurus* species; does not appear to have an eclipse plumage. Male: crown, ear-coverts, lores and broad malar stripe black; broad supercilium stripe sky blue; back mid tawny-brown; tail mid-bluish grey-brown, narrowly tipped buff-white; long, pointed turquoise-blue ear-tufts; underparts from chin to crissum delicate powdery sky blue; iris blackish-brown; bill black; tarsi deep flesh-grey. Female similar, but the lower breast is white, crissum pale tawny-rufous, and ear-tufts shorter and less pointed.
Habitat Forested foothills, swampy secondary forest with Sago Palm *Metroxylon sagu* and rattan *Calamus* and other vines, at 800m, although the forest zone where the bird has been found extends from 300–1,100 m (Schodde 1982).
Food and feeding Few data; forages low down; insectivorous.
Breeding Nest and eggs undescribed.

Voice Undescribed; Broad-billed Fairy-wren has a reeling song of 2–3 seconds duration.
Movements Presumably sedentary.
Range Currently known from Mt Bosavi, a massif isolated from the main New Guinean cordilleras by lower terrain; Schodde (1982) speculates that it may extend into eastern Irian Jaya.
Conservation status Presently unclassified by BirdLife International; doubtless protected by the very remote and inaccessible nature of its habitat. Appears to be uncommon.
Etymology Named in honour of Rob Campbell, who mist-netted the first examples (Schodde & Weatherly 1983).

GREY GRASSWREN
Amytornis barbatus

A distinctive, especially long-tailed, grasswren; genetic evidence suggests that it is rather distant from other members of the genus, interpreted as being an older, more primitive, form (Christidis 1999).

Type information *Amytornis barbatus* Favaloro & McEvey 1968, Teurika, Balloo River, NW New South Wales, Australia.
Discovery In September 1942 N. Favaloro and A. Storer were travelling by car in a remote area of outback New South Wales when a notably pale, long-tailed grasswren was seen. They were not able to investigate this further, and it was not until 25 years later, in July 1967, that Favaloro was able to return. Then, on 7 July, he collected five specimens and found three nests with eggs, thereby allowing the description of a new species. A second race, *A. b. diamantina*, was described by R. Schodde and L. Christides from a disjunct area in SW Queensland and NE South Australia in 1987.
Description 18–21 cm. In form differs from all other grasswrens in having a notably long, rather narrow, tail. Nominate male has white-streaked black crown, white face with contrasting black markings including a black eyestripe, a black V-mark on white ear-coverts; upperparts pale cinnamon, laterally streaked with black and white; tail black; underparts white, the chest with grey-black streaks, the flanks buff; iris dark brown; bill blackish; tarsi dark grey-brown to greyish-black. Female similar but smaller and duller. Race *diamantina* is larger than the nominate, less densely streaked and more cinnamon below.
Habitat Unusually in grasswrens, it inhabits swampy floodplains rather than arid areas; lignum (*Muelenbeckia cunninghamii*) and swamp canegrass (*Eleocharis pallens*) flats on inland-flowing rivers (Carpenter 2002). In drought years seems to concentrate in areas of tall dense lignum (Hardy 2002).
Food and feeding Secretive; runs on the ground with tail cocked. Forages in the open and in shrubs;

▼ Grey Grasswren *A. b. barbatus*, adult at the nest. Caryapundy Swamp, NW New South Wales, Australia, 1968 (*Len Robinson*).

▲ Grey Grasswren *A. b. diamantina*, adult. Goyder Lagoon, Warburton Creek floodplain, northern South Australia, August 2012 (*Chris Tzaros*).

diet includes seeds, insects and sometimes water snails (Favaloro & McEvey 1968).

Breeding Nesting may be irregular in response to variable rainfall; in dry years may not breed. Nest is a bulky, semi-domed construction, loosely woven from grass stems, lined with fine grass, plant down and a few feathers, situated in the centre of a lignum or canegrass clump, usually from 0.3–0.6 m up. Eggs 2–3, creamy-white, densely marked with reddish-brown spots, heavier on blunt end. Incubation by female; young tended by both parents. Laying recorded in winter and spring (early July, mid-August, September) but possibly variable according to conditions (Rowley & Russell 1997).

Voice Song is simpler than those of most grasswrens; a series of high-pitched metallic ringing notes *trip-ip-ip*. Alarm call is a single, high-pitched piercing *eep* (Rowley & Russell 1997).

Movements Non-migratory, but appears to move locally in response to climatic conditions; in times of drought seems to concentrate in patches of high lignum (Hardy 2002).

Range Endemic. Central Australia. *A. b. barbatus*, extreme SW Queensland and NW New South Wales; *A. b. diamantina*, drainage of Diamantina River, Goyder Lagoon and the Warburton System, SW Queensland, with 54 sites identified. A population in the Cooper Creek area is undetermined as to subspecies (Black *et al.* 2012). Some potential areas of range are inaccessible and not yet investigated.

Conservation status Nominate race occurs in one (quite extensive) location and is classified Vulnerable; *A. b. diamantina* occurs in at least five discrete populations with 54 identified sites and is classified as Least Concern. Main threats are habitat changes from fire, grazing by introduced mammals including feral goats, pigs and horses, and changes to flow of internal rivers for irrigation. No major areas of habitat are currently protected.

Etymology *Barbatus*, Latin, 'bearded', referring to the striking facial pattern.

KALKADOON GRASSWREN
Amytornis ballarae

Formerly regarded as a race of Dusky Grasswren *Amytornis purnelli*, but biochemical evidence suggests full species status. Shares a lineage with four other grasswren species, Thick-billed, Dusky, Eyrean and Black Grasswrens *A. textilis*, *A. purnelli*, *A. goyderi* and *A. housei* (Christidis *et al.* 2010).

▼ Kalkadoon Grasswren, male. Mt Isa, NW Queensland, Australia, July 2014 (*Lindsay Hansch*).

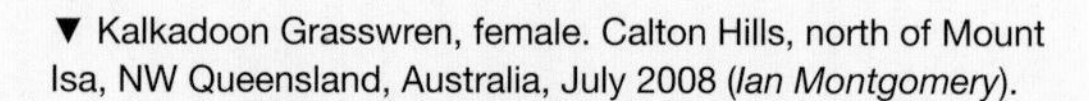

▼ Kalkadoon Grasswren, female. Calton Hills, north of Mount Isa, NW Queensland, Australia, July 2008 (*Ian Montgomery*).

Type information *Amytornis textilis ballarae* Condon 1969, 10 km south of Mary Kathleen, near Ballarae copper mine, SE of Mount Isa, NW Queensland, Australia.

Alternative names Grey-bellied Grasswren.

Discovery Grasswrens of the genus *Amytornis* are characterised by geographically isolated populations, which lend themselves to speciation. Consequently specimens taken from the Mount Isa region of western Queensland were referred for examination by H. T. Condon of the South Australian Museum, who described them as a race of Thick-billed Grasswren *A. textilis,* a species with two populations in central and western Australia, both highly disjunct. However, Parker (1972) reassessed the situation, classifying *ballarae* as a race of *purnelli.* More recent authors (e.g. Higgins *et al.* 2001; Rowley & Russell 2007 HBW 12) have treated *ballarae* as a valid species in its own right, a view supported by biochemical studies.

Description 15–16 cm. Monotypic. Male: face dusky rufous with white streaks; upperparts bright rufous-brown, streaked black and white on head and neck; upperwing and tail grey-brown; throat and breast straw-colour with white streaks; belly grey; iris dark brown; bill dark grey to blackish with dark blue-grey base to lower mandible; tarsi dark grey. Female has dark rufous flanks.

Habitat Spinifex (*Triodia*) on arid rocky hillsides and in gullies.

Food and feeding Forages in pairs or small groups, mostly on ground or low bushes; food items comprise insects and seeds in about equal proportions. Young fed on insects and spiders (Higgins *et al.* 2001).

Breeding Nest is a partly-domed construction, often with a pronounced roof and projecting hood, made of soft grass stems, lined with shredded grass and fine fibres; located in tussocks of spinifex, or dead bushes surrounded by spinifex. One nest at a height of 0.45 m. Eggs 2–3; dull white, heavily marked with spots and blotches of light to dark brownish and purplish-red, especially around blunt end. Incubation probably wholly or entirely by female. Young fed by both parents but brooded by female alone. Laying season July–September or October (Carruthers *et al.* 1970).

Voice Little known. Stated to differ from that of Thick-billed Grasswren *A. textilis*; a series of high-pitched, varied short trills, *tei-tei-tit-t-tewi* (Chapman 1996). Contact calls are single high squeaks; alarm call a loud squawk.

Movements Probably sedentary.

Range Endemic. NW Queensland; around Mount Isa, east Barkly Tableland and Selwyn Range, north to Thorntonia and south to Boulia, an area about 300 km north to south and somewhat less east to west.

Conservation status Classified by BirdLife International as Least Concern. Although the range is geographically restricted, the population appears to be stable.

Etymology *Ballarae,* from the Ballarae copper mine, near Mount Isa, Queensland; Kalkadoon, from the aboriginal tribe which formerly inhabited the region.

EUNGELLA HONEYEATER
Lichenostomus hindwoodi

Sometimes placed in a separate genus *Bolemoreus,* together with another *Lichenostomus* species. Alternatively, sometimes placed in *Caligavis.*

Type information *Meliphaga hindwoodi* Longmore & Boles 1983, Massey Creek, Clarke Range, eastern Queensland, Australia.

Discovery This taxon was first mentioned and illustrated by J. S. Robertson (1962) who believed it to represent a disjunct population of Bridled Honeyeater *M. frenata,* a species resident in northern Queensland. Noting differences between the southern population and *frenata,* N. W. Longmore and W. E. Boles collected specimens between 1975 and 1980, as well as making field observations and recording vocalisations. It was shown that *frenata* did not respond to taped calls of the new taxon, although the reverse experiment was not performed. Consistent and significant differences in plumage and vocalisations allowed the new form to be described as a new species (Longmore & Boles 1983). This was accepted by Christidis & Boles (2008).

Description 17–20 cm. Monotypic. In form very similar to other *Lichenostomus* honeyeaters, a medium-sized bird with prominent facial markings. Above dark grey-brown, the forehead and crown scalloped with pale grey, and pale grey streaking on mantle, back and scapulars; underparts paler grey-brown with fine white streaks; facial pattern with black in front of eye, gape and moustachial stripes off-white; small off-white patches behind eye and below ear-coverts; iris blue-grey to greenish-blue; bill black; tarsi bluish-grey. Sexes similar in plumage. Male larger than female. Differs from *frenata* in having a solid black (not bicoloured) bill, a paler crown and by having ventral streaks.

▲ Eungella Honeyeater. Eungella National Park, Queensland, Australia, September 2014 (*Vik Dunis*).

Habitat Rainforest, usually above 900 m, occasionally down to 150 m; casuarinas, eucalypts, sometimes in gardens.

Food and feeding Nectarivorous and frugivorous; flowers, also arthropods. Feeds mostly in canopy.

Breeding Nest is a deep cup of fine plant fibres, covered with moss outside, at heights of 4–20 m; in horizontal branch of foliage, sometimes in mistletoe (Cooney *et al.* 2006), built by both sexes. Chicks fed by both parents. Nesting season late August–November.

Voice Song is a short series of metallic notes, beginning with a harsh rattle. Calls *chip, churr* etc.

Movements Non-migratory, but some birds appear to move altitudinally.

Range Endemic. Clarke Range, Queensland.

Conservation status Classified by BirdLife International as Least Concern. Although the overall range is small, populations appear to be stable, and most of the known range is in protected areas. Quite common in some parts of its range (Morcombe 2000).

Etymology Named in honour of Keith A. Hindwood (1904–1971), formerly President of the Royal Australasian Ornithologists' Union, in appreciation of his lifetime study of Australian ornithology. The common name 'Eungella' comes from the township of Eungella, a word of aboriginal origin meaning 'mountain of mists' (Longmore & Boles 1983).

WATTLED SMOKY HONEYEATER
Melipotes carolae

Forms a superspecies with Spangled, Arfak and Common Smoky Honeyeaters (*M. ater, M. gymnops* and *M. fumigatus*). Not accepted as a valid species by the BirdLife Taxonomic Working Group; Diamond & Bishop (2015) comment that the species limits within the *Melipotes* of western New Guinea are uncertain and require more specimen material for clarification.

Type information *Melipotes carolae* Beehler *et al.* 2007, Bog Camp, Sarmi District, Papua Province, Indonesia.
Alternative names Fakfak Honeyeater.
Discovery The Foja Mountains, located north of the main central spine of Irian Jaya (West Papua) are completely isolated geographically from other highland areas, and are consequently of great potential biological interest. Nevertheless, the sheer inaccessibility of the area, largely uninhabited and lacking tracks and even trails, prevented comprehensive ornithological investigation until the use of helicopters allowed J. Diamond to make the first visits to higher altitudes in 1979 and 1981. However, the area remained largely unknown until November 2005 when a team from Conservation International and the Indonesian Institute of Sciences made a Rapid Assessment of animal and plant life in the western part of the range, arriving by helicopter in one of the few feasible landing sites, a peat bog. Within a few minutes of landing party members noticed a distinctive wattled honeyeater foraging unwarily in forest at the edges of the bog. Several specimens were mist-netted, providing the holotype and paratypes of a new species.
Description Male 22 cm, female smaller. Monotypic. Overall largely sooty-black, darker on the crown and nape, paler on the underparts, which are scaled with paler grey; sides of face, including area above and behind eye, covered by a fleshy orange area of bare skin, with a large loose pendant wattle at the lower edge; iris dark brown or red-brown; bill and tarsi black.
Habitat Closed humid tropical forest, with annual rainfall about 3–4 m, both interior and edge; type specimens taken at 1,150 m, but upper and lower levels of distribution not defined.
Food and feeding Forages in the middle and upper levels of vegetation, especially on plants producing small fruits. Predominantly frugivorous (Beehler *et al.* 2007).
Breeding Nest and eggs unknown. However, Diamond (1985) described a *Melipotes* nest in the Foja Mountains which was assumed to be that of *M. fumigatus*; since that species does not appear to occur in the Foja range, it would seem likely that it belonged to *carolae*. It was described as a bundle of debris 20 cm in diameter, woven around the trunk of a sapling at a height of 9 m. Specimens collected in late November and early December were not in breeding condition (Beehler *et al.* 2007).
Voice No data.
Range Endemic. Confined to the Foja Mountains, northern Papua Province, NW New Guinea. Due to the inaccessibility of the area, full limits of its range are not defined.
Conservation status Not currently assessed by BirdLife International. Restricted range species. However, given that the known range appears to be pristine, and not even visited by local people (partly due to inaccessibility, but also due to the fact that the summits of the range are considered sacred), it seems likely that no threats presently exist.
Etymology Named for Carol, wife of the senior author of the type description, "acknowledging...her personal commitment to biodiversity studies in New Guinea" (Beehler *et al.* 2007).

▼ Wattled Smoky Honeyeater. Foja Mountains, New Guinea, Indonesia, November 2008 (*Tim Laman/Getty Images*).

BISMARCK HONEYEATER
Melidectes whitemanensis

A distinctive high-altitude honeyeater that was originally placed in a monotypic genus *Vosea*; however, Diamond (1971) suggests that it is best placed in the mostly New Guinea genus *Melidectes*. Possibly closest to Sooty Honeyeater *M. fuscus*.

Type information *Vosea whitemanensis* Gilliard 1960a, Camp 12, Wild Dog Range, Whiteman Mountains, central New Britain.
Alternative names Gilliard's Honeyeater.
Discovery In 1958 and early 1959 Thomas and Margaret Gilliard made a collecting expedition to the Whiteman Mountains in central New Britain. In December 1958 a new and distinctive honeyeater was collected.
Description 22.5 cm, female slightly smaller. Monotypic. In form typical of the genus *Melidectes*, a long-billed slim bird with a long and graduated tail. Overall olive-brown on head and neck, becoming more brownish-olive on back and more yellowish-olive on lower back, rump and uppertail-coverts; underparts olive-brown, becoming yellow or paler on lower belly and vent; conspicuous large yellow-olive patch on spread wing; central rectrices yellowish-olive; iris dark brown; bare skin below and behind eye pale grey, tinged pale yellow; bill fine, decurved, black; tarsi dark grey. Female similar in plumage but somewhat duller.
Habitat Primary humid montane forest, 1,200–1,800 m, occasionally down to 850 m; sometimes in gardens.
Food and feeding Moves vigorously and abruptly, sometimes hanging head-downwards, in middle and upper levels of forest. Diet arthropods and nectar (Diamond 1971).
Voice Song is a series of 3–8 soft, mellow, whistled pairs of notes, the first note simple, the second trilled, lower pitched and downslurred (Diamond 1971); similar to that of *M. fuscus* but in shorter sequences.
Movements Apparently sedentary.
Range Endemic. Confined to median and higher altitudes on the island of New Britain (Nakanai Mountains, Whiteman Mountains and Mount Talawae).
Conservation status Currently classified by BirdLife International as Near Threatened, upgraded from Least Concern. Widespread, but uncommon over much of its range (Dutson 2011). Forest clearance for agriculture, especially oil-palm plantations, has in recent years resulted in losses of about 20% at lower altitudes though less severe higher up, where the topography discourages plantations; however, increased logging is a concern. Population estimated in the range of 3,500–15,000 mature individuals (BLI SF 2014).
Etymology The obsolete generic name *Vosea* honours Charles R. Vose, "a generous friend in exploration who recently lost his life in Alaska" (Gilliard 1960). The specific name refers to the type locality in the Whiteman Mountains.

HALL'S BABBLER
Pomatostomus halli

A large and distinctively plumaged babbler, quite widely distributed, which somehow managed to escape discovery until 1963.

Type information *Pomatostomus halli* Cowles 1964, Tyrone Station, SW Queensland, Australia.
Alternative names Black-bellied/Dark-bellied/White-breasted/White-throated Babbler.
Discovery A series of five expeditions to rebuild Australian bird collections was sponsored by Major Harold Hall. On the first of these, in 1963, P. R. Colston of the Natural History Museum collected a new babbler, obviously of the genus *Pomatostomus*, in south-central Queensland.
Description 19–21 cm. Monotypic. Diagnostic and quite unlike any other Australian babbler. Largely dark sooty-brown, the lores jet black, underparts more dark brown; broad white supercilium, extending to base of bill; chin, throat and upper breast white; rectrices very dark brown, tipped with white, narrowly on central feathers, broadly on remainder; iris brown; bill black with pale grey base to lower mandible; tarsi black.
Habitat Arid mulga (*Acacia*) woodland, especially areas with more than 70% tree cover, with canopy height 7–11 m (Brown & Balda 1977); also, more rarely, in more sparse woodland, with scattered trees. In some areas sympatric with White-browed Babbler *P. superciliosus*, but seems to prefer more arid habitat (Ford 1977).

► Hall's Babbler. Bowra Sanctuary, SW Queensland, Australia, May 2016 (*David Cook*).

Food and feeding Forages in groups, of about ten birds on average; in better habitat group sizes are larger (Brown & Balda 1977). Groups appear to maintain exclusive home ranges. Feeds mostly on the ground or low down; food items comprise invertebrates, mostly insects.

Breeding A cooperative breeder; groups consist of a breeding pair and one or two auxiliaries (Dow 1980, 1982); pairs may breed without helpers, but in these cases rarely successfully (Higgins & Peter 2002). Several nests may be built, but at any one time only one is used for breeding; others may be used for roosting. Nest is smaller and neater than those of other babblers (Andrew & Rogers 1993), dome-shaped with a side-entrance, made of twigs, the inner chamber lined with grass, hair and feathers; usually in mulga trees, 3–8 m up (Brown & Balda 1977). Eggs creamy light brown or beige, with light brown markings, heavier at blunt end. Clutch probably two. Incubation probably by female. Fledging period greater than ten days (Gill & Dow 1983). Auxiliaries assist with feeding young and communal nest-defence. May breed through much of year (Brown & Balda 1977; Higgins & Peter 2002).

Voice Noisy and demonstrative. A variety of *chirp* calls, loud buzzes and staccato *chit-chit*; a sharp whistled call in alarm. Contact call of group is a soft clucking.

Movements Apparently sedentary, but local dispersal, including break-up of larger groups at breeding season, may occur.

Range Endemic. Interior eastern Australia, mostly in Queensland, south of 22° S, south to northern New South Wales (upper Western Region) (Higgins & Peter 2002; Hooper 1974; Ford 1977).

Conservation status Classified by BirdLife International as Least Concern. Has a very large geographic range (487,000 km², BLI SF 2014). May be declining locally; in New South Wales classed as Vulnerable (NSW NPWS 1996), due to grazing by livestock.

Etymology Named in honour of Harold Wesley Hall (1888–1964). Major Hall, who made his fortune in the Mount Morgan goldfields and as a promoter of 'Hall's Fortified Wines', sponsored the five 'Harold Hall Expeditions'. A book documenting the results of these was edited by Mrs Pat Hall (no relative) of the Natural History Museum (Beolens & Watkins 2003).

DARK BATIS
Batis crypta

Formerly included in the nominate race of Short-tailed Batis *Batis mixta*, but clearly separated morphologically and biochemically; forms a superspecies with that form, as well as with Cape Batis *B. capensis* and several other East African batis species.

Type information *Batis crypta* Fjeldså *et al.* 2006, Mdandu Forest, Kipengere Range, west of Njombe, Ludewa District, Iringa Region, Tanzania.

Discovery As a result of a comprehensive collecting effort in the Eastern Arc mountains in Tanzania, Fjeldså *et al.* (2006) noticed a clear-cut morphological

▲ Dark Batis, female. Above Maringo, Malawi, March 2005 (*Johannes Ferdinand*).

◄ Dark Batis, male. Luala Camp, Udzungwa Mountains, Tanzania, November 2015 (*Markus Lagerqvist*).

change in the populations of Short-tailed Batis *B. mixta* in central Tanzania. Subsequent biochemical analysis showed that the two population types appear to be genetically isolated from each other.

Description 9.5–10 cm. Monotypic. A typical batis in form and plumage; small, rather large-headed and short-tailed, strikingly-plumaged and sexually dimorphic. Male: crown and back dark grey, the back with dark spots; conspicuous black mask, contrasting with white throat; wings blackish with conspicuous white wing-stripe; tail black with white tips and white edges to outer feathers; iris red or reddish-brown; bill and tarsi black. Female: crown grey; upperparts olivaceous; throat and pectoral-patch brick-red with white lower throat; wing-patch cinnamon; blackish mask; lower belly white.

Habitat Evergreen montane forest, 540–2,160 m, most abundant at 1,500 m. Outside breeding season found in miombo (*Brachystegia*) woodland and in wooded foothill areas (Louette 2006 HBW 11).

Food and feeding Forages in lower and middle levels of trees; insectivorous, including termites.

Breeding Nest undescribed; reported to breed in December and January, and to lay two eggs (Louette 2006 HBW 11).

Voice A variety of whistling and harsh churring notes; song of male is a series of short, low whistles; wings make a loud mechanical whirr in flight.

Movements There appear to be local or altitudinal post-breeding movements.

Range Endemic. Eastern and southern Tanzania (regions of Dodoma, Morogoro and Iringa); extreme NW Malawi (Misuku Hills).

Conservation status Currently classified by BirdLife International as Least Concern. Occurs over a wide geographic range, in parts of which it appears quite common; also appears to be able to tolerate significantly degraded forest (Fjeldså *et al.* 2006).

Etymology *Crypta*, Greek, meaning 'hidden'.

WEST AFRICAN BATIS
Batis occulta

A widespread and well-known bird which remained undescribed due to taxonomic confusion.

Type information *Batis occultus* (sic), Lawson 1984, Mt Nimba, Liberia.
Discovery A species of batis was described in 1903 from the island of Bioko (formerly the Spanish possession of Fernando Pó) in the Gulf of Guinea, West Africa, and named *Batis poensis*. A similar batis was widespread on mainland West Africa, but due to a lack of good quality material was not named. This was rectified by the fieldwork of M. Eisentraut on Bioko in 1963 and A. Forbes-Watson in Liberia in 1977. In 1984 W. J. Lawson, taking advantage of the new material, rather diffidently described the mainland form as a new species, *Batis occultus* (sic). Other authorities (e.g. Erard & Fry 1997) have treated the mainland birds as a race of *poensis*, and the latest IOC World Birdlist (Version 7.1) follows this treatment, but the two taxa are considered separate species in the latest HBW/BirdLife International checklist (del Hoyo & Collar (2016).
Description 12 cm. Monotypic. A typical batis, small, dumpy and short-tailed, with a proportionately large head. Sexually dimorphic. Male: crown blackish, separated from blackish ear-coverts by a white supercilium behind the eye; white supraloral spot; upperparts blackish or blackish-grey, with concealed white spots on the back; tail black with white edgings to outermost rectrices; wings black with conspicuous white wing-stripe; underparts white, except for broad black chest-band and black thighs; iris yellow; bill and tarsi black. In the female the chest-band is chestnut.
Habitat Secondary growth and degraded forest, shrubby areas, sea level to 2,500 m.
Food and feeding Forages in forest canopy, often in mixed flocks; insectivorous, including hard-shelled prey such as Coleoptera, Hemiptera and Hymenoptera, which are broken up by being beaten against twigs; also caterpillars (Harris & Franklin 2000).
Breeding Nest is a cup of plant material, including bark flakes, situated up to 30 m high at extremities of branches. Nesting season, in Cameroon, January–March; nest-building in October in Nigeria, incubating bird in January in Gabon. Eggs undescribed; incubation probably by female alone (Harris & Franklin 2000). Few other data.
Voice Song is a long trill, *tee-tee-tee-trrruuu tu-tu-tu-tu* etc, and a long series of very high-pitched penetrating notes, *heet-heet-heet* (Borrow & Demey 2001). Also bill-snapping and 'wing-fripping' (Harris & Franklin 2000).
Movements Apparently sedentary.
Range West Africa, disjunctly from Sierra Leone, Liberia, Ivory Coast, Ghana, SW and SE Nigeria, Gabon and Congo.
Conservation status Classified by BirdLife International as Least Concern; widely distributed, and appears to be able to adapt well to modified habitats.
Etymology *Occulta*, Latin, 'hidden' or 'concealed'.

▼ West African Batis, female. Kakum NP, Ghana, October 2013 (*Yann Kolbeinsson*).

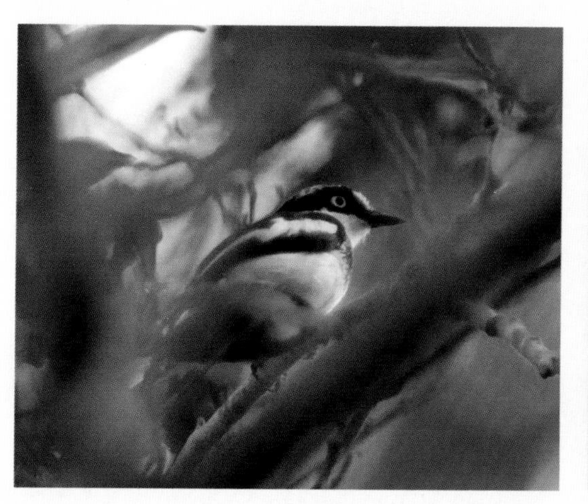

▼ West African Batis, male. Kakum NP, Ghana, November 2015 (*Laval Roy*).

WILLARD'S SOOTY BOUBOU
Laniarius willardi

An overlooked cryptic species with a very restricted geographic distribution.

Type information *Laniarius willardi* Voelker *et al.* 2010b, Nteko, Kisoro District, southern Uganda.

Alternative names Willard's Boubou.

Discovery In 1997, T. P. Gnoske and colleagues collected a series of four boubous (genus *Laniarius*) on privately owned land adjacent to Bwindi Impenetrable Forest National Park in southern Uganda. These were ascribed to the race *holomelas* of Mountain Boubou *L. poensis*, a species with an unusually disjunct range, with two other races occurring in West Africa, separated from *holomelas* by over 2,000 km. However, the fact that the Ugandan specimens had distinctive grey or blue-grey eyes – as opposed to black or reddish-black – raised suspicions as to their true specific identity. Careful morphological examination and comparison with specimens of other black members of the genus revealed significant if subtle differences in measurements, albeit nothing of use in the field, apart from eye-colour. Unfortunately, many older museum specimens are of no comparative use since collectors frequently neglected to record soft-part colours on the labels.

Description 18–19 cm. Monotypic. A typical *Laniarius* bush-shrike in form; medium-sized, with a fairly long tail and heavy powerful hook-tipped bill. Sexes similar. Plumage black, slightly iridescent; iris grey or blue-grey; bill and legs black.

Habitat Montane forest, probably mostly 1,600–1,900 m; apparently at lower elevations than *L. poensis holomelas*.

Food and feeding No data. Other members of the genus prey mainly on large insects, caterpillars and snails.

Voice Song is a quick series of four hooting notes on same pitch *hoo-hoo-hoo-hoo* (XC 166364); also harsh scolding notes (XC 113291).

Movements No data; presumably sedentary.

Range Imperfectly known. SW Uganda (Nteko) and NW Burundi (Kibara); recordings made in Rwanda (Nyungwe NP, XC 194594). Possibly also eastern Democratic Republic of the Congo.

Conservation status Not currently classified by BirdLife International but probably Near Threatened. The overall range, while uncertain, seems to be small; forest destruction at the altitudes preferred by the species is very prevalent.

Etymology Named in honour of Dr David Willard (b. 1946), under whose direction the Field Museum of Natural History in Chicago has become one of the world's foremost centres of ornithological studies.

◄ Willard's Sooty Boubou. Bwindi Impenetrable Forest National Park, Uganda, July 2014 (*David Hoddinott*).

RED-SHOULDERED VANGA
Calicalicus rufocarpalis

Closely related to the more widely distributed Red-tailed Vanga *C. madagascariensis*, from which it seems to be allopatric.

Type information *Calicalicus rufocarpalis* Goodman *et al.* 1997, La Table 20 km south-east of Toliara, province of Toliara, Madagascar.

Discovery In June 1948 Col. P. Milon collected two female vangas from a mixed-species flock near a stone quarry in Toliara province, SW Madagascar. The specimens, which had good anciliary data, were ascribed to *C. madagascariensis*, and remained undetected in the collection of the Muséum National d'Histoire Naturelle in Paris for almost half a century. However, S. Goodman and colleagues examined the two Milon specimens in comparison with numerous examples of *madagascariensis* held in a number of collections in Europe and North America and concluded that a new species was involved (Goodman *et al.* 1997). No male specimens were available at that time. Previously, in March 1991, one of the authors of the type description (C. A. Domergue) had photographed a male vanga attending a nest at a site about 10 km from Milon's type locality; this allowed comparison of the male plumage of the two *Calicalicus* species.

Description 14–15 cm. Monotypic. The smallest of the vangas, with striking plumage; sexually dimorphic. Male: crown and back pale grey; forehead and front supercilium white, contrasting with black eyestripe and throat; flight feathers and tertials brown, with an obvious brick-red patch on the shoulders; tail pale brick-red with brownish central feathers; chest to vent white with pinkish on chest and flanks; iris pale lemon-yellow; bill brownish-black with grey-blue base; tarsi pinkish-grey. Female generally duller, lacking the striking facial pattern and with a reduced shoulder-patch. Differs from *madagascariensis* in larger size, yellow, not dark brown, eye and greater extent of shoulder-patch.

Habitat Dense, semi-arid *Euphorbia* scrub, below 100 m.

Food and feeding Forages in low dense bushes by gleaning from leaves and by sally-gleaning (Hawkins *et al.* 1998); often in family groups. Prey is insects including beetles and grasshoppers.

Breeding Few data recorded. In early March 1991 C. Domergue found a nest but few details were published; the Milon holotype, taken in June, had granular ovaries with microscopic follicles, presumably indicating post-breeding condition.

Voice Distinctive from that of *C. madagascariensis*. Alarm or contact call, a peevish *karr-trkkk* or *kwoiroikk*, ending in a rattle; a dropping, whistling cadence, *ksisisisisisusususu*. Male song, a loud *tyub-tee* or *pu-teer* (Hawkins *et al.* 1998).

▼ Red-shouldered Vanga, male. La Table, Toliara province, Madagascar, October 2016 (*Ken Behrens*).

▲◄ Red-shouldered Vanga, female (top) and juvenile (bottom). La Table, Toliara province, Madagascar, January 2011 (*John C. Mittermeier*).

Movements Presumably sedentary.

Range Endemic. SW Madagascar. Toliara, south to the Mahafaly Plateau; recently found further south of the Linta River (Sim & Zefania 2002).

Conservation status Currently classified by BirdLife International as Vulnerable, with a note that a change to Threatened is possible. Fairly restricted geographically, and of patchy occurrence, although population densities may be locally high (Sim & Zefania 2002). Present estimates are in the range of 350–1,500 individuals. Main threats are habitat clearance for timber, charcoal and goat grazing.

Etymology *Rufocarpalis*, Latin, referring to the diagnostic brick-red shoulders.

CHOCO VIREO
Vireo masteri

A distinctive vireo species, probably most closely related to Yellow-winged Vireo *Vireo carmioli* which is endemic to the highlands of eastern Costa Rica and western Panama.

Type information *Vireo masteri* Salaman & Stiles 1996, Alto de Pisones, 7 km NW of Geguardas, Municipio de Mistrató, Dpto. Risaralda, Colombia.

Alternative names Spanish: Vireo de Chocó.

Discovery Between July and September 1991 a team of ornithologists from the Anglia Polytechnic University, UK, and the Universidad de Vallé, Colombia, conducted fieldwork in western Nariño, Colombia, in an area potentially impacted by highway construction. In August a small bird was mist-netted and later prepared as a study skin; unfortunately this specimen was partially destroyed by ants and was then temporarily misplaced in Bogotá. However, from photographs and notes the bird was announced as a probable undescribed vireo at the Neotropical Ornithological Congress in Quito in November 1991 (Salaman 1996). The following year G. Stiles observed a small canopy-feeding passerine which he could not identify in Dpto. Risaralda, Colombia, some 520 km north-east of the Nariño location; when a specimen was taken it proved to be a new vireo, on which basis the type description was written.

Description 11–11.5 cm. Monotypic. A relatively small, wing-barred vireo; generally olive-green above, brighter on the back; conspicuous whitish supercilium with dark lores and eyestripe; wings dull dark brown with two obvious slightly yellowish-white bars formed by tips to the median and greater coverts; chin and throat dull whitish, tinged yellow; chest brighter ochraceous-yellow, becoming paler and clearer below; iris dark brown; bill pale horn-colour, shading to whitish at the base; tarsi bluish lead-grey.

Habitat In Colombia, primary epiphytic rainforest, usually on steep slopes with rather broken canopy 20–25 m tall, with gaps caused by tree fall, 1,200–1,900 m. In Ecuador, apparently inhabits a wider spectrum of forest types including forest-edge, 800–1,500 m (Jahn *et al.* 2007).

Breeding Nest and eggs undescribed. In Colombia, adults feeding fledged young on 6 and 15 August; a 25 August specimen has skull only 30% ossified.

Voice Song is higher in pitch and more rapid in delivery than most other vireos, and quite distinct from that of the sympatric Brown-capped Vireo *Vireo leucophrys*; a rapid series of about ten notes, lasting about two seconds, starting with three short, high-pitched syllables, four lower-pitched ones and a slurred ending. Calls include a sharp brief *chip* and a nasal *zhree-zhree-zhree* alarm call.

Movements Presumably sedentary.

Range Currently known from a limited number of locations in Colombia (Risaralda and two locations in Nariño) and more recently in NW Ecuador (Alto Tambo and Cristal, Esmeraldas (Jahn *et al.* 2007); and the Mashpi Road, Pichincha (Brinkhuizen 2014)). Apparently absent from some suitable areas in between the geographic extremes.

► Choco Vireo. Rio Ñambi Nature Reserve, Nariño Department, Colombia. Photograph taken on 25 August 1991, the day of the species' discovery (*Carl Downing*).

Conservation status Currently classified by BirdLife International as Endangered. Although the potential global population is quite substantial – based on the area of suitable forest, it is placed at 78,000 plus or minus 7,000 mature individuals – extensive searches in many apparently suitable areas have failed to locate it, leading to a more precautionary estimate of 15,600 plus or minus 1,400 mature individuals (Jahn *et al.* 2007), or a total population of 20,000–25,000 birds. Habitat destruction and fragmentation remains the greatest threat. Vigorous action is being taken to preserve habitat, with, for example, the creation of the Río Ñambí Community Nature Reserve (Salaman 1996; Salaman & Stiles 1996).

Etymology Named for Dr Bernard Master (b. 1941), of Minneapolis, USA, in recognition of his major financial contributions which helped enormously in preserving important areas of habitat for this and other species. Although there was some controversy in ornithological circles about this procedure, the end result has been wholly admirable.

FLORES MONARCH
Monarcha sacerdotum

One of several highly restricted monarch species of SE Indonesia, with distribution limited to small island ranges.

Type information *Monarcha sacerdotum* Mees 1973, Sesok, Flores, Lesser Sundas, Indonesia.

▲ Flores Monarch. Puarlolo, Flores, Indonesia, August 2015 (*Dubi Shapiro*).

Discovery A single specimen of a monarch (*Monarcha*) was sent to G. F. Mees of the Rijksmuseum van Natuurlijke Historie in Leiden, the Netherlands, by two Catholic priests, J. A. J. Verheijen and E. Schmutz. It was collected by Fr Schmutz on 25 September 1971. The plumage was sufficiently distinctive in comparison to all other monarch species for Dr Mees to describe a new species based on the one unique specimen.

Description 15.5 cm. Monotypic. Morphologically a typical *Monarcha* monarch, medium-sized with a powerful hook-tipped bill, strong rictal bristles and fairly long tail. Generally dark grey above, the wings somewhat darker, and pure white below; conspicuous black facial mask, covering the forehead, area around eye, ear-coverts and throat; tail black with conspicuous white outer rectrices; iris dark; bill pale grey; tarsi grey or dark grey. Sexes similar. Differs from the Spectacled Monarch *M. trivirgatus*, which also occurs on Flores, in lacking any rufous below.

Habitat Medium-altitude moist evergreen forest, 350–1,000 m, mostly 700–900 m. Also old secondary and partially degraded forest (Clement *et al.* 2006 HBW 11). *M. trivirgatus* is probably separated altitudinally, being a lowland species (Butchart *et al.* 1996).

Food and feeding Actively forages in forest trees. Diet not well known; seen to feed caterpillars (Lepidoptera), lacewings (Neuroptera) and crickets (Orthoptera) to fledgling.

Breeding Nest and eggs undescribed. Recently fledged chick seen in August. Other members of the genus build conical cup-shaped nests in vertical tree forks.

Voice A nasal *schr-schr-schr* followed by a rich, flute-like whistle rising towards the end; also a single upwardly-inflected whistle repeated three or four times, and various buzzing and chattering notes (Clement 2006 HBW 11).

Movements Presumably sedentary.

Range Endemic. Confined to the island of Flores, Nusa Tenggara, Indonesia; all records, recent and historic, come from the western approximately one-eighth of the island, with no records east of Puarlolo and some recent records north to Bari (Anon. 2002a,b; BLI SF 2014).

Conservation status Classified by BirdLife International as Endangered. The overall range is small and within most of it the species appears to be fairly uncommon. Outright loss of forest and degradation of habitat is "rampant", for agriculture, firewood and lumber. No semi-evergreen forest below 1,000 m is presently adequately protected (BLI SF 2014).

Etymology *Sacerdotum*, Latin, 'of the priests', recognising the contribution of Frs Verheijen and Schmutz in providing the first specimen.

CAMPINA JAY
Cyanocorax hafferi

A distinctive *Cyanocorax* jay, probably most closely related to Azure-naped Jay *C. heilprini*, from which it is widely allopatric, and somewhat more distantly to Black-chested Jay *C. affinis*, even more distant geographically.

Type information *Cyanocorax hafferi* Cohn-Haft *et al.* 2013, 65 km NNW of Manicoré, Campo do Matupiri-Amapá, Amazonas, Brazil.

Alternative names Portuguese: Cancão-da-campina.

Discovery In August 2002 M. Cohn-Haft hiked into a remote area of natural savanna (known locally as *campina*), some 140 km south of Manaus, Amazonian Brazil. He briefly sighted a *Cyanocorax*-type jay which he could not identify; since no jays of this genus were known from the area, this observation was extremely interesting; the sighting indicated either a major range extension of a known species or an unknown species. Resolution of this mystery was obviously a high priority; however, it was not possible to visit the area for over two years. In 2005 Dr Cohn-Haft and A. Fernandes again hiked into the area, where they encountered a flock of jays; unfortunately they were not equipped to take specimens or record the birds. Afterwards they were unable to find any more. Fortuitously, the botanical crew of their group was able to obtain one specimen, which became the holotype. Over the next several years the species was encountered at several sites and a series of specimens was taken, allowing the description of the new species. Opinion in the SACC (Proposal 635) is mixed, with some members voting for acceptance, while others, pointing to the similarities in vocalisations between *hafferi* and *heilprini*, preferred a subspecific treatment.

Description 33–36 cm. Monotypic. A typical *Cyanocorax* jay, with a stiff, bristly frontal crest covering the nostrils. Facial mask including forecrown, orbital area and throat black, with three small blue patches – supra-orbital, sub-orbital and malar; mid-crown, nape and upper mantle sky blue; back, wings and most of tail dull brownish-grey with a faint purplish wash; breast azure-blue, becoming dull brownish-grey on upper belly, dull dirty white on lower belly and tinged yellowish on vent; tip of tail dull white, much broader on underside; iris pinkish-white; bill black; tarsi dull black.

Habitat Very specific; edge of Amazonian savannas, usually close to the forest edge, especially in islands of low forest. Absent from areas without a transition zone between tall forest and savanna. Areas flood shallowly after heavy rain.

Food and feeding Usually found in small flocks, feeding at all levels in forest; food items are both vegetable (seeds and fruits) and animal (arthropods).

▲ Campina Jay. Left bank of Purús River, Canutama, Amazonas, Brazil, September 2010 (*Dante Buzzetti*).

Breeding Nest is a large, bulky cup, mostly of twigs, located in the canopy of low woods, almost always in small, isolated forest islands. Probably a communal breeder, with groups of 3–8 birds, presumably involving offspring of previous year. Breeds during the dry season. No data on eggs, incubation period, etc.

Voice Noisy and vocally varied. Calls include sharp whistles, toots, mews and soft burbling sounds. May not be distinguishable from calls of *C. heilprini*.

Movements May disperse locally into *terra firme* forest outside the breeding season.

Range Endemic. Restricted to a very small area in the Purus-Madeira interfluve, from north of Novo Aripuanã, south to near Boca do Acre, all SW of Manaus in Amazonas state.

Conservation status Not classified by BirdLife International, but undoubtedly worthy of Vulnerable at least. Total known habitat area is only a little over 1,000 km^2. Main threats are habitat destruction (especially burning, which the species appears not to be able to tolerate) for agriculture and other purposes. The paving of the main road running through the species' range can only increase these pressures.

Etymology Named in honour of Jürgen Haffer (1932–2010), a pioneer of South American biogeography, and originator of the 'Pleistocene Refuge Hypothesis' of Amazonian bird distribution and species richness (Haffer 1969, 1982). Dr Haffer was, happily, made aware prior to his death of the intent to name a species in his honour.

LITTLE RAVEN
Corvus mellori

A widespread and common Australian species, which was finally recognised as such only in 1967.

Type information *Corvus mellori* Rowley 1967, based on a 1901 specimen, Angas Plains, South Australia. *Corvus marianae mellori* Mathews 1912.

Discovery The taxonomy of Australasian corvids has had a complicated and confusing history. In a 1912 study G. M. Mathews described ten subspecies; there was, however, continuing dispute (e.g. Vaurie 1958). For a concise history of this confusing issue, which is beyond the scope of this book, see Rowley (1967). In this paper Rowley also

◄ Little Raven. Deniliquin, New South Wales, Australia, October 2016 (*David Cook*).

argued for full specific status for a population extending from eastern South Australia to New South Wales.

Description 48–50 cm. Monotypic. A medium-sized crow, entirely glossy-black, the concealed feather bases on the head and neck grey; throat hackles bifurcated, bill somewhat slender; iris white; bill and tarsi black. A proposed race *halmaturinus* confined to Kangaroo Island is not considered valid.

Habitat Highly varied, though absent from dense unbroken forest. Open areas, farmland, scattered woodland with open patches, suburban areas, semi-arid areas and shorelines; sea level to 2,000 m.

Food and feeding Omnivorous, but specialises in insects. Other food items include small vertebrates, nestlings and eggs of small birds, carrion and some plant material including seeds and fruit (Higgins *et al.* 2006). Will cache surplus food (Lewis 1978).

Breeding Nest is a substantial bowl of sticks and twigs, lined with grass, bark, feathers etc., built by both sexes, usually in an upright tree-fork, height from ground 3–12 m (in a New South Wales study) averaging about 7.5 m (Rowely 1973). Eggs 1–6, usually 3–5, pale green to bluish-green, spotted and blotched with blackish-brown, umber etc.; incubation by female alone 19–20 days (Rowely 1973). Young brooded by female alone but fed by both sexes, fledging period 34–39 days; young may continue to be fed by parents for up to three months.

Voice Calls varied, including a harsh cawing, a series of guttural territorial notes *kar-kar-kar*, longer wailing notes, a low conversational *caa* etc.

Movements Sedentary while breeding; fledged birds join mobile flocks. Adults seem to be largely sedentary, but immatures are more mobile. May move in response to availability of ephemeral food supplies, such as mouse plagues. Longest ringing recovery about 350 km (Higgins *et al.* 2006).

Range Endemic. Southern Australia, from west of the Eyre Peninsula (South Australia), all of Victoria and much of New South Wales almost to the Queensland border; Kangaroo Island.

Conservation status Least Concern. Abundant and widespread. May have benefited from human activity.

Etymology Named in honour of John White Mellor (1868–1931), a founder member of the Royal Australasian Ornithologists' Union and its President from 1911 to 1912.

RED SEA SWALLOW
Petrochelidon perdita

One of the least known and most enigmatic recent discoveries to come out of Africa (indeed, the species' name in Norwegian is 'Enigmasvale'), known from one incomplete specimen. Described as a *Hirundo* but DNA evidence shows it to belong to the genus *Petrochelidon*, sister species to South African Cliff Swallow *P. spilodera*.

Type information *Hirundo perdita* Fry & D. A. Smith 1985, Sanganeb lighthouse, Red Sea coast NE of Port Sudan, Sudan.

Discovery In May 1984 D. A. Smith discovered a highly decomposed swallow corpse underneath the lighthouse on Sanganeb Reef, some 20 km north-east of Port Sudan on the western shore of the Red Sea. The body had fortuitously come to rest in a crevice protected from the attentions of the omnipresent scavenging land crabs but was in a very poor state. The corpse was photographed, but only the wings and tail, currently housed in the Natural History Museum, Tring, could be preserved. These and the photographs are the only material available.

Description Probably ca 14 cm. Monotypic. Based on a partial specimen of unknown sex. Crown blue-black; back glossy deep blue; rump grey-white; forehead and lores blackish; wings and tail blackish-brown; tail square-ended; small white patch on chin, remainder of throat and upper chest bluish-black, rest of underparts white except for buffy undertail-coverts.

Habitat Breeding habitat unknown.

Voice Undescribed.

Movements No data, although the location of the unique specimen clearly indicates some tendency to move.

Range The breeding location is unknown. Butchart (2007b) surmises that it probably breeds in the Red Sea hills of Sudan or Eritrea. However, another possibility is the Saudi Arabian coast north of Jeddah, since unidentified pale-rumped swallows have been seen flying out over the Red Sea towards Jeddah (Madge & Redman 1989). There have been numerous observations of unidentified swallows in several locations in Ethiopia (Madge & Redman 1989; Turner & Rose 1989; Atkins & Harvey 1994; Vermeulen 2000), although since these did not correspond precisely in plumage to the Sanganeb specimen they might refer to a yet further unknown species, other than *perdita* (Butchart 2007b).

Conservation status Currently listed as Data Deficient; clearly, locating the breeding area is critical for any conservation measures.

Etymology *Perdita,* Latin, 'lost', an entirely appropriate name given the circumstances.

WHITE-EYED RIVER MARTIN
Pseudochelidon sirintirae

One of the most charismatic discoveries to come out of Asia in recent years, a spectacularly plumaged and taxonomically unexpected swallow. Originally placed in the genus *Pseudochelidon* for anatomical reasons, sharing this genus with the otherwise unique African River Martin *P. eurystomina*, whose range is restricted to forested rivers and coastal savanna in central Africa (mostly the Congo Basin). It has also been placed in a new monotypic genus *Eurochelidon* (Brooke 1972b), though the validity of this has been challenged (Zusi 1978).

Type information *Pseudochelidon sirintirae* Thonglongya 1968, Bung Boraphet, Nakhon Sawan province, central Thailand.
Alternative names Thai: Nok Ta Phong.
Discovery During ringing operations in a large marshland complex in central Thailand in January–February 1968, nine examples of a unique and unusual swallow were obtained among very large numbers of other swallow species (Thonglongya 1968). The birds were apparently caught by professional trappers, who go out in boats after dark and throw fishing-nets over the reedbeds where the swallows are roosting; the precise location of the roosting birds is apparently not known, since the market trappers who were hired to catch the birds then brought them to a hotel in Nakhon Sawan for ringing (Tobias 2000).

Description 15 cm, with two central rectrices elongated a further 8.5 cm, with narrow tip rackets 3.5 cm long and 2 mm wide. Monotypic. Plumage generally black with a green sheen; narrow silvery-white rump; wings black; tail black with a green gloss; iris white, with white eyelids; bill very broad, greenish-yellow tipped black; tarsi flesh-coloured. Sexes similar.
Habitat Known only from what is apparently the wintering area; seen to feed and drink over an extensive marsh. No other data.
Food and feeding No data. The very large, wide bill obviously suggests the capture of flying arthropods; the very large eye has led to suggestions that it is primarily a crepuscular feeder.
Breeding Nest and eggs undescribed.
Voice Not recorded.
Range Known only from one location in central Thailand, some 300 km north of Bangkok, during the winter (October or November to March, according to local knowledge). The breeding area is not known and might well be in southern China or Myanmar (Burma), or other South-East Asian countries. It was tentatively suggested, on the basis of some rather imaginative, undated Chinese scroll paintings of unknown provenance, that the species may breed in China (Dickinson 1986), although a more credible identification of the subjects might be Oriental Pratincoles *Glareola maldivarum* (Parkes 1987).

◄ White-eyed River Martin, caught in November 1968 in the Bung Boraphet region of Nakhon Sawan province, central Thailand, and brought alive to Bangkok where it was photographed (*H. E. McClure*).

Conservation status Classified by BirdLife International as Critically Endangered; however, it is very probably extinct. It was probably always very restricted in distribution. On the wintering grounds it was, along with other reedbed roosting species, subject to enormous pressure from professional bird-trappers who catch large numbers of birds for sale as food and as subjects for release by devout Buddhists, who gain merit from such actions; however, it is likely that mortality among released examples of an obligate aerial insectivore kept in captivity even for a short time would be very high. After the initial capture of nine specimens in early 1968, a tenth was caught in November 1968 and brought to Bangkok where it was photographed (Thonglongya 1969). Two birds were caught in early 1971 and later died in Dusit Zoo, Bangkok (Sophasan & Dobias 1984). The first actual field observations, of six birds, were made by B. King and S. Kanwanich in February 1978, with further probable observations in January 1980 and an unconfirmed report of one trapped by local people in 1986. It is suggested that the species came under special pressure from bird-trappers because its royal associations may have made it a marketable commodity, with rumours of up to 120 individuals being sold locally. The area where the original specimens were taken has nominally been protected since 1972, but enforcement has been difficult and in fact one reserve ranger was murdered in 1987 by bird poachers (Tobias 2000). There have been no authenticated records in recent years.

Etymology The generic name *Pseudochelidon* comes from the Greek, *pseudos* 'false' and *chelidon* 'a swallow'. The proposed generic name *Eurochelidon* also comes from the Greek, *euros,* the 'south wind', and *chelidon,* a swallow. The specific scientific name *sirintarae* is a dedication to HRH Princess Sirindhorn Thepratanasuda, daughter of HM the late King Bhumibol Adulyadej of Thailand.

ASH'S LARK
Mirafra ashi

A little-known and endangered species occurring only in an area of permanent political instability.

Type information *Mirafra ashi* Colston 1982, Uarsciek, south Somalia.

Alternative names Ash's Bushlark.

Discovery In July 1981 J. S. Ash collected eight specimens of an unfamiliar lark near Uarsciek (variously spelt Uarshek or Warshiiki), some 80 km NE of Mogadishu (Mogadiscio or Muqdisho), Somalia. Dr Ash suspected that a new species might be involved, but since there was insufficient material available at the National Museum of Kenya for a valid comparison he forwarded the specimens to P. R. Colston at the Natural History Museum, Tring, who described the new species.

Description 17 cm. Monotypic. In shape and form typical of the genus *Mirafra*; a medium-sized lark with a heavy, quite thick bill. Upperparts greyish-brown with paler edgings on the dorsal feathers which have darker centres; underparts buff, paler on vent and belly, the throat buffy-white with fine grey spots, becoming larger and browner on the chest; slightly crested with whitish or pale buffy supercilium; wings with an obvious rufous patch; tail dusky-brown, the outer rectrices narrowly edged white; iris dark brown; bill dark grey above, pale bluish-grey below; tarsi creamy.

Habitat Short, open grassland in stabilised dune systems with a few scattered bushes and fossil coral outcrops (Ryan 2009 HBW 9).

Food and feeding Terrestrial; runs across open ground, sometimes perches on tops of grass tufts. Diet unknown.

Breeding Nest and eggs undescribed. A female specimen had enlarged ovaries (but also active flight feather moult) in July.

Voice Reported to sing from tops of small bushes; song undescribed.

Movements No data.

Range Endemic. Imperfectly known. Occurs in coastal central Somalia; the northern limits of its range are not precisely defined.

Conservation status Classified by BirdLife International as Endangered. The proven range is small (estimated at 346 km², although this could be an underestimate). Described as locally common (Ash & Miskell 1998) in this area. Population estimates in the range of 3,500–15,000 individuals (BLI SF 2014). Threats include coastal development; the ecological relationship of the species to grazing is unknown. Given the perennial political unrest of the area, conservation work may prove to be challenging.

Etymology Named in honour of Dr John S. Ash (1925–2014), an English ornithologist who lived in Somalia for a number of years and co-authored the *Birds of Somalia* (1998). He also lived and worked in Ethiopia where he discovered the Ankober Serin *Crithagra ankoberensis* in 1976 (q.v). His work in Ethiopia resulted in the publication of *Birds of Ethiopia and Eritrea* (2009).

TANA RIVER CISTICOLA
Cisticola restrictus

An enigmatic bird, possibly not justifiable as a valid species; it has been suggested that it might be an aberrant example of Ashy Cisticola *C. cinereolus* or a hybrid between that species and Rattling Cisticola *C. chiniana*. There do not appear to have been any comparative DNA studies.

Type information *Cisticola restricta* Traylor 1967, Karawa, lower Tana River, Kenya, based on a specimen taken in 1932.

Discovery The species, if such it is, was described by M. A. Traylor in 1967 based on his discovery of three specimens, in the collection of the Field Museum, Chicago, of a cisticola that evidently did not belong to any known species. A further three specimens were located in the Los Angeles County Museum. All had been collected in the region of the lower Tana River, Kenya.

Description 13 cm. Monotypic. A typical cisticola in form, small, slight, long-legged and with a long, graduated tail. Overall tone of plumage washed-out grey above, with darker markings on coverts and flight feathers; crown rusty-brown; tail brown with darker subterminal band and buffy tips; below creamy, the flanks buffy-grey; iris warm brown; bill dark brown with paler base; tarsi dull pink. Differs from *cinereolus* by rufous wash on crown, narrower streaking on underparts and in lacking white bases to nape feathers; from *chiniana* by paler appearance, smaller size and lack of rufous wing-edgings.

Habitat Flat sandy areas or black 'cotton' soils with semi-arid *Acacia* scrub (Zimmerman *et al.* 1996).

Food and feeding Diet insects, including grasshoppers (Orthoptera).

Breeding Nest and eggs unknown.

Movements No information.

Voice Song is stated to be similar to that of *C. chiniana* (A. Forbes-Watson, label data).

Range Endemic. All specimens are from the lower Tana River area of Kenya; suggestions that it occurs in adjacent southern Somalia are presently unsubstantiated.

Conservation status Classified by BirdLife International as Data Deficient. If the form is a valid species, it is possibly extinct; the last record was in 1972, and more recent searches by competent ornithologists have failed to find it (D. J. Pearson, pers. comm.).

Etymology *Restrictus*, Latin, referring to the very limited range of the species.

RIVER PRINIA
Prinia fluviatilis

A species with an extraordinary disjunct range, with populations in small areas from the Atlantic coast of West Africa through to NW Kenya. Forms a superspecies with Plain, Tawny-flanked and Pale Prinias (*P. inornata, subflava* and *somalica*).

◄ River Prinia. Diffa, Nigerian border, September 2005 (*Jan F. Rasmussen*).

Type information *Prinia fluviatilis* Chappuis 1974, N'Djamena, Chad.
Alternative names French: Prinia Aquatique.
Discovery The taxonomy of *Prinia* warblers has caused considerable confusion over the years. C. Chappuis, over several years' observations, noted that birds in areas of Chad were vocally distinct from the locally widespread Tawny-flanked Prinia *P. subflava*; furthermore, there were differences in habitat preference, singing habits and in plumage. The situation was complicated by the absence of a specified type specimen (Chappuis 1974), necessitating the selection of a lectotype (i.e. a specimen chosen out of a series) at a later date (Chappuis *et al.* 1989).
Description 11–12 cm. Monotypic. A small, rather long-tailed and slender unstreaked prinia. Generally grey above, with an obvious pale supercilium; tail graduated, more grey-brown with buffy margins, rectrices except central pair with blackish subterminal spots on underside and creamy-buff tips; underparts from chin whitish, more buffy on flanks, thighs and vent; iris pale brown; bill black: tarsi pinkish-brown. Birds from Chad have shorter tarsi than Senegal birds. In a species with widely disjunct populations subspeciation might be expected. Differs from *P. subflava* by smaller size, colder grey tone to upperparts and whiter underparts.
Habitat Grass, shrubland and swamp vegetation in the vicinity of rivers, but absent from reedbeds. *P. subflava* tends to occur in more arid habitat.
Food and feeding Forages in grasses, shrubs and reeds but apparently it does not fly-catch aerial prey. Diet insects.
Breeding Literature misleading due to confusion with *P. subflava* (e.g. Bannerman 1939); however, several unequivocally identified nests have been found (de Naurois & Morel 1995). Nest is a deep oval ball with an entrance high up on the side, made of fine grasses with a deep internal pocket, preventing loss of eggs in high wind; usually located at end of hanging vegetation over water. Eggs four, blue, unmarked (unlike of those of *P. subflava*). No data on incubation or fledging periods.
Voice Song is a distinctive series of plaintive down-slurred *pleeu* notes, the notes longer and higher-pitched than those of *P. subflava* and lacking an initial *chic* note (Chappuis 1974).
Movements Apparently sedentary.
Range Highly disjunct. West Africa: Senegal (Borrow & Demey 2001; Rodwell *et al.* 1996); Gambia (Barnett & Emms 2001); Mauritania (Salewski *et al.* 2005); Cameroon, Mali, Chad and Niger, and recently in NW Kenya (Finch 2005). Nesting proven in Guinea-Bissau (Barlow *et al.* 1997).
Conservation status Currently classified by BirdLife International as Least Concern; widespread and seems to be quite common in some locations.
Etymology Both the common and scientific names reflect the species' preference for riverside habitat.

CAMBODIAN TAILORBIRD
Orthotomus chaktomuk

The discovery of a new and distinctive tailorbird in and around Phnom Penh, the capital city of Cambodia with a population of over 2 million, was one of the most surprising in the recent ornithological history of SE Asia; aptly described as "hiding in plain sight" (Anon. 2013c).

► Cambodian Tailorbird. Phnom Penh, Cambodia, March 2015 (*Ross Gallardy*).

Type information *Orthotomus chaktomuk* Mahood *et al.* 2013, Bateay District, Kompong Cham province, Cambodia.

Discovery In January–March 2009, in the course of a routine programme to monitor wild bird populations for avian influenza, four tailorbirds were mist-netted at two sites not far from Phnom Penh, Cambodia. All were photographed and released; based on photographs, complicated by some incorrect location information, these birds were mistakenly identified as Ashy Tailorbirds *Orthotomus ruficeps*. In January 2012 Hong Chamnan found a similar bird at Prek Ksach, about 15 km from Phnom Penh, which was again identified as *ruficeps*. A. J. I. John photographed birds at this site in June 2012; these photographs aroused the interest of S. P. Mahood, and subsequent discussions with J. W. Duckworth, P. Round and other experienced ornithologists raised the possibility that an undescribed taxon might be involved. Further fieldwork in June 2012 and April 2013 found more than 100 individual birds at 10 locations. Examination of the photographs taken in 2009 confirmed that these also corresponded to the putative new taxon. In August 2012 five specimens, of differing ages and sexes, were taken. Comparison of this material with specimens of all other relevant *Orthotomus* species allowed the description of a new species (Mahood *et al.* 2013).

Description 11–12 cm. Monotypic. In form a typical *Orthotomus* tailorbird; a dainty, rather large-headed bird with a long graduated tail and long fine bill. Crown, nape and lores rich cinnamon-rufous, abruptly demarcated from the white cheeks and mid-grey upper neck, mantle and rump; wings slightly darker grey than mantle; tail rounded, mid-grey with blackish-grey subterminal band and whitish tips; chin and throat whitish, the lower throat and upper chest contrastingly marked with blackish-grey; remainder of underparts greyish-white with diffuse darker streaks on lower chest and upper belly; iris orange-brown; bill with culmen decurved toward tip, slightly hook-tipped, dark horn above, pink-horn below with paler base; tarsi pinkish. Female somewhat duller overall, with less heavily streaked underparts.

Habitat Very dense evergreen scrub, with bushes 2–6 m tall, usually seasonally flooded; absent from unbroken forest. All locations 3–25 m above sea level (Mahood *et al.* 2013).

Food and feeding Forages by gleaning and sally-gleaning in dense vegetation, from ground level to the canopy. Observed prey items: flies (Diptera), small spiders, small caterpillars and a small katydid (Tettigoniidae).

Breeding Nest and eggs undescribed but presumably similar to other members of the genus, which are deep cups of down, spider-webs etc., located in a pouch made by sewing (usually) two leaves together.

Voice Song is lengthy, often lasting a minute or more, consisting of multiple phrases made up of 2–5 trilled strophes of varied types. Female's trill is higher than that of the male. Mahood *et al.* (2013) give a detailed analysis of vocalisations.

Movements Presumably sedentary.

Range Endemic. Incompletely known. Confined to seasonally flooded dense scrub within the floodplain of the Tonle Sap, Mekong and Bassac rivers in Cambodia. Extreme limits of presently known distribution about 140 km north to south and 120 km east to west, but not present in all apparently suitable habitat within this area. Total potential range probably less than 10,000 km^2 (Mahood *et al.* 2013).

Conservation status Mahood *et al.* (2013) suggest a classification of Near Threatened as appropriate. The overall range is small and patchy and the habitat is liable to degradation and change; invasion of scrub by forest, caused by the absence of wild or domestic ungulates, could be a negative factor. At least one site, where the original birds were mist-netted in 2009, appears to have been lost.

Etymology *Chaktomuk*, a Khmer word meaning 'four faces', referring to the low-lying area forming an 'X' where the three major rivers come together; Phnom Penh was historically known as 'Krong Chaktomuk' (literally, the City of Four Faces).

BARE-FACED BULBUL
Pycnonotus hualon

An unusual and distinctive bulbul from limestone karst areas of central Laos; its position within the large genus *Pycnonotus* requires further clarification. The bare face, unique in the genus, is not presently deemed to be adequate reason for the erection of a new monotypic genus. The first new bulbul species from Asia in more than a century.

Type information *Pycnonotus hualon* Woxfold *et al.* 2009, Pha Lom, south of Ban Nonsomphou, Vilabouli district, Savannakhet province, central Laos.

Discovery In May 1995 R. J. Timmins observed a small group of distinctive and unfamiliar bulbuls in karst country at the northern end of the Khammouan limestone massif in central Laos. Because of the limited plumage detail that he was able to provide, his observations were not documented in the survey report (Duckworth *et al.* 1998), and he "subsequently weathered a fair amount of good-natured ribbing on relating the sighting to sceptical colleagues" (Woxfold *et al.* 2009). In February 1999 T. D. Evans was able to obtain good views and described an unidentified bird at a site about 16 km from the location of the 1995 observation, but did not publish his description. Finally, in December 2008, J. W. Duckworth and I. A. Woxfold saw and recorded several birds near the village of Ban Nonsomphou, some 165 km SE of the earlier observations. Two specimens were taken, allowing the description of a new species.

Description 19 cm. Monotypic. Plumage largely dull olive-grey; chin and throat off-white; undertail-coverts more olive; striking bare skin on face bluish around eyes and on lores, orange-pink on ear-coverts, behind and below eye to base of bill; behind the bare skin on the face an area of whitish, upswept feathers; iris very dark brown; bill black; legs mainly black. Female similar but smaller and paler, with paler and more sparse crown feathering.

Habitat Sparse vegetation on almost soil-less limestone karst country, with small deciduous trees and shrubs. However, also apparently common in evergreen forest valleys in between karst blocks, which may be the most preferred habitat (Pilgrim *et al.* 2009). Lowlands, probably below 500 m.

Food and feeding Limited data; probably mostly fruits and berries (*Ficus, Bridelia*) (Woxfold *et al.* 2009). Also seen to sally for aerial prey, and to search bare rock faces for food, presumably invertebrates (Pilgrim *et al.* 2009).

Breeding Nest and eggs undescribed. Vocally active in December, much less so in August, when birds appear to be moulting.

Voice Rather vocal. A series of dry, bubbled whistles, sometimes followed by a longer series of notes accelerating towards the end; also various trilling calls.

▲ Bare-faced Bulbul. Na Hin, Khammouan province, Laos, December 2009 (*James Eaton*).

Movements No data; presumably sedentary.

Range Endemic. Known from Savannakhet province, Laos, but may also be in Bolikhamxai and Khammouan provinces; much apparently suitable habitat has not been investigated.

Conservation status Presently classified by BirdLife International as Least Concern. The overall range is quite extensive and might well be further extended with more fieldwork. The inaccessible nature of the habitat does not encourage human exploitation, although quarrying and firewood gathering might have local impacts. Several areas of potential occurrence are protected.

Etymology *Hualon*, 'bald-headed' in the Lao language, also applied to a bald- or shaven-headed person.

PRIGOGINE'S GREENBUL
Chlorocichla prigoginei

An almost unknown species, with the first, rather scanty, field observations published only in 1997. May form a species-pair with Joyful Greenbul *C. laetissima.*

Type information *Chlorocichla prigoginei* de Roo 1967, Maboya, on Upper Luhule River, Democratic Republic of Congo (DRC).

Alternative names French: Bulbul de Prigogine.

Discovery Described by A. de Roo in 1967, based on material taken in an area north-west of Lake Edward, DRC.

Description 19–20 cm, female smaller than male. Monotypic. In form a typical greenbul, medium-sized. Upperparts yellowish olive-green, darker brown on wings, tail olive with yellowish olive-green margins; lores and area in front of eye pale grey; greyish-white eye-ring; chin white, throat bright yellow; remainder of underparts greenish-yellow, darker and more green on flanks and sides of chest; undertail-coverts yellowish-brown; iris dark red; bill black; tarsi dull bluish-grey, dark horn or blackish.

Habitat Isolated forest patches in savanna and gallery forest along upper courses of rivers, 1,350–1,800 m. Absent from mature lowland forest. Most recently seen in slightly degraded damp montane forest (Pedersen 1997).

Breeding Nest and eggs undescribed.

Voice Not recorded; apparently not as vocal as *C. laetissima* when in mixed flocks (Pedersen 1997).

Movements Presumably sedentary.

Range Endemic. NE DRC, on the Lendu Plateau between Nioba and Djugu and in scattered localities between Beni and Butembo, NW of Lake Edward.

Conservation status Currently classified by BirdLife International as Endangered, although upgrading to Critically Endangered is possible. The suitable habitat has been severely impacted by logging and clearing for agriculture, a process which is continuing; in fact recent reports (BLI SF 2016) suggest that all forest on the Lendu Plateau has now been destroyed. No area of occurrence has any protection. Population is (speculatively, based on little data) estimated in the range of 2,500–10,000 individuals.

Etymology Named in honour of A. Prigogine (see under Prigogine's Nightjar *Caprimulgus prigoginei*).

LIBERIAN GREENBUL
Phyllastrephus leucolepis

Previously thought to be an almost unknown, highly restricted and endangered greenbul, occurring in an area of rampant habitat destruction, uncontrolled population growth and continuous political instability. Work currently in progress (Collinson *et al.*, in prep.) suggests that the only specimen is, in fact, an abnormally plumaged example of the Icterine Greenbul *P. icterinus*.

Type information *Phyllastrephus leucolepis* Gatter 1985, 20 km NW of Zwedru, near Cavalla River, Grand Gedeh County, Liberia.

Alternative names Spot-winged Greenbul, White-winged Greenbul.

Discovery During fieldwork in the rainforests of SE Liberia, conducted between 1981 and 1984, what was believed to be a distinctive and clearly undescribed bulbul species was seen by W. Gatter, always during the dry season (December to March). The unique specimen was taken in January 1984; it was, however, in very poor condition, having been lodged for a day up in the branches of a tree where it was extensively damaged by ants, and had to be preserved as a 'mummy'; consequently internal examination was not possible.

Description 16 cm. Monotypic. Distinctive. In form a typical greenbul, rather small, fairly long-tailed with a stout powerful bill. Head and upperparts olive-green, with inconspicuous yellow eye-ring; wings more brown, with an obvious whitish wing-bar formed by tips to the greater coverts, pale subterminal area on primaries, and conspicuous pale tips to the tertials, obvious on the closed wing. Throat bright yellow, remainder of underparts drab yellowish with strong olive wash on breast and flanks; tail brownish-chestnut; iris colour undescribed; bill black above, dark horn below; tarsi pale grey. Frequently flicks open wings, showing pale areas. Distinguished from sympatric Icterine Greenbul *P. icterinus* by wing markings.

Habitat Lowland forest, in transition zone between evergreen and semi-deciduous zones, ca 200 m.

Food and feeding Forages singly or in pairs, often in mixed-species flocks, in mid-levels of forest (4–8 m) (Borrow & Demey 2001). No data on diet.

Breeding Nest and eggs undescribed.

Voice No information.

Movements Presumably sedentary.

Range Endemic. Known only from two forest patches near Zwedru, SE Liberia.

Conservation status Classified by BirdLife International as Critically Endangered; might well be extinct. Habitat destruction for logging and clearance for smallholdings is very prevalent; during the ruinous civil wars of 1989–1996 and 1999–2003 the situation apparently worsened. Surveys of both known sites in 2010 failed to re-find the species (Molubah & Garbo 2010).

Etymology *Leucolepis*, Greek, 'with white scales'.

NEW BRITAIN THICKETBIRD
Megalurulus grosvenori

An almost unknown species. Described by Gilliard as a species in the genus *Cichlornis*, since subsumed into *Megalurulus*. Ripley treated this and *M. llaneae* as races of Melanesian Thicketbird *M. whitneyi*, a course followed by several authorities. However, the IOC World Bird List follows Sibley & Monroe (1990) in treating it as a full species.

Type information *Cichlornis grosvenori* Gilliard 1960b, Wild Dog Range, Whiteman Mountains, New Britain.
Alternative names New Britain Thicket Warbler, Bismarck Thicketbird.
Discovery E. T. and M. Gilliard conducted an extensive expedition in 1958 and 1959 to the island of New Britain, east of Papua New Guinea. The first specimens were taken on 18 and 22 December 1958 by Tesako and Rambur, native hunters from the Sepik River, New Guinea. The Gilliards themselves never saw the bird, and it could be that no Western ornithologist has ever done so.
Description 19 cm. Monotypic. Upperparts dark olive-brown, becoming darker on the rump; ochraceous-tawny supercilium; broad black mask running from base of bill through lores, eyes and malar region to posterior ear-coverts; underparts ochraceous-tawny, paler on chin, throat and central abdomen; flanks darker; wings blackish, the upper coverts as mantle; tail sharply graduated, dark brown; iris dark brown with a dull green ring; tarsi dark smoky-brown.
Habitat Upland wet forest; specimens taken at about 1,575 m.
Food and feeding Few data; stomach contents of specimens comprised insect parts and a small snail.
Breeding No data.
Voice Song were not described. Call-note "a single note, emitted only occasionally".
Movements Presumably sedentary.
Range Endemic. Known only from the type locality.
Conservation status Not classified, but would appear to be restricted and uncommon. Extensive searches in New Britain (Nakanai Mountains, Mt Talawe) failed to find it (Bishop & Jones 2001).
Etymology Named in honour of Dr Gilbert Grosvenor (1875–1966), editor of the magazine of the National Geographic Society for more than 50 years; the Society supported the Gilliards' fieldwork. Dr Grosvenor also has a sandstone arch in Utah named in his honour.

BOUGAINVILLE THICKETBIRD
Megalurulus llaneae

Treated as a race of Melanesian Thicketbird *M. whitneyi* by Ripley (1985) and in the *Handbook of the Birds of the World* (Vol 11, 2006); given full specific status in the IOC World Bird List.

Type information *Megalurulus llaneae* Hadden 1983, Crown Prince Range, central Bougainville Island, North Solomons Province, Papua New Guinea.
Alternative names Bougainville Thicket Warbler.
Discovery In June 1979 D. Hadden was trying to track down the nesting sites of shearwaters on Bougainville Island, east of Papua New Guinea. He was informed by a local of an apparently promising hole in a bank high up in the Crown Prince Range; after a prolonged climb the hole was located, but it clearly belonged to a passerine of some kind, not a shearwater. Mist-nets were

► Bougainville Thicketbird. Top Camp, above Panguna, Bougainville Island, Papua New Guinea, June 1980 (*Bruce Beehler*).

erected in the forest and one specimen of an unfamiliar warbler was caught, which became the holotype of a new species. In fact, the unique example of the nest of *M. llaneae* was in a rock crevice, not a hole (see below).
Description ca 18 cm. Monotypic. Head, wings, back and rump sooty-olive; rich cinnamon-rufous supercilium; small dark mask before, behind and below eye; lesser wing-coverts black with brownish-olive tips; throat and abdomen cinnamon-rufous, shading into dull brown flanks and lower abdomen; tail black; iris very dark brown; bill and tarsi black.
Habitat Wet, rather open, montane forest, canopy height about 15 m, with much epiphytic moss growth, ca 1,200 m.
Food and feeding Feeds low down; stomach contents insects.
Breeding Only one nest so far described, in June. It was a cup of dark vegetation, lined with pale fibres, placed ca 2 m above a stream-bed in a niche in the vertical rock wall of a narrow gulley. Eggs two. The same nest was used in subsequent years (Ripley 1985).
Voice Song not described; distress-call of mist-netted bird *Shreed-shreed.*
Movements Presumably sedentary.
Range Endemic. Apparently confined to the mountain range where the type specimen was taken.
Conservation status Presently unclassified. Would appear to be fairly uncommon.
Etymology Named by the discoverer for his wife, Llane Hadden, "as a token of gratitude for her support and interest" (Hadden 1983).

TANIMBAR BUSH WARBLER
Horornis carolinae

A little known and rather restricted species, of uncertain affinities; possibly most closely related to Fiji Bush Warbler *H. ruficapilla*. Formerly placed in the widespread genus *Cettia*; the genus *Horornis* is confined to SE Asia and the western Pacific (Alström *et al.* 2006).

Type information *Cettia carolinae* Rozendaal 1987, Yamdena, Tanimbar Islands, Indonesia.
Alternative names Yamdena Bush Warbler.

▲ Tanimbar Bush Warbler. Tanimbar Islands, Maluku, Indonesia, October 2014 (*James Eaton*).

Discovery Between August and November 1985, F. G. Rozendaal, while conducting fieldwork on Yamdena, Tanimbar Islands, Indonesia, mist-netted several specimens of an unknown sylviid warbler. The birds were so secretive that observations of its habits were rather limited, but Dr Rozendaal was able to make recordings of the characteristic vocalisations as well as noting some of the feeding behaviour. Comparison of the collected material with several other members of the genus as well as comparative sonogram studies established that a new species was involved.
Description 11–13 cm. Monotypic. A medium-sized, rather short-tailed bush warbler. Generally rather nondescript brown, the cap rufous, bordered by a conspicuous cinnamon-buff supercilium; upperparts and tail dull warm brown; chin, throat and upper chest greyish-white, olive-grey on lower breast with browner flanks; iris brown; bill blackish, lower mandible dull greyish-pink; tarsi dull pinkish-flesh.
Habitat Undergrowth of primary monsoon forest and secondary forest, regenerating forest and roadside bamboo thickets (Rozendaal 1987); also selectively logged semi-evergreen forest (Bishop & Brickle 1998).
Food and feeding Skulking, feeding low down in dense undergrowth, clinging to tree trunks and lianas and saplings, or hopping along horizontal branches of rattan palms; stomach contents of specimens comprised invertebrates (ants, true bugs, beetles) and small snails.
Voice Song, a long, drawn-out, penetrating whistle, rising slightly in pitch and volume, with a pleasing terminal flourish *tuuuuuuuuuuuu-chirrup*, lasting just

over one second (Kennerley & Pearson 2010). Particularly vocal in the wet season (Bishop & Brickle 1998).

Movements Presumably sedentary.

Range Endemic. Confined to Yamdena Island, Tanimbar group, Indonesia; presently known from two locations in the southern part of the island, but may be more widespread.

Conservation status Currently classified by BirdLife International as Near Threatened, mainly because of its restricted geographic range on an island where there has been significant habitat loss. May be able to tolerate some degradation of forest habitat.

Etymology Named by the discoverer, Frank G. Rozendaal, for his wife Caroline.

BOUGAINVILLE BUSH WARBLER
Horornis haddeni

A bird that first came to the attention of ornithologists because of its arresting song but which resisted discovery for a considerable time; ironically, well known to local people who had several dialect names for it. Recent biochemical studies (Alström *et al.* 2011) created a new taxonomy in which *haddeni*, along with several other species in western Oceania, are widely separated from *Cettia* of Eurasia and should belong in the new genus *Horornis*.

Type information *Cettia haddeni* LeCroy & Barker 2006, between Kupei and Moreni, Crown Prince Range, Bougainville Island, Papua New Guinea.

Alternative names Odedi (Nasioi language), Kopipi (Rotokas language).

Discovery In 1972, J. Diamond heard a distinctive song in montane forest on the island of Bougainville. Although the singer was probably "not uncommon", Dr Diamond was never able to see it and made no guesses as to its identity (Diamond 1975). Between 1977 and 1980 D. Hadden heard the song on numerous occasions, but did not see its author, nor was he able to capture a specimen. In 1979, B. Beehler visited the area, also hearing the song but failing to glimpse the singer; in a thoughtful if somewhat speculative paper (Beehler 1983) he argued that the bird – already known by one of its onomatopoeic native names, Odedi – was most probably a *Cettia* warbler. During most of the 1990s the area was off-limits due to civil disorder, but with the deployment of Australian and New Zealand troops in 1999 the situation became more stable, although some areas remained under rebel control. D. Hadden was able, with the help of a local man, John Toroura, to mist-net three birds in steep, undisturbed forest, proving Beehler's prescient opinion that the mystery bird was indeed a *Cettia*. The formal description of the species was based on three specimens, taken in 2000 and 2001, and currently in the American Museum of Natural History (LeCroy & Barker 2006).

Description 13 cm. Monotypic. In form a typical but notably large *Cettia*-type warbler, rather large-headed with a flat forehead, rounded wings and tail. Head dark chestnut-brown with a pale central streak; upperparts very dark chestnut-brown, the wings and tail more bright chestnut; dark lores and eyestripe; chin whitish-grey, underparts mottled greyish with olive-brown flanks; iris dark, bill dark brown with yellow-ochre base to lower mandible; tarsi yellow-brown to yellow-ochre.

Habitat Wet montane forests, 700–1,500 m. May have a fairly restricted altitudinal distribution, being absent from higher areas.

Food and feeding No data. Appears to feed low down or on the ground, possibly associates with Island Thrush *Turdus poliocephalus* (Dutson 2011).

Breeding Nest and eggs undescribed. An adult male taken on 16 September had enlarged testes (LeCroy & Barker 2006).

Voice Song is distinctive and arresting; two- and three-note rising pure whistled phrases at intervals of a few

▲ Bougainville Bush Warbler. Panguna, Bougainville, July 2009 (*Ashley Banwell*).

seconds, notes occasionally trilled, reminiscent of the song of North American Hermit Thrush *Catharus guttatus* (Beehler 1983).

Movements Presumably sedentary.

Range Endemic. Confined to limited mountainous areas on the island of Bougainville.

Conservation status Classified by BirdLife International as Near Threatened, mainly because of its restricted range. Within this, there may be some habitat loss due to clearance for smallholdings; the effect of introduced species, especially Black Rats *Rattus rattus*, is not known. Based on song, may be quite common in some locations, but more fieldwork is required.

Etymology Named in recognition of the contributions of Don Hadden to the ornithology of Bougainville Island, "whose long search for the bird (was) finally rewarded" (LeCroy & Barker 2006).

MOHELI BRUSH WARBLER
Nesillas mariae

One of several isolated Comoro species or races of the endemic Malagasy genus *Nesillas*. Recent genetic studies (Fuchs *et al.* 2016) indicate that there was more than one colonisation of the western Indian Ocean by ancestral forms of the genus.

▲ Moheli Brush Warbler. Djwayezi, Mohéli, Comoros, December 2015 (*Ken Behrens*).

Type information *Nesillas mariae* Benson 1960, Bandamale, Mohéli (Mwali), Comoro Islands, Indian Ocean.

Alternative names Moheli Tsikirity, Mrs Benson's Warbler. French: Nésille de Moheli.

Discovery The type specimen was taken by the British Ornithologists' Union 1958 Centenary Expedition to the Comoro Islands, in September of that year, and described by C. W. Benson in 1960. A total of five specimens was acquired.

Description 15–16 cm. Monotypic. Upperparts dull olive-green, brighter on rump; below yellowish-buff, paler on chin, warmer brown on flanks, the upper breast and throat with soft greyish streaks; iris dark brown, narrow bare eye-ring reddish; bill blackish; tarsi grey.

Habitat Upland forested areas, especially with almost complete cover of herb and shrub layers; at lower elevations, in woody vegetation with small trees and streamside scrub; mostly 500 m upwards, but also down to 150 m (Kennerley & Pearson 2010; Cheke & Diamond 1986).

Food and feeding Forages mostly at mid- or upper levels of shrubs and lower levels of trees; appears to be ecologically separated from the other *Nesillas* species of the island, the Madagascar Brush Warbler *N. typica moheliensis,* which forages lower down (Louette *et al.* 1988). Food items are invertebrates including beetles, ants and spiders; occasionally seeds.

Breeding Few data. One nest under construction, by both sexes, was a small cup of moss and grass with untidy bits of grass dangling below, situated 9 m up at the end of a horizontal branch of a Mango tree *Mangifera indica*. Eggs undescribed (Cheke & Diamond 1986).

Voice Song is a series of loud, high-pitched scratchy notes, somewhat resembling the song of Barn Swallow *Hirundo rustica*. Alarm call is a rapid rattle; contact calls of foraging pairs are a clear *chinkachoyit-chetwie* (Kennerley & Pearson 2010).

Movements Presumably sedentary.

Range Endemic. Confined to Mohéli, Grand Comoro, Indian Ocean, mostly in forested mountain areas.
Conservation status Currently classified by BirdLife International as Least Concern. Although geographically restricted, it appears to be common in suitable habitat (Kennerley & Pearson 2010).
Etymology Named in honour of Florence Mary Benson (1909–1993), wife of the discoverer, C. W. Benson (see under Forbes-Watson's Swift).

ALDABRA BRUSH WARBLER
Nesillas aldabrana

Fuchs *et al.* (2016), basing their conclusions on comprehensive biochemical studies, show that the colonisation of the western Indian Ocean islands by warblers of the genus *Nesillas* was quite complex, with two main hypotheses as to the actual sequence of events. Since Aldabra Atoll was probably submerged some 0.125 million years ago, and would not then have supported any terrestrial fauna, the Aldabra Brush Warbler is likely to be of recent lineage. Perhaps surprisingly, it is a sister species to *N. lantzii*, the Subdesert Brush Warbler, which is geographically the most distant of any species of the genus, being found in semi-arid habitats in southern Madagascar.

Type information *Nesillas aldabranus* Benson & Penny 1968, western extremity of Middle Island, Aldabra Atoll, Indian Ocean.
Alternative names Aldabra Tsikirity.
Discovery An adult female was collected, along with a nest and eggs, by M. J. Penny in December 1967, during the Royal Society Aldabra Expeditions of 1967–1968. A second male specimen was taken by C. W. Benson at a nearby location in January 1968.
Description 18–20 cm. Monotypic. A dingy-coloured, long-tailed bird, pale brown above, brighter cinnamon on the rump and tail; well-marked greyish-white supercilium; greyish-white below, brown on the breast; flanks pale tawny-brown; iris mid-brown to reddish-brown; bill dark horn above, paler below and at base; tarsi grey.
Habitat Dense closed canopy scrub, up to 5 m in height, with good leaf-litter. Prŷs-Jones (1979) gives a detailed study of the habitat.
Food and feeding Skulking, foraged low down. Stomach contents of male specimen were small spiders, moths, winged ants, beetles, bugs and a caterpillar (Benson & Penny 1968).
Breeding One active nest described; a cup made of shredded leaves, lined with fine grass stems, ca 0.6 m up in the base of a *Pandanus* shrub. On 11 December it held three freshly laid eggs, very pale purplish with dark brown blotches and speckles on underlying pale lilac blotches. Two other inactive nests were at heights of 2.6 m and 10 m in shrubs.
Voice Call was a nasal, trisyllabic *chinkachoy*; alarm call a harsh *chirrr*. Prŷs-Jones (1979) gives a more detailed description of vocalisations.
Range Endemic. Formerly in a narrow strip of scrub, about 50 m wide and 2 km long (R. Prŷs-Jones, pers. comm), on the north coast of Middle Island (le Malabar), Aldabra Atoll. Absent from other islands in the group.
Conservation status Extinct. The total distribution was always highly restricted and the species appeared to be uncommon even within that. A study in 1974–1975 located only five birds, with a maximum extrapolated population of only 25 (Prŷs-Jones 1979). By 1977 only two males were known to have survived, and no birds have been seen since 1983. The causes of extinction are not entirely certain but may relate to predation by rats and habitat destruction by introduced goats and the native giant tortoise *Geochelone gigantea*.
Etymology *Aldabrana*, from Aldabra.

▲ Aldabra Brush Warbler. Aldabra, Seychelles, 1974 (*Robert Prŷs-Jones*).

TAIWAN BUSH WARBLER
Locustella alishanensis

A little-known species endemic to Taiwan, previously the source of considerable taxonomic confusion.

Type information *Bradypterus alishanensis* Rasmussen *et al.* 2000, Ta-Kuan, Hua Lian Hsien, Taiwan, based on a specimen taken in March 1960.

Alternative names Chinese: Taiwan Duan Chi Ying.

Discovery It has been known for almost a century that bush warblers (*Bradypterus*, presently *Locustella*) occur on Taiwan; specimens were collected in 1917 (Hachisuka & Udagawa 1951) but were tentatively identified as Brown Bush Warbler *B.* (= *L.*) *luteoventris*, the species breeding from the eastern Himalayas to south China and north Vietnam. Later, Delacour (1952) placed Taiwan specimens in the taxon *B. (mandelli) idoneus*, known only from southern Vietnam, while later authors (e.g. Meyer de Schauensee 1984) referred Taiwan birds to the eastern Chinese race *melanorhynchus* of Russet Bush Warbler *B. mandelli*. However, more recently several ornithologists realised that Taiwan birds had vocalisations very different from those of populations on the adjacent mainland, prompting a study of morphometrics and voice which led to the recognition of the Taiwan population as a distinct new species.

Description 13–14 cm. Monotypic. Medium-sized, rather short-winged with a long, rounded tail. Upperparts dark brown, with an indistinct pale region above the lores and eye-ring; sometimes a pale supercilium; throat whitish; in some individuals obvious darker spotting on chin or throat, often absent; breast pale brown, flanks dull russet-brown; belly greyish-white; undertail-coverts russet-brown, the feathers tipped with pale brown; iris brown; bill black, sometimes, perhaps in immature birds, with a pale base to lower mandible; tarsi pale pink. A melanistic example has been reported (Kuroda 1938).

Habitat A variety of habitats, but requires a luxuriant understorey of thick grasses, tangled shrubs and ferns; at lower elevations, slopes of both coniferous and deciduous woodland; at higher elevations, hillsides covered in tall grass and bamboo; 1,200–3,000 m (Kennerley & Pearson 2010). Maturing of secondary forest that results in a diminution of understorey is deleterious to the species (Yuang *et al.* 2005).

▲ Taiwan Bush Warbler, juvenile. Yushan National Park, Taiwan, June 2004 (*Liao Pen-shing*).

◄ Taiwan Bush Warbler. Note this bird lacks the characteristic throat-spotting. Nantou County, Taiwan, July 2005 (*Liao Pen-shing*).

Food and feeding Forages low down in vegetation; no data on diet.
Breeding Few data. Breeding may occur from late March at lower elevations, while birds have been seen carrying food in August. Nest is placed low down, often in a *Miscanthus* clump; two eggs.
Voice Song is distinctive and quite unlike that of other members of the genus. Each sequence starts with a short monotone whistle, perhaps rising slightly in pitch, followed by a series of three or four flat, hollow-sounding clicks, and finishes with a monotone whistle (Kennerley & Pearson 2010). Calls include a series of raspy *ksh-ksh-ksh* notes, speeding up when alarmed.
Movements Mainly resident, though some altitudinal movement may occur.
Range Endemic to Taiwan. Fairly well distributed at suitable altitudes, over most of the eastern mountain range.
Conservation status Classified by BirdLife International as Least Concern. Although the geographic range is not great, the species seems to be quite common in some locations, and furthermore appears to be able to adapt to quite a wide variety of habitats including some modified by human activity.
Etymology *Alishanensis*, from Mount Ali-Shan, on which the 1917 specimens were collected.

SAKHALIN GRASSHOPPER WARBLER
Locustella amnicola

Treated as a race of Gray's Grasshopper Warbler *L. fasciolata* by many authorities (e.g. Cramp *et al.* 1992; Kennerley & Pearson 2010); however, DNA studies (Drovetski *et al.* 2004, 2015) suggest that full species status may be justified, and it is treated as such in the IOC World Bird List.

Type information *Locustella amnicola* Stepanyan 1972, Tonino-Anivsky Peninsula, Sakhalin Island, eastern Russian Federation.
Alternative names Stepanyan's Warbler. Russian: Sakhalinskiy Sverchok; Japanese: Ezo-sennyu.
Discovery In May and June 1972 L. S. Stepanyan was conducting an ornithological investigation of the central part of the Tonino-Anivsky Peninsula of southern Sakhalin Island, eastern Russian Federation. On 10 June he obtained two adult specimens of a *Locustella* warbler, which he deemed sufficiently distinct from Gray's Grasshopper Warbler of the eastern mainland of the Russian Federation to warrant description as a new species.
Description 16–17 cm. Monotypic. In form a typical large *Locustella* warbler, rather large-billed, flat-headed with a rounded tail. Generally brown, richer above, paler below, with an indistinct brownish supercilium; iris brown; bill brownish above, paler below; tarsi pinkish-brown. Differs from Gray's Grasshopper Warbler in lacking the grey tone of the ear-coverts and upper chest, and in minor details of wing-formula.

► Sakhalin Grasshopper Warbler, adult. Hokkaido, Japan, June 2009 (*Pete Morris*).

▲ Sakhalin Grasshopper Warbler, juvenile. Yehliu, Taiwan, October 2009 (*Liao Pen-shing*).

Habitat In damp areas, including woodland edge and dense, shady alder and willow thickets (Brazil 1991); on wintering grounds, thickets, forest-edge and tall grass, up to 1,000 m.

Food and feeding Extremely secretive, foraging low down in vegetation. Diet arthropods.

Breeding Nest is a cup, 1.5–4.5 m up in a bush or tree; eggs 1–5, laid from early June to late July (Brazil 1991).

Voice Song is a very loud, explosives *chot-pin,chot-pin kake taka*; sings by day and night (Brazil 1991); slightly slower, more mellow and relaxed than that of Gray's Grasshopper Warbler (Kennerley & Pearson 2010).

Movements Migratory. Migrates through Taiwan and winters in the Wallacean region, probably overlapping with the wintering range of Gray's Grasshopper Warbler, which winters from the southern Philippines to eastern Indonesia and western New Guinea (Irian Jaya). Rheindt *et al.* (2014) noted birds on Ternate Island (Halmahera) whose song seemed to resemble that of *amnicola* rather than *fasciolata* but were not willing to make a definite species assignment based on their observations.

Range Breeds over most of the island of Sakhalin and northern Japan (Hokkaido); also southern Kuril Islands.

Conservation status Not classified by BirdLife International, but in the Japanese part of the range locally common, so presumably Least Concern.

Etymology *Amnicola*, Latin, 'dwelling by the riverside'.

SICHUAN BUSH WARBLER
Locustella chengi

A cryptic species found in several provinces in central China, suspected for a considerable period but only recently confirmed.

Type information *Locustella chengi* Alström *et al.* 2015, Qinling Mountains, Sichuan, China.

Alternative names Chinese: Sichuan Duan Chi Ying.

Discovery In May 1987 P. Alström and U. Olsson were investigating the avifauna of the Emei Mountains in Sichuan province, China, when they heard a distinctive, and to them unfamiliar, song emanating from an area of dense herbs near a trail. The singer was extremely secretive and allowed only brief glimpses, which revealed it to be a cryptically-plumaged warbler, resembling the widespread Russet Bush Warbler *Locustella (Bradypterus) mandelli*, a species breeding from the eastern Himalayas, Assam, northern Myanmar (Burma) across much of southern China to northern Indochina and Thailand. However, the song heard was quite distinct from that species. At this point both observers had to leave the matter in abeyance while they pursued other work, while not forgetting their 'mystery bird'. Returning to China in 2011, they found the unknown warbler to be common in the Qinling Mountains, and collected one specimen (subsequently designated the holotype). They later learnt that birds with the song-type of the new taxon as well as the usual *mandelli* song had been heard at the same time in the breeding season at one place in Sichuan province; they went there in 2014 and found them to be breeding fairly commonly in sympatry. Subsequently other specimens were captured, measured and photographed; DNA samples were also taken. The species was formally described by Alström *et al.* in 2015. The DNA analysis revealed that the species was closely related to *mandelli* and had probably diverged from that species some 850,000 years ago.

Description 13 cm. Monotypic. In form a typical *Locustella* warbler, with rather short, rounded wings and a rounded, graduated tail. Upperparts warm grey-brown; centre of throat whitish, brownish-grey faintly tinged yellow at sides; belly whitish with pale dingy yellowish tinge, vent feathers cold grey-brown, tipped greyish-white; iris brown; bill blackish; tarsi pinkish.

► Sichuan Bush Warbler. Laojun Shan, Sichuan, China, May 2014 (*Per Alström*).

Habitat Breeding habitat: slopes and flat ground with a dense vegetation of herbs and ferns, often in association with trees, such as in forest clearings and in secondary forest; also in tea plantations. 1,000–2,275 m. Habitat on wintering grounds not known. Generally allotopic from *mandelli*, which tends to occur at higher elevations, with only limited overlap.

Food and feeding No data; forages low down in dense vegetation.

Breeding Nest and eggs undescribed. From vocal activity, breeding season probably begins in May.

Voice Song is distinctive, described as insect-like, a "low-pitched drawn-out buzz, followed by a short click, repeated in series"; markedly lower-pitched than that of *mandelli*. Each species reacts vigorously to playback of its own song but very little to that of the other species.

Movements Migratory, but winter range unknown.

Range Breeding endemic. Central China, at suitable altitudes in Shaanxi, Sichuan, Guizhou, Hubei, NW Hunan and NW Jiangxi.

Conservation status Not currently listed by BirdLife International, but doubtless of Least Concern. Appears to have a quite wide breeding distribution and to be not uncommon in many locations; also appears to tolerate some habitat modification and to inhabit secondary growth. Habitat requirements on wintering grounds unknown.

Etymology Named in honour of Professor Cheng Tsohsin (Zheng Zuoxin) (1906–1998), one of the foremost Chinese ornithologists of the 20th century, in recognition of "his unparalleled contributions to Chinese ornithology" (Alström *et al.* 2015).

RUBEHO WARBLER
Scepomycter rubehoensis

A cryptic species, very similar in plumage to Winifred's Warbler *S. winifredae*, itself a little known and very restricted species only described in 1938 (Moreau 1938).

Type information *Scepomycter rubehoensis* Bowie *et al.* 2009, Ukwiva Forest, Rubeho Mountains, Tanzania.

Discovery Winifred's Warbler was believed to have disjunct populations in several ranges of the Eastern Arc mountains of Tanzania (Pearson 2006 HBW 11). In view of the apparent isolation of these populations, Bowie *et al.* (2009) undertook a genetic analysis which showed that birds from the Rubeho-Ukaguru Mountains differed significantly from those from the Uluguru Mountains; examination of specimens also showed consistent if small plumage differences between the two populations.

Description 13–14 cm. Monotypic. Crown chestnut, remainder of head, face, chin, throat and upper chest deep rufous; nape and mantle dark mouse-grey; wings and tail darker, the tail strongly graduated; remainder of underparts hair-brown with pale mottling in centre;

▲ Rubeho Warbler. Ukaguru Mountains, Tanzania, November 2015 (*Markus Lagerqvist*).

iris dark; bill black; tarsi dark grey to dark brown. Differs from *S. winifredae* in having deeper reddish head, more conspicuously mottled underparts and a longer bill.

Habitat Dense vegetation in wet forest, on steep slopes, with broken canopy caused by landslides, logging or clearing by African Elephants *Loxodonta africana*; 1,500–1,900 m.
Food and feeding Forages in pairs in low vegetation; no data on diet.
Breeding Nest and eggs undescribed.
Voice Duet song a long series of ringing calls alternating between two pitch levels (Pearson 2006 HBW 11).
Movements Presumably sedentary.
Range Endemic. Confined to highest parts of Ukaguru Mountains and Ukwiva Forest in the Rubeho Mountains, NE Tanzania.
Conservation status Not assessed by BirdLife International, but probably deserves status of Endangered or similar. Appears to be rather uncommon; the area has a dense human population, who frequently burn areas for agricultural purposes.
Etymology Both the vernacular and scientific names refer to the Rubeho Mountains.

MANGAREVA REED WARBLER
Acrocephalus astrolabii

In 1978 D. T. Holyoak and J.-C. Thibault described several new taxa of *Acrocephalus* warblers from the Pacific, based on old museum specimens that had been overlooked or inadequately analysed. Among these were two specimens, presently in the Muséum d'Histoire Naturelle in Paris, which they described as a new race (*astrolabii*) of Nightingale Reed Warbler *A. luscinius* of the northern Mariana Islands. Cibois *et al.* (2011) argue for full specific status for *astrolabii*, a treatment followed by several authorities including the IOC World Bird List. The two specimens are rather distinctive, being large, with a relatively short bill and extraordinarily stout tarsi, feet and claws (Holyoak & Thibault 1978b). The origin of the two specimens is highly debatable. Both, each inadequately labelled, were collected at some time during the 1838–1839 Second Antarctic Expedition under the command of Captain J. Dumont D'Urville, who travelled widely in Polynesia and Melanesia (Dumont D'Urville, 1841–1846). One specimen is labelled "Mangareva" (in the Gambier Islands, French Polynesia), the second "Nouheva" (= Nuku Hiva, in the Marquesas Islands), thereby putting the provenance of both in doubt. Holyoak & Thibault (1978b) argue that Yap Island, in the southern Marianas, is a more plausible origin. No ornithological exploration of Yap occurred for another 30 years (Hartlaub & Finsch, 1872), when no examples of *astrolabii* were encountered. The species is, in all probability, extinct; however, given the large number of islands visited by Captain Dumont D'Urville, the vagueness of the label data, and the recent discovery of a new *Acrocephalus* at a site as well known as the Cook Islands, a definite classification of Extinct might, just, be premature.

Type information *Acrocephalus luscinius astrolabii* Holyoak & Thibault 1978b, based on two specimens of disputed provenance.
Alternative name Astrolabe Reed Warbler, Gambier Islands Warbler.
Etymology The scientific name is derived from the French corvette *Astrolabe*, Captain Dumont D'Urville's ship.

COOK ISLANDS WARBLER
Acrocephalus kerearako

Two races, described originally as races of Pitcairn Reed Warbler *A. vaugheni*, a treatment that seems implausible given the large geographic separation (over 2,500 km), the very different plumage and the existence of numerous other species and races of *Acrocephalus* on intervening archipelagos.

Type information *Acrocephalus vaugheni kerearako* Holyoak 1974, Mangaia, Cook Islands.
Acrocephalus vaugheni kaoko Holyoak 1974, Mitiaro, Cook Islands.

Alternative names Cook Reed Warbler. Local names: Kerearako (Mangaia Island) and Kaoko (Mitiaro Island).

Discovery Two races, each endemic to one island, described (along with several other new bird species and subspecies) as a result of fieldwork by D. T. Holyoak in July–September 1973; curiously, the existence of such birds on the island of Mangaia had been known for 50 years prior to Dr Holyoak's work (Christian 1920).

Description 16 cm. Nominate race: Generally olive above, the wings darker with tawny-olive feather edgings; underparts yellowish-white; narrow bright buff supercilium; dark line from lores through eye; ear-coverts warm brown; iris dull brown; bill dark grey, paler below with pink base; tarsi pale blue-grey. *A. k. kaoko*: larger than nominate, upperparts duller with only slight tawny tinge; underparts paler yellow, with indistinct light grey-brown streaking on throat and breast.

Habitat On Mangaia Island, common in wooded and bushy areas around the island periphery, but also occurs in scrubby areas and in introduced pine forests; on Mitiaro, found in trees and thick bushy scrub in coastal coral zone, in *Casuarina* thickets and in village gardens.

Food and feeding Forages in bushes, trees and on ground; prey items include small beetles, dipterous larvae, grasshoppers, moths and spiders.

▲► Cook Islands Warbler *A. k. kerearako*. Top: male, Mangaia, Cook Islands, May 2006. Bottom: female at nest, Mangaia, Cook Islands, December 2007. (*Gerald McCormack*).

◄▼ Cook Islands Warbler *A. k. kaoko*, male (top) and female (bottom). Mitiaro, Cook Islands, April 2011 (*Gerald McCormack*).

Breeding Few data; nest of nominate race apparently undescribed. On Mitiaro, three occupied nests found in September. Nest is a deep cup, made of fine dry stems and fibres, at heights of 1 m to canopy height. The three nests described each had one egg, but local inhabitants state that clutches of two and (rarely) three occur. The specimens collected by Dr Holyoak in August and September were in breeding condition.

Voice Songs of both races have a simple structure with distinct single loud whistled phrases; on Mangaia, a rapid delivery given in shorter bursts of 5–7 seconds; on Mitiaro, less varied and slower. Both songs are interspersed with harsher notes.

Range Endemic. Nominate race, Mangaia Island; race *kaoko*, Mitiaro Island, southern Cook Islands. Whether birds ever occurred on the intervening islands of Atiu and Mauke is not known.

Conservation status Currently classified by BirdLife International as Near Threatened. Although both races appear to be quite common, each has a restricted distribution (Mangaia is about 52 km², with a human population of about 700; Mitiaro is 22 km² with about 200 inhabitants). Potential threats on Mangaia are predation by rats, both Black and Polynesian, and habitat destruction by goats.

Etymology Both scientific names are taken from the names of the birds in the local dialect.

CRYPTIC WARBLER
Cryptosylvicola randrianasoloi

A somewhat nondescript but taxonomically fascinating species. Originally placed in the Old World warblers (Sylviidae); recent biochemical work (Cibois *et al.* 1999; Cibois *et al.* 2001; Cibois *et al.* 2010; Beresford *et al.* 2005) has resulted in the creation of a new family (Bernieridae; Malagasy warblers) which encompasses a number of species previously placed, unsatisfactorily, in families such as Timaliidae (Old World babblers) and Pycnonotidae (bulbuls).

Type information *Cryptosylvicola randrianasoloi* Goodman *et al.* 1996, east ridge of Volotsangana River, Réserve Naturelle Intégrale No 5, 38 km south of Ambalavao, Fianarantasoa province, Madagascar.

Alternative names French: Fauvette à pattes noires, Sylvicole cryptique.

Discovery In November 1992 B. M. Whitney and J. Pierson, while on a field trip in the Maromiza Forest, east of the capital, Antananarivo, Madagascar, heard an unfamiliar song, which they recorded without seeing the bird. Later the same day they saw a distinctive warbler-like bird which responded vigorously to playback of the song earlier recorded. The observers immediately realised that an unknown species was involved. Over the next couple of days the species was found to be quite common in the area, with at least eight singing birds along 2 km of trail. The observers returned to the capital and discussed their observations with O. Langrand and S. Goodman, who visited the area, confirmed that the species was fairly common and took the first specimen in December 1992. In November 1992 Whitney observed and recorded the same species at Ranomafana, some 275 km south of the original location. The recordings were played to L Wilmé, who realised that she had recorded the same song at Ranomafana in 1987. Subsequent fieldwork showed the species to be widely distributed in suitable habitat in a band of eastern Madagascar almost 1,000 km in length. Detailed morphological examination of the four specimens now available and comparison with other relevant genera showed that the new species could not be allocated to any known genus, requiring the erection of a new one (Goodman *et al.* 1996).

Description 12 cm. Monotypic. Forehead, crown, nape, hindneck and upperparts olive-green, becoming yellower on rump; narrow greenish-yellow supercilium; dark olive-green eyestripe; cheek and ear-coverts mottled greenish-yellow; chin light yellow; throat, breast, flanks and belly dirty greyish-yellow; undertail-coverts and thighs olive-green; rectrices yellowish olive-green above, dull olive-green below; iris dark brown with greyish cast; bill black above with dull orange cutting-edge, dull orange below with darker tip; tarsi brownish-black. Sexes similar.

Habitat Rainforest, of quite varied types, including primary forest, areas of bamboo and in quite modified habitat. At the upper altitudinal range, open grassland with clumps of ericacious vegetation, 900–2,100 m (Goodman *et al.* 1996).

Food and feeding Forages actively in the canopy and subcanopy, often in mixed-species flocks; food items are invertebrates; observed prey includes crickets, spiders, caterpillars and moths.

Breeding Breeding season October–December. So far only one nest described; it had three eggs on 24 October, but no other details were published.

Voice Song, delivered from tree crowns or open branches lower down, is a series of raspy notes, 5–12 in number, slightly slurred, on roughly the same pitch *chick-tss-tss-tss-tss*. Call, a distinctive scolding *tsick*.

Movements Presumably sedentary.

Range Endemic. Occurs along much of the length of eastern Madagascar, from about 14° 40' S to at least 24° S (Hawkins & Pearson 2006 HBW 11). With more fieldwork, further range extensions are possible.

Conservation status Currently classified by BirdLife International as Least Concern. The species has a long, if narrow, geographic distribution and appears to be able to tolerate moderate habitat disturbance. The montane forest habitat seems to be less liable to the slash-and-burn agriculture which has devastated other areas of Madagascar (Goodman *et al.* 1996).

Etymology The generic name *Cryptosylvicola* is a mixture of roots meaning 'hidden forest inhabitant'; the specific epithet honours Georges Randrianasolo (1930–1989), a Malagasy ornithologist who contributed much to the study of Madagascar birds; he was also an expert on lemur distribution. The newly created family name, Bernieridae, recognises Chevalier J. A. Bernier, a French naval surgeon who collected a number of bird specimens between 1831 and 1834. Two Madagascar species, Bernier's Teal *Anas bernieri* and Bernier's Vanga *Oriolia bernieri* are also named for him.

► Cryptic Warbler. Mantadia National Park, Madagascar, October 2012 (*Markus Lilje*).

APPERT'S TETRAKA
Xanthomixis apperti

Originally described as a member of the bulbul (Pycnonotidae) family; however, more recent biochemical work suggests that it is more properly placed in the endemic Malagasy family Bernieridae (see under Cryptic Warbler *Cryptosylvatica randrianasoloi* and references therein). Sometimes placed in the genus *Bernieria*.

Type information *Xanthomixis apperti* Colston 1972, 40 km SE of Ankazoabo, SW Madagascar.

Alternative names Appert's Bulbul. French: Tetraka d'Appert, Bulbul d'Appert.

Discovery In the early 1970s P. R. Colston and C. W. Benson were asked to examine two specimens of what appeared to be an unfamiliar bulbul, collected by the Rev. Otto Appert in SW Madagascar in 1962. Careful examination and comparison of these specimens with other bulbul species resulted in the description of a new species, placed by Colston in the predominantly mainland African genus *Phyllastrephus*. At this point the two specimens were the only ones known; two other reported specimens, reputedly taken by G. Randrianasolo, could not be located (Colston 1972).

Description 15 cm. Monotypic. Crown, nape and ear-coverts blue-grey, remainder of upperparts including tail greenish olive-grey; supercilium buffish-white to pale grey; eye-ring pale whitish-grey above and below, dark grey in front of and behind eye; chin and throat white; underparts warm yellow, centre of belly whitish; iris dark brown; bill mostly dark with dirty pale pink base to lower mandible; tarsi pale horn or pale reddish.

Habitat Undisturbed deciduous forest at 500–800 m and a tract of relict evergreen forest on an isolated massif at 900–1,330 m (Fishpool & Tobias 2005 HBW 10; Langrand 1990).

Food and feeding Forages low down or on the ground, sometimes in flocks of up to eight birds, also in mixed-species flocks. Diet arthropods.

Breeding Nest and eggs undescribed; subadult recorded in April.

Voice Quite vocal; call, a high shivering *tsee* or *tseetsee*; alarm call a high-pitched *tsirr*.

Movements May undertake local altitudinal movements but hard data lacking.

Range Endemic. Restricted to three locations in SW Madagascar: Zombitse-Vohibasia National Park and Anavelona Classified Forest, and recently near Salary Bay (Langrand & von Bechtolsheim 2009).

Conservation status Currently classified by BirdLife International as Vulnerable. Very restricted in distribution. Of the two main locations, one (Zombitse-Vohibasia) is protected and the second (Anavelona) is relatively remote; only two birds were seen at Salary Bay. In some parts of the range appears to be fairly common, in others rare (Mustoe *et al.* 2000). Population believed to be in the range of 600–1,700 mature individuals.

Etymology Named in recognition of Otto Appert (1930–2012), a Swiss-German Catholic missionary who collected the first two specimens. Dr Appert was perhaps better known as a palaeobotanist rather than an ornithologist, having published extensively on fossil cryptograms.

◄ Appert's Tetraka. Zombitse Forest, Zombitse-Vohibasia National Park, Atsimo-Andrefana region, Madagascar, December 2010 (*Dubi Shapiro*).

ALPINE LEAF WARBLER
Phylloscopus occisinensis

Treated as a race of Tickell's Warbler *P. affinis* by some authors.

Type information *Phylloscopus occisinensis* Martens *et al.* 2008, western shore of Lake Donggi Cona, Qinghai, China.

Alternative names West Chinese Leaf Warbler. Chinese: Hua Xi Liu Ying.

Discovery Martens *et al.* (2008) made a comprehensive study of the bush-dwelling *Phylloscopus* warblers of the western Chinese mountains, better defining relationships and describing two new taxa; a new race of Tickell's Warbler *P. affinis* (*P. a. perflavus*) and a population clearly enough differentiated on vocal, morphological and molecular genetics to merit full specific status as a new species, *P. occisinensis*. This form was previously included in Tickell's Warbler, a widespread montane species.

Description 10–11 cm. Monotypic. A medium-sized *Phylloscopus* with prominent yellow supercilium but no wing-bars. Upperparts dark grey-brown, tinged olive-green; eyestripe dark brownish, contrasting with supercilium; underparts bright yellow with darker breast-band and flanks; iris dark; bill dusky-brown with orange base to lower mandible; tarsi dusky-brown, soles rich yellow.

Habitat In breeding season, various types of alpine forest, including *Juniperus* forest with *Salix* and *Caragana* bushes, birch (*Betula*), alpine steppes and open meadows, 3,000–4,500 m (Martens *et al.* 2008). Wintering habitat not known.

Food and feeding Forages singly or in pairs, or in small flocks in winter. Diet insects.

Breeding Nest and eggs undescribed. Adults feeding dependent young on 24 July.

Voice Song is reported to be lower in pitch than that of *P. affinis*, with rapid double or triple trills.

Movements Few data; may undertake altitudinal movements; speculated to winter in Myanmar (Burma).

Range Breeding endemic. West-central China (Qinghai, Sichuan, Gansu); recorded in non-breeding season in Yunnan. Wintering range not certainly known.

Conservation status Not currently classified by BirdLife International, but probably Least Concern. Quite widely distributed and apparently fairly common in some locations.

Etymology *Occisinensis*, an abbreviated Latin complex for 'western China'.

▼ Alpine Leaf Warbler. Koko Nur, Qinghai, China, July 2014 (*James Eaton*).

EMEI LEAF WARBLER
Phylloscopus emeiensis

Another *Phylloscopus* species whose existence was first suspected as a result of careful observations of distinctive vocalisations.

Type information *Phylloscopus emeiensis* Alström & Olsson 1995, Emei Shan, Sichuan, China.
Alternative names Chinese: Emei Liu Ying.
Discovery In May 1987 P. Alström and U. Olson heard and saw a *Phylloscopus* warbler in the Emei Shan range of mountains, Sichuan province, China. The song and call were unfamiliar to them and highly distinctive. In form the bird appeared essentially identical to the locally common Blyth's Leaf Warbler *P. reguloides* (now Claudia's Leaf Warbler *P. claudiae*). No further individuals of this song-type were found then or in subsequent visits in May 1989 or June 1990, and the observers concluded that the 1987 sighting probably involved aberrant vocalisations of *P. reguloides*. However, they later learned that A. Goodwin had heard birds with that same song-type in a different area of Emei Shan. This prompted them to visit Goodwin's site in April 1992, where they found at least 20 individuals in song within a very limited area, allowing for much more detailed observations of vocalisations and morphology. It was found that the new form did not respond to playback of the songs of two sympatric species, *P. reguloides* (now Claudia's Leaf Warbler *P. claudiae*) and White-tailed Leaf Warbler *P. davisoni* (now Kloss's Leaf Warbler *P. ogilviegranti*). Three males were captured, measured and photographed; blood samples were taken from all three and one was collected as a specimen, allowing the description of a new species.

▼ Emei Leaf Warbler. Tang Jia He, Sichuan, China, June 2014 (*James Eaton*).

Description 11–12 cm. Monotypic. A medium-sized *Phylloscopus* with well-marked plumage. Crown greenish-grey with pale median stripe; supercilium yellowish-white; eyestripe dark grey-green; ear-coverts grey-green, mottled with indistinct pale yellowish; upperparts dull green; upperwing brown-grey with one distinct pale greenish-yellow wing-bar on greater coverts, a less defined one on median coverts; throat and underparts whitish with indistinct yellow streaks, pale greyish-green flanks and pale yellow undertail-coverts; iris dark brown; bill blackish above, pale orange below; tarsi pinkish-grey. In plumage very similar to *P. reguloides* (now *P. claudiae*), but crown pattern less distinct, rear of lateral crown-stripes dark greenish-grey, not blackish, and pale areas on outer rectrices reduced.
Habitat In breeding season, temperate deciduous of broadleaved forest, sometimes mixed with spruce (*Picea*) and fir (*Abies*). Winter habitat not well known.
Food and feeding Little information. Insectivorous.
Breeding Nest and eggs undescribed. Presumably begins breeding in mid to late April.
Voice Diagnostic, the best distinction from other species in the *P. reguloides* complex. Song is a clear, slightly quivering straight trill, lasting 3–4 seconds, among Eurasian *Phylloscopus* most resembling that of Arctic Warbler *P. borealis*. Call is also distinctive, a soft *tu-du-du, tu-du* or *tu-du-du-du*.
Movements Migratory, but little information.
Range Breeding endemic. Central China: SE Sichuan, NE Yunnan, northern Guizhou, NW Guangdong and S Shaanxi; wintering area little known, with one record from SE Myanmar (Burma) (Robson 2005) and a second (November 2015) from Hong Kong (Robson *et al.* 2016).
Conservation status Classified by BirdLife International as Least Concern. Occurs in several widely scattered locations, in some areas not uncommon.
Etymology *Emeiensis,* Latinised form of Emei.

HAINAN LEAF WARBLER
Phylloscopus hainanus

A distinctive *Phylloscopus* warbler from Hainan Island, specimens of which had previously been unrecognised or misidentified.

Type information *Phylloscopus hainanus* Olsson *et al.* 1993, Diao Lou, Lingshui county, Hainan, China.
Alternative names Chinese: Hainan Liu Ying.
Discovery In 1986 U. Olsson, as part of an extensive study of the *Phylloscopus* warblers of SE Asia, had occasion to examine four specimens in the collection of the Academia Sinica, Kunming, China, taken on Hainan, a tropical island off the southern coast of China. These had previously been identified as the race *goodsoni* of Sulphur-breasted Warbler *P. ricketti*, a species of wide distribution in mainland China. However, he was immediately struck by the distinctiveness of the Kunming specimens and considered the possibility of an undescribed species. In 1988 U. Olsson and P. Alström visited central Hainan, but despite searches in apparently suitable habitat they were not able to find any examples of the bird in question. However, in April 1988 S. Jensen found the bird to be common in the Jian Feng Ling Forest Reserve and made recordings of the song. Olsson and Alström noted that the song differed significantly from that of mainland *P. ricketti.* The following year they visited Jensen's site and observed the bird, confirming its morphological distinctiveness. To confuse matters, examination of specimens from Hainan in the American Museum of Natural History, previously labelled as *P. ricketti goodsoni*, turned out to be no such thing, but instead were Blyth's Leaf Warbler *P. reguloides*, a species breeding from southern India to eastern China and wintering on Hainan; hence *P. ricketti goodsoni* is a synonym of *P. reguloides*, and the Hainan specimens represent a new species. This was described by Olsson *et al.* (1993), based on a 1962 specimen located in the collection of the Academica Sinica in Beijing. (*P. reguloides* was later split into three species, treating *P. goodsoni* as a distinct species.)
Description 11 cm. Monotypic. A medium-sized, brightly plumaged *Phylloscopus.* Green above, bright yellow below; median crown-stripe pale yellow, lateral crown-stripes vivid green; ear-coverts, throat and underparts deep warm yellow; a broad yellowish wing-bar formed by tips to the greater coverts and a narrower one on the median coverts; two outermost rectrices are entirely white on the inner webs; iris dark brown; bill dark above, pale below; tarsi medium dark, the soles paler.
Habitat Broadleaved forest, including primary forest, selectively logged areas and secondary growth, at 600–1,500 m (Chan Bosco Pui Lok *et al.* 2005). May previously have occurred at lower altitudes but little of this forest now remains.
Food and feeding Insectivorous, but few data.
Breeding One nest found, dome-shaped with a side entrance, located on the face of a steep roadside bank 170 cm up, made of slender panicles and broad grass leaves, lined with fibres, other fine vegetable matter and feathers. In late April it held well-fledged young; several other pairs with fledged young seen at the same time (Olsson *et al.* 1993).
Voice Distinctive. Song resembles that of White-tailed Leaf Warbler *P. davisoni* and even more Kloss's Leaf Warbler *P. ogilviegranti,* now split from *P. davisoni*; short varied phrases *tsitsitui...titsu-titsui-titsui* etc. Olsson *et al.* (1993) provide sonograms of varied examples. Call is *pitsitui, pitsui* etc.
Movements Not clearly known, but may be sedentary. Breeding probably complete by May, when birds found in mixed flocks; adult photographed in same breeding area in November (J. & J. Holmes, Oriental Bird Images 2003).
Range Endemic. Confined to Hainan Island.
Conservation status Currently classified by BirdLife International as Vulnerable. Much presumed habitat has been destroyed; however, known from eight locations, some of which are protected (Chan Bosco Pui Lok *et al.* 2005); in some areas rather common or "one of the most common bird species" (Olsson *et al.* 1993). Population estimates in the range of 1,500–7,000 mature birds, probably declining overall.
Etymology *Hainanus,* Latinised from the island of Hainan, an Endemic Bird Area of critical priority (BirdLife International 1998).

▼ Hainan Leaf Warbler. Jianfengling, Hainan, China, December 2014 (*Robert Hutchinson*).

LIMESTONE LEAF WARBLER
Phylloscopus calciatilis

Probably forms a superspecies with Yellow-vented Warbler *P. cantator* and Sulphur-breasted Warbler *P. ricketti*. Although probably closest genetically to the former, it more resembles the latter in plumage. Placed in the genus *Seicercus* by some authorities.

Type information *Phylloscopus calciatilis* Alström *et al.* 2009, Phong Nha-Ke Bang National Park, Quang Binh province, Central Annan, Vietnam.

Alternative names Chinese: Hui Yan Liu Ying.

Discovery Sulphur-breasted Warbler *P. ricketti* is a species with a widespread breeding distribution across central China, wintering in Thailand, Laos and Vietnam (Robson 2002; Robson 2005). However, the observation by several ornithologists during the mid-1990s of birds apparently identical in plumage to *ricketti* in Laos and Vietnam during the breeding season clearly indicated that further investigation was warranted. Several recordings of songs were made. Subsequently in June 1996 two specimens were taken in Phong Nha-Ke Bang National Park; in February 1998 the putative new form was found to be common and territorial in the Hin Namno National Protected Area. Some birds were trapped, measured and photographed. In April 2004 Phong Nha-Ke Bang was revisited, resulting in the songs of nine territorial birds being recorded, and detailed data including blood samples obtained from two trapped individuals. Careful comparison of all data, morphometric and genetic, with specimen material of *P. ricketti* and *P. cantator*, which appear to be allopatric or parapatric, established sufficient differences to allow Alström *et al.* to describe a new species. A specimen taken in 1929 by Jean Delacour in West Tonkin, Vietnam, and presently in the Field Museum, Chicago, was also assigned to *P. calciatilis*.

▼ Limestone Leaf Warbler. Ba Be National Park, Vietnam, March 2016 (*James Eaton*).

Description 10–11 cm. Monotypic. A strikingly marked *Phylloscopus*. Median crown-stripe greenish-yellow, bordered by a blackish lateral crown-stripe; supercilium greenish-yellow; eyestripe blackish; cheeks and ear-coverts greenish-tinged yellow; narrow yellow crescent below eye; upperparts rather bright greyish-green, brighter on rump; wings greenish with two narrow wing-bars formed by yellowish-white tips to greater and median coverts; underparts bright yellow with green tinge on flanks; iris dark grey-brown; bill blackish above, pale orange below; tarsi light greyish-buff. Sexes similar. In plumage probably indistinguishable in the field from *P. ricketti*; marginally colder yellow below and more greyer-tinged above, and lateral crown-stripes slightly greyer. May differ from *ricketti* in wing-formula and bill measurements, but sample sizes are small.

Habitat Appears to be restricted to broadleaved evergreen and semi-evergreen forests, sometimes in secondary growth, on limestone karst mountains, at 80–1200 m.

Food and feeding No data.

Breeding Nest and eggs undescribed. Singing birds in mid-February in east-central Laos and in May in north-central Laos; adults carrying food seen in mid-April in north-central Vietnam.

Voice The best distinction from other *Phylloscopus* species in the area. Call note seems to be diagnostic, a short, soft *pi-tsiu, pi-tsu* or similar. Alström *et al.* (2009) give a detailed analysis of songs in comparison with those of *ricketti* and *cantator*. The song is 7–9 softly whistled notes of varying pitch, duration and structure and on a slightly falling scale (2013 HBW SV).

Movements Apparently sedentary.

Range Limestone karst hills in north and central Vietnam and North and central Laos, from about 17° 20' N to about 23° N; also occurs in adjacent areas of southern China (Guangxi).

Conservation status Alström *et al.* (2009) point out that it is widely distributed and appears to be quite common in some locations; some areas of occurrence are protected in both Laos and Vietnam. The nature of the karst country where the species occurs does not lend itself to agricultural exploitation; however, some clearance for firewood and timber does occur, so the overall population may be declining.

Etymology *Calciatilis,* Latin, 'dweller on limestone'.

ALSTRÖM'S WARBLER
Seicercus soror

One of two newly described species to emerge from the highly complex Golden-spectacled Warbler *Seicercus burkii* group.

Type information *Seicercus soror* Alström & Olsson 1999, based on a 1902 specimen from Hopiachen, Sichuan, China, presently in the Natural History Museum, Tring, UK.
Alternative names Plain-tailed Warbler.
Discovery The Golden-spectacled Warbler *S. burkii sensu lato* is a widely distributed form, breeding from northern India to eastern China. Earlier systematists had classified it as a polytypic species with several described races (Ali & Ripley 1973); however, extensive fieldwork by several ornithologists drew attention to the inconsistencies in song of several geographic populations. Consequently in the 1980s and 1990s P. Alström and U. Olsson undertook an extensive and comprehensive study of the genus, involving the examination of some 700 museum specimens, recordings of vocalisations and playback tests, and other field observations. They concluded that no fewer than seven valid species are involved; some of these are 'splits' of previous subspecies, but one form which they named *S. soror* is a newly described species. Although some authorities do not accept these conclusions, *soror* is treated as a good species by, for example, *Handbook of the Birds of the World* (Alström 2006, HBW 11).
Description 11–12 cm. Monotypic. Generally greyish-green above, yellow below with a variable greenish tinge on sides of breast and flanks; median crown-stripe dull pale grey, lateral crown-stripes greyish-black; face and ear-coverts greyish-green, the lores slightly suffused with yellowish; pale yellow eye-ring; bill dark with paler tip and paler below; tarsi pale.
Habitat On breeding grounds, warm temperate evergreen broadleaved forest, in mature forest and in secondary growth; 700–1,300 m in Sichuan and 600–1,500 m in Shanxi. Appears to be sharply demarcated altitudinally from other sympatric *Seicercus* species. In winter (in Thailand) understorey of broadleaved evergreen forest, sometimes mangroves, below 1,000 m.
Food and feeding Forages in understorey; diet insects but no specific data.
Breeding Nest and eggs undescribed. Breeding season probably May–June.

▼ Alström's Warbler. Tai Po Kau, Hong Kong, China, October 2012 (*Michelle & Peter Wong*).

Voice Song is simple, lacking trills, *chip-chu-se-sis-chu-se-ris* etc.; Alström & Olsson (1999) give a very comprehensive comparative study of *Seicercus* vocalisations.
Movements Migratory, abandoning the breeding range completely.
Range Breeding endemic. Breeding range: disjunct populations in east-central and SE China (Shanxi, SE Sichuan, Guizhou, Fujian). On migration elsewhere in southern China. Winters in Thailand, Laos, Vietnam and Cambodia (Dymond 2008).
Conservation status Classified by BirdLife International as Least Concern, having a very large if disjunct breeding range and a wide distribution on the wintering grounds.
Etymology *Soror,* Latin, 'sister', referring to its close relationship with other *Seicercus* species. The common name, suggested by Rheindt (2006), not by the original authors, recognises the outstanding contributions of Professor Per Alström to the ornithology of SE Asia.

MARTENS'S WARBLER
Seicercus omeiensis

The second of two new species to be disentangled from the notoriously complex species swarm of Asian *Seicercus* warblers.

Type information *Seicercus omeiensis* Martens *et al.* 1999, Emei Shan, Sichuan, China.
Alternative names Omei Spectacled Warbler, Emei Shan Warbler. Chinese: Emei Wong Ying.
Discovery *Seicercus* is a complex and puzzling genus of conservatively plumaged warblers, with a group of species usually referred to as the '*S. burkii* complex', stretching from the western Himalayas to the eastern Chinese coast, between the approximate latitudes of 35° N and 20° N. The species tend to be very similar in appearance, in plumage differing in relatively minor details of the amount of white on the tail feathers and in crown patterns. During June and July 1997, on the Tabai Shan range of Sichuan, China, Prof. J. Martens and colleagues noticed that there was a clear altitudinal differentiation between *Seicercus* warblers, based on morphological and vocal characters; subsequent more detailed analysis, of vocalisations and tissue samples,

▼ Martens's Warbler. Wawu Shan, Sichuan, China, July 2011 (*Michelle & Peter Wong*).

showed that no fewer than four separate species were involved. The following year investigations in the Emei Shan range of Sichuan revealed that, again, four species of *Seicercus* were present; however, only three species were common to both mountain ranges. This led to a very comprehensive review of the members of the genus found in western China, resulting in the description of a cryptic species from the Emei Shan range.

Description 11–12 cm. Monotypic. Extremely similar to Bianchi's Warbler *S. valentini*. Upperparts green, upperwing and tail brown-grey with green edges to feathers; head conspicuously striped, the median-stripe pale to bright grey; lateral crown-stripes black or greyish-black; side of crown and nape mainly green; ear-coverts and lores green; obvious yellow eye-ring; underparts from chin to vent bright yellow, the flanks suffused with greenish; white triangles on webs of two outermost rectrices; iris dark brown; bill blackish above, pale orange below; tarsi pale greyish-pink. Differs from *valentini* by purer grey median crown-stripe, blacker and more extensive lateral crown-stripes, brighter green upperparts and more saturated yellow underparts.

Habitat Breeds in montane forest. Upper part of warm temperate evergreen forest zone and lower part of cool temperate zone with lush deciduous broadleaved forest; also secondary growth with bushes and trees. 1,200–2,300 m. Winters in evergreen broadleaved forest, mostly above 1,000 m but sometimes down to 400 m (Alström 2006 HBW 11).

Food and feeding Insectivorous, but few data. Forages mainly in understorey; prey often caught by short flights (Alström 2006 HBW 11).

Breeding Nest and eggs undescribed. Probably breeds in May–June.

Voice Song consists of short abrupt strophes of varied whistled notes *chu-si-tsu-chu-si-tsu* etc. Martens *et al.* (1999) give very detailed sonogram comparisons of the songs of this and other members of the genus.

Movements Totally migratory; leaves breeding range by September, returning in April.

Range Breeding endemic. Breeding range is east-central China (Sichuan, Shaanxi, probably south Gansu). Winters in Thailand and Cambodia.

Conservation status Currently listed by BirdLife International as Least Concern. Small breeding range, but appears to be locally abundant in some areas.

Etymology *Omeiensis*, from the Omei Mountains, China.

NAUNG MUNG SCIMITAR BABBLER
Jabouilleia naungmungensis

A form with a varied taxonomic history, described as a full species, then classified as a race of Indochinese or Short-tailed Scimitar Babbler, but reclassified as a full species in the current (version 7.1) IOC World Bird List. The genus *Jabouilleia* is merged with *Rimator* by some authorities.

Type information *Jabouilleia naungmungensis* Rappole *et al.* 2005, Naung Mung, Kachin state, Myanmar.

Alternative names Naung Mung Wren-babbler.

Discovery In February 2004 a joint expedition of the Myanmar Nature and Wildlife Conservation Division and the Smithsonian Conservation and Research Center, while conducting investigations in the vicinity of Naung Mung village in Kachin state, Myanmar, mist-netted two unfamiliar long-billed babblers. A third individual was caught a couple of days later. The birds were tentatively assigned to the genus *Jabouilleia*, with some misgivings, since the closest population of this genus (*J. danjoui*, the Indochinese or Short-tailed Scimitar Babbler) was some 600 km to the south-east, in east Tonkin. However, comparison of two of the specimens with material from several institutions led to the conclusion that the Naung Mung birds were indeed a new species of *Jabouilleia*. Subsequently, Collar (2011) argued that *naungmungensis* was insufficiently different from the Vietnamese populations in both voice and plumage to warrant full species status. However, using the criteria for species delimitation advanced by Tobias *et al.* (2010), the IOC World Bird List (version 7.1) restores it to a full species, as does the HBW/BirdLife Illustrated Checklist (del Hoyo & Collar 2016), the latter using the name Naung Mung Wren-babbler *Rimator naungmungensis*.

Description 18–19 cm. Monotypic. A plump-bodied, short-tailed babbler with a spectacular, long and decurved bill. Upperparts generally brown, the crown hair brown, paler on forecrown, lores, ear-coverts and supercilium; chin and throat whitish with a black malar stripe; breast whitish, somewhat mottled; sides of breast cinnamon-brown, belly whitish; undertail-coverts cinnamon; iris dark; bill notably long and decurved, about 33 mm, slightly shorter in the one male specimen, sepia at base, distally more pale; tarsi burnt umber.

Habitat Understorey of premontane temperate rainforest; specimens collected at 540 m.

◀ Naung Mung Scimitar Babbler. Hkakaborazi region, Myanmar, March 2006 (*Christopher Milensky/ Smithsonian Institution*).

Food and feeding No direct observations; all evidence suggests that it feeds on or near the ground, probing into soil with the long bill. Stomach contents of one specimen contained cicada nymphs, which feed on roots in soil.

Breeding Nest and eggs unknown. Gonadal data from specimens suggests that February is near the start of the breeding season; a male specimen taken in mid-March had somewhat enlarged testes.

Voice Song is a single, whistled note on the key of D, lasting about 0.5 s., repeated at 2-second intervals; slightly higher in pitch in response to playback. Also a raspy *chur* (Rappole *et al.* 2008).

Movements Presumably sedentary.

Range So far known only from the type locality and a second site about 24 km south.

Conservation status BirdLife International suggested classification as Vulnerable or Endangered, based on an apparently small range without formal protection. However, Rappole *et al.* (2008) speculate that the total potential range may be several hundred square kilometres of pristine forest, with no current habitat destruction.

Etymology *Naungmungensis*, from the village of Naung Mung, Myanmar.

NONGGANG BABBLER
Stachyris nonggangensis

A distinctive babbler, most closely related to Sooty Babbler *S. herberti* of central Laos and Vietnam; both species are placed in the genus *Nigravis* by some authorities.

Type information *Stachyris nonggangensis* Zhou & Jiang 2008, Nonggang, Longzhou County, Guangxi Zhuang, China.

Discovery On numerous occasions in 2005 and 2006 an unfamiliar babbler was observed in the Nonggang Natural Reserve of Guangxi, southern China. In January 2006 two individuals, which became the holotype and a paratype, were captured in mist-nets in karst rainforest at an altitude of 290 m.

Description 17 cm. Monotypic. Plumage is mostly sooty-brown, slightly tinged olive on upperparts; white crescent at rear of ear-coverts; feathers of middle throat and upper chest white with sooty-brown tips; iris pale blue to light greyish-green; bill black, slightly browner towards tip; tarsi quite sturdy, reddish-black. Sexes similar. The allopatric *S. herberti* lacks the distinctive ear-crescent and throat markings.

Habitat Seasonal rainforest on limestone karst country with numerous outcrops, probably mostly 150–650 m.

Food and feeding Usually seen walking on rocks, foraging in vegetation in gaps; in winter in small flocks, occasionally with other species, in pairs during the breeding season. Diet probably arthropods (Zhou & Jiang 2008).

Breeding Six nests so far discovered. These were open bowls about 15 cm in external diameter and 8 cm in internal diameter, made of dried leaves and twigs, in open cavities in limestone bedrock outcrops; eggs 4–5, pure white, incubated apparently by female alone for 18 days. Nesting period April; nest predation apparently rather high (Jiang *et al.* 2013).
Voice Call a distinctive downward-inflected whirring (Eames & Nguyen 2016).
Movements Presumably sedentary.
Range Presently known from three locations in Guangxi province, China; might also be expected in similar habitat in Yunnan province. Recently discovered in Cao Bang province, extreme northern Vietnam (Eames & Nguyen 2016).
Conservation status Currently listed by BirdLife International as Vulnerable. The presently known range is quite small and the habitat requirements appear to be quite specific. Li *et al.* (2013) estimate the population size at about 1,500 individuals, although with recently discovered range extensions into Vietnam this might be an underestimate. Two of the known Chinese locations as well as the Vietnamese one are protected.
Etymology *Nongganggensis*, from the type locality.

▲ Nonggang Babbler. Nonggang NNR, Guangxi, China, January 2009 (*James Eaton*).

CORDILLERA GROUND WARBLER
Robsonius rabori

Described originally in the genus *Napothera*, widespread in SE Asia, now placed along with two other species (all recently described) in the endemic Philippine genus *Robsonius*, based on the lack of rictal bristles, the partially-feathered nostril, some other plumage features and its walking habits (Collar 2006; see also Hosner *et al.* 2013). The taxonomic position of the genus is somewhat fluid; some authorities include it in the family Locustellidae (Oliveros *et al.* 2012). Hosner *et al.* (2013) suggest a wholesale renaming of the three species of *Robsonius* to remove inaccuracies found in older literature based on incomplete knowledge of plumage and distribution, while incorporating the new taxonomic position of the genus; thus *R. rabori* would become the Cordillera Ground Warbler, *R. sorsogonensis* the Bicol Ground Warbler and *R. thompsoni* the Sierra Madre Ground Warbler.

Type information *Robsonius rabori* Rand 1960, Tabbug, Pagupud, Ilocos Norte, Luzon, Philippines.
Alternative names Rabor's Ground Warbler, Rabor's Wren-babbler, Rusty-faced Babbler. Spanish: Ratina de Rabor.
Discovery Originally described in 1960 from a single juvenile specimen taken by D. S. Rabor in northern Luzon in 1959. It has since been found more widely in northern Luzon and the adult plumage is now well known.
Description 20–22 cm. Monotypic. A plump terrestrial bird with powerful legs and a rather long bill. Head and cheeks dull rusty-chestnut, becoming rufescent-olive on the back with blackish scaling on the shoulders; rectrices more rusty; wings olive with blackish subterminal area on greater coverts and two obvious wing-bars formed by bold white tips on greater and median coverts, the remainder of closed wing more chestnut-brown, the two outer primaries more grey with white tips; lores white, chin and throat whitish with prominent black malar stripe and black spots on lower throat; chest and belly grey with dark rufescent-brown vent; iris hazel to brown; bill dark horn-brown above, greyish-pearl below; tarsi pearly brownish-horn. Sexes similar.
Habitat Lowland evergreen forest and second-growth, sea level to 1,300 m; often in the vicinity of small streams.

Breeding Nest and eggs undescribed; the nest described by Sánchez-Gónzalez *et al.* (2010) was in fact that of *R. thompsoni* (Hosner *et al.* 2013).

Voice Song is an "extremely high, wispy, slurred, insect-like *tsui-ts-siliuu-ee* etc., rising at the end" (Collar & Robson 2007 HBW 12).

Movements Presumably sedentary.

Range Endemic. Confined to northern Luzon, Philippines (Cordillera Central of Ilocos Norte).

Conservation status Currently classified by BirdLife International as Vulnerable, although Hosner *et al.* (2013) suggest Endangered as more appropriate. The overall range is quite limited, and habitat destruction fuelled by rapid population growth is very prevalent. Generally fairly uncommon – indeed there appear to be no field observations from the original collection site to 1990 (Harrap & Mitchell 1994) – if perhaps overlooked; in a couple of locations apparently quite common. Part of the range is within the Northern Sierra National Park.

Etymology The genus is named after British ornithologist Craig Robson, tour leader and author of various field guides to the birds of SE Asia, to "pay tribute to [his] expertise and contribution...with respect to the study of Asian birds and Asian babblers in particular". The specific name honours Dioscoro S. (Joe) Rabor (1911–1996) who collected the first specimen. Dr Rabor, "preeminent Philippine zoologist and conservationist for more than half of the 20th century", was born in Cebu City, Philippines, and received a PhD under the direction of S. Dillon Ripley. His numerous (more than 50) expeditions led to a huge expansion of the knowledge of Philippine ornithology, with eight new species and 61 new races described from his collections. In addition to the present species, a sunbird (*Aethopyga linaraborae*, q.v.) is named for his wife Lina. Incidentally, his four daughters, all of whom became physicians, are all named after birds (Kennedy & Miranda 1998), while his habit of wearing a firearm in a holster in the field reputedly "kept his many students in line".

BICOL GROUND WARBLER
Robsonius sorsogonensis

Treated as a race of Cordillera Ground Warbler *R. rabori* by some authorities; the race *mesoluzonicus* was also initially described as a subspecies of *rabori* (duPont 1971), but this is now usually regarded as a synonym of *sorsogonensis*.

Type information *Robsonius sorsogonensis* Rand & Rabor 1967, Mount Bulusan, San Roque, Bulusan, Sorsogon Province, southern Luzon, Philippines.

Alternative names Sorsogon Wren-babbler, Grey-banded Babbler. Spanish: Ratina de Rand.

Discovery Described on the basis of an adult female specimen taken by D. S. Rabor in May 1961 on Mount Bulusan, Luzon. The race *mesoluzonicus* was also described from a unique specimen from central Luzon in 1971 (duPont 1971).

Description 20–22 cm. Monotypic. In form very similar to *R. rabori*, a plump terrestrial bird with stout bill and heavy legs. Forehead greyish-olive, crown and mantle olive with faint dark scales; lower back brown, paler on rump; wings brown with two conspicuous white bars formed by white tips to the greater and median coverts; tail dark brown; lores white, cheeks grey, narrow malar stripe black, bordered above and below by white; throat white; chest grey, becoming white on central belly; undertail-coverts dark red-brown; iris brown; bill blackish-brown with pale grey base to lower mandible; tarsi light brown. Differs from *rabori* in having much greyer, less rusty, crown and in lacking black spotting on the lower throat.

Habitat Broadleaved evergreen forest, including forest edge and second growth, sea level to 1,000 m.

Food and feeding Terrestrial, foraging by walking on the forest floor, flipping over leaf debris. Food items are insects.

Breeding One nest described; dome-shaped with side-entrance, placed 0.6–0.75 m up in a rattan; eggs two, white with reddish-brown speckles; nestlings fed by both parents. Breeding season February–August (Collar & Robson 2007 HBW 12).

Voice Song is a very high-pitched *tit'tsu-tsuuuu-tsuui* etc.

Movements Presumably sedentary.

Range Endemic. Nominate, south Luzon and Catanduanes Island; race *mesoluzonicus*, central Luzon.

Conservation status Hosner *et al.* (2013) suggest that a classification of Vulnerable is appropriate.

Etymology *Robsonius*, see under Cordillera Ground Warbler. *Sorsogonensis*, from Sorsogon Province; *mesoluzonicus*, central Luzon.

SIERRA MADRE GROUND WARBLER
Robsonius thompsoni

The description of this species and the details in the type description paper should finally bring to an end the confusion surrounding the taxonomy and distribution of the genus *Robsonius.*

Type information *Robsonius thompsoni* Hosner *et al.* 2013, 12 km SW of Baler, San Luis municipality, Aurora Province, Luzon, Philippines.
Alternative names Spanish: Ratina de Sierra Madre.
Discovery The population of ground warblers (to use the name suggested by Hosner *et al.* 2013) found in an extensive area of eastern Luzon, east and south of the Cagayan Valley is the best known of the genus, being frequently observed by visiting birdwatchers. The first member of the genus to be named, *R. rabori*, is by contrast little known, initially having been described from one juvenile specimen. In June 2011 an adult *Robsonius* was salvaged from a mammal trap near the village of Adams; this proved to be quite distinct in plumage from other *Robsonius* specimens in adult plumage. This prompted a detailed biochemical examination that revealed a previously unrecognised species that is quite widely distributed in eastern Luzon, while the Adams specimen was attributable to *rabori.* Since no name had been given to the new population, Hosner *et al.* (2013) named it as a new species, based on a specimen taken in June 2009 near Baler, Luzon.
Description 15 cm. Monotypic. Crown, nape and ear-coverts amber; lores white; back cinnamon brown, the feathers tipped with dusky brown; concealed white rump patch; wings chestnut, with several bars of white spots formed by tips on the wing coverts; thin white eye-ring; throat white with black tips; malar stripe black; necklace of black spots on lower throat and upper chest; breast grey; flanks dusky brown; iris dark brown; bill dark brown above, pale grey below; tarsi light brown. Differs from *rabori* (the closest species in plumage) by having a necklace of black spots and black feather tips on the throat.
Habitat Broadleaved lowland and lower montane forest, including primary and secondary forest, forest-edge and logged secondary growth, sea level to at least 1,300 m (Hosner *et al.* 2013).
Food and feeding Largely terrestrial; walks slowly across forest floor, flipping over litter in search of invertebrates. Stomach contents of four specimens comprised unidentified insect remains.
Breeding The nest described by Sánchez-Gonzáles *et al.* (2010) (as that of *R. rabori*) was, based on its geographic location, *R. thompsoni.* It was a large clump of dry sticks, branches and leaves, domed with a front entrance, located 50 cm up in a bank of rock and mud. On 18 June it held one egg, whitish-cream, blotched with grey and light brown (a second egg was in the oviduct of the collected female). No data on incubation or fledging periods etc.
Voice Song is very similar to the other members of the genus, a very high-pitched series of three or four variable phrases with ascending and descending notes. Probable alarm call is a rapid trill (Hosner *et al.* 2013).

► Sierra Madre Ground Warbler, juvenile.
Mt Cagua, Gonzaga, Luzon, Philippines, July 2011 (*Peter Hosner*).

Movements Presumably sedentary.
Range Endemic. Eastern Luzon, from about 14° 30' N. to about 18° 10' N., east of the Cagayan Valley.
Conservation status Not currently classified by BirdLife International. Known from 21 localities. Hosner *et al.* (2013) suggest a designation of Vulnerable.
Etymology *Robsonius,* see under Cordillera Ground Warbler. Specific name in honour of Professor Max C. Thompson of Southwestern College, Winfield, Kansas," for his decades of contributions to natural history collections and ornithology in particular" (Hosner *et al.* 2013).

NEPAL WREN-BABBLER
Pnoepyga immaculata

Formerly included in the Timaliidae, a large assemblage of rather diverse species; however, recent molecular work (Gelang *et al.* 2009) shows that the genus (of at least four species, possibly more – Päckert *et al.* 2013) is closely related to neither the Timaliidae nor the Old World warblers Sylviidae, but instead merits its own family, Pnoepygidae.

Type information *Pnoepyga immaculata* Martens & Eck 1991, below Lete, Thakkola, Nepal.
Alternative names Spotless Wren-babbler.
Discovery In May 1985 S. Harrap observed some singing wren-babblers in the Langtang Valley of central Nepal. On the basis primarily of song he concluded, on the information then available, that his observations referred to the Scaly-breasted Wren-babbler *P. albiventer,* while recognising that extreme caution is required in identifying birds of this genus (Harrap 1989). Only later did he realise that he was one of the first ornithologists to observe what later transpired to be an undescribed species (Harrap 2011). The recognition that *immaculata* was indeed a new species, initially alerted by distinctive vocalisations, came from the work of Martens & Eck (1991); in fact, the holotype had already been taken, in 1980, and a paratype now in the Field Museum in Chicago dates back to 1947.
Description 8.5–10 cm. Monotypic. A small, almost tail-less, terrestrial bird, occurring in two morphs. Pale morph has brownish-olive upperparts with obscure darker scaling; underparts whitish on chin, becoming darker olive-grey on the flanks and belly, the feathers with obvious darker scaling; iris brown; bill horn-brown above, paler below; tarsi sturdy, with large feet, horn-brown or flesh-brown. Dark morph is darker olive-grey above, deep fulvous-brown below with darker scaling.
Habitat Understorey and edges of rocky broadleaved evergreen forest, often close to streams and rivers, also secondary growth, degraded forest and areas used intensively by humans; scrub and gullies outside forest. 1,730–3,100 m. Winters as low as 250 m, also in semi-

◀ Nepal Wren-babbler, pale morph. Himachal Pradesh, India, May 2015 (*James Eaton*).

▼ Nepal Wren-babbler, dark morph. Dehradun, Uttarakhand state, India, December 2012 (*Nitin Srinivasa Murthy*).

deciduous and deciduous forests and tall grass (Collar & Robson 2007 HBW 12; Chantler & Rye 2008).

Food and feeding Forages low down or on the ground or on boulders. Diet presumably invertebrates.

Breeding Actual nest and eggs undescribed. One adult was seen carrying nest material into a hole, about 3 m up in a stream bank in June (Choudhary 2000).

Voice Song is a series of sharp thin whistles, 5–8 in number, *si-su-si-si-swi-si-si-si* etc. Call a sharp repeated *tchit* or *tslick* (Harrap 2011).

Movements Incompletely known, but short distance and altitudinal movements occur; occurs at lower altitudes down to 250 m in winter.

Range. Breeding range is southern foothills of Himalayas, mostly Nepal, marginally into India (Himachal Pradesh) in west, probably into Sikkim in east and marginally into southern Tibet. Winters mostly within Nepal.

Conservation status Currently classified by BirdLife International as Least Concern, *contra* the opinions of Martens & Eck (1991). Fairly widely if apparently spottily distributed. Appears to be able to adapt, at least to some degree, to degraded and goat-grazed habitat (Chantler & Rye 2008).

Etymology *Immaculata,* Latin, 'lacking spots'.

PANAY STRIPED BABBLER
Zosterornis latistriatus

Treated as conspecific with Negros Striped Babbler *Z. nigrorum* by some authorities, but vocally distinct. Previously placed in the genus *Stachyris*, widespread in SE Asia; *Zosterornis* is endemic to the Philippines.

Type information *Stachyris latistriata* Gonzales & Kennedy 1990, 1.1 km SSW of peak of Mount Baloy, Barangay San Augustin, municipality of Valderrama, Antique Province, Panay, Philippines.

Alternative names Spanish: Timalí Rayado.

Discovery In February and March 1987 a team from the National Museum of the Philippines visited Mt Baloy on the island of Panay to determine the status of the endangered Visayan Spotted Deer *Cervus alfredi*. During this visit a small number of birds were mist-netted in montane forest; among the 77 specimens taken was an unidentified babbler clearly of the (then) genus *Stachyris*. Comparison of this specimen with two other *Stachyris* species from nearby islands (Luzon and Negros Striped Babblers *S. striata* and *S. nigrorum*) indicated that a new species was involved. In October–November 1989 a further, much larger, series of specimens was taken, allowing the description of the Panay birds as a new species in 1990.

Description 15 cm. Monotypic. A strikingly marked babbler. Forehead blackish; remainder of upperparts dull olive-green, more greyish-olive on wings and tinged rusty on the tail; face creamy-white with a conspicuous dark narrow eye stripe, moustachial line and malar stripe; chin and upper throat creamy-white, becoming pale sulphur-yellow on lower throat, buff-yellow on breast and belly with greenish-olive flanks; conspicuous broad black streaks on lower throat, chest and belly, becoming more diffuse and dark olive-green on lower belly and undertail-coverts; iris bright rusty, outer edge paler, with creamy eye-ring; bill dark or horn, paler at base of lower mandible; tarsi bluish-olive.

Habitat Montane forest, especially forest with abundant epiphytic mossy growth, 1,400–1,900 m; sometimes down to 1,000 m (Gonzales & Kennedy 1990).

Food and feeding Feeds in lower to upper layers of canopy, 1.5–12 m up, singly, in pairs or small flocks; does not apparently joined mixed-species flocks. Food items insects (beetles and cicadas); probably also seeds (Gonzales & Kennedy 1990).

▼ Panay Striped Babbler. Mount Madja-as, Panay, Philippines, February 2008 (*Edward Vercruysse*).

Breeding One nest so far described. It was a cup, in branches in the crown of a small tree (*Ficus* sp.), 5 m up, made of loosely woven mosses, lined with hair-like epiphytic fern fibres. On 14 October it contained two well-grown young. Breeding season probably coming to an end in October based on the gonadal condition of specimens (Gonzales & Kennedy 1990).
Voice A noisy, active bird. Song is a series of slightly ascending trills, consisting of 11–17 notes, lasting 1.2 seconds; call a sharp *tsik* (Gonzales & Kennedy 1990).
Movements Presumably sedentary.
Range Endemic. Confined to mountainous areas in west-central Panay Island, Philippines; Mt Baloy, Mt Madja-as and Dagsalan, upper Aklan River.
Conservation status Classified by BirdLife International as Near Threatened. The overall range is quite small; suitable undamaged habitat is about 700 km² in extent. In some locations abundant; described by Gonzales & Kennedy (1990) as the most common species in high-elevation forests on Mt Baloy. Although forest destruction at lower altitudes on Panay has been very extensive, the rugged nature of much of the species' habitat has so far protected it.
Etymology *Latistriatus,* Latin, from the characteristic broad black streaks on the underparts.

PALAWAN STRIPED BABBLER
Zosterornis hypogrammicus

Formerly placed in the genus *Stachyris*; may form a superspecies with Luzon, Panay and Negros Striped Babblers (*Z. striatus, latistriatus* and *nigrorum*).

Type information *Stachyris hypogrammicus* Salomonsen 1962, Mt Mataling, Mantalingajan Range, Palawan, Philippines.
Alternative names Palawan Tree Babbler, Buff-capped Babbler. Spanish: Timalí de Palawan.

Discovery During 1961 and 1962 the small two-masted Danish vessel *Noona Dan* made an expedition to the Western Pacific and landed a party of biologists on Palawan Island in the Philippines in August 1961. They moved inland and set up camp in early September on Mt Mataling at an altitude of about 1,100 m. Among the birds collected in thick mossy forest were seven specimens of a babbler unfamiliar to the ornithologists of the expedition; one specimen was sent to

▼Palawan Striped Babbler. Mount Victoria, Palawan, Philippines, May 2016 (*Ross Gallardy*).

the eminent Danish ornithologist Finn Salomonsen who described it as a new species.

Description 14–15 cm. Monotypic. Crown dull orange-buff, upperparts olive-green; sides of face buffy mid-grey; tail greyish olive-brown; chin and throat whitish-buff to yellowish-grey; breast and belly olive-yellow, boldly streaked black; flanks dull olive-green; iris brown or chestnut-brown; bill blackish-horn, paler below; tarsi olive grey-horn. Sexes similar.

Habitat Montane evergreen forest with abundant epiphytic moss growth, 1,000–2,030 m.

Food and feeding Forages mostly in canopy but also at lower levels, in pairs or small groups. Food items insects, seeds and vegetable matter.

Breeding Nest and eggs undescribed. Birds in breeding condition April–May.

Voice Call, a distinctive *zeep-zeep-zeep-zuup*, the second note highest, the last lowest; also "bubbling noises" (Collar & Robson 2007 HBW 12).

Movements Presumably sedentary.

Range Endemic. Upper levels of southern Palawan Island, Philippines (Mts Mataling, Borangbato, Mantalingajan and at Magtagguimbong) (BLI SF 2014).

Conservation status Currently classified by BirdLife International as Near Threatened. Total range is quite small, but it appears to be fairly common in suitable habitat. Rates of deforestation on Palawan are increasing.

Etymology *Hypogrammicus*, Greek, 'streaked underneath'.

VISAYAN MINIATURE BABBLER
Micromacronus leytensis

A tiny, very distinctive and virtually unknown babbler whose discovery required the erection of a new genus. Previously treated as conspecific with Mindanao Miniature Babbler, itself only described six years later.

Type information *Micromacronus leytensis* Amadon 1962, Dagami, Barrio of Patok, eastern shoulder of Mt Lobi (1,500 feet), Leyte, Philippines.

Alternative names Leyte Tit-babbler, Miniature Tit-babbler. Spanish: Timalí Enano de Visayan.

Discovery On a collecting expedition to the island of Leyte, eastern Philippines, in August 1961, sponsored by the Philippine National Museum, G. Alcasid and M. Celestino encountered a group of birds which they initially took to be white-eyes (*Zosterops*). On taking a specimen they noticed that it was "different"; they tried to get more, "but succeeded in downing only four; in about two minutes they were gone" (Amadon 1962).

Description 7–8 cm. Monotypic. Tiny, with a wing-length shorter than any bird other than some hummingbirds and 10% shorter than that of the other smallest passerines, the kinglets (*Regulus*), with unique, almost barbless specialised back feathers. Sexually dimorphic. Male bright olive-green above, the crown with indistinct black tips; underparts and supercilium bright yellow; lower back, rump and flanks with almost bare white-shafted feathers with brownish tips, 4–4.5 cm in length, projecting backwards; iris red, bill black, legs greenish-grey with straw-yellow feet. Female duller yellow, with buffy-white supercilium.

Habitat Undergrowth and canopy of montane broad-leaved evergreen forest and forest-edge. Type specimen taken at 500 m.

Food and feeding No data on diet. Moves actively, in upper levels of forest, often with other species.

Breeding Nest and eggs undescribed.

Voice Not recorded.

Movements Presumably sedentary.

Range Endemic. Samar, Biliran and Leyte islands, Philippines.

Conservation status Classified by BirdLife International as Data Deficient. However, given the very few observations it would appear to be very rare.

Etymology The genus *Micromacronus* is derived from the generic name of the tit-babblers *Macronus* with the Greek prefix *micro* meaning small; *leytensis* from the island of Leyte.

▼ Visayan Miniature Babbler. Samar Island Natural Park, Samar, Philippines, May 2016 (*Michael Kearns*).

MINDANAO MINIATURE BABBLER
Micromacronus sordidus

Initially described as a race of Visayan Miniature Babbler *M. leytensis*; now given full specific status by most authorities (e.g. Collar 2006).

Type information *Micromacronus leytensis sordidus* Ripley & Rabor 1968, mountains of south Mindanao, Philippines.
Discovery Described from an adult male specimen taken by D. S. Rabor in February 1964 on Mt Matutum, Tupi, Cotabato Province, Mindanao, at an altitude of between 3,000 and 4,300 feet (ca 1,000–1,325 m).
Description 7–8 cm. Monotypic. In form very similar to *M. leytensis,* extremely small with extraordinary almost barbless long feathers on the back, rump and flanks, though these are somewhat shorter than in *leytensis.* Overall much duller olive-green above and below, without any bright yellow tones. Iris red, bill horn-grey, tarsi greenish-grey.
Habitat Undergrowth and canopy of montane broad-leaved evergreen forest and forest-edge, 600–1,670 m.
Food and feeding No data on diet. Associates with other species in forest canopy.
Breeding Nest and eggs undescribed; birds in breeding condition in May–June, recently fledged young in June.
Voice Not recorded.

▲ Mindanao Miniature Babbler. Mount Matutum Protected Landscape, South Cotobato, Mindanao, Philippines, September 2016 (*Tonji Ramos*).

Movements Presumably sedentary.
Range Endemic. Southern Mindanao, Philippines.
Conservation status Not currently assessed by BirdLife International, but appears to be rare. Occurs in two national parks (Mt Apo and Mt Kitanglad). Elsewhere doubtless threatened by forest clearance.
Etymology *Sordidus,* Latin, 'dirty', referring to the very dull plumage in comparison to *M. leytensis.*

CHESTNUT-EARED LAUGHINGTHRUSH
Ianthocichla konkakinhensis

Previously included in the large Oriental genus *Garrulax.*

Type information *Garrulax konkakinhensis* Eames & Eames 2001, Mt Kon Ka Kinh, Gia Lai province, Vietnam.
Discovery During survey work which was part of a project between BirdLife International and the Forest Inventory and Planning Institute, in April 1999, J. C. Eames mist-netted and collected an unfamiliar laughingthrush *Garrulax* sp. along the summit ridge of Mt Kon Ka Kinh in the southern part of the Kon Tum Plateau, Vietnam. Subsequently two further specimens, an adult and a juvenile, were taken at different sites in the same locality. Comparison with specimen material of several races of Rufous-chinned and Barred Laughingthrushes *G.* (*I.*) *rufogularis* and *G.* (*I.*) *lunulatus,* both highly disjunct, showed the new taxon to be a distinctive species (Eames & Eames 2001).
Description 24 cm. Monotypic. Crown and nape grey, tipped black; mantle and upperwing-coverts olive-buff; back more rufous, heavily marked with black; tail long, yellow-tinged rufous-brown, with broad black subterminal band and white tips, conspicuous when seen from below; ear-coverts chestnut; throat white, becoming olive-buff and grey-buff lower down with conspicuous black markings on the chest and upper flanks; iris brown or dark brown; bill dark horn above, pale horn below; tarsi flesh-horn.
Habitat Broadleaved forest, including secondary growth and forest edge, often with bamboo; 1,200–1,750 m.
Food and feeding Forages in understorey; no data on diet.

► Chestnut-eared Laughingthrush. Den, Vietnam, January 2016 (*Robert Hutchinson*).

Breeding Nest and eggs undescribed; breeding season March–April.
Voice Song is a "sweet *Turdus*-like rambling series of fairly well-spaced and stressed notes, and some mimicry, lasting about 4–6 seconds"; very different from that of *I. rufogularis*. Calls include a low grumbling *rreeek-rreeek-rreeek* (Collar & Robson 2007 HBW12).
Movements Presumably sedentary.
Range Initially described from Mt Kon Ka Kinh, Gia Lai province, Vietnam. Has since been found in a number of locations in the highlands around Kon Tum and recently in Laos (Mahood *et al.* 2012). Some areas of potential habitat have not been investigated due to their inaccessibility.
Conservation status Currently listed by BirdLife International as Vulnerable, due to its limited geographic range; population estimated in the range of 1,500–4,000 individuals. It does, however, seem to be quite common in some parts of its range, several of which are protected as national parks. At the Laos location (Xe Sap) it may be abundant (Gray 2012, in BLI SF 2014).
Etymology The specific name is derived from the type locality, Mt Kon Ka Kinh.

GOLDEN-WINGED LAUGHINGTHRUSH
Garrulax ngoclinhensis

Placed in the genus *Trochalopteron* by some authorities. Sometimes treated as a race of Chestnut-crowned Laughingthrush *G.* (*T.*) *erythrocephalus*, but appears to be distinct in several respects.

Type information *Garrulax ngoclinhensis* Eames *et al.* 1999a, Mt Ngoc Linh, Kon Tum province, Vietnam.
Alternative names French: Garrulaxe de Ngoc Linh.

Discovery In 1996 an expedition was mounted under the auspices of BirdLife International and the Forest Inventory and Planning Institute to examine the avifauna of Mt Ngoc Linh in the Western Highlands of southern Vietnam. The height, and isolation from areas of similar altitude, suggested that this area would be a productive site for investigation. In May 1996 a strikingly plumaged laughingthrush was

▲ Golden-winged Laughingthrush. Ngoc Linh Kon Tum Nature Reserve, Annamite Range, Vietnam, April 2016 (*Thang Nguyen*).

observed on several occasions; it clearly resembled Chestnut-crowned and Collared Laughingthrushes *G. erythrocephalus* and *yersini* but was obviously distinct from both. Consequently one specimen (an adult male) was collected on 15 May, and two paratypes, one male and one female, in March 1998. From these the species was described by Eames *et al.* in 1999.

Description 27 cm. Monotypic. A medium-sized dark grey laughingthrush, faintly scaled on the breast; crown chestnut; forehead dark grey, tinged maroon; back grey, tinged olive; greater coverts broadly tipped chestnut; primary coverts black; primaries and secondaries broadly fringed golden; rectrices slaty, the outer ones fringed dull ochre-golden; face blackish; ear-coverts and throat pinkish dark grey; breast dark grey, scalloped broadly with mid-grey; mid-belly dark olive-grey; flanks, thighs and vent dark grey; iris dark brown to blackish; bill blackish-horn; tarsi dark brown.

Habitat Primary upper montane broadleaved evergreen forest with canopy height of 10–15 m and bamboo understorey.

Food and feeding Forages in herb and shrub layers of undergrowth, usually singly or in pairs; once in association with Red-tailed Laughingthrush *G. milnei*.

Breeding Nest and eggs undescribed. The holotype male, taken on 15 May, had enlarged testes.

Voice Call is described as a two-noted, rather cat-like mewing *Rr-raow* with emphasis on the second note (Eames *et al.* 1999a).

Movements Presumably sedentary.

Range Endemic. Currently recorded from Mts Ngoc Linh and Ngoc Boc in the Kon Tum Plateau, and in the Kon Piong Forest Complex (Eames 2001). May possibly also occur in suitable habitat in adjacent Laos (Eames *et al.* 1999a).

Conservation status Currently classified by BirdLife International as Vulnerable. The known range is small and has been subject to deforestation for agriculture. Parts of its range are protected by the Ngoc Linh (Kon Tum) Nature Reserve and the Ngoc Linh proposed reserve.

Etymology Named for Mt Ngoc Linh, which in Vietnamese means ‘sacred precious stone’.

BUGUN LIOCICHLA
Liocichla bugunorum

A spectacular and extremely rare liocichla, whose discovery "was perhaps the greatest ornithological sensation of 2006", taking "the world's media by storm" (Collar & Pilgrim 2007). The decision to describe the species without taking a type specimen attracted great attention and has stimulated widespread discussion on the ethics of collection.

Type information *Liocichla bugunorum* Athreya 2006, Lama Camp, near Eaglenest Wildlife Sanctuary, Arunachal Pradesh, India.

Discovery In January 1995 R. Athreya had brief views of a pair of liocichlas which he was unable to identify from standard reference guides. A short reference was made online in 1996, but no further evidence ensued until early 2005, when Dr Athreya saw a flock at the same location on three occasions. Field sketches suggested a similarity to Grey-cheeked Liocichla *L. omeiensis*, a species disjunct by more than 1,000 km, though whether the Eaglenest birds were identical or merely similar was unclear. After very considerable effort, a probable male specimen was mist-netted in May 2006. In view of its obvious rarity it was not collected; it was instead measured, comprehensively photographed and released, after some rectrices (a distinction from *omeiensis*) and one secondary feather were taken. The new species was described on this basis, a procedure acceptable under the rules of the International Commission of Zoological Nomenclature. A second bird, similar but duller in plumage, had been netted a few days earlier at the same location, but had escaped after only a few photographs had been taken. It was assumed that this was a female.

Description About 22 cm. Monotypic. Presumed male: erectile crown feathers black; triangular loral patch orange-yellow; post-ocular streak yellow; narrow area above eye blackish; ear-coverts grey; upperparts greyish-olive; shoulders greyish; on closed wing a broad golden-yellow patch on the base of the primaries; remainder of primaries and secondaries slaty-black, the secondaries having broad conspicuous crimson inner webs, all remiges with narrow whitish tips; chin yellow-olive, below a grey band, breast brighter yellow-olive; tail graduated, mostly blackish above with indistinct darker bars, each feather ending in a spray of orange-red barbs; underside of tail flame-coloured on outer rectrices, becoming more olive, then black, towards centre; undertail-coverts black with flame-red tips and the yellow sides; iris dark reddish-brown; bill pale horn at tip, darker at base; tarsi flesh-coloured. Presumed female has duller crown, yellowish patch on wing much duller, underside of tail copper-red, no red on undertail-coverts, rectrices with broad yellow tips, orange-yellow loral patch possibly bigger.

▼ Bugun Liocichla, male. Eaglenest, Arunachal Pradesh, India, April 2013 (*James Eaton*).

▲ Bugun Liocichla, female. Eaglenest, Arunachal Pradesh, India, March 2008 (*János Oláh/Birdquest*).

Habitat So far found in heavily-disturbed, logged forest with dense shrubbery and small trees, on ravines; once on edge of primary forest; 2,060–2,340 m.

Food and feeding Seen to eat berries. Forages at all levels from ground to canopy, sometimes in monospecific flocks of up to six birds, sometimes in association with other species.

Breeding Nest and eggs undescribed. Breeding season probably May when birds respond to playback and are seen in pairs.

Voice Song is fluty, including a descending series of notes, slightly slurred and inflected at end.

Movements Presumably sedentary.

Range Endemic. So far found only in a very small area around the Eaglenest Sanctuary and around Lama Camp, near Tenga in Arunachal Pradesh, in NE India.

Conservation status Classified by BirdLife International as Critically Endangered. The total population may be as few as 14 birds, including three breeding pairs (Athreya 2006). Since the original observations the bird has been seen a number of times by several observers (e.g. Allen & Catsis 2007). However, given that it is very distinctive in plumage and vocally conspicuous it seems unlikely that substantial undetected populations occur in other areas such as adjacent Bhutan. Vigorous conservation efforts are underway, involving the local Bugun people (Sykes 2006); a potential further threat includes a plan to build a road through some of the known habitat. The proposed Tsangyang Gyatso Biosphere Reserve would protect habitat for this and several other species of conservation concern (Mazumdar *et al.* 2011).

Etymology *Bugunorum*, genitive plural, derived from the Bugun people, in whose hands the survival of the species now rests.

BLACK-CROWNED BARWING
Actinodura sodangorum

A large and distinctively plumaged barwing, so far found only in limited areas of eastern Laos and central Annam, Vietnam.

Type information *Actinodura sodangorum* Eames *et al.* 1999b, Ngoc Linh, Kon Tum province, Vietnam.

Alternative names French: Actinodure à calotte noir, Actinodure à tête noir.

Discovery Ngoc Linh is a highland area of central Vietnam that until the mid-1990s was largely uninvestigated ornithologically. Consequently it was selected as a survey site by a team from BirdLife International and the Forest Inventory and Planning Institute. Fieldwork started on 28 April 1996 near the hamlet of Dak Rit and almost immediately an unfamiliar barwing (*Actinodura*) was observed by J. C. Eames and R. Eve. Birds were relocated the following day and on numerous occasions over the next 20 days; however, efforts to trap them were unsuccessful, although recordings were made. Finally a female specimen was obtained on 15 May 1996; a second, a male, was procured on 17 March 1998, allowing the description of a new species.

Description 24 cm. Monotypic. A rather large barwing, with diagnostic black cap and throat streaks. Sides of face grey, lores darker, white eye-ring; upperparts mostly olive-brown, indistinctly barred darker; scapulars with fine bars; primaries and secondaries blackish-brown with chestnut or orange-buff bars; tail long, graduated, blackish-brown with chestnut bases and white tips, broadly barred with black; throat, breast and belly rich rufous-orange; iris dark brown; bill horn, with flesh-coloured base; tarsi grey-horn. Sexes similar.

Habitat Primary and modified lower and upper montane forest, including forest-edge adjacent to cultivation; most common in closed canopy primary evergreen forest, 1,000–2,400 m; also in tall damper grassland with scattered bushes at 1,150 m (Oláh 2006).

Food and feeding Forages singly or in pairs, mostly in the canopy, sometimes lower, and on tree boles. No data on diet.

Breeding Nest and eggs undescribed. del Hoyo & Collar (2011) describe an extraordinary mating display in late April involving both birds swinging in a head-first full circle around a branch, possibly with brief cloacal contact.

Voice Sings and calls antiphonally. A series of two or three drawn-out catlike *wa-wa* or *wa-wa-wa*, in couplets or triplets. The second bird joins in with a descending phrase of five or six short piping notes. Birds sometimes engage in allopreening after duetting (Eames *et al.* 1999b).

Movements Presumably sedentary.

Range So far known from about seven locations in Vietnam and three in adjacent Laos.

Conservation status Formerly classified as Vulnerable, but downgraded to Near Threatened in view of recent observations in new locations; does not appear to be in decline. Appears to be able to tolerate, or even take advantage of, some clearance of primary forest. Part of range protected by Ngoc Linh Nature Reserve, Vietnam. Road-building plans may be a future issue.

▲ Black-crowned Barwing. Lo Xo Pass, central Vietnam, March 2015 (*James Eaton*).

Etymology Named in honour of the Södang tribe, an ethnic group inhabiting areas of Vietnam, Cambodia and Laos.

TOGIAN WHITE-EYE
Zosterops somadikartai

A distinctively plumaged white-eye, forming a superspecies with several other *Zosterops* species found in eastern Indonesia, New Guinea and New Britain.

Type information *Zosterops somadikartai* Indrawan *et al.* 2008, Pulau Malenge, Togian Islands, Sulawesi, Indonesia.

Discovery The Togian (or Togean) Islands, in the Gulf of Tomini, Sulawesi, have been relatively unexplored ornithologically, with brief visits in 1871 and 1939. However, longer-term fieldwork took place in 1996, 1997 and 2001, resulting in the discovery of a new hawk-owl *Ninox burhani* (q.v.). During this, observations were made of a *Zosterops* white-eye. A specimen was collected in July 2003; this is so far unique. Comparison with examples of Black-crowned White-eye *Z. atriceps*, which occurs in Sulawesi adjacent to the Gulf of Tomini as well as on islands further east, showed significant plumage differences, while recordings of the vocalisations also seem to differ from those of *atriceps*. On this basis a new species was described.

Description 11 cm. Monotypic. Forehead and area in front of eye jet-black, contrasting sharply with citrine-green crown and olive-green ear-coverts; back citrine, rump more yellowish; wing-coverts olive-green; rectrices dark brownish-black with olive edgings; throat yellow, chest grey, belly and flanks white; vent yellow; iris dark red, bare eye-ring dark greyish; bill black with pale flesh base to lower mandible; tarsi pale metallic horn.

◀ Togian White-eye. Togian Islands, Central Sulawesi province, Sulawesi, Indonesia, November 2015 (*James Eaton*).

Habitat Mangrove and secondary vegetation and in mixed gardens, below 100 m (Indrawan *et al.* 2008).
Food and feeding Forages in twos and threes, gleaning from branches and below leaves, often in dense shrubs. Food items insects and caterpillars.
Breeding Nest and eggs undescribed. Copulation of birds believed to be of this species observed in January.
Voice Song is a thin, sweet warble, less modulated than that of *atriceps* (see Indrawan *et al.* 2008 for detailed, comparative sonograms). Call consists of twittering chirrups (Eaton *et al.* 2016b).
Range Endemic. Confined to the Togian Islands, Gulf of Tomini. Present on Malenge, Talatakoh and Batudaka Islands, but not so far on Togian Island itself.
Conservation status Currently classified by BirdLife International as Near Threatened. Restricted-range species, apparently rather uncommon. Habitat on the islands is probably decreasing.
Etymology Named in honour of Professor Soekarja Somadikarta, Indonesia's leading avian taxonomist.

VANIKORO WHITE-EYE
Zosterops gibbsi

A dull-plumaged, large-billed isolated island population of a very widespread genus; possibly forms a superspecies with Santa Cruz White-eye *Z. sanctaecrucis.*

Type information *Zosterops gibbsi* Dutson 2008, Lavarka, Vanikoro, Santa Cruz Islands, Solomon Islands.
Discovery In 1994, while on a brief visit to the island of Vanikoro (Vanikolo) in the Santa Cruz Islands, D. Gibbs observed several white-eyes which he realised were of an unknown species; he was not, however, able to collect specimens and confined himself to describing it as *Zosterops* sp. (Gibbs 1996). In November 1997 G. Dutson visited Vanikoro and quickly found examples of the new taxon, as well as an active nest. Four specimens were taken, allowing the description of a new species (Dutson 2008). Curiously, personnel from the Whitney South Seas Expeditions spent a total of 14 days on Vanikoro in 1925–1927 but failed to detect any white-eyes (Mayr 1933).
Description 12–13 cm. Monotypic. Overall yellowish-green, upperparts citrine, underparts paler except on flanks; throat and chin pale yellow; naked skin around eye powder-grey, but no white feathers; iris deep rufous-orange; bill long, black with pale pink base to lower mandible; tarsi pale orange. Differs from Santa Cruz White-eye (*Z. sanctaecrucis*) in having a longer, stouter bill, pale orange legs and naked grey eye-ring.
Habitat Most abundant in old-growth forest between 350–750 m, but also occurs in degraded habitat down to sea level.
Food and feeding Forages in small groups, in a slow, methodical manner, on tree trunks, the undersides of branches and clumps of dead leaves; food items small fruit and insects.
Breeding Possibly a cooperative breeder. One nest described (Dutson 2008). Nest was a deep pensile bowl made from tightly woven grass stems, in a fork in a horizontal branch, 4 m up; in early November it contained two nestlings which were fed by at least three adults.
Voice Full song consists of various phrases of usually 6–20 melodic notes, longer and slower than the corresponding notes of *Z. sanctaecrucis*; a subsong of about

five notes, a quiet buzzing *vruh*, a high-pitched flight call and a harsh nasal scold.

Movements Presumably sedentary.

Range Endemic. Confined entirely to Vanikoro (Vanikolo) Island, Santa Cruz Group, Solomon Islands.

Conservation status Classified by BirdLife International as Least Concern. Although confined to a small range (the total area of the Vanikoro group, including smaller islands, is 173 km²), the population, which may number in the tens of thousands, appears to be stable and in suitable forest habitat it seems to be the most abundant species. The human population is quite small (1,300 people, speaking five different languages) and much of the upland habitat is little touched, despite earlier logging.

Etymology Named in honour of D. Gibbs, who first observed the bird and recognised it as a new taxon.

NAVA'S WREN
Hylorchilus navai

Originally described as a race of Sumichrast's Wren, itself a highly restricted Mexican endemic.

Type information *Hylorchilus sumichrasti navai* Crossin & Ely 1973, 26 km north of Ocozocoautla, Chiapas, Mexico.

Alternative names Crossin's Wren. Spanish: Cuevero de Nava, Cucarachero de Nava.

Discovery In December 1969 Santos Farfán noted an unusual wren in limestone karst country on the Caribbean slope of Chiapas, Mexico. Allan R. Phillips, who was present at the time, suspected that it was a Sumichrast's Wren, the nearest known population of which at that time being some 400 km to the north-west. In December 1970 and January 1971 a short series of specimens was taken from one location in humid forest; comparison of these with the limited material available of *H. sumichrasti* led to the description of the new taxon as a race of that species, *H. s. navai*. However, Phillips (1986), basing his opinions on morphological features, suggested that a full species might be involved. A detailed study of vocalisations by Atkinson *et al.* (1993) found major differences, resulting in the classification of the two populations as separate species, a view now widely accepted (e.g. Kroodsma & Brewer 2005 HBW 10; Whittingham & Atkinson 1996).

Description 16 cm. Monotypic. In overall form unique among wrens, only resembling *H. sumichrasti*; fairly large and bulky, with a notably long, fine and slightly decurved bill. Upperparts dark, rich brown, the remiges with faint blackish bars; rectrices brown, very faintly barred darker; throat and upper chest whitish, becoming lightly scalloped grey on lower chest and dark sooty-brown on flanks; iris brown; bill blackish with pale yellow base to lower mandible; tarsi dark grey. Sexes similar, although the female has a shorter bill. Differs from *H. sumichrasti* in having a white, not orange-brown, chest as well as in vocalisations.

Habitat Forest, probably requiring undisturbed forest, on limestone karst outcrops, 75–800 m; does not, apparently, colonise disturbed secondary habitat.

Food and feeding Gleans invertebrates from lichens on boulders and in cracks of rock-faces (Atkinson *et al.* 1993).

Breeding Nest and eggs undescribed.

Voice Both sexes sing. Song of male is a varied, often slightly jerky warble of rich notes, sometimes introduced by a few soft slightly upslurred accelerating notes and ending with a strong upslurred note (Howell & Webb 1995); countersong of female is an introductory note followed by eight or more loud, shrill whistles,

▼ Nava's Wren. Reserva de la Biosfera Selva el Ocote, Chiapas, Mexico, March 2007 (*Christopher L. Wood*).

ending abruptly (Gómez de Silva *et al.* 2004; Gómez de Silva 1997b), who give comparative sonograms with *H. sumichrasti*.

Movements Presumably sedentary.

Range Endemic. Confined to southern Mexico, in western Chiapas, SE Veracruz and very marginally eastern Oaxaca.

Conservation status Currently classified by BirdLife International as Vulnerable. The total range was originally estimated to be in the area of 4,800–4,900 km^2 (Gómez de Silva 1997a) but may be smaller; however, within this it is confined to isolated patches of habitat on limestone outcrops. These are usually unsuitable for agriculture but are used as sources of firewood. There are six known sites in Veracruz, two in Chiapas and one in Oaxaca. Population estimates are in the range 1,500–7,000 mature individuals (BLI SF 2015). Some areas are protected.

Etymology Named in honour of Juan Nava "who, through his devotion to learning the birds of his native Mexico, has earned the respect and admiration of numerous American ornithologists" (Crossin & Ely 1973).

INCA WREN
Thryothorus eisenmanni

A distinctively marked *Thryothorus* wren, closely related to and probably forming a superspecies with Plain-tailed Wren *T. euophrys*. Sometimes placed in the genus *Pheugopedius* (e.g. Mann *et al.* 2006).

Type information *Thryothorus eisenmanni* Parker & O'Neill 1985, San Luis, Cuzco, Peru.

Alternative names Spanish: Cucarachero Inca.

Discovery As early as 1965, J. O'Neill and G. Lowery repeatedly observed wrens near the spectacular Inca ruins of Machu Picchu, Peru, that resembled Plain-tailed Wren *T. euophrys*, but in a location quite distant from the nearest known range of that species. However, it was not until 1974 that specimens were obtained, which established the clear-cut morphological differences from that species.

Description 16 cm. Monotypic (but see below). Male: crown, nape, up to half of the ear-coverts and lores dull black; supercilium contrastingly white; back and rump bright russet, the tail duller with obscure darker bars; throat white; submoustachial streak dull black; chest and upper belly white, boldly streaked black; flanks and undertail-coverts dull yellow-brown; iris chestnut or reddish-brown-brown; bill dark brown above, blue-grey below; tarsi greyish-horn or grey-black. Female has fewer streakings on belly, no bars on tail, darker streaking on underparts and a charcoal-grey crown. A currently undescribed wren, obviously closely related to this species or to *euophrys*, has been seen by numerous observers, including this author, in the area between the (disjunct) ranges of *eisenmanni* and *euophrys*; the taxonomic status of this population

◄ Inca Wren. Abra Malaga, Cusco, Peru, March 2012 (*Nick Athanas*).

is presently uncertain (G. Engblom, pers. comm: see essay on Future New Species). *Eisenmanni* is immediately distinguishable from all races of *euophrys* by much heavier and darker streaking on the underparts.

Habitat Open subtropical or tropical montane forest especially with thickets of *Chusquea* bamboo; not in dense forest. 1,700–3,350 m.

Breeding Nest and eggs undescribed.

Voice Noisy and vociferous. Both sexes sing in well-synchronised antiphonal duets; these consist of sequences of rapidly repeated chortling whistles, rising and falling within a sequence of ten notes; female contribution apparently higher in pitch. Call a rich *tchp* or *tchp-er*, sometimes in a chattered series (Schulenberg *et al.* 2007).

Movements Sedentary, apart from local movements according to die-off of bamboo patches.

Range Endemic. Peru, in Dpto. Cuzco, approaching border of Junín; marginally into Apurímac (Schulenberg *et al.* 2007).

Conservation status Currently classified by BirdLife International as Least Concern. Notwithstanding the relatively small range, it appears to be quite common and perhaps increasing, since *Chusquea* bamboo is a primary colonist of disturbed areas. Readily seen around the ruins of Machu Picchu.

Etymology The scientific name honours Eugene Eisenmann (see under Azuero Parakeet).

ANTIOQUIA WREN
Thryophilus sernai

Placed in the large, mainly Neotropical, genus *Thryothorus* by some authorities (e.g. 2013 HBW SV); however, recent opinions (e.g. Mann *et al.* 2006) suggest that the genus is polyphyletic, involving at least four clades. Hence the genus *Thryophilus* (Baird 1874) is resurrected as a home for this and two other South American species.

Type information *Thryophilus sernai* Lara *et al.* 2012, vereda El Espinal, 3.2 km SSW of Santa Fé de Antioquia, Antioquia, Colombia.

Alternative names Spanish: Cucarachero Antioqueño, Cucarachero Paísa.

Discovery In February 2010 C. E. Lara found a "*Thryothorus*-type" wren in the canyon of the Cauca River in Antioquia, Colombia. Since this was outside the then known range of any member of this genus this was obviously of great interest. In March 2010 a type specimen (an adult male) was taken after the recording of its vocalisations, becoming the holotype; two further paratypes, adult male and female, were taken in August 2010. Careful comparisons of plumage and vocalisations, along with molecular data, allowed the description of a new species, distinctive from the two other Colombian species (Rufous-and-white Wren *T. rufalbus* and Niceforo's Wren *T. nicefori*). This conclusion was accepted by the majority of the SACC (Proposal 562).

Description 14 cm. Monotypic. A medium-sized wren; upperparts brown, varying from earth-brown on the forehead to cinnamon-brown on the rump; flight feathers with dark transverse bars; conspicuous white supercilium; post-ocular stripe earth-brown; narrow white crescent below eye; cheeks whitish with dark greyish mottling; chin, throat and central abdomen white, the flanks and lower belly light brown, undertail-coverts white, barred brownish-black; iris deep brown or chestnut, bare eye-ring light yellowish; bill grey to blackish above, whitish below; tarsi pale grey or grey-flesh. Sexes similar.

▼ Antioquia Wren. Dry Cauca River Valley, Dpto. Antioquia, Colombia, February 2013 (*Steve Bird*).

Habitat Semi-deciduous dry forest with open understorey, dry scrub and vegetation along watercourses, so far recorded at 250–800 m.

Food and feeding Forages actively, usually in pairs, in dead vegetation, vine tangles etc., from ground level to subcanopy; food items include a variety of insects (beetles, moths, grasshoppers etc.).

Breeding Few data; an inactive nest (undescribed) was found in December, and the two August paratypes were in breeding condition (Lara *et al.* 2012).

Voice Both sexes sing; song is melodious and flute-like, that of the male being longer in extent. Lara *et al.* (2012) give a detailed study of vocalisations and comparisons with those of *T. rufalbus* and *T. nicefori*, emphasising the distinctions.

Movements Presumably sedentary.

Range Endemic. So far found in six locations in the northern Cauca Valley, from Ituango south to the Concordia/Salgar border. This is a narrow strip of land, only 150 km in extent, and within 7 km of the river itself. Not all potential habitat has been fully explored.

Conservation status Not currently classified by BirdLife International, but clearly under considerable threat, notwithstanding some apparent ability of the species to adapt to some habitat modification. Much of the habitat in the Cauca Valley has already been heavily modified or destroyed, for agriculture, urban development and recreational use. Planned hydroelectric development would also inundate a significant portion of some of the best habitat for this and several other species of concern.

Etymology Named in honour of the late Marco Antonio Serna Díaz (1936–1991), whose "often unappreciated legacy has been foundational for the ongoing rise of ornithological and herpetological research in Antioquia" (Lara *et al.* 2012).

MUNCHIQUE WOOD WREN
Henicorhina negreti

Probably most closely related to Grey-breasted Wood Wren *H. leucophrys*, a widely distributed highland species found from Mexico to western Bolivia.

Type information *Henicorhina negreti* Salaman *et al.* 2003, Tambito Nature Reserve, municipality of El Tambo, Cauca, Colombia.

Alternative names Spanish: Cucarachero de Munchique.

▼ Munchique Wood Wren. Cerro Montezuma, Colombia, February 2013 (*Robert Lewis*).

Discovery The Chocó Endemic Bird Area of western Colombia has been a very productive area for the discovery of ornithological novelties, with three new species described in the 1990s alone. However, the difficult terrain, wet climate and, until recently, chronic civil unrest have resulted in the region receiving less than its full share of attention. In 1978 and 1984, in Munchique National Park, S. Hilty observed birds closely resembling Grey-breasted Wood Wren *H. leucophrys* in plumage, but with quite distinct vocalisations. Based on his data, it was suggested, incorrectly, that these birds might be the *brunneiceps* race of *leucophrys* (Brewer 2001). Nothing further was done until 1996, when P. Coopmans visited the area, noting once again the distinctive song, in close proximity to birds of typical *leucophrys* song-pattern; on this basis he postulated that a new, reproductively and ecologically isolated taxon was involved. Further work was done, in Munchique National Park, Dpto. Nariño, from 1997 to 2000. In August 2000 a group including P. Salaman and colleagues studied the distribution of wood wrens in the area, discovering that *H. l. leucophrys* was present in drier habitats, *H. l. brunneiceps* in wetter ones, both below 2,250 m, while above this altitude birds of the new song-type were exclusively found. Several birds, both adults and juveniles, were captured along with examples of *brunneiceps* from only 400 m away, allowing direct comparison. A total of four specimens, two adult males, one adult female and a juvenile, was collected; these specimens and the recordings of vocalisations allowed the description of a new species.

Description 10.8–11.7 cm. Monotypic. In adult plumage very similar to *H. leucophrys*; a small, large-headed, short-tailed wren. Crown dark brown, more black on sides; remainder of upperparts slightly brighter brown; narrow white supercilium; cheeks and malar area blackish, mottled with dull white and greyish; throat indistinctly streaked black; breast grey, paler below with brown flanks and vent; iris hazel-brown; bill black, bluish lead-grey below at base; tarsi bluish slaty-grey. Darker overall than *leucophrys*; tail shorter and tarsi longer, with different soft-part colours. Juveniles are quite distinctive, being dark sooty-blackish below and dull brown above.

Habitat Stunted, wet, epiphytic montane cloud-forest, 2,250–2,640 m. Apparently more specific in habitat requirements than the more catholic *H. leucophrys*.

Food and feeding Forages busily in pairs or family parties, usually low down (below 2 m) in dense wet vegetation. No data on diet.

Breeding Nest and eggs undescribed. The female specimen, taken on 25 July 2000, had an old brood-patch; recently-fledged juveniles seen in late July.

Voice Song is distinctive, consisting of repeated phrases of 6–12 pure notes, often 10 phrases each lasting about two seconds. Salaman *et al.* (2003) give detailed sonograms and comparisons with the two adjacent races of *H. leucophrys*.

Movements Presumably sedentary.

Range Endemic. Confined to very limited areas in western Colombia. Originally described from two areas on the uppermost Pacific slope of the western Andes in Dpto. Cauca. Has since been found in southern Antioquia (Krabbe 2009), some 350 km north of the type locality; and in Dpto. Chocó, about 230 km north of the Cauca site (van Oosten & Cortes 2009).

Conservation status Classified by BirdLife International as Critically Endangered, *contra* Kroodsma & Brewer (2005). The population is confined to a narrow ecological zone and appears to be declining. The main threat is habitat destruction for agricultural purposes.

Etymology Named in honour of Alvaro José Negret (1949–1998), one of Colombia's finest naturalists. Among his lasting contributions to conservation was the foundation of the Tambito Nature Reserve, home to more than 300 species of birds (Salaman *et al.* 1999).

BAR-WINGED WOOD WREN
Henicorhina leucoptera

A distinctively plumaged wood-wren confined to southern Ecuador and northern Peru.

Type information *Henicorhina leucoptera* Fitzpatrick *et al.* 1977, Cordillera del Cóndor, above San José de Lourdes, Dpto. Cajamarca, Peru.

Alternative names Spanish: Cucarachero Aliblanco, Soterrey-Montes Alibandeado.

Discovery During a 1975 survey of the avifauna of the Cordillera del Cóndor, an isolated mountain range shared between northern Peru and southern Ecuador, J. W. Fitzpatrick and colleagues discovered a distinctively plumaged *Henicorhina* wren living in dense cloud-forest on the Peruvian section of the range. A series of specimens was taken at the type locality, above San José de Lourdes, Dpto. Cajamarca, which is located 20 km from the Ecuadorian border, and one further specimen from a location in Dpto. San Martín, about 150 km to the south-east, across the valley of the Rio Marañón. These, and detailed studies of vocalisations, allowed the description of a new species.

Description 11 cm. Monotypic. In form a typical *Henicorhina* wren, very small, large-headed, with powerful legs and a tiny tail. Above rich brown, crown more greyish-brown; obvious white supercilium, cheeks mottled black and whitish; wings darker brown, with two prominent white bars formed by tips of the greater and lesser coverts; tail tiny, brown with darker bars;

▼ Bar-winged Wood Wren. Abra Patricia, Peru, September 2011 (*Fabrice Schmitt*).

chin, throat and upper breast nearly white, often with darker streaks at sides of chest; underparts more grey on belly, with cinnamon-brown on lower belly, flanks and crissum; iris dark reddish-brown, bill black above, pale grey or whitish at base of lower mandible; tarsi dull grey. Distinguished from sympatric race of Grey-breasted Wood Wren by wing-bars, dingier tone above and whiter underparts.

Habitat Mostly in somewhat impoverished forest growing on quartz sand, with stunted trees and a canopy height of 6–9 m; usually with a heavy understorey of ericaceous shrubs and bromeliads, the ground carpeted with mosses and lichens. Also, in San Martín, in more varied habitats including tall, moist hill forest to fern-covered slopes on white sandy soil savanna (Mazar Barnett & Kirwan 1999). At the original location, syntopic with Grey-breasted Wood Wren *H. leucophrys*, but with a clear altitudinal preference; at 1,950 m both species were present in, apparently, equal numbers; at 2,200 m *leucoptera* was abundant and *leucophrys* absent (Fitzpatrick *et al.* 1972). Overall altitude range is 1,350–2,600 m.

Food and feeding Forages busily, alone or in pairs, in lower levels of forest; stomach contents insects (Davis 1986).

Breeding Nest and eggs undescribed. Specimens from San Martín in August and September were not in breeding condition (Davis 1986).

Voice Very vocal; both sexes sing in syncopation. Song is a series of rich, warbled phrases, sometimes with trills; higher, more ringing and rapid and with more trills than that of *leucophrys*. Fitzpatrick *et al.* (1977) give detailed sonogram comparisons of the two species. Alarm call is a rapid, high-pitched chatter.

Movements Presumably sedentary.

Range Southern Ecuador in the Cordillera del Cóndor, Dpto. Zamora-Chinchipe (Krabbe & Sornoza 1994). Northern Peru (Cajamarca, and several disjunct locations in San Martín) (Schulenberg *et al.* 2007; Davis 1986; Hornbuckle 1999).

Conservation status Currently classified by BirdLife International as Near Threatened, although Ridgely & Greenfield (2001), refer to the Ecuadorian population as "truly at risk". In several locations described as common (Davis 1986) or fairly common (Stotz *et al.* 1996). The poor quality of the soil in most areas of occurrence gives some measure of protection.

Etymology *Leucoptera*, Greek, 'white-winged'.

IQUITOS GNATCATCHER
Polioptila clementsi

The most restricted, and certainly the rarest, of all the gnatcatchers, occurring in a tiny area of very specialised habitat in western Peru. Most closely related to Guianian, Para, Rio Negro and Inambari Gnatcatchers (*P. guianensis, P. paraensis, P. facilis* and *P. attenboroughi*), all forming a sister group to Slate-throated Gnatcatcher *P. schistaceigula*.

Type information *Polioptila clementsi* Whitney & Álvarez 2005, Zona Reservada Allpahuayo-Mishana, approximately 25 km WSW of Iquitos, Dpto. Loreto, Peru.

Alternative names Spanish: Perlita de Iquitos.

Discovery In September 1997, J. Álvarez Alonso observed what appeared to be Guianian Gnatcatcher

◀ Iquitos Gnatcatcher. Allpahuayo Mishana National Reserve, Peru, 1999 (*José Álvarez Alonso*).

(*P. guianensis*) in *varillal* forest (a tall woodland of canopy height 15–30 m, *Catalpa* being the dominant species), in the proposed Allpahuayo-Mishana Reserve, near Iquitos, Peru. This was some 800 km from the nearest known area of occurrence of *guianensis*. He did, however, note that the songs differed distinctly from those of *guianensis*. Subsequently he and B. Whitney obtained three specimens and made numerous recordings of vocalisations, establishing that the population in the reserve was an undescribed species.

Description 11 cm. Monotypic. In form a typical *Polioptila* gnatcatcher, tiny (5–6 g), waif-like, with a long, graduated tail. Upperparts fairly uniform slaty-grey; narrow broken white eye-ring; throat and breast slightly paler grey than back; belly and vent white; central rectrices black, becoming increasingly white on each side-feather; iris brown; bill black above, greyish-horn below; tarsi bluish-grey, soles of feet whitish. All the specimens were juveniles.

Habitat Appears to be very specific and demanding in its requirements, being found in very restricted areas of the Allpahuayo-Mishana Reserve, in tall, humid *varillal* forest on impoverished white quartz-sand soils.

Food and feeding Forages in canopy and subcanopy, usually in mixed-species flocks, energetically checking leaf surfaces. Stomach contents of specimens comprised insect fragments and arthropod eggs (Whitney & Álvarez 2005).

Breeding Nest and eggs undescribed. Most vocal in September–December (Shany *et al.* 2007b). The specimens, taken in December and April, were juveniles with incompletely ossified skulls.

Voice Diagnostic. Song is similar to, but immediately distinguishable from, that of *P. guianensis*, with three introductory notes followed by a series of evenly spaced notes at a faster tempo than those of *guianensis*. Whitney & Álvarez (2005) give detailed sonogram comparisons with the songs of geographically adjacent gnatcatcher species.

Movements Presumably sedentary.

Range Endemic. Confined entirely to the Allpahauyo-Mishana Reserve, Dpto. Loreto, Peru.

Conservation status Critically Endangered (indeed, prior to the publication of the type description, Lowen (2002) lamented that the species was "in danger of disappearing before being described"). The total area of occurrence is tiny; Whitney & Álvarez estimated a total population of no more than 50 pairs in six locations totalling 2,000 ha, while Shany *et al.* (2007b), despite intensive searching, found only 10–15 pairs in an area of 250 ha. The reserve status does not totally protect the habitat from logging for lumber and charcoal production, clandestine road-building and clearing for agriculture. The nature of the forest, on very poor soil, means that regeneration is extremely slow and areas already destroyed may not recover in the near future. Vigorous efforts are being made to enlist local public opinion to the cause of conservation in the reserve, which is also home to several other new or rare species; the gnatcatcher has been designated the official bird of the city of Iquitos, which has not, however, blunted the opinions of some local people who regard the reserve as a hindrance to the development of the city.

Etymology Named in honour of James S. Clements (1927–2005). Dr Clements, who supported efforts to create the Allpahuayo-Mishanha Reserve, was the author and founder of the enormously influential *Birds of the World: a Checklist*, published in six print editions and now available online.

INAMBARI GNATCATCHER
Polioptila attenboroughi

A member of the Guianian Gnatcatcher *P. guianensis* group, which includes four species in northern South America and Amazonia (see comments under Iquitos Gnatcatcher *P. clementsi*).

Type information *Polioptila attenboroughi* Whittaker *et al.* 2013, Tupana Lodge, municipality of Careiro, Amazonas, Brazil.

Alternative names Portuguese: Balança-rabo-do-inambari.

Discovery In their 2005 paper on the taxonomy of the Guianian Gnatcatcher complex, Whitney and Álvarez predicted that there might be an undetected population inhabiting the region west of the Rio Madeira in western Amazonia, which should be looked for especially on white sand and weathered clay *terra firme* forests. Subsequently, in July 2007, A. Whittaker and A. Aleixo tape-recorded and collected two specimens from west of the Madeira, in upland *terra firme* on sandy-soil forest. These were phenotypically and genetically distinct from all other taxa in the genus, justifying the description of a new species (Whittaker *et al.* 2013). This proposal received majority, but not unanimous, assent from the SACC (Proposal 619).

Description 11 cm. Monotypic. A rather uniform plumbeous-grey gnatcatcher, in form typical of the genus, very small and long-tailed. Upperparts from head to tail-coverts plumbeous; throat and chest plumbeous, lightest on throat, becoming white on lower belly and vent; narrow broken white eye-ring; tail long and graduated, entirely black on inner feathers, with increasing amounts of white on lateral rectrices; iris light brown or creamy; bill blackish-grey, paler below; tarsi bluish-grey.

Habitat Canopy of upland *terra firme* forest on sandy soils, below 200 m.

Food and feeding Forages alone or in pairs, apparently always in mixed-species flocks. Gleans prey (presumably arthropods) from leaf surfaces, sometimes catching it in short aerial sallies.

Breeding Nest and eggs undescribed. The type specimen taken in early July had enlarged testes.

Voice The loud song is a series of about six notes, most resembling that of the Para Gnatcatcher *P. paraensis* but the notes are perceptibly slower; also a more complex song which does not appear to have equivalents in nearby species. Whittaker *et al.* (2013) give relevant spectrograms.

Movements Presumably sedentary.

Range Endemic. Presently known from western Amazonian Brazil, west of the Rio Madeira and south of the Rio Solimões, in the states of Amazonas and Rondônia; the locations of documented occurrences are in an area about 350 km east to west and 650 km north to south. The most south-westerly location is about 50 km from the Bolivian border, but there are as yet no records from that country.

Conservation status Not presently classified by BirdLife International. The total area of actual or potential occurrence is quite extensive, although it appears to be at low densities in known locations. Whittaker *et al.* (2013) identify the increased clearing of forest for soyabean production as the worst potential threat.

Etymology Named in honour of Sir David Attenborough OM, CH, FRS (b. 1926). For more than 60 years Sir David has advanced public awareness of nature and conservation issues by countless superb television programmes, perhaps most famously the Life series produced with the BBC Natural History Unit. The specific epithet *attenboroughi* has also been appended to organisms as diverse as a long-beaked echidna, a flightless weevil, a goblin spider, an alpine hawkweed, a ghost shrimp, a pitcher-plant, an Ecuadorian flowering tree, a fossil grasshopper and a fossil armoured fish, while the genera *Attenborosaurus* (a plesiosaur-like reptile) and, perhaps more poetically, *Sirdavidia* (a Gabon custard-apple) have also been erected. The English and Portuguese names recognise the main centre of distribution of the new species.

ALGERIAN NUTHATCH
Sitta ledanti

The discovery of the Algerian Nuthatch, the only new species to be described from the Western Palaearctic mainland in the 20th century, caused a press campaign unequalled in the history of zoology (Vielliard 1978). A member of a species group, treated by some authorities as a superspecies, which includes Corsican and Krüper's Nuthatches in the Mediterranean Basin, Chinese and Yunnan Nuthatches in East Asia and Red-breasted Nuthatch of North America (*Sitta whiteheadi, krueperi, villosa, yunnanensis* and *canadensis*) (see Pasquet 1998).

Type information *Sitta ledanti* Vielliard 1976, Djebel Babor, Algeria.

Alternative names Kabyle or Kabylian Nuthatch. French: Sitelle kabyle.

Discovery In October 1975, the botanist J.-P. Ledant and colleagues were examining an endemic population of the endangered Algerian Fir *Abies numidica* in the Djebel Babor, Algeria, when a woodpecker-like tapping was heard. On investigation the sound was found to be caused by a small nuthatch, which was observed for some 15 minutes; the following day, the species was seen by several observers. On consulting standard text books, M. Ledant realised that there were no reliable records of nuthatches from Algeria. Consequently he wrote to J. Vielliard of the Société d'Études Ornithologiques de France. Professor Vielliard admitted that, at the time, his reaction was one of incredulity; he "regularly received more or less serious reports from uninformed amateurs"; furthermore, the name Ledant was unfamiliar to him. Nevertheless he encouraged M. Ledant to return to the site to gather more information. A visit to the Djebel was attempted by Ledant and two companions in December 1975, but they were unable to climb to the required habitat due to deep snowdrifts. In April 1976 they returned and in the thaw were able to see and describe two nuthatches.

Independently and without any knowledge of the observations of Ledant, in June 1976 Dr E. Burnier discovered the species and made a number of sketches (Burnier 1976). A month later Ledant returned on an expedition with Prof Vielliard and other colleagues. On arrival in suitable habitat they quickly found the

▲ Algerian Nuthatch, male (left) and female (right). Tamentout Forest, Djemila, Petite Kabylie, Algeria, June 2009 (*David Monticelli*).

nuthatch, also observing broods and locating nests. Two specimens, a male holotype and a female paratype, were taken and the species was rapidly described in 1976.

Curiously, Heim de Balsac & Mayaud (1962) mention unsubstantiated reports dating from the 19th century of nuthatches in the mountains of Algeria; these were assumed to refer to the European Nuthatch *Sitta europea* but, in retrospect, might well have been the first observations of *S. ledanti* (Vieilliard 1976a, 1976b).

Description 13.5 cm. Monotypic. A medium-sized nuthatch with blue-grey upperparts, the forehead black in the male, a smaller diffuse black area in the female (Jacobs *et al.* 1978); prominent white supercilium, thin blackish eyestripe; throat and underparts creamy-pink; vent paler; central rectrices blue-grey, outer rectrices blackish with narrow blue-grey tips and subterminal white bar; iris blackish-brown with narrow whitish eye-ring; bill slaty-black, basal half of lower mandible blue-grey; tarsi grey or dark grey.

Habitat Relict montane forest, largely dominated by the endemic Algerian fir *Abies numidica* and other coniferous species, with epiphytic mosses and lichens; also in lower altitude areas dominated by oaks; 350–2,000 m, but apparently most abundant at upper levels, especially above 1,900 m (Ledant 1981; van den Berg 1982; Ledant & Jacobs 1977).

Food and feeding Feeds in typical nuthatch fashion on trunks, branches and twigs of trees, often hammering pieces of bark and wood-chips (Gather & Mattes 1979). In summer probably largely insectivorous; in winter feeds largely on seeds and nuts. Caches seeds in lichen-covered branches. Young fed on insects, spiders, conifer seeds and mosses. Territorial in winter.

Breeding Nests in cavities, in rotted trees or crevices created by snow damage to trees, sometimes in old woodpecker holes; usually 3–15 m up; may excavate own hole. One example of mud-plastering around hole entrance in the manner of a European Nuthatch *S. europea* (Gatter & Mattes 1979). Eggs undescribed, but clutches probably 2–4. Nest made of wood-chips, leaves, feathers, animal fur etc. Incubation probably by female alone, young fed by both sexes, fledging period probably 22–25 days. Probably single-brooded.

Voice Foraging and flight call a nasal *kua*, reminiscent of call of Red-breasted Nuthatch *S. canadensis*; other calls include harsh rasping notes and nasal notes (Monticelli & Legrand 2009a, 2009b). Song is a repetition of 7–12 nasal or fluty notes. For comparative sonograms of the songs of the three Mediterranean species (*ledanti, kruperi and whiteheadi*) see Matthysen (1998).

Movements Largely sedentary, but some degree of altitudinal movement may occur.

Range Endemic. Confined to four sites in northern Algerian mountains; Taza National Park, Mt Babor, Tamentout and Djimla (Vielliard 1976b; Bellatreche & Chalabi 1990).

Conservation status Currently classified by BirdLife International as Endangered. The total known range is in the region of 240 km², with an estimated population of 250–1,000 mature individuals. Main threats include loss of habitat due to fire, degradation by livestock grazing and illegal deforestation (Harrap & Quinn 1996, Ledant *et al.* 1985).

Etymology Named in honour of its first discoverer, Jean-Paul Ledant. M. Ledant, a Belgian, was trained as a botanist and forest specialist; nevertheless, he published numerous papers on the ecology of "his" new bird (Ledant 1977, 1978, 1981; Ledant & Jacobs 1977; Ledant *et al.* 1985). The French and alternative English common names refer to the Little Kabyle region of Algeria.

SICHUAN TREECREEPER
Certhia tianquanensis

Originally described as a race of Eurasian Treecreeper *C. familiaris* with which it is sympatric; however, genetic and morphological studies suggest that its closest relative is in fact the allopatric Rusty-flanked Treecreeper *C. nipalensis* (Teitze & Martens 2009; Teitze *et al.* 2006).

Type information *Certhia familiaris tianquanensis* Li 1995, Tianquan county, Sichuan, China.

Alternative names Chinese: Sichuan Xuan Mu Que.

Discovery In 1995 Li, on the basis of nine specimens collected in two locations in Sichuan, China, described a new race of Eurasian Treecreeper *C. familiaris*, although his paper (Li 1995) went largely unnoticed. However, in May 2000 J. Martens and Y.-H. Sun observed treecreepers that they immediately recognised as distinctive in plumage and song. Comparison with Li's original seven specimens, taken in 1991 and 1992, and two others, one dating back to 1940 and misidentified, showed major differences from the local race of *familiaris*, which was in any case syntopic without any apparent hybridisation. Further studies of vocalisations and cytochrome b supported the status of *tianquanensis* as a full valid species (Martens *et al.* 2002, 2003).

Description 14 cm. Monotypic. A typical *Certhia* treecreeper in form and plumage, but notably short-billed and long-tailed. Distinguished from other treecreepers of the general geographic area by the contrasting white throat and dusky, not rusty, flanks and belly, by the lack of bars on the rectrices, and by vocalisations.

Habitat Apparently a relict species, dependent on open stands of old-growth coniferous forest, especially Emei Fir *Abies fabri* with dense bamboo understorey (Rheindt 2004). At 2,500–2,900 m in the breeding season, descending to at least 1,600 m in winter.

Food and feeding Foraging behaviour as other *Certhia* species; little data on diet but presumably mainly arthropods.

Breeding Five nests so far discovered (Sun *et al.* 2009). Nests were located in cracks and fissures of dead fir stems. Nests are made of mosses, lined with hair and feathers, the upper wall with a few dried bamboo leaves. Eggs white with dense red spots concentrated at blunt end; one clutch was of four eggs. Incubation apparently by female, fed by male. Nesting period May and June.

▲ Sichuan Treecreeper. Wawu Mountain, Sichuan, China, July 2007 (*Li Liwei*).

Voice Song is distinctive, a loud, rapid, high-pitched trill, starting explosively but tailing off and falling in pitch, and introduced by a higher, sweeter note (Harrap 2008 HBW 13). Longer and more slurred than that of *C. nipalensis*. Martens *et al.* give comparative sonograms of the songs of *tianquanensis* and geographically adjacent taxa.

Movements Short-range altitudinal migrant, moving down to 1,600 m in winter.

Range Endemic. Confined to a restricted area of Sichuan, western China; originally known from five sites only, more recently found in other areas, extending the accepted range, including the Shaanxi range (Rheindt 2004; BLI SF 2015).

Conservation status Currently classified as Near Threatened. Rather specific in its habitat requirements and geographically highly restricted. Much logging has occurred in what may have been a more extensive range; some habitat is protected by the rugged nature of the terrain. Some areas of habitat are formally protected. Total population estimated to be in the range of 700 mature individuals.

Etymology Both scientific and vernacular names reflect the limited geographic range of the species.

HIMALAYAN FOREST THRUSH
Zoothera salimalii

A widespread but cryptic form whose specific identity was only established by a meticulous study of plumage, measurements, vocalisations and DNA.

Type information *Zoothera salimalii* Alström *et al.* 2016, based on a specimen taken in January 1954 in the Khasi Hills, Meghalaya, India, by Rupchand.

Alternative names Himalayan Thrush. Chinese: Xi Shan Lin Dong.

Discovery The genus *Zoothera* occurs from northern India, through SE Asia to Australia and the western Pacific. In the Himalayas and western Chinese, two species, *Z. mollissima* (Plain-backed Thrush, with two or three subspecies) and *Z. dixonii* (Long-tailed Thrush), have always been regarded as closely related and sometimes conspecific. In 2009, P. Alström and S. Dalvi noticed that in Arunachal Pradesh, India, two populations of Plain-backed Thrush, separated completely by habitat, altitude and song, existed. This led to a very comprehensive review of the entire complex, in northern India and western China. These studies showed that the Plain-backed Thrush complex comprises at least three distinct species: *Z. mollissima*, Alpine Thrush; *Z. griseiceps*, formerly *Z. m. griseiceps*, Sichuan Forest Thrush; and one population that had not been previously described and hence is named as a new species, the Himalayan Forest Thrush *Z. salimalii*. There may be a fourth cryptic species, the 'Yunnan Thrush', although this requires further study. Although the differences between *Z. salimalii* and other members of the group are subtle, in plumage and measurements, the new species is clearly distinct; genetic and vocal evidence backs this up.

Description 25 cm. In form a typical *Zoothera* thrush, medium-sized, quite long-tailed, bulky, with strong legs. Upperparts from forehead to rump uniform dark russet-brown; narrow buff-tinged eye-ring; lores and area below eye blackish, supraloral area pale rufous; pale buff sub-moustachial and malar areas; chin buffy-white, throat pale buff; breast strongly buffy, with heavy blackish-brown scaly markings which cover the underparts, becoming less prominent on central lower breast; rectrices brown, paler on central pair, with sharply demarcated white tips; bill dark blackish-brown without obvious pale base; iris dark brown; tarsi pink. Sexes similar.

Habitat In the breeding season, occurs in montane, mostly coniferous forest, to the upper tree limit, at 3,430–4,200 m. Sympatric, but not syntopic, with Alpine Thrush *Z. mollissima sensu stricto*, which occurs above 4,000 m.

Food and feeding Plain-backed Thrush *Z. mollissima sensu lato* feeds on insects, snails and leeches, also berries and seeds. Forages on the ground, turning over leaves and probing in the earth. In winter, sometimes in flocks (Collar 2005).

Breeding No information specifically referable to *Z. salimalii*.

Voice The song of Himalayan Forest Thrush is a mixture of rich, drawn-out clear notes and shorter, thinner ones, with hardly any harsh scratchy notes. It differs from that of Alpine Thrush in being more musical and slower, with more pronounced variations in pitch. Alström *et al.* give a very comprehensive comparative study of the songs of all the members of the genus in the geographic study area, showing that the vocalisations of the species are truly distinctive.

► Himalayan Forest Thrush. Dulongjiang, Yunnan, southern China, June 2014 (*Per Alström*).

Movements Migratory or partially so, moving to lower altitudes in winter, with non-breeding season records in Meghalaya and Manipur states, India; eastern Myanmar (Burma); Yunnan province, China; and northern Vietnam.
Range Occurs from Sikkim and Darjeeling, India, to NW Yunnan. In addition, the authors of the species description have examined a single specimen (IOZ 35463) from south Sichuan, China, collected during the breeding season (10 May 1960), and a 7 May 1972 bird (KIZ 72183) from Luchun county, SE Yunnan; these birds may have been migrants or on their breeding grounds. It is not known whether the distribution is continuous or patchy.
Conservation status Alström *et al.* suggest a classification of Least Concern, since the species is locally common or abundant in parts of its breeding range, while the habitat does not appear to be under imminent threat.
Etymology Named in honour of Sálim Moizuddin Abdul Ali. Born in Bombay (Mumbai), India in 1896, Dr Ali was surely the doyen of 20th century Indian ornithology. In a long and productive life he made enormous contributions to the study and conservation of the birds of the subcontinent; perhaps his most outstanding accomplishment was, with S. Dillon Ripley, the ten-volume *Handbook of the Birds of India and Pakistan,* but he also published other regional guides as well as making voluminous contributions in refereed journals. His awards included the Gold Medal of the British Ornithologists' Union, as well as recognition from learned societies in the United States, the Soviet Union, the Netherlands and his native India. He died at the age of 90 in June 1987.

BOUGAINVILLE THRUSH
Zoothera atrigena

A distinctive thrush originally described as a race of New Britain (or Black-backed) Thrush *Z. talaseae,* but treated as a full species by many authorities (e.g. Dutson 2011, Clements 2014).

Type information *Zoothera talaseae atrigena* Ripley & Hadden 1982, 15 km SSW of Arawa, Bougainville Island, North Solomon Islands, Papua New Guinea.
Discovery Described on the basis of a specimen mist-netted by D. Hadden and B. Beehler in montane forest on Bougainville Island.
Description 20-23 cm. Monotypic. A typical *Zoothera* thrush in form. Upperparts blackish-grey; two wing-bars formed by whitish edgings to the greater and lesser coverts; rectrices black with white tips; underparts white, the flanks with prominent blackish scales; iris dark, bill blackish, tarsi black in male, probably brownish-black in female. Differs from *Z. talaseae* in having black cheeks, generally blacker above, smaller white covert tips and more prominent flank scaling.
Habitat Wet montane forest; type specimen was taken at 1,500 m.

◄ Bougainville Thrush. Top Camp, above Panguna, Bougainville Island, Papua New Guinea, June 1980 (*Bruce Beehler*).

Food and feeding Few data; forages low down or on the ground.
Breeding Nest and eggs undescribed.
Voice Undescribed.
Movements Presumably sedentary.
Range Known only from one region of the Crown Prince Range, Bougainville Island.
Conservation status Apparently rather rare and very localised, although possibly overlooked. IUCN classification is Near Threatened.
Etymology *Atrigena*, from Latin, 'dark-cheeked'.

VARZEA THRUSH
Turdus sanchezorum

A cryptic taxon whose existence as a valid species has only recently been elucidated.

Type information *Turdus sanchezorum* O'Neill *et al.* 2011, south of the Amazon, ca 10 km SSW of the mouth of the Río Napo on the east bank of Quebrada Vainilla, Loreto, Peru.
Alternative names Spanish: Zorzal de Várzea. Portuguese: Sabía-da-várzea.
Discovery In 1961, on his first field trip to Peru, J. P. O'Neill collected a *Turdus* thrush which was identified as Hauxwell's Thrush *T. hauxwelli*, a widespread South American species ranging over much of lowland western Amazonia. However, as more specimen material became available, Dr O`Neill noticed specimens with plumage features consistently differing from *hauxwelli*; these became known as a 'grey-tailed morph' and were illustrated as such in Schulenberg *et al.* (2007) although these authors presciently commented that a separate species might be involved. In 2003, in Dpto. San Martín, Peru, B. J. O'Shea and B. Walker noted a *Turdus* thrush with a diagnostic "mewing" call. Study of recordings made by several ornithologists, including some made by T. Parker in the 1980s, revealed more birds with 'mewing' calls with the plumage characters of the 'grey-tailed morph'. A detailed study by O' Neill *et al.* (2011) demonstrated a number of subtle but consistent differences in plumage and soft parts of this 'morph' from proven *hauxwelli*, documenting a widespread distribution and showing in some cases sympatry and syntopy with *hauxwelli*. On this basis the new species was described; extensive genetic data provide further support (O'Neill *et al.* 2011). In fact, authors as early as Hellmayr (1934) and, later, Gyldenstolpe (1945a) had noticed the variability of soft-part colours of specimens, but by the time they received the material (collected in the case of Count Gyldenstolpe by the Olalla brothers), colours would undoubtedly have changed.
Description 22–24 cm. Monotypic. A dull-brown *Turdus* thrush, rather large, generally sepia-brown above, the tail greyish-brown; chin and throat buffy-white with dusky-brown streaks; breast and flanks camel-brown; centre of belly and vent whitish; iris brown, eye-ring dull orange; bill dull yellow at tip, darkening to blackish at base; tarsi olive-grey. *Hauxwelli* has no orange orbital ring and a blackish bill; the greyish rectrices of *sanchezorum* are also diagnostic.
Habitat Mostly *várzea* forest; also in white-sand forest and savanna, though adjacent to intermittently flooded forest. Mostly lowlands around 100 m, although the white-sands population was at 800–1,100 m.
Food and feeding Probably omnivorous; one specimen had stomach contents containing blackish fruit pulp with seeds; also seen to eat small palm fruit.

▼ Varzea Thrush. Moyobamba, San Martín, Peru, October 2014 (*Daniel Lane*).

Breeding Nest and eggs undescribed. Two specimens from northeastern Loreto in December and January were in breeding condition.

Voice Most distinctive call is a rising, querulous mewing note, as well as several other short calls. Song is a slow, mellow caroling with slurred phrases (O'Neill *et al.* 2011).

Movements Probably sedentary.

Range Western Amazonia. Currently recorded from Peru in San Martín, Loreto, Cuzco and Ucayali; Colombia, in Dpto. Amazonas; Brazil, in Amazonas and Acre states (Guilherme 2013).

Conservation status Not currently listed by BirdLife International, but has a wide distribution and appears to be locally common.

Etymology Named in honour of Manuel Sánchez and Marta Chávez de Sánchez, in recognition of their tireless help to the ornithologists of the Louisiana State University Museum of Zoology over many years. Sr Sánchez was, in fact, responsible for capturing the first specimen of the iconic Long-whiskered Owlet (*Xenoglaux loweryi*) (qv) in 1976. The vernacular names in English, Spanish and Portuguese reflect the habitat requirements of the species.

OLIVE-BACKED FOREST ROBIN
Stiphrornis pyrrholaemus

Treated as a race of the Forest Robin *S. erythrothorax* complex by some authorities; molecular studies suggest that *pyrrholaemus*, and possibly other races, should be treated as full species. Recently discovered sympatry of *erythrothorax* and *pyrrholaemus* in NE Gabon (Boano *et al.* 2015) supports this view. Genetics also suggest that the genus should be in the family Muscicapidae (Beresford & Cracraft 1999).

▼ Olive-backed Forest Robin, male. Moukalaba, Doubou NP, Gabon, April 2003 (*Brian Schmidt/Smithsonian Institution*).

▼ Brian Schmidt, a research ornithologist at the Smithsonian Institution, makes notes of a female specimen of Olive-backed Forest Robin in Gabon, 2003 (*Carlton Ward Jr*).

Type information *Stiphrornis pyrrholaemus* Schmidt *et al.* 2008, N'dodo Lagoon, Monkalaba-Doudou National Park, Ogooné-Maritime province, Gabon.
Alternative names French: Rouge-gorge à dos olive.
Discovery In 2000 a team from the Smithsonian Institution, Washington DC, while conducting fieldwork in Gabon to investigate ways of ameliorating the effects of resource development and extraction, took specimens of a *Stiphrornis* forest robin from two sites in the southwest region of the country. Comparison with museum material of other forest robins, followed by DNA analysis, indicated that a new species was involved. A previous undetected specimen (an unsexed juvenile), taken in November 1953, was found in the Muséum d'Histoire Naturelle de Paris. The taxonomy of other members of the genus is presently controversial; Beresford & Cracraft (1999) favour treating the races of *S. erythrothorax sensu lato* as four separate species (the Smithsonian birds being the fifth), while the IOC World Bird List currently lumps all five under *erythrothorax*. Collar (2005 HBW 10) regards the four-way split of *erythrothorax* as perhaps premature, while *pyrrholaemus* is given full specific status by del Hoyo *et al.* (2013 HBW SV), a course of action which we have followed.
Description 12 cm. Monotypic. A typical forest robin in form, fairly small, large-headed and with short legs and robust feet. Sexually dimorphic. Male, crown, nape and ear-coverts olive-washed grey; face blackish with obvious white spot in front of eye; back to tail olive-green; throat and chest fiery orange becoming yellow below with grey-tinged flanks; iris dark brown; bill black; tarsi pinkish. Female has less intense colour on throat and chest, less black on face.
Habitat Wet lowland primary forest below 200 m, with moderate understorey, less common or absent when undergrowth is reduced by grazing.
Food and feeding Forages unobtrusively on or near ground; food items are small arthropods.
Breeding Nest and eggs undescribed.
Voice Two types of vocalisations; a single chirp, followed by a series of modulated notes, and continually repeated phrases. Differs from corresponding vocalisations of *S. erythrothorax* subspecies.
Movements Presumably sedentary.
Range Endemic. Initially recorded only from southern Gabon, south of the R. Ogooné, but more recently at a second location some 600 km NE of the type locality, in the Makokou region of NE Gabon (Boano *et al.* 2015).
Conservation status Not currently assessed by BirdLife International. Limited geographic range, although it appears to be fairly common in optimum habitat; less common in habitat with heavy grazing by native animals, which would doubtless also apply to domestic ungulates.
Etymology *Pyrrholaemus*, Greek, 'fire-throated'.

RUBEHO AKALAT
Sheppardia aurantiithorax

Molecular studies indicate that this species' closest relative is Iringa Akalat *S. lowei*, from which it is geographically isolated by the lowlands of the Great Ruaha River in Tanzania.

► Rubeho Akalat, adult feeding young. Ukaguru Mountains, Tanzania, November 2015 (*Markus Lagerqvist*).

Type information *Sheppardia aurantiithorax* Beresford *et al.* 2004, Mafwemiro Forest, Rubeho Mountains, Tanzania.

Discovery In 1989 T. D. Evans and G. Q. A. Anderson mist-netted an akalat in the Ukaguru Mountains, Tanzania; since they were unable to take specimens, the mist-netted birds were identified as the Iringa Akalat *S. lowei*, a range extension. When specimens were finally collected in the Rubeho Mountains by J. Linne in 2000, comparison with authentic specimens of adjacent akalats immediately revealed that a new species was involved.

Description 14 cm. Monotypic. In form typical of the genus, a small terrestrial chat with robust legs and an upright posture. Above dark slate, the rump with a coppery wash; forehead to nape tawny-olive, pale loral area; throat and upper chest deep orange, becoming dark ochre on breast and tawny-olive on flanks; lower breast and belly creamy-buff, vent deep orange; iris brown; bill black; tarsi grey with ochraceous soles. Distinguished from *S. lowei* by absence of white mid-throat and richer coppery tinge to upperparts.

Habitat Understorey of montane forests, at 1,800–3,200 m (although once at 400 m), often modified by natural or artificial forest fires.

Food and feeding Forages on the ground or in lower levels of vegetation; no data on diet. Presumably follows swarms of driver ants (*Dorylus*).

Breeding Nest and eggs undescribed. Probably nests in March (Beresford *et al.* 2004).

Voice Call, a series of dry nasal rattles, *trrr-rrr-rrr* etc., lower in pitch than call of *S. lowei*. Song not recorded.

Movements Data lacking; however, one record in August at 400 m in the foothills of the Muguru Mountains suggests some local movement during the dry season.

Range Endemic to the Eastern Arc mountains of Tanzania: Rubeho Highlands, Wota, Kiboriani, Muguru and Ukaguru Mountains (Kahindo *et al.* 2007).

Conservation status Currently classified by BirdLife International as Endangered. Although apparently fairly common in some parts of its range, this is quite small (550 km^2) and often degraded; some parts of the range are protected, although the effectiveness of protection is variable (BLI SF 2015).

Etymology *Aurantiithorax*, a combination of Greek and Latin roots, 'golden-throated'. The vernacular name refers to one of the most important locations for this species.

NIMBA FLYCATCHER
Melaenornis annamarulae

A local but very characteristic flycatcher of lowland rainforests in West Africa.

▼ Nimba Flycatcher. Atewa, Ghana, April 2009 (*Nik Borrow*).

Type information *Melaenornis annamarulae* Forbes-Watson 1970, Gracefield, Mt Nimba, Liberia.

Alternative names Liberian/West African/Mrs Forbes-Watson's Black Flycatcher. French: Gobe-mouche du Libéria.

Discovery While taking part in a survey of the Mt Nimba area, Liberia, in 1967, initiated by the International Union for the Conservation of Nature, A. D. Forbes-Watson encountered a new and distinctive flycatcher; a series of 11 specimens was taken, allowing the description of a new species.

Description 19 cm. Monotypic. A large, robust flycatcher, uniformly blackish-plumbeous above, the remiges and rectrices more blackish, the throat slightly paler; iris dark-brown; bill black; tarsi black with dull creamy soles. Sexes similar.

Habitat Lowland tropical rainforest on flat terrain, also in clearings and disturbed forest.

Food and feeding Forages alone, in pairs or in small groups, in the canopy or upper levels, catching aerial prey by short sallies, also by searching in crevices. Prey items are insects, including flying ants, beetles and caterpillars.

Breeding Nest and eggs undescribed; birds with enlarged ovaries at the end of July (Forbes-Watson 1970).
Voice Song consists of short, varied phrases of pleasant, melodious whistles. Calls include a thin, soft *wheep-wheep* and loud harsh strident notes (Borrow & Demey 2001).
Movements Presumably sedentary.
Range West Africa. Disjunct locations (Klop *et al.* 2010; Alport *et al.* 1989) in eastern Sierra Leone, southern Guinea, northern and eastern Liberia, SE and central Ivory Coast, and recently a small, highly disjunct population in eastern Ghana (Demey & Hester 2008; Borrow 2010).
Conservation status Classified by BirdLife International as Vulnerable. Althought the total geographic range is quite extensive, habitat destruction is extremely serious and continuing, driven by expanding human numbers (the population of Liberia increased by 410% between 1960 and 2015, and that of Ivory Coast by 580%). Déné Forest Reserve in Guinea was 90% destroyed by 2003, and other nominally protected areas are under intense pressure from smallholdings, logging and mineral exploitation (BLI SF 2015).
Etymology The scientific name honours Anna Marula Forbes-Watson (1941–2006), the wife of the discoverer.

SULAWESI STREAKED FLYCATCHER
Muscicapa sodhii

A species observed by many ornithologists and birdwatchers over a period of 15 years prior to its final collection and description. Although closely resembling Grey-streaked Flycatcher *M. griseisticta* in plumage, a migrant breeding in far-eastern Russia, North Korea and China, DNA evidence shows a closer relationship to Asian Brown Flycatcher *M. dauurica.*

Type information *Muscicapa sodhii* Harris *et al.* 2014, Baku Bakulu, Sulawesi, Indonesia.
Discovery In 1997, B. King and colleagues observed a *Muscicapa* flycatcher in Lore Lindu National Park, Sulawesi, Indonesia, which they realised was an undescribed taxon (King *et al.* 1999), although a formal description was not possible in the absence of specimen material. The bird most resembled Grey-streaked Flycatcher *M. griseisticta*, but since the observations were made during the boreal summer, when that species should be on its distant NE Asian breeding grounds, it appeared unlikely that it could be that species. Furthermore, the Sulawesi bird appeared to differ from *griseisticta* in some details of plumage and structure, being obviously shorter-winged (and by implication, non-migratory). In 2012 an international group of ornithologists from the United States and Indonesia visited the site of the original observations and were able to record vocalisations of the new bird, but were unable to capture one. However, on 23 June

► Sulawesi Streaked Flycatcher, adult and juvenile. Banti Murung, Sulawesi, Indonesia, September 2009 (*Nigel Voaden*).

they were presented with a specimen by a local hunter who had just shot it with an air gun. Two days later the same hunter, unsolicited, gave them a second specimen obtained in the same manner. These became the holotype and a paratype.

Description 12 cm. Monotypic. In form a typical *Muscicapa* flycatcher, large-headed with a hooked bill. Plumage drab grey-brown above, with indistinct facial patterning. Underparts brown, heavily streaked on the breast, less so on the belly; centre of lower belly and vent unmarked white; iris dark brown; bill black with paler yellowish-cream base to lower mandible; tarsi brownish-black.

Habitat Secondary broadleaved forest with some residual large trees and patchy roadside forest habitat; at least 250 m to 1,050 m.

Food and feeding Forages at all levels but mainly in mid-level, catching insects by aerial sallies; observed prey items include large damselflies and katydids. May sometimes join mixed-species flocks.

Breeding Nest and eggs undescribed. Adults observed feeding juveniles in late September and early October, i.e. at the start of the monsoon season (Harris *et al.* 2014).

Voice Song consists of thin, very high-pitched whistles, chirps, trills and twitters, higher in pitch than other members of the genus (Harris *et al.* 2014, who give detailed sonograms).

Movements Presumably sedentary.

Range Endemic. Widespread on the island of Sulawesi, including several locations in the NE peninsula (Sulawesi Utara), central (Sulawesi Tengah), the SE Peninsula (Sulawesi Tenggara) and the SW peninsula inland from Makassar. Doubtless other locations will emerge now that the species has been described.

Conservation status Not currently listed by BirdLife International, but undoubtedly of Least Concern. Has a wide geographic distribution, and appears to be able to adapt to substantial habitat modification, so long as some large trees remain.

Etymology Named in honour of the late Prof. Navjot S. Sodhi (1962–2011) "for his monumental contributions to conservation biology and ornithology in SE Asia" (Harris *et al.* 2014).

ASHY-BREASTED FLYCATCHER
Muscicapa randi

Probably forms a superspecies with Asian Brown Flycatcher *M. dauurica* and Sumba Brown Flycatcher *M. segregata*.

▼ Ashy-breasted Flycatcher. Makiling, Los Baños, Laguna, Luzon, Philippines, November 2012 (*Sylvia Ramos*).

Type information *Muscicapa latirostris randi* Amadon & duPont 1970. Dalton Pass, Nueva Vizcaya, Luzon, Philippines.

Discovery Several specimens of a flycatcher, taken in August 1959 by Filipino collectors at Dalton Pass, Luzon, Philippines, were sent to J. E. duPont at the Delaware Museum. Comparison of these with a long series of the migratory Asian Brown Flycatcher *M. latirostris* (sic, a junior synonym for *dauurica*) showed consistent differences in plumage and bill morphology on which basis the discoverers described them as a race of '*latirostris*'. Dickinson *et al.* (1991) "provisionally" treat it as a full species, a course of action followed by Taylor (2006 HBW 11). In fact, one unrecognised specimen was taken in August 1877 (Dickinson *et al.* 1991).

Description 12.5–14 cm. Monotypic. A small or medium-sized flycatcher with a large head and short rounded wings. Upperparts grey-brown, with paler lores and a narrow pale eye-ring; throat whitish; chest light grey-brown, lacking obvious streaking; belly whitish; iris dark; bill dark horn above, yellowish-horn below; tarsi blackish.

Habitat Mountain forest, mostly below 1,000 m, occasionally up to 1,200 m; also in selectively logged areas.

Food and feeding Forages alone or in pairs, mostly in lower levels of forest; no data on diet.

Breeding Nest and eggs undescribed. Recently fledged birds in June and August.
Voice Song is a very quiet *wee-tit* or *zeeet-tipp*; also a high trill and short high warbled phrases.
Movements No data; may perform short post-breeding movements (Dickinson *et al.* 1991).
Range Endemic. Negros, Samar and Luzon, Philippines; breeding proven on Luzon (Hornskov 1995).
Conservation status Currently classified by BirdLife International as Vulnerable. Although geographically quite widely distributed (further fieldwork is desirable), it does not appear to be common; habitat destruction by logging and slash-and-burn agriculture is major and ongoing.
Etymology Named in honour of Austin Rand (1905–1982), a Canadian ornithologist who conducted extensive fieldwork in New Guinea and Madagascar. Curator of birds at the Field Museum, Chicago, and later President of the American Ornithologists' Union (1962–64).

FURTIVE FLYCATCHER
Ficedula disposita

Treated as conspecific with Cryptic Flycatcher *F. cryptica* and Lompobattang Flycatcher *F. bonthaina* by some authorities, but differs substantially in several characters.

Type information *Ficedula bonthaina disposita* Ripley & Marshall 1967, Zambarles Mountain, above Crow Valley, Tarlac Province, Luzon, Philippines.
Discovery In January 1966 J. T. Marshall mist-netted a single female specimen of a small flycatcher in the Zambarles Mountains, west-central Luzon. This was obviously related to Cryptic Flycatcher *F. cryptica* of Mindanao, itself only known at that time by three specimens. On the basis of the limited material available, Ripley & Marshall (1967) described the Luzon bird as a race of the widely disjunct Lompobattang Flycatcher *F. bonthaina* of southern Sulawesi, Indonesia, in which species they also included, as a race, *cryptica*. However, more recent observations, including adult male birds (e.g. Dutson 1993), show significant differences from both *cryptica* and *bonthaina*, suggesting that full specific status is warranted for all three forms.
Description 11–11.5 cm. Monotypic. A small, large-headed flycatcher generally greyish-brown above, more olive on head and nape and more rufous on rump; chin and throat whitish; an indistinct olive-greyish breast-band; belly to vent white; tail has pale orange-rufous sides with broad dark brown tips to all feathers; iris dark brown; bill dark horn above, paler and with greyish base below; tarsi greyish-pink.
Habitat Lowland dense secondary growth forest, including disturbed forest; mostly below 700 m.
Food and feeding Forages singly or in pairs, usually below 5 m. No data on diet.
Breeding Nest and eggs undescribed. Recently fledged young seen in May.
Voice Song is a quiet, high-pitched two- or three-note whistle, *wan-he* or *wanhe he*; calls include a repeated sharp *zeet zeet* (Taylor 2006 HBW 12).
Movements Presumably sedentary.
Range Endemic to Luzon. Recent records have extended the known range considerably (Taylor 2006; Poulsen 1995); Western Luzon, in the Zambales Mountains west of the central plain, and disjunctly on the east coast, in the region of Aurora.
Conservation status Currently classified by BirdLife International as Near Threatened, although with further fieldwork this might be downgraded. Appears to be able to adapt to some habitat modification and in some locations may be quite common in secondary growth.
Etymology *Disposita*, Latin, referring to the furtive, secretive and skulking habits of the bird; Dutson (1993) suggested the English name to reflect the same characteristics.

▼ Furtive Flycatcher. Sierra Madre, Luzon, Philippines, March 2012 (*Robert Hutchinson*).

WHISKERED FLOWERPECKER
Dicaeum proprium

A distinctive flowerpecker whose relationships to the remainder of the genus *Dicaeum* are not entirely clear.

▲ Whiskered Flowerpecker. Davao City, Mindanao, Philippines, February 2014 (*Sylvia Ramos*).

Type information *Dicaeum proprium* Ripley & Rabor 1966, Mt Mayo, Limot, Mati, Davao Province, Mindanao, Philippines.
Discovery The type specimen, a male, was collected by D. S. Rabor in Davao Province, Mindanao, in July 1965. One other specimen was taken at the same time.
Description 9 cm. Monotypic. Unlike any other flowerpecker. Upperparts glossy and blue-black, the primaries dark brown; chin, throat and moustachial stripe greyish-white; prominent thick blue-black malar stripe; remainder of underparts from lower throat to lower belly pale sepia-brown; darker on the flanks, with white pectoral tufts and greyish-white crissum; bill blackish with pale base to lower mandible; tarsi dark brownish. Female similar but less glossy above and slightly darker below (Cheke & Mann 2001).
Habitat Forest, forest-edge and secondary growth, above 800–900m.
Food and feeding No data recorded; presumably mistletoe berries and other fruit.
Breeding Nest and eggs undescribed. Seen carrying nest material to the top of 20 m trees in late April/early May (Robson & Davidson 1995).
Voice Four or five high-pitched notes, first two higher than others. High insect-like trills, with lower *tsenk* notes possibly made by a second bird. High *swink* and *chenk* notes with buzzing quality. These are run into a song, with rising and falling notes (Cheke & Mann 2001).
Movements Presumably sedentary.
Range Endemic. Restricted to Mindanao, Philippines, where it is known from about six discrete locations, from the base of the Zamboanga Peninsula in the west to north of Mayo Bay in the east.
Conservation status Currently classified by BirdLife International as Near Threatened, downlisted from Vulnerable. Generally uncommon or locally common (Cheke & Mann 2001). Occurs in some nominally protected areas; habitat destruction is an issue.
Etymology *Proprium*, Latin, 'special' or 'characteristic'.

RUFOUS-WINGED SUNBIRD
Cinnyris rufipennis

A distinctive Tanzanian endemic sunbird. Placed in the genus *Cinnyris* by several authorities (e.g. Cheke & Mann 2001; Cheke & Mann 2008 HBW 13), a treatment followed here.

Type information *Nectarinia rufipennis* Jensen 1983, Ndzungwa Mountains, Tanzania
Discovery The Mwanihana Forest Reserve in the Udzungwa Mountains of Tanzania was little explored ornithologically until quite recently; it has proven to harbour several endemic races of birds and one new primate. In August 1981 F. P. Jensen conducted a ten-day visit to the reserve, in the course of which he mist-netted a male sunbird, obviously distinct in plumage from other members of the family. In September 1981 a further visit by several other ornithologists resulted in the collection of a female specimen, as well as field observations, allowing the description of a new species.

▲ Rufous-winged Sunbird, female. Nyumbanitu, Udzungwa Mts, Tanzania, December 2002 (*Louis A. Hansen*).

◄ Rufous-winged Sunbird, male. Udzungwa Scarp Proposed NR, Tanzania, November 2015 (*Markus Lagerqvist*).

Description 12 cm. Monotypic. Sexually dimorphic. A medium-small sunbird with a quite long, strongly decurved bill. Male: upperparts and face iridescent violet or turquoise-blue; chin, throat and upper chest glossy bronze; lower chest iridescent violet, below this a chestnut band; two small yellow pectoral tufts; upper belly grey, tinged olivaceous, becoming more citrine on lower belly; bright rufous patch on closed wing; iris dark brown; bill long (23.5 mm), decurved, black; tarsi black. Female: greyish-olive on crown, face and nape, becoming olive-green on back and rump; throat olive-yellow, more sulphur-yellow on chest and belly with blackish spotting; edgings of flight feathers with tawny appearance. Male non-breeding plumage has white patches on chest-band and head.

Habitat Moist epiphytic mountain forest, mostly 1,500–1,700m, occasionally down to 600 m.

Food and feeding Usually feeds at 2–8 m, occasionally up to 30 m. Nectarivorous, seen feeding in *Achyrospermum* sp., *Leucas, Tecoma* and mistletoes. Presumably also insects. Defends clumps of flowers against other sunbird species.

Breeding Only one (used and empty) nest so far described; a typical sunbird pensile pouch, 5 m up (Dineson *et al.* 2001; Jensen & Brøgger-Jensen 1992). Dependent juvenile seen in January (Cheke & Mann 2001).

Voice Song is a soft, high-pitched trilling; female adds a high-pitched chirping (Stuart *et al.* 1987). Calls include squeaking and fizzing noises, a loud *tyew* etc.

Movements Appears to wander locally according to availability of nectar sources, moving downwards in colder months, although always present at preferred locations.

Range Endemic. Eastern escarpment of the Udzungwa Mountains, Tanzania.

Conservation status Classified by BirdLife International as Vulnerable. Overall range is small and within this has been recorded at eight known locations. Although some parts of range have protected status, illegal logging and forest clearance still occurs (Dinesen *et al.* 2001). Population esimated in range 6,000–15,000 mature individuals (BLI SF 2015).

Etymology *Rufipennis*, Latin, 'rufous-winged'.

LINA'S SUNBIRD
Aethopyga linaraborae

Forms a superspecies with the allopatric Apo Sunbird *A. boltoni.*

Type information *Aethopyga linaraborae* Kennedy *et al.* 1997, near peak of Mt Pasian, Davao del Norte Province, Philippines.

Discovery The Philippine Biodiversity Inventory was initiated in 1989 to study forested areas in the Philippines, in response to growing concern about the rampant deforestation occurring in the archipelago. In 1993 surveys were conducted on three Mindanao mountains, resulting in the collection of several sunbird specimens on Mt Puting Bato. Comparison of these with specimens of Apo Sunbird *A. boltoni* from Mt Apo, about 120 km to the west, showed that they were strikingly different. Attempts to return to Mt Puting Bato in 1994 were frustrated by civil disorder; consequently the team went to Mt Pasian, 45 km to the north, where the new sunbird was observed and some specimens taken. In fact, several further specimens (labelled *A. b. boltoni*)

◄▼ Lina's Sunbird, male (top) and female (bottom). Compostela Valley, Mindanao, Philippines, November 2015 (*Tonji Ramos*).

were found in the collections of the Field Museum, Chicago, and the United States National Museum, Washington DC, obtained by the pre-eminent Philippine ornithologist Dioscoro Rabor in 1965 on Mt Mayo, a further 35 km to the south of Mt Puting Bato.

Description 11 cm. Monotypic. Sexually dimorphic. Male: forehead, cheeks, wing-coverts and uppertail-coverts metallic green; head and tail blackish-green; back olive-green; rump yellow; underparts bright yellow, more orange on breast; small scarlet pectoral tufts. Female: no metallic hues above, rump more olive; underparts olive-green with diffuse darker streaking; iris blood red; bill long (males, 22–26 mm, females 19–23 mm), fine and decurved, black; tarsi blackish-grey with pale yellow-ochre soles.

Habitat Wet mossy montane forest, about 975–1,980 m.

Food and feeding Forages singly or in pairs, sometimes in mixed-species flocks.

Breeding Nest and eggs undescribed. Specimens taken in May were in breeding condition.

Voice Song is a long series of high-pitched twittery notes; calls include a high-pitched *suweet* (Cheke & Mann 2008 HBW 13).

Movements No data.

Range Endemic. Confined to three mountains, Pasian, Puting Bato (or Batu) and Mayo, eastern Mindanao, Philippines.

Conservation status Currently classified by BirdLife International as Near Threatened. The overall range is very small; however, it is to a degree protected by the rugged and inaccessible nature of its habitat, which is uninviting for agriculture and contains few trees of commercial interest.

Etymology Named in honour of Lina N. Florendo Rabor (d. 1997), wife of Dioscoro S. Rabor (q.v.), who accompanied her husband on more than 40 scientific expeditions throughout the Philippines between 1936 and 1975; "a remarkable woman" who contributed to some 80 papers published by Dr Rabor (Kennedy *et al.* 1997).

KILOMBERO WEAVER
Ploceus burnieri

A very restricted *Ploceus* weaver of central Tanzania.

Type information *Ploceus burnieri* N. E. Baker & E. M. Baker 1990, Kilombero River, Tanzania.

Discovery In December 1986 a visit was made by N. E. Baker and E. M. Baker to the Morogoro region of east-central Tanzania, as part of the fieldwork for the Tanzanian Bird Atlas; however, the Bakers had previously been informed by Dr E. Burnier of the presence in the area of a *Ploceus* weaver species which he had not been able to identify. The unknown bird proved to be quite common, and using mist-nets five female and two male birds were captured; one specimen of each sex was taken, allowing the description of a new species.

Description 13 cm, female somewhat smaller. Monotypic. In overall appearance a typical *Ploceus* weaver, rather small, and sexually dimorphic. Male (breeding) largely yellow above and below, with a fairly restricted black mask edged with chestnut, the crown tinged orange; mantle and back olive-green, flight-feathers with yellow edgings and yellow wing-bar formed by covert tips; rump lemon-yellow; iris dark brown without reddish tinge; bill black; tarsi dark horn. Female: forehead, crown and nape grey-green, the crown darker; sides of face, ear-coverts, throat and upper breast pale buff; mantle grey-green, with darker feather centres and pale yellowish edgings; iris brown, bill greyish above, pale horn below; tarsi dark horn.

Habitat Seasonally flooded grasslands in river floodplains, below 300 m; occurs in extensive riverside swamps with tall fringing beds of *Phragmites* reeds, generally away from trees (Craig 2010 HBW 15). Seems to be the only *Ploceus* species in this habitat.

Food and feeding Forages in groups in reed stems, sometimes on the ground. Stomach contents included seeds, also dried fish and domestic refuse.

▲ Kilombero Weaver, male. Ifakara ferry causeway, Tanzania, February 2012 (*Paul Oliver*).

▲ Kilombero Weaver, females. Kilombero floodplain, Ifakara, Tanzania, November 2015 (*Markus Lagerqvist*).

Breeding Nest, built initially by male, and finished by female, is oval-shaped, made of grass with a side-entrance, attached to a *Phragmites* stem, usually 2–3 m up, lower over water. Nests in colonies of up to 20 nests, rarely 30, occasionally singly. Eggs 1–2, olive-brown to turquoise with light brown vermiculations. Probably polygynous. Breeding season December–February. Baker & Baker (1990) provide a sketch of a typical 'male' nest prior to being finished.

Voice Song is described as rambling, including chips and squeaks. Alarm call *tjack*.

Movements Presumably sedentary.

Range Endemic. Apparently restricted to the floodplain of the Kilombero River in central Tanzania, an area not exceeding 1,500 km^2.

Conservation status Currently classified by BirdLife International as Vulnerable, due mainly to the small known range, not all of which is occupied. However, in some locations it is common. Main threats are destruction of habitat for agriculture, including grazing with attendant burning to maintain pasture, rice farming and sugarcane plantations. Parts of the habitat seem to be unattractive to agricultural exploitation, however. No part of the range is protected.

Etymology Named in recognition of E. Burnier, a Swiss national, who first drew the species to the attention of the Bakers. Dr Burnier, a medical practitioner, attended to the people of the area for many years; he was, incidentally, one of the first persons to observe the Algerian Nuthatch (q.v.) (Burnier 1976).

LUFIRA MASKED WEAVER
Ploceus ruweti

Formerly regarded as a superspecies with several other *Ploceus* species of central and southern Africa, and treated by some authorities as a race of Tanganyika Masked Weaver *P. reichardi*. Molecular studies suggest that the closest relationship is with Golden-backed Weaver *P. jacksoni*, while supporting full specific status for *P. ruweti* (Craig *et al.* 2011).

Type information *Ploceus ruweti* Louette & Benson 1982, Lake Lufira, Democratic Republic of Congo.

Alternative names Lake Lufira Weaver, Ruwet's Masked Weaver. French: Tisserin de Ruwet.

Discovery In about March 1960 J.-C. Ruwet was investigating the avifauna of the environs of the Lufira River in Katanga province, SE Democratic Republic of Congo

◄ Lufira Masked Weaver, male. Lake Tshangalele, Democratic Republic of Congo, January 2011 (*Michel Hasson*).

▲ Lufira Masked Weaver. Left: female, near Kiubo Falls, Lufira River, Democratic Republic of Congo, February 2010. Right: juvenile, Lake Tshangalele, Democratic Republic of Congo, March 2009. (*Michel Hasson*).

(DRC), a large area of marshland bush originally created by a dam built in 1926, and since largely silted up. He encountered a weaver, common in the area, which he identified as Black-headed Weaver *P. melanocephalus*. One specimen, a male, was taken. Over the years it was examined by several ornithologists without consensus; finally, Louette & Benson (1982) described it as a new species. It was not, however, observed again until M. Hasson, acting on the suggestion of Dr Louette, visited Lake Lufira in early 2009 and found several nesting colonies (Ruwet himself had found several colonies, of up to 20 nests, without giving much further information). In their 2009 paper, Louette & Hasson give details of nests, eggs and juvenile plumage.

Description 13 cm. Monotypic. Sexually dimorphic. Male (breeding): black face-mask extending over most of crown, face, ear-coverts and upper throat; rear crown and nape bright yellow; back dark olive-yellow with obscure darker streaking; rump pale yellow; tail and upperwing olive-green, with pale yellow margins to remiges and coverts; breast and flanks chestnut-brown, the remainder of underparts yellow with a rufous tinge on crissum; iris reddish; bill black; tarsi grey-brown. Non-breeding plumage imperfectly known but throat becomes yellow and top of head greenish. Female: lacks black and most of the rufous of the male, generally more greenish; bill brown above, horn-brown below; eyes dark.

Habitat Lowlands near rivers and swamps, but not, apparently (*contra* Louette & Benson 1982, and Cotterill 2004) exclusively in swamp vegetation; nesting colonies also found in acacia trees over dry ground (Craig *et al.* 2011).

Food and feeding Food items comprise insects and seeds. Young fed on insects.

Breeding Nests in loose colonies of 3–20 nests in bushes above water (Louette & Benson 1982) and over dry ground (Craig *et al.* 2011). Nest construction starts in February. Nest woven by male, a somewhat untidy hanging oval ball of woven fibres with a spoutless bottom entrance. Sometimes sited next to wasps' nests. Eggs two, pale green with brown spots, especially at blunt end. May be double-brooded. Probably polygynous.

Voice Vocal. Song of male is a series of extended wheezing notes, some ending in short *tat-tat-tat* sounds; also short notes and a territorial trill (Craig 2010 HBW 15).

Movements Breeding areas apparently deserted at end of wet season; local knowledge states that birds are seen away from the lake in the dry season (Louette & Hasson 2009).

Range Endemic. SE DRC in the region of Lake Lufira (Tshangalele or Shangalele), Katanga. Recently found 120 km downstream near Kiubo Falls. Intervening habitat not explored, but might hold other populations (Craig *et al.* 2011).

Conservation status Currently listed by BirdLife International as Data Deficient. However, it appears to be not uncommon in at least two locations, with substantial potential areas of occurrence not yet investigated.

Etymology Named in honour of its discoverer, the ethologist and animal behaviourist Professor Jean-Claude Ruwet (d. 2007), of the University of Liège, Belgium.

GOLA MALIMBE
Malimbus ballmanni

A striking and distinctive malimbe which may form a superspecies with the allopatric Rachel's Malimbe *M. racheliae* of SE Nigeria to Gabon, the only other member of the genus to have yellow in its plumage.

Type information *Malimbus ballmanni* Wolters 1974, between Cavally and Keibli Rivers, NW of Tai, SW Ivory Coast.

Alternative names Ballmann's Malimbe, Tai Malimbe. French: Malimbe de Ballmann.

Discovery In December 1971 and again in 1972 G. D. Field observed an unfamiliar malimbe in the Gola Forest in eastern Sierra Leone on numerous occasions. From field notes taken at the time a detailed description was prepared, on which basis a tentative type description, under the name *M. golensis* was published, appropriately in the first issue of *Malimbus*, the Journal of the West African Ornithological Society "though, I suppose, no official name can be bestowed on the new bird without a type specimen" (Field 1979). In the meantime, in 1974, H. E. Wolters published a brief description of a new malimbe which had been collected by P. Ballmann, a soil scientist, in SW Ivory Coast at some date in 1972; Dr Ballmann passed the alcohol specimen to the Alexander Koenig Museum, Bonn, Germany. Dr Wolter's description, and consequently the specific epithet *ballmanni*, has priority (Prigogine 1981).

▼ Gola Malimbe. Gola, Sierra Leone, February 2008 (*Nik Borrow*).

Description 15–17 cm. Monotypic. Sexually dimorphic. Male: plumage largely black, with a golden-yellow nape, washed with tawny-orange at margins; large cadmium-yellow breast-patch; crissum bright yellow; iris dark red; bill black; tarsi grey. Juvenile male has a yellow throat and orange-brown on crown. Female similar to male but lacking nape-patch. Borrow & Demey (2001) show a complete golden-yellow collar on the male.

Habitat Wet lowland evergreen forest, including very old secondary forest and moderately or heavily-logged forest (Gatter & Gardner 1993). Mostly below 400 m, but one sighting at 1,000 m in Guinea.

Food and feeding Forages singly or in pairs, sometimes in mixed-species flocks, from 5 m to 30 m up, but mostly around 10–20 m (Gatter & Gardner 1993). Food items comprise insects, especially grasshoppers and mantids.

Breeding Main period of nest building is from September to October, in second half of rainy season; may also nest in the intermediate dry season in July–August. Nest is a free-hanging "inverted sock", with a vertical tubular spout entrance, suspended from the ends of vertically-hanging vines, 8–21 m up, mostly at 10–20 m; total length of nest including attachment about 115 cm. Eggs undescribed (Gatter & Gardner 1993).

Voice Both sexes sing. Male song is *cheg chig cheg-chega-zzzzzzzz*; female song lacks the terminal wheeze (Gatter & Gardner 1993).

Movements Presumably sedentary.

Range West Africa; western Liberia and adjacent eastern Sierra Leone; southern Guinea; eastern Liberia and adjacent Ivory Coast.

Conservation status Currently classified by BirdLife International as Endangered. The main threat is rampant deforestation, driven by an ever-increasing human population, logging (much of it uncontrolled and illegal) and clearing for smallholdings. The population is believed to be declining seriously, and the bird now seems to be rare in most locations; conservation efforts have been severely hampered by civil wars. Rediscovered after a 30-year hiatus in Gola Forest in 2007 (Dowsett-Lemaire & Dowsett 2008). One recent population estimate for all countries combined is in the region of 10,000–20,000 individuals (Niemann 2015). The most recent estimate (BLI SF, April 2017) is 6,000–15,000 individuals.

Etymology The scientific and vernacular names recognise Dr Peter Ballmann (b. 1941), a German geoscientist who has also described several fossil birds including barbets and owls.

ROCK FIREFINCH
Lagonosticta sanguinodorsalis

A very restricted firefinch which is the usual host of the, also newly described and also highly restricted, Jos Plateau Indigobird *Vidua maryae*.

Type information *Lagonosticta sanguinodorsalis* Payne 1998, Taboru, Jos Plateau, Nigeria.
Alternative names Rock Finch.
Discovery The brood-parasitic indigobirds of the genus *Vidua*, which are widespread in sub-Saharan Africa, have developed a strategy that involves mimicking the songs of their estrildid (waxbill family) hosts. During fieldwork in the Jos Plateau of central Nigeria, R. B. Payne noticed that indigobirds which were at that time called Quailfinch Indigobirds *V. nigeriae* mimicked a firefinch then called African Firefinch *L. rubricata* (Payne 1968a, b). However, the Jos indigobirds were larger and brighter than authentic *V. nigeriae*, and were initially described as a subspecies (*maryae*) of the Dusky Indigobird *V. funerea*, the species that mimics and parasitises *rubricata* over much of eastern and southern Africa. The songs of the Jos firefinches were clearly different from those of *Lagonosticta* species, having a unique descending whistled *treeee*, while the combination of blue-grey bill in the adult, red back and grey head in the male and bright reddish-brown back in the female is distinctive and allows the new species to be distinguished from other firefinches (Payne 1998).

Description 10–11 cm. Monotypic. Sexually dimorphic. Male: face and underparts bright pinkish-red, the central belly to vent black; crown and nape grey; upperparts brownish, rump and base of tail deep carmine-red, the remainder of the tail blackish; small white spots on upper flanks and sides of breast. Female: generally less brightly coloured and paler.
Food and feeding Forages on the ground; food items include small grass seeds.
Breeding Breeds August–November. Nest is a spherical structure made of coarse grasses on the outside and finer ones inside, hidden in tufts of grass between boulders from ground level to 1 m up, occasionally up to 5 m. Clutch 2–5 eggs. The predation rate of nests is very high (Brandt & Cresswell 2008), with only 29% of nests producing chicks, just about enough to compensate for adult mortality; birds may attempt a second brood following failure of the first.
Voice Payne (1998) gives an exhaustively comprehensive study of vocalisations. Most characteristic call is a rapid dissenting trill *treeee*, given by both sexes. Numerous other calls.
Movements Some local movements, as during the dry season birds will move from the rocky inselberg breeding habitat to nearby gallery forest and savanna where water is more readily available.
Range Central Nigeria in the Jos Plateau, and in the Mandara Mountains in eastern Nigeria/Cameroon.

► Rock Firefinch, pair. Mora, northern Cameroon, April 2012 (*Werner Suter*).

Sightings at other locations in northern Nigeria, ascribed to other firefinches, may refer to this species.

Conservation status Currently classified by BirdLife International as Least Concern. Despite its small overall range, the species is locally common. Human pressure on habitat for firewood, fence posts and creation of pasture has resulted in the destruction of some areas of previous occurrence. However, some parts of the range are protected as formal reserves or by local custom (one area is a sacred grove, formerly used as a site for ritual circumcisions – Wright & Jones 2005).

Etymology *Sanguinodorsalis*, Latin, having a blood-red back.

RED-EARED PARROTFINCH
Erythrura coloria

Sometimes placed in the genus *Trichroa*. Probably forms a superspecies with Tricoloured and Blue-faced Parrotfinches *E. tricolor* and *E. trichroa*.

Type information *Erythrura coloria* Ripley & Rabor 1961, Mt Katanglad (Kitanglad), Malaybalay, Bukidnon, Mindanao, Philippines.

Alternative names Mindanao/Red-collared/Mount Kitanglad Parrotfinch.

Discovery In 1960 the Peabody Museum, Yale University, Connecticut, and the Siliman University, Dumaguete, Philippines, jointly sponsored an expedition to investigate the avifauna of Mt Katanglad (also spelled Kitanglad), an isolated massif in central Mindanao. In late March R. B. Gonzales, a field assistant of Professor Dioscoro Rabor, obtained an adult male specimen of a parrotfinch *Erythrura* which was clearly quite distinct from any other members of the genus known from the Philippines. Subsequently four other specimens, three male and one female, were taken, allowing the description of a new species.

Description 10–10.5 cm. Monotypic. Sexually dimorphic. Male: overall green above and below, the face cobalt-blue; large red crescent on sides of neck and rearmost ear-coverts; uppertail-coverts red, red sides to brown rectrices, giving a red appearance on spread tail; iris brown; bill black; tarsi pinkish-brown. Female with less blue and less red on head. However, Ziswiler *et al.* (1972) and Goodwin (1982) state that in adult plumage, the sexes are alike.

Habitat Montane forest and forest-edge, secondary growth and grassy clearings, 1,000–2,250 m. Not strongly dependent on bamboo.

Food and feeding Forages singly or in small groups, low down or on the ground. Food items a variety of seeds, also small insects.

Breeding Few data on wild population. Birds in breeding condition in January–April. Frequently bred in captivity (e.g. Kühn 1994), where it readily takes to half-open nest-boxes, building an unlined open nest out of fibres and grasses. Clutch 2–3, incubation period 12–13 days, fledging period 21–23 days.

◄ Red-eared Parrotfinch. Mt Kitanglad, Mindanao, Philippines, February 2010 (*Adrian Constantino/www.birdingphilippines.com*).

Voice Calls include a ticking trill, a flight alarm call *prrrt* and a sharp repeated *tik* (Kennedy *et al.* 2000).
Movements Presumably sedentary.
Range Endemic. Mindanao, Philippines in seven or more disjunct highland areas including Mt Hilong-hilong, Mt Pasian, Mt Pulong Bato, Mt Kitanglad, Mt Apo, Mt Busa, and several other locations.
Conservation status Currently classified by BirdLife International as Near Threatened. However, its range is now known to be more extensive than previously believed (it is rather quiet and unobtrusive) and in some locations (e.g. Mt Kitanglad) it seems to be quite common, leading Peterson *et al.* (2008) to suggest that a downgrading to Least Concern might be appropriate.
Etymology *Coloria*, Latin, referring to its striking plumage.

JAMBANDU INDIGOBIRD
Vidua raricola

One of a large number of indigobirds, essentially identical in plumage, which in the field are only safely distinguished vocally.

Type information *Vidua raricola* Payne 1982, Banyo, Cameroon.
Alternative names French: Combassou jambandou.
Discovery In a monumental study of the indigobirds (*Vidua*) of western Africa, R. B. Payne described several new taxa, including Jambandu Indigobird *V. raricola*, Barka Indigobird *V. larvaticola* and Jos Plateau Indigobird, originally described as a race (*maryae*) of Dusky Indigobird *V. funerea*. These species are extremely similar in plumage, to the extent that some field guides (e.g. Borrow & Demey 2001) do not bother to illustrate them separately; the specific characteristics which give rise to the descriptions rest mainly on vocalisations, which include mimicry of the songs of the host species, and the pattern of the insides of the mouths of nestlings, which have also evolved to mimic the patterns of young host birds. In the case of the Jambandu Indigobird *V. raricola* the situation is somewhat more complicated; the species was given the specific name *raricola* because it was believed that its host was Black-bellied Firefinch *Lagonosticta rara*. However, the type series of specimens, it transpired, included material of two species. The holotype of *V. raricola* in fact parasitises Zebra Waxbill *Amandava subflava*, while the birds whose song resembles that of Black-bellied Firefinch were actually Cameroon Indigobird *V. camerunensis* (Payne 2010 HBW 15).
Description 11.5 cm. Monotypic. Sexually dimorphic. Breeding male largely uniform black with a green gloss, a concealed white flank-spot, the wing feathers mostly slightly browner; iris dark brown; bill white; legs light purplish. Non-breeding male more resembles female, which is generally brown with an obvious crown-stripe and supercilium, upperparts brown with blackish shaft-streaks and buffy wing-bars; chin whitish, underparts whitish-grey, flanks buffy-grey; bill grey or greyish-brown above, whitish below. Female and non-breeding male indistinguishable from other indigobirds in range.
Habitat Floodplains and edges of cultivation, fields, marshes and grassy areas; type specimen taken at 1,050 m.
Food and feeding Food items mainly small grass seeds; forms mixed flocks with other species after breeding.

▼ Jambandu Indigobird, male. Bumbuna, Sierra Leone, December 2009 (*John Caddick*).

Breeding Obligate nest-parasite of Zebra Waxbill *Amandava subflava*; eggs white, unmarked, in clutch of three, laid one per day. Breeding season probably September–January.
Voice Song is churring, grating and scratchy, some parts in imitation of the song of the host species, other parts not (Payne & Payne 1994).
Movements No data; presumably sedentary.
Range Disjunct in a number of locations in Sierra Leone, Ghana, Nigeria, Cameroon, west and south Sudan and western Ethiopia.
Conservation status Currently classified by BirdLife International as Least Concern. The species has a wide geographic range (and, due to identification difficulties, may well have been undetected in other locations).
Etymology The specific name *raricola* refers to the then (incorrectly) identified host-species, the Black-bellied Firefinch *Lagonosticta rara* (see above). The vernacular name derives from a local greeting in the Fulani language of Banyo, translated as 'How's your body?'

BARKA INDIGOBIRD
Vidua larvaticola

Sometimes treated as conspecific with Wilson's, Quailfinch and Cameroon Indigobirds *V. wilsoni*, *V. nigeriae* and *V. camerunensis*.

Type information *Vidua larvaticola* Payne 1982, Zaria, Nigeria.
Alternative names Barka Indigobird. French: Combassou barka.
Discovery Described in the same paper as Jambandu and Jos Plateau Indigobirds *V. raricola* and *V. maryae* (q.q.v.).

▼ Barka Indigobird, male. Poli, Cameroon, October 2007 (*Nigel Voaden*).

Description 10–11 cm. Monotypic. Sexually dimorphic. Breeding male mostly black with a blue or green-blue gloss, with a concealed white flank-spot; most of primaries, secondaries and tail brown; iris dark brown; bill white; tarsi pinkish-white to light purplish or whitish-mauve. Non-breeding male similar to female; brown above, with prominent crown-stripe and whitish supercilium, back brown with darker streakings and some pale edgings; chin whitish; chest and flanks more buffy; belly whitish.
Habitat Open secondary growth, abandoned farmland, savanna with scattered bushes, riverine thickets.
Food and feeding Forages alone, in small groups and in company of other seed-eating species. Food items: seeds.
Breeding Obligate brood-parasite, in northern Nigeria on Black-faced Firefinch *Lagonosticta larvata*, possibly also on Mali Firefinch *V. virata* (Mills 2010). Eggs unmarked white.
Voice Song similar to those of other indigobirds, a series of churring scratchy notes; mimics song of host species (in Nigeria, Black-faced Firefinch *Lagonosticta larvata*; in Mali, Mali Firefinch *V. virata* – Payne & Barlow 2004). Payne (1982) gives detailed analyses of indigobird vocalisations and mimicry.
Movements Presumably sedentary.
Range Not entirely known, due to similarity with other members of the genus. Disjunctly across a wide swathe of sub-Saharan Africa; NW & W Ethiopia, W & E Sudan, Central African Republic, southern Chad, central Cameroon, Nigeria, NE Ivory Coast, southern Mali, NW Ghana, Guinea-Bissau.
Conservation status Currently classified by BirdLife International as Least Concern. Has a wide distribution which may be extended with further study.
Etymology *Larvaticola*, derived from the specific epithet of the best-known host-species, Black-faced Firefinch.

JOS PLATEAU INDIGOBIRD
Vidua maryae

Originally described as a race of Dusky Indigobird *V. funerea*, but more recent work (Payne 1998) suggests that a full species is involved.

Type information *Vidua funerea maryae* Payne 1982, Panshanu, 50 km east of Jos, Nigeria.
Alternative names French: Combassou du Plateau de Jos.
Discovery Described (as a subspecies) by Payne (1982), in the same paper as the preceding two species.
Description 10–11 cm. Monotypic. Sexually dimorphic. Male largely black with a blue-green gloss; tail, outermost secondaries and primaries brownish; concealed flank-spot white; iris dark; bill white; tarsi light purplish. Female indistinguishable from females of other indigobirds, brown above, pale supercilium, chest buff-brown, belly whitish. Non-breeding male resembles female.
Habitat Semi-arid, relatively undisturbed or lightly grazed savannas, grasslands with scattered bushes, around bases of rocky outcrops; 800–1,200 m.
Food and feeding Forages on the ground; diet is small grass seeds.
Breeding Obligate brood-parasite. Probable host is Rock Firefinch *Lagonosticta sanguinodorsalis*. Eggs undescribed; breeding season probably September/October to December.
Voice Song much like other *Vidua* species, churring scratchy notes, mixed with imitations of song of the host, Rock Firefinch.

▲ Jos Plateau Indigobird, male. Amurum Forest, near Jos, Nigeria, November 2002 (*Tasso Leventis*).

Movements Presumably sedentary.
Range Central and northern Nigeria (Panshanu Pass, Taboru, Kagoro) (Borrow & Demey 2001); disjunctly, in northern Cameroon (Mills 2010).
Conservation status Currently classified by BirdLife International as Least Concern. Although the known range is fairly small, it does not appear to be in significant decline; some parts of range (e.g. Magama Forest Reserve) are protected.
Etymology Named in recognition of the contributions of the Canadian ornithologist Mary Gartshore (b. 1950) to the knowledge of the species.

MEKONG WAGTAIL
Motacilla samveasnae

Probably forms a superspecies with White, White-browed, Japanese and African Pied Wagtails (*M. alba*, *M. maderaspatensis*, *M. grandis* and *M. aguimp*). Despite the very close resemblance to the last-named species, genetic studies (Alström & Mild 2003) suggest that the Mekong Wagtail is more closely related to the Asian species than to *aguimp*.

Type information *Motacilla samveasnae* Duckworth *et al.* 2001, San River, Stung Teng province, Cambodia.

► Mekong Wagtail. Kratie, Cambodia, February 2011 (*James Eaton*).

Discovery In 1972 Kitti Thonglongya collected two black-and-white wagtails in Ubon Ratchani province, Thailand, which were then ascribed to the *alboides* race of White Wagtail *M. alba*. Numerous observations of 'white wagtails' were made by many observers in subsequent years in a variety of locations in Indochina and Thailand, but little interest was generated since they apparently belonged to a known taxon. However, in 2000, a new guide to the birds of the area (Robson 2000) illustrated very accurately examples of *alboides*, making it obvious that birds seen over a wide area of Indochina could not be of this taxon. Examination of specimens in several institutions confirmed this fact. Consequently, in February 2001 several ornithologists visited the Mekong, San and Kong rivers in north-eastern Cambodia, where they observed over 100 individual birds of the new form. A total of eight specimens was taken, in various plumages, allowing the description of a new species (Duckworth *et al.* 2001).

Description 17–17.5 cm. Monotypic. In form a typical *Motacilla* wagtail, slim and long-bodied, with a long tail. Above largely black in adult male, paler and greyer in female, with a conspicuous white supercilium, black ear-coverts, and extensively white median and greater coverts and on the bases of the inner primaries and secondaries; broad black chest-band; remainder of underparts white; tail black with conspicuous white markings; iris dark brown; bill black; legs greyish-black.

Habitat Apparently rather specific in habitat requirements; found almost exclusively on small rocky and often bushy islets in broad, braided, swift-flowing rivers with numerous channels and gravel shoals, below 110 m, and only rarely on the sides of a river (Duckworth *et al.* 2001; Davidson *et al.* 2001).

Food and feeding Feeds mainly on the ground, often along the water's edge but, unlike most wagtails, regularly feeds from bushes in water, walking along branches and picking items from leaves. Little data on diet (Alström & Mild 2003).

Breeding One nest only so far described. It was in a natural rock cavity on a rocky islet in the river, constructed of rootlets. On 11 March it contained three nestlings aged about 5–6 days; fed by both parents. Independent juveniles seen on the same date, suggesting possible multi-brooding (Handschuh & Packman 2010).

Voice Song is a brief, rapid series of thin, high-pitched, often rather harsh notes; flight call is a short, sharp, harsh *dzeet* (Duckworth *et al.* 2001). Alström & Mild (2003) give a very detailed comparative analysis of vocalisations.

Movements Probably sedentary, although local movements may occur according to varying river conditions.

Range Indochina. NE Thailand, NE Cambodia, south Laos, extreme western Vietnam (Le Con Trai & Craik 2008).

Conservation status Currently classified by BirdLife International as Near Threatened. The population in some areas is stable and healthy; however, the specialised habitat could be very vulnerable to change, especially dam construction and dredging to improve navigation, both of which have been proposed.

Etymology The species is named in honour of Sam Veasna (1966–1999), one of Cambodia's leading ornithologists who died at a tragically early age from malaria contracted during fieldwork in northern Cambodia (Duckworth *et al.* 2001).

LEMON-BREASTED CANARY
Serinus citrinipectus

May belong to a genus separate from most of the other African canaries (*Ochrospiza*) (Arnaiz-Villena *et al.* 2008).

Type information *Serinus citrinipectus* Clancy & Lawson 1960, near Panda, Inkambane, Sul do Save, southern Mozambique.

Alternative names Lemon-breasted Seedeater.

Discovery In October 1959 C. H. Scheepers trapped about 40 individuals of a canary in Sul do Save, Mozambique. These were kept in aviaries until 1960, when they were examined by staff of the Durban Museum on an ornithological expedition. Based on information provided by Mr Scheepers, museum staff collected nine specimens from the original location, with a tenth captive specimen provided by Mr Scheepers. Based on this series the new species was described in 1960.

Description 12 cm. Monotypic. In form a typical canary, fairly small, with a conical bill. Sexually dimorphic. Male: grey-brown above, the head finely, and the back heavily, streaked; upper ear-coverts greyish, whitish patch on lower ear-coverts; two narrow whitish wing-bars; rump bright yellow; chin to breast deep lemon-yellow; sides of breast and flanks buffish, remainder of underparts white or pale peach-buff; iris dark brown; bill greyish-horn above, paler below; legs pinkish-brown. Female has less prominent facial features and yellow on underparts replaced by pale pinkish or peach-buff.

▲ Lemon-breasted Canary, male (left) and female (right). Inhassoro, Inhambane province, Mozambique, January 2011 (*Niall Perrins*).

Habitat Lowland palm savannas; especially associated with the Ilala Palm *Hyphaene natalensis* (Chittenden 1998). Various types of open woodland, usually below 750 m.

Food and feeding Outside breeding season occurs in flocks of up to 250 individuals. Food items, seeds and other vegetable matter, also insects including termites.

Breeding Most early information came from captive breeding (Clemlow 1993; Jolliffe 2000), and it is possible that the species was bred in aviaries even before its recognition as a new species (Lawson 1970). In the wild, nests in December–May, mostly January–February. Nest is an open cup of fibrous vegetable material including petals and bark strips, bound with cobwebs, located 1.5–6 m up, usually in Ilala Palm. Eggs 3–4, plain white or with sparse red-brown speckles. Incubation by female alone, 12–14 days; young fledge in 14–16 days (Robson 1990; Chittenden 1998).

Voice Song is a brief, rapid and fairly tuneless series of short, slightly rising and falling twittering notes (Clement 2010 HBW 15).

Movements Largely sedentary, with short nomadic movements according to food availability.

Range Several apparently disjunct populations in southern Malawi, SE Zimbabwe, southern Mozambique, Zambia and South Africa (Zululand to northern Natal).

Conservation status Currently classified by BirdLife International as Least Concern. Although patchy in occurrence, the overall range is quite extensive. Threats include apparently substantial trapping for cagebirds (Parker 1999) and the use of the Ilala Palm for furniture. In some cases, man-made habitat change may be helpful. Occurs in the Vilanculos Coastal Wildlife Sanctuary, Mozambique (Cizek 2008; Read *et al.* 2014).

Etymology *Citrinipectus*, Latin, 'lemon-breasted'.

ANKOBER SERIN
Crithagra ankoberensis

Placed in the genus *Carduelis* by some authorities; recent genetic work shows that the genus *Serinus* is polyphyletic (i.e. has different evolutionary origins), requiring the resurrection of the genus *Crithagra* (Zuccon *et al.* 2012; Gill & Donsker 2015). Sometimes treated as a race of Yemen Serin *S. menachensis*, although the bill shape differs markedly from that species.

► Ankober Serin. Ankober Escarpment, Ethiopia, February 2010 (*Ken Behrens*).

Type information *Serinus ankoberensis* Ash 1979, Ankober, Shoa province, Ethiopia.
Alternative names Ankober Seedeater.
Discovery In November 1976 J. S. Ash encountered a flock of 14 small serin-type birds on a highland massif north of Ankober, Ethiopia. These were unknown to him; he took detailed notes, and after consultation with the limited literature then available, and with other ornithologists, began to consider that they might be an unknown species. Consequently, in January 1977 two specimens were taken, and in February a further example was captured, examined and released, allowing the description of a new species.
Description 12–13 cm. Monotypic. A medium-small, rather drab and heavily streaked finch. Head, face and underparts streaked with blackish-brown, with a small, paler suborbital crescent and paler cheeks; back brown, paler on rump; upperwing dark brown; tail dark brown, outer rectrices edged pale grey; iris black; bill greyish-horn, paler on lower mandible; tarsi flesh-coloured or pinkish-brown. Female similar but cheeks less pale.
Habitat Windswept rocky grassland, cliff tops and cliff faces, with many boulders and very low vegetation, sometimes with stunted tree-heath (*Erica arborea*), 2,620–4,250 m.
Food and feeding Forages on ground, frequently in flocks; food items seeds of grasses and herbs (*Rumex* and *Sida*).
Breeding One nest only so far described (Ash 1979). This was under construction in late February; bird incubating on 5 March. Nest was a deep cup of fine roots, animal hair and wool, located in a hole in an overhanging earth bank. Eggs three, white.
Voice Song, often given by several birds in concert, is a series of musical chirps and chirrups; various other calls including a chattering *chee-chachachacha*, a double *treet-treet* etc (Clement 2010 HBW 15).
Movements May be slightly nomadic, since birds may be absent from previously consistent locations.
Range Endemic. Highlands of central and northern Ethiopia. Known from four locations in northern Shoa province, around Ankober; south of Debra Sine; Simien Mountains and Abuna Yosef Mountains (Shimelis 1999; Atkins 1992; Ash & Atkins 2009). Recently found at other locations in western highlands and may be more widespread than previously known.
Conservation status Currently classified by BirdLife International as Vulnerable, although with further work a downgrading to Near Threatened might be appropriate. Total range is rather restricted; does appear to have tolerance for some habitat modification. Main threats are intensive grazing and planting of non-native *Eucalyptus*.
Etymology Both scientific and vernacular names commemorate the location of its discovery.

SILLEM'S MOUNTAIN FINCH
Carpodacus sillemi

One of the least known and most enigmatic species of the Palaearctic; collected (and misidentified) in 1929, recognised as a new species only in 1992, then, after an absence of more than 80 years, finally rediscovered in 2012.

◀ Sillem's Mountain Finch, male (left) and presumed female (right). Kunlun Mountains, Qinghai, China, June 2014 (*Mark Beaman*).

Type information *Leucosticte sillemi* Roselaar 1992, 5,125 m, Kushku Maidan, west Tibetan Plateau.
Discovery In 1929, the Netherlands Karakoram Expedition was exploring some ornithologically unknown high-altitude areas in western Tibet. On 7 September, J. A. Sillem, working under very demanding physical conditions, collected three birds, all of which were identified as Brandt's Mountain Finch *Leucosticte brandti* (although Sillem himself, in his diary, referred to the third specimen as "a single of another species", the first two being indeed *brandti*) (Roselaar 1994). The following day Sillem collected a bird in juvenile plumage. These two specimens, both now in the Zoological Museum of Amsterdam, are to this day the only ones collected.

In 1991, while looking through the museum's collection of snowfinches *Montifringilla*, C. S. Roselaar came upon these specimens, labelled *L. brandti*, which he recognised as having numerous significant plumage differences from that species. Based on a careful analysis of plumage characters involving examination of more than 400 specimens from a variety of museums it was possible to describe a new species unequivocally.

In early June 2012 Y. Muzika photographed many finches while on an arduous trek in the Yeniugou Valley in western Qinghai, China (Muzika 2014). One individual was photographed but not identified. On returning home he submitted the photographs to K. Kazmierczak, who suggested that the mystery bird was the long-lost Sillem's Mountain Finch. Examination of other photographs of female-type birds revealed several of a bird which, structurally (notably the very long wings in relation to the tail), resembled the male, but was quite different from females of other species. This, it seemed highly probable, was the undescribed female plumage of *sillemi*. The fact that the female plumage differs from that of the male (see Kazmierczak & Muzika 2012 and Anon. 2012c) has taxonomic implications: in *Leucosticte*, the sexes do not differ greatly from each other. A recent study of mitochondrial DNA from the type specimen (Sangster *et al.* 2016) shows that Sillem's Mountain Finch belongs to the rosefinch genus, *Carpodacus*, although unique in having no trace of reddish in the male plumage.

Description 15 cm. Monotypic. Sexually dimorphic. Male: head and neck bright tawny-cinnamon; upperparts drab grey; rump and uppertail-coverts sandy-white; tail dark drab grey; chin, breast and sides of breast pale cinnamon-buff, remainder of underparts white with pale yellowish-buff wash. Soft parts undescribed. Female lacks the bright head colours and is diffusely streaked on the chest.

Habitat High altitude plateau, above 5,000 m.

Food and feeding Feeds on the ground in company of other species; no data on diet.

Breeding Nest and eggs undescribed.

Movements No data.

Range So far described from two locations; the type locality in the western Tibetan Plateau, and in the Yeniugou Valley, Qinghai, China. These locations are about 1,500 km apart.

Conservation status Currently classified by BirdLife International as Data Deficient. Much of the intervening habitat between the two known locations remains to be explored ornithologically. It would appear that the species is uncommon, although until the recent rediscovery it may have been overlooked simply because birdwatchers and ornithologists did not know what to look for.

Etymology Named in recognition of its discoverer, the Dutch explorer Jerome Alexander Sillem (1902–1986), who served on the Netherlands Karakoram expedition of May 1929 to August 1930 as ornithologist. Sadly, Sillem did not live to see the naming of 'his' mountain finch.

PO'OULI
Melamprosops phaeosoma

A unique and very aberrant honeycreeper, with no close extant relatives, requiring the erection of a new genus, and having a unique diet based on molluscs. The relationship to other members of the Drepaniidae is very difficult to determine, due in part to the many extinctions that have occurred in the family since human colonisation of the Hawaiian Archipelago. It was probably most closely related to two extinct members of the genus *Xestospiza*, both known only from subfossil remains (James & Olson 1991). Shortly after discovery the species suffered a catastrophic decline and is now extinct.

Type information *Melamprosops phaeosoma* Casey & Jacobi 1974, Koolau Forest, NE slopes of Haleakala Volcano, Maui, Hawaii.

► Po'ouli, male. Hana Rainforest, eastern Maui, Hawaii, January 2002. At that time there were only two known Po'ouli remaining and both were males (*Jack Jeffrey Photography*).

Alternative names Black-faced Honeycreeper.

Discovery As part of a survey of the Hana Rainforest, an almost inaccessible area on the NE slope of Haleakala Volcano on the island of Maui, Hawaii, in the summer of 1973, T. L. C. Casey, J. D. Jacobi and colleagues were able to observe a honeycreeper, in appearance totally unlike any known member of that family. The bird had previously escaped detection because of the extreme difficulty of working in its tiny range – precipitous slopes, frequent torrential rain and fog – and access was achieved only by the use of a helicopter. At first, the possibility of some introduced alien species seemed likely, but the very remote and pristine nature of the habitat argued against this. It became clear that material in-hand was essential, so, with considerable reluctance, the authors took two specimens (Powell 2008); this allowed the description of the new species, while a study of the anatomy, including the detailed tongue structure, emphasised its unique taxonomic position (Pratt 1992; Bock 1978).

Description 14 cm. Monotypic. A plump, dumpy, very short-tailed bird. Upperparts brown, underparts pale brown to off-white with deep cinnamon thighs and crissum; conspicuous black mask, including forehead, lores and throat, separated from the crown by a narrow grey band and from the back by a striking white cheek-patch; iris brown; bill, which is thick, elongated and slightly downcurved, black; tarsi black. Female basically similar, but more drab and grey below; juvenile has a smaller mask than adult.

Habitat In recent times, confined to wet, epiphyte-covered precipitous slopes, from about 1,440–2,070 m, with the dominant tree species Ohi'a-lehua *Metrosideros collina*; prior to human settlement apparently in drier forest (see below under Conservation status).

Food and feeding Foraged in understorey and sub-canopy; food mostly invertebrates, including insects and spiders and especially snails (Baldwin & Casey 1983). Small snails were crushed and eaten whole; the fleshy parts were extracted from larger snails. The alien garlic snail *Oxychilus alliarus*, though abundant, was not taken. Some fruit (Pratt *et al.* 1997).

Breeding Two nests only described, sequential, of the same pair of birds, found in 1986. Nest was an open cup among leafy branchlets, made of twigs, the interstices stuffed with coarse moss, about 8 m up in 15 m trees. The first nest was built in early March; after its failure, a second nest was built in the third week of April. Eggs probably two, not observed, but egg fragments suggest colour was whitish, with fine dense grey-brown speckling. Incubation by female alone; male fed both chicks and female. Most food was carried internally and fed by regurgitation. Fledging period 21 days (Engilis *et al.* 1996; Kepler *et al.* 1996).

Voice Call a sharp *chit-chit-chit*; alarm call a more whistled *chee-up*. Song, a jumbled series of similar notes, accelerating and rising in pitch.

Movements Sedentary.

Range Endemic. Since discovery, only in a limited area of eastern Maui, Hawaii.

Conservation status Extinct. In a part of the world which has seen – and continues to see – far more than its fair share of ornithological tragedy, the saga of the Po'ouli is particularly heartbreaking. Based on palaeontological material, it appears that prior to human settlement, the Po'ouli was more widespread, both geographically and in habitat; subfossil remains have been found at elevations as low as 300 m, in habitat that was quite different to the present, a xeric and mesic forest. The early Polynesian settlers cleared off virtually all such forest below 1,500 m, while forest clearance after European settlement intensified, leaving only rainforest (Cuddihy & Stone 1990). It seems likely that, by the mid-1800s, the range of the Po'ouli was already much smaller than prior to human settlement, since none of the many assiduous collecting ornithologists of the 19th century encountered it.

Shortly after discovery, field surveys estimated a population density of about 76 birds/km^2; between 1975 and 1981, this had dropped to 15 birds/km^2; and by 1985, to 8 birds/km^2 (Scott *et al.* 1986; Mountainspring *et al.* 1990). The overall population in 1986 was estimated at 140 birds (with a wide margin of error). Surveys in 1994/1995 found six birds at four locations; in 1997/2000 only three of these birds could be located (Baker 1998, 2001). The last evidence of wild breeding was in 1995 (Reynolds & Snetsinger 2001), while the last wild sightings were in 2004 (US Fish and Wildlife Service 2006). Efforts to create a breeding pair by relocating a single female into the breeding territory of a male were frustrated when the female returned to the place of capture (Groombridge *et al.* 2004).

The causes of this catastrophic decline were undoubtedly several, while the relative importance of each is the subject of considerable spirited debate. Certainly, introduced species had a deleterious effect; feral pigs destroyed understorey and exacerbated soil erosion, while rats and cats are potential predators. The role of the introduced Garlic Snail *Oxychilus alliarius*, which invaded Po'ouli habitat but which was not eaten by that species, may have been significant.

The question as to whether Po'ouli could have been saved is again the source of much vigorous debate. Powell (2008) makes no bones in stating that "pigs, rats and bureaucratic dithering would take another eighteen years to kill the last Po'ouli, essentially finishing the job begun by the Polynesians 1,500 years earlier". Despite much effort and vigorous work by many dedicated conservationists, it is generally accepted that efforts to preserve habitat from degradation by alien species came far too late, as did attempts to maintain the species by captive breeding; had these efforts been 20 years earlier, it is possible that the end result would have been happier.

Etymology The generic and specific names are from the Greek, meaning respectively 'black forehead' and 'brown body'. The vernacular name, suggested by Mrs Mary Kawena Pukui, a noted authority on Hawaiian language and culture, means 'black-faced'.

Footnote The battle for the Po'ouli has been lost. Several other Hawaiian species, unless prompt and effective action is taken, are poised to follow in its footsteps very shortly. Two, both on the island of Kaua'i, are in quite desperate straits: Akikiki *Oreomystis bairdi* and Akeke'e *Loxops caeruleirostris*. Organisations which are in the forefront of this effort are the Kaua'i Forest Bird Recovery Project (www.kauaiforestbirds.org) and the American Bird Conservancy (www.abcbirds.org); on the island of Maui, home to several other Critically Endangered species, the corresponding organisation is the Maui Forest Bird Recovery Project (www.mauiforestbirds.org).

MINDANAO SERIN
Chrysocorythus mindanensis

A little-known species whose taxonomic position has been the source of much debate; originally placed in the genus *Serinus*, then *Crithagra* and finally *Chrysocorythus* (Zuccon *et al.* 2012).

Type information *Serinus mindanensis* Ripley & Rabor 1961, Mt Katanglad, Mindanao, Philippines.

Discovery Mountain Serin *Chrysocorythus estherae* has an extraordinary distribution, being found, in various races, in widely disjunct alpine and montane habitats in Indonesia (east and west Java, northern Sumatra, south Sulawesi and north-central Sulawesi). In 1961 S. D. Ripley and D. Rabor described a form from Mt Katanglad, Mindanao, Philippines. In view of its geographic separation from all races of *estherae*, as well as its rather distinctive plumage, it was given full specific status. However, a majority of more recent authorities (e.g. Clement 1993; Clement in HBW15, 2010; IOC World Bird List Version 7.1) have treated it as a race of *estherae*. Most recently, it has been resurrected as a full species by the *HBW/BirdLife International Illustrated Checklist of the Birds of the World* (del Hoyo & Collar 2016).

Description 11cm. Monotypic. Upperparts dark olive-brown, edged greenish-olive on mantle, back and scapulars; rump bright yellow; two prominent yellow wing-bars; yellow edgings to tertials; face, excluding dark lores but including forecrown, cheeks, throat and upper breast yellow, becoming whitish on lower chest and belly, with diffuse darker streaks on flanks; iris dark brown or black, bill conical, olive-brown above, pale brown below; tarsi brown to dark brown.

Habitat Alpine forest; montane rainforest or dwarf ericaceous forest (Clement 1993).

Food and feeding Few data; forages low down, singly or in small flocks. Diet probably mostly seeds.

Breeding Nest and eggs undescribed.

Voice No information specific to *mindanensis*; Mountain Serins have a short, tinkling song, with a dull metallic chittering call.

Movements Probably sedentary.

Range Endemic. Mountains of Mindanao (Mt Katanglad, Mt Apo).

Conservation status Not classified by BirdLife International. Restricted-range species, apparently rather uncommon.

Etymology *Mindanensis*, from the island of Mindanao.

ELFIN WOODS WARBLER
Setophaga angelae

The discovery of a very distinctive new species in a national park in the US territory of Puerto Rico, at a site visited by numerous birdwatchers and ornithologists, caused a major sensation among the North American ornithological community. It probably forms a superspecies with Plumbeous Warbler *S. plumbea* and Arrow-headed Warbler *S. pharetra*.

Type information *Dendroica angelae* Kepler & Parkes 1972, Sierra de Luquillo, Puerto Rico.

Alternative names Puerto Rico Warbler. Spanish: Reinita de Angela, Reinita de Bosque Enano.

Discovery In 1968, C. B. and A. Kepler established residence in the Sierra de Luquillo, an extensive forest area in eastern Puerto Rico, in order to study the endemic Puerto Rican Parrot *Amazona vittata* and

▲ Elfin Woods Warbler. Left: male, Maricao State Forest, Maricao, Puerto Rico, October 2011. Right: female, Maricao State Forest, Maricao, Puerto Rico, April 2012. (*Mike Morel*).

Puerto Rican Tody *Todus mexicanus*. While setting up census routes in elfin forest, a habitat confined to the higher peaks of the Sierra, they occasionally observed a black-and-white warbler which they were unable to identify. By March 1971 they had come to the conclusion that an unknown species must be involved. This was confirmed by the taking, by C. B. Kepler and K. C. Parkes, of four specimens in May and July 1971.

Description 13.5 cm. Monotypic. Largely black and white. Upperparts black with diagnostic white facial markings including an incomplete white eye-ring, lores, post-ocular region and sides of nape. Two prominent white wing-bars; tail black with narrow white margin on outer web of outer rectrices; underparts white, profusely streaked with black on throat and chest, the streaks becoming coarser on sides of lower belly, which is white; iris very dark fuscous brown; bill grey, almost black; tarsi dark bluish-grey, soles greyish-yellow. Female very similar, streaking on underparts probably less prominent.

Habitat Montane forest, including elfin forest and *palo colorado* forest; in some locations, commonest in *Podocarpus* woodland; 250–950 m, most abundant 600–800 m. More recently found in modified habitats, including shade coffee plantations and timber plantations; may have some ability to accept modified habitat when prime habitat has been affected by tropical storms (Delannoy 2007; Gonzales 2008).

Food and feeding Forages at heights of 1.5–16 m, mostly by foliage-gleaning, sometimes by sally-gleaning (Cruz & Delannoy 1984a, 1984b; Delannoy 2007). Prey: arthropods. Frequently occurs in mixed-species flocks.

Breeding Breeding season is March–June (Arroyo-Vásquez 1992; Raffaele *et al.* 1998). Nest is a cup of rootlets and plant fibres, lined with grass leaves and down, placed 1.3–7.6 m up in aerial leaf litter trapped in vegetation or vines (Arroyo-Vásquez 1992); one nest, apparently unusual, was located in a cavity in a vertical rotten tree stump (Rodríguez-Mojica 2004). Eggs 2–3, one record of 4, dull white, spotted with reddish-brown, especially at blunt end. Young fed by both sexes on insects.

Voice Song is a rapid series of notes on one pitch, getting louder and ending with a series of double notes; calls also include a metallic *chip* note.

Movements Apparently sedentary.

Range Endemic to Puerto Rico. Main areas of occurrence are the Luquillo National Forest at the eastern end of the island and in the Maricao State Forest, about 150 km to the west (Gochfield *et al.* 1973). Populations also reported in Carite State Forest (1977) and in Toro Negro State Forest (late 1970s), but these two appear to have been extirpated (Delannoy-Julia 2009).

Conservation status Currently classified by BirdLife International as Vulnerable but recent recommendations suggest upgrading to Threatened (Delannoy-Julia 2009). There is strong evidence of a recent significant decline in the Luquillo population (Arendt *et al.* 2013). Main threat is habitat destruction for roads, communication towers, picnic sites and campgrounds (Delannoy-Julia 2009). The two populations are essentially genetically isolated; the largest of these is in the Maricao State Forest, with about two-thirds of the total estimated population of about 1,800 individuals.

Etymology Named for Angela Kay Kepler (b. 1943), wife and co-worker of one of the authors of the type description. This maintains a tradition in the nomenclature of the Parulidae, which includes Lucy's, Adelaide's, Virginia's and Grace's Warblers, celebrating, respectively, two daughters, a wife and a sister of prominent ornithologists. Dr Kepler, Australian by birth, is an accomplished biologist and author in her own right; the extinct Hawaiian rail *Porzana keplerorum* is named for her and her husband.

SELVA CACIQUE
Cacicus koepckeae

One of the least known icterids, with a very limited distribution. Probably most closely related to Ecuadorian Cacique *C. sclateri* and sometimes regarded as conspecific; however, there are significant differences in vocalisations.

▲ Selva Caciques. Santuario Nacional Megantoni, Dpto. Cusco, Peru, April 2004 (*Daniel Lane*).

Type information *Cacicus koepckeae* Lowery & O'Neill 1965, Balta, SE Dpto. Ucayali, Peru.
Alternative names Spanish: Cacique de Koepcke.
Discovery Two specimens of a yellow-rumped icterid were obtained by J. P. O'Neill in a remote area of Amazonian Peru in 1963 and 1965. Initially the specimens did not attract much attention until they were later compared to material obtained in Bolivia, when it was realised that they did in fact represent a new species, which was described in November 1965. The species then disappeared from view for more than 30 years, until rediscovered in two new locations in Peru (Gerhart 2004). It has now been observed at a number of locations (e.g. Tobias 2003).
Description 23 cm. Monotypic. Essentially all black with a striking yellow rump. Iris pale blue; bill pale bluish-grey; tarsi black. Sexes alike.
Habitat Apparently confined to dense riparian vegetation and transitional forest, often with *Guadua* bamboo patches, along the margins of high-gradient streams (Gerhart 2004); 300–700 m.
Food and feeding Feeds in small groups, possibly family parties. Diet nectar, seed pods etc.
Breeding Nest is a typical cacique structure, a pendant bag about 50–70 cm in length, built mostly of rhizomorphs of the fungus *Marasmius* (Grilli *et al.* 2012), suspended from the tip of a tree branch from 5–18 m up. Eggs undescribed; in late July an adult was visiting the nest frequently, presumably feeding young. May be a cooperative nester, with several birds seen visiting and guarding the nest (Grilli *et al.* 2009, Grilli *et al.* 2012). Does not apparently nest colonially.
Voice A rapid series of loud, quick, explosive paired notes *chick-pouw* or *chick-pouw-pouw* (Gerhart 2004).
Movements Presumably sedentary.
Range Has now been recorded at several locations in Amazonian Peru, in Ucayali, Cuzco and Madre de Dios; the north to south and east to west limits are each about 300 km. Recently reported from Acre, Brazil (Grilli *et al.* 2009).
Conservation status Currently classified by BirdLife International as Endangered. Apparently very local, occurring in low densities. Some observations from protected areas (Tobias 2003), but main threat is habitat destruction.
Etymology Named in honour of Dr Maria Koepcke (1924–1971); see under Koepcke's Screech Owl.

PALE-EYED BLACKBIRD
Agelasticus xanthophthalmus

Originally described as a member of the widespread genus *Agelaius*; subsequently placed into *Chrysomus* and then *Agelasticus* (Lowther *et al.* 2004). Probably most closely related to Unicolored Blackbird *A. cyanopus*.

Type information *Agelaius xanthophthalmus* Short 1969, Tingo María, Huánuco, Peru.
Alternative names Spanish: Varillero Ojipálido.
Discovery In August 1968, while studying woodpeckers in a small marsh near Tingo María, Peru, L. L. Short noticed a pair of icterids, entirely black in plumage. Being aware that no all-black blackbird was known from the Amazonian drainage of central Peru, he collected them; the new species was described on the basis of these two specimens.
Description 20.5 cm. Monotypic. A relatively small

▲ Pale-eyed Blackbird. Cocha Camungo, Madre de Dios, Peru, July 2011 (*Daniel Lane*).

blackbird, entirely black in plumage with a greenish-blue gloss. Iris straw-yellow or orange-yellow; bill and tarsi black. Sexes alike.

Habitat Lowland marshes, including oxbow lakes, especially areas with floating marsh grasses and scattered bushes (Parker 1982).
Food and feeding Forages mostly low down in marsh vegetation, gleaning insects from leaf surfaces and clumps of dead leaves; also takes aerial termites.
Breeding One nest so far described. It was an open cup, made of strips of dead leaves, situated about 45 cm up in marshside vegetation. In early November two eggs, pale bluish-white, the blunt end solid dark brown, the remainder blotched dark brown. No other data (Orians & Orians 2000).
Voice Song is a loud, fairly rapid series of notes all on the same pitch, given when perched and in flight. Also trills and single *chee* or *tew* notes (Orians & Orians 2000).
Movements Presumably sedentary.
Range Disjunctly in Amazonian Ecuador (Napo), Peru (San Martín, Ucayali, Huánuco and Madre de Dios) and extreme northern Bolivia.
Conservation status Currently classified by BirdLife International as Least Concern, although Ridgely & Greenfield (2001) believe that, for Ecuador at least, Near Threatened might be more appropriate. Appears to be generally uncommon, but further work might reveal other locations within its quite wide range.
Etymology *Xanthophthalmus*, Greek, 'yellow-eyed'.

TROPEIRO SEEDEATER
Sporophila beltoni

May form a superspecies with the widely distributed Plumbeous Seedeater *S. plumbea*.

Type information *Sporophila beltoni* Repenning & Fontana 2013a, Rio Tainhas Valley, Jaquirana municipality, Rio Grande do Sul, Brazil.
Alternative names Portuguese: Patativa-tropeiro.
Discovery Plumbeous Seedeater *S. plumbea* is a species with wide distribution in South America, from northern Colombia to SE Brazil. In plumage it is generally plumbeous-grey with a black bill. For many years it had been recognised that some Brazilian individuals had yellow bills – indeed, a series of specimens of this type was collected by the Austrian naturalist Johann Natterer in 1820–1821. Later ornithologists referred to a 'yellow-billed form' of *plumbea* (with further confusion caused by the misidentification of some specimens as Temminck's Seedeater *S. falcirostris*, which does indeed have a yellow bill but which is otherwise quite distinct). Finally, in 2005, M. Repenning and C. S. Fontana discovered a population of yellow-billed birds in shrub-grassland habitat in the states of Santa Catarina and Rio Grande do Sul, which they were unable to ascribe to any known species. There followed eight years of intensive fieldwork, during which time populations were studied and numerous individuals colour-banded. This established the very distinctive nature of the yellow-billed form in the morphology, colour pattern, vocalisations and habitat preferences; this, along with evidence of reproductive isolation from populations of *plumbea* in the narrow zone of sympatry, allowed the description of a new species.
Description 12.4 cm. Monotypic. An essentially grey seedeater with a distinctive yellow bill. Above, dark neutral grey, inconspicuously streaked on the head and neck; tail darker; white patch at base of primaries and secondaries, conspicuous on closed wing and in flight; throat, breast and ear-coverts bluish neutral grey; centre of belly and undertail-coverts white; iris chestnut; bill bright yellow; tarsi blackish-grey. Takes some four years to acquire totally adult plumage. Female plumage more drab grey or yellowish-buff, bill

◀▲ Tropeiro Seedeater. Left: male, Bom Jesus, Rio Grande do Sul, Brazil, December 2010. Right: immature female, São Joaquim, Santa Catarina, Brazil, November 2009. (*Márcio Repenning*).

pale yellowish. Differs from *plumbea* in bill colour, the darker grey, with a bluish-grey tone, of the breast, and frequent absence of the white malar streak found in *plumbea*.

Habitat Apparently rather specific in its requirements, breeding in grasslands with dense tall shrubs associated with *Araucaria angustifolia* forests (Repenning & Fontana 2013a, who give a more detailed analysis of the habitat requirements).

Food and feeding Nestlings are fed predominantly on seeds, but also on spiders and termites (Repenning & Fontana 2016). No other data.

Breeding Repenning and Fontana (2016) conducted an intensive study over several years, during which time some 133 nests were monitored. Breeding occurred from early November to early March, with the peak of breeding activity from the second week of November to the end of December (coinciding with the peak of winter grass seeding). Later nestings may have involved replacement clutches for failed earlier attempts. Eggs 1–3, with 2 usual (91% of nests). Incubation by female alone, 11–13 days, usually 12; young fed by both sexes, but male involvement increases as the nestling become larger. Interestingly, in seven cases where the nestlings had been sexed and colour-ringed, the male parent fed exclusively the male chick, the female parent, the female chick.

Voice Has a very large repertoire of call types. Song is complex, including an introductory portion involving clear and whistled notes. Repenning & Fontana (2013a) give a detailed comparative analysis of the song and that of *plumbea*.

Movements Austral migrant. Arrives on breeding territory mid-October, leaving in late January, disappearing completely by early March.

Range Endemic. Breeds in SE Brazil south from NE Paraná through Santa Catarina to NE Rio Grande do Sul; limits of breeding distribution from about 24° 10 S to 28° 45 S. Wintering range from eastern Goias and western Minas Gerais south to SW Bahia, some 1,500 km from the southern limit of the breeding range. On wintering grounds associates with several other *Sporophila* species.

Conservation status Not currently classified by BirdLife International, but Repenning & Fontana (2013) believed that Endangered would be appropriate, pointing out that it is the least common of all 35 species of *Sporophila* seedeaters. There are several threats, principally the destruction of its rather specialised breeding habitat by the planting of exotic *Pinus* trees. A further serious threat is the (probably unsustainable) drain on the population by trapping for the cagebird trade, which has already caused some extirpations (Repenning 2012; Repenning & Fontana 2013b). It no longer occurs in the locations where Natterer collected it in 1820–1821.

Etymology Named in honour of Dr W. Belton (1914–2009), an American diplomat who was involved in the study and conservation of the birds of Rio Grande do Sul (Belton 1974, 1985). The vernacular names refer to the Rota dos Tropeiros, a herders' trail which for 200 years was used for the transportation of stock in Brazil, and which mirrors the species' migration route.

IBERA SEEDEATER
Sporophila iberaensis

A species of confused nomenclature, having been described twice in a short period of time.

Type information *Sporophila iberaensis* Di Giacomo & Kopuchian 2016, Estancia San Alonso, Esteros del Iberá, Dpto. Concepción, Corrientes, Argentina.
Alternative names Spanish: Capuchino del Iberá; Portuguese: Caboclinho-do-Iberá.
Discovery In October 2001 A. S. Di Giacomo observed two puzzling *Sporophila* seedeaters in Corrientes, northern Argentina; he was not able to identify them, since they appeared not to be in full plumage. Characteristically they were a creamy-pallid colour underneath, with a grey crown, black collar and brown wings and shoulders. In studies conducted in 2007 and 2011, the creamy-pallid birds were again encountered, and were quite abundant in some locations. It was also noted that the song was distinctive, quite different from other syntopic *Sporophila* species. In 2010 B. López-Lanús joined the investigation, collecting a holotype in February and two paratypes in December 2011. There appears to be some disagreement between Dr Di Giacomo and Sr López-Lanús on the subject of publication; the latter published a detailed type description in 2015 with himself as sole author, designating the new species as *Sporophila digiacomoi*. Subsequently, Di Giacomo & Kopuchian published an abbreviated description, based on the same three specimens, using the name *S. iberaensis*, while promising a more complete disclosure in the near future. The SACC (Proposal 715) has given *iberaensis* priority over *digiacomoi*, as the former is the only name that fulfils ICZN requirements for valid publication.
Description 10 cm. Monotypic. In form a typical *Sporophila* seedeater, small with a conical bill. Crown grey; lower face, chin, ear-coverts and nape blackish-grey; remainder of underparts and rump creamy-pallid; back mid-brown, the wing-coverts and remiges darker brown with paler edgings; small white patch at base of primaries; rectrices dusky brown, edged olive-brown; iris dark; bill blackish; tarsi blackish-grey. All three specimens were males; immature birds appear to have an incomplete nuchal collar.
Habitat Flat pastureland, mostly subject to annual flooding, on sandy soils; apparently absent from dry-soil pastures and savannas; all records below 100 m.

▼ Ibera Seedeater, adult male (left) and adult female (right). San Nicolás Portal, Iberá Wetlands, Corrientes, Argentina, October 2009 (*Carlos Figuerero*).

Food and feeding Seed-eating; identified food plants include *Andropogon*, *Paspalum* and *Hymenachme* (López-Lanús 2015).

Breeding Nest and eggs undescribed. Fledged young being fed (by both parents) in February; female visiting nest (in tall pasture adjacent to a swamp) in November. The three male specimens, taken in February and December, had enlarged testes.

Voice Song is stated by López-Lanús (2015), who provides extensive comparative sonograms, to be diagnostic and distinguishable from other *Sporophila* species in the area.

Range Northern Argentina (Corrientes), southern Brazil (Mato Grosso do Sul) and three Paraguayan departments.

Conservation status In some locations apparently common or abundant. Di Giacomo & Kopuchian (2016) suggest a classification of Endangered for Argentina and Vulnerable worldwide.

Etymology The specific epithet *iberaensis* refers to the type location; *digiacomoi* recognises the original discoverer, Dr Adrián S. Di Giacomo.

SAO FRANCISCO SPARROW
Arremon franciscanus

Possibly most closely related to Half-collared Sparrow *A. semitorquatus* (Raposo 1997), but genetic evidence lacking (Schulenberg & Jaramillo 2012).

Type information *Arremon franciscanus* Raposo 1997, Rio São Francisco, Minas Gerais, Brazil.

Alternative names San Francisco Sparrow. Portuguese: Tico-tico-do-são-francisco.

▼ Sao Francisco Sparrow. Palmeiras, Bahia, Brazil, September 2014 (*Ciro Albano*).

Discovery During fieldwork in *caatinga* habitat near Mocambinho, Minas Gerais, Brazil in December 1995, M. A. Raposo encountered an undescribed *Arremon* sparrow. Four specimens were collected after their vocalisations had been recorded. Examination of the collection in the Museu Nacional, Rio de Janeiro, revealed a further undated specimen taken by an unknown collector across the border in Bahia state. On this basis a new species was described.

Description 15 cm. Monotypic. A strikingly marked sparrow obviously related to other *Arremon* species. Head black, with narrow white central crown-stripe and white supercilium; nape grey; back and tail greenish-olive, brightest in carpal region; throat and upper breast white, with a black mark at sides of upper chest, not continued across chest as a collar; centre of breast and belly white; flanks pale grey; iris dark reddish-brown; bill striking yellow-orange with black ridge along culmen; tarsi greyish. Female plumage undescribed but presumably similar.

Habitat Thick xerophytic *caatinga* scrub, also in the transition zone between *caatinga* and more humid forest (Schulenberg & Jaramillo 2012). Can tolerate some substantial habitat modification (Kirwan *et al.* 2004).

Food and feeding Forages on the ground; no data on diet.

Breeding Nest and eggs undescribed; male specimens taken in mid-December had enlarged testes.

Voice Song is rather variable, thin and high-pitched. One type consists of a repeated single note followed by a trill; or a more varied introduction, with an up-down pattern for the introductory notes. Call a sharp, ringing *tiip* (Schulenberg & Jaramillo 2012).

Movements Presumably sedentary.

Range Endemic. East-central Brazil in southern and central Bahia (Parrini *et al.* 1999) and northern Minas Gerais; more recently found further south in Minas Gerais (D'Angelo Neto & Ferreira de Vasconcelas 2003).

Conservation status Classified by BirdLife International as Near Threatened, although described by Kirwan *et al.* (2004) as "reasonably common in suitable habitats". The main threat is the impact of fire, agriculture and cattle ranching on the habitat. Part of the known range may be the subject of a large irrigation scheme which may cause further loss of habitat.

Etymology *Franciscanus*, from the São Francisco River.

BLACK-SPECTACLED BRUSHFINCH
Atlapetes melanopsis

One of several highly restricted *Atlapetes* species found in the NW Andes. Although, in plumage, it most closely resembles Apurimac Brushfinch *A. forbesi*, García-Moreno & Fjeldså (1999) suggest, based on DNA evidence, that it is genetically nearer to certain forms currently referred to as Bolivian and Slaty Brushfinches *A. rufinucha* and *A. schistaceus.*

Type information *Atlapetes melanops* (sic), Valqui & Fjeldså 1999, 4 km south-east of Huachocollpa, Provincia Tayacaja, Dpto. Huancavelica, Peru.

Alternative names Spanish: Atlapetes de Anteojos.

Discovery Several examples of a new *Atlapetes* brushfinch were seen in central Peru in 1996. Initially these were thought to be Apurímac Brushfinches *A. forbesi* and little attention was paid to them; however, on 21 July it was noted that the birds differed in several respects from that species. A type specimen (acquired by a local resident using a slingshot) was taken on 24 July, allowing the description of the species.

Description 19 cm. Monotypic. Crown ochraceous-tawny; back dark olive; wings blackish, the coverts tinged olive; rump lighter olive than back; tail black; forehead black; narrow white area extending above loral region; area around and above eye black; malar area white, submalar stripe blackish; chin greyish, throat off-white; remainder of underparts neutral grey, darker on flanks with yellowish wash; vent dark olive; iris dark reddish-brown; bill black; tarsi grey.

Habitat Dry open bushy areas (but in an area of high seasonal rainfall), often with dense thickets in small ravines, also near edges of humid elfin forest, 2,480–3,600 m.

Food and feeding Forages from ground level to about 3 m up, singly or in groups of up to three birds. Food items seeds and insects.

Breeding Nest and eggs undescribed.

Voice Song begins with a pair of introductory notes then chipping notes running together into a trill; calls include a liquid *twiit*, repeated when agitated (Jaramillo 2011 HBW 16).

Movements Presumably sedentary.

Range Endemic. Five locations in Dptos Junín and Huancavelica, both north and south of the Río Mantaro, Peru; more recently, in the Apurímac Valley, west of Cuzco, a substantial range extension (Hosner *et al.* 2015a).

Conservation status Currently classified by BirdLife

▲ Black-spectacled Brushfinch. Chilifruta, Junin region, Peru, September 2013 (*Carlos Calle*).

International as Endangered. The overall range is small and not all apparently suitable habitat is occupied. Habitat destruction, especially burning for agricultural clearances, has been going on since historic times, though the local human population may be declining due to migration to urban centres. Population estimates in the range of 1,500–7,000 mature individuals.

Etymology *Melanopsis*, Greek, referring to the black orbital area. Valqui & Fjeldså, in their original description, used the specific epithet *melanops* (1999) but due to potential confusion with *Buarremon melanops*, itself a *nomen dubium* probably referring to the nominate race of Bolivian Brushfinch *A. rufinucha* (see Sclater, 1886), it was renamed *melanopsis* (Valqui & Fjeldså 2002).

ANTIOQUIA BRUSHFINCH
Atlapetes blancae

Presumably most closely related to Slaty Brushfinch *A. schistaceus*.

Type information *Atlapetes blancae* Donegan 2007, based on a misidentified specimen taken prior to 1971 from San Pedro, Antioquia, Colombia.

Alternative names Spanish: Atlapetes Antioqueño, Gorrión-Montés Paísa.

Discovery During museum work related to the description of a new race (*yariguierum*) of Yellow-breasted Brushfinch *A. latinuchas* (Donegan & Huertas 2006), T. M. Donegan noticed three specimens, in the collections of three different Colombian institutions, of a bird closely resembling, but differing from, Slaty Brushfinch *A. schistaceus*. The accompanying data on these specimens was somewhat incomplete and fragmentary. What became the holotype specimen was unsexed and undated, with the location of San Pedro,

Antioquia; a second, one of the paratypes, was also unsexed, undated, and with no location other than "Antioquia", while the third, the last paratype, was a male in breeding condition, taken at San Pedro on 10 June 1971, the only firm date. All three had been identified by their collectors as *A. schistaceus*. The species was described on the basis of plumage differences consistent among all three specimens (see below). This treatment has been accepted by a majority of the SACC (Proposal 322).

Description 17 cm. Monotypic. In form a typical *Atlapetes*, a fairly large, stout-bodied finch. Crown rufous, contrasting with greyish-black face; remainder of upperparts mid-grey, darker on wings and tail; white flash on wing formed by narrow white edgings on middle primaries; throat and chin greyish-white with sharp, narrow dark malar stripe; remainder of underparts greyish-white, slightly darker on flanks. Soft parts undescribed. Differs from *A. schistaceus* in narrow malar stripe and much whiter underparts.

Habitat No firm data; probably forest-edge at 2,400–2,800 m.

Food and feeding No information.

Breeding Nest and eggs undescribed. One male specimen in late July had enlarged testes.

Voice No information.

Movements Presumably sedentary.

Range Uncertain; endemic. Specimens probably came from a Universidad de la Salle retreat called La Lana, near San Pedro de los Milagros, in the northern Andes in Antioquia, Colombia.

Conservation status Classified by BirdLife International as Critically Endangered; in fact, there is a distinct probability that it is extinct. Known only by the three specimens discovered by Donegan. In January 2007 he and several colleagues conducted extensive searches in the presumed type locality, as did several other ornithologists in subsequent years, all without success. Since the presumed dates of the original specimens, much habitat destruction has occurred in the region.

Etymology Named for Blanca Huertas, the wife of the discoverer, in recognition of her many contributions to Colombian biology, especially with regard to lepidoptera; coincidentally, in a different grammatical case, *blanca* describes the greatest distinction in plumage from *A. schistaceus*, namely the whitish underparts.

VILCABAMBA BRUSHFINCH
Atlapetes terborghi

Probably most closely related to Black-faced Brushfinch *A. melanolaemus*, despite the quite different plumage.

Type information *Atlapetes rufinucha terborghi* Remsen 1993, Cordillera Vilcabamba, Dpto. Cuzco, Peru.

Discovery The Cordillera de Vilcabamba is a spur of the eastern Peruvian Andes, largely isolated from other nearby high-altitude habitats by deep river valleys and lowlands, and as such is a rich source of endemic highland taxa, both species and races. In July 1967 J. W. Terborgh and J. S. Weske collected five specimens of an *Atlapetes* brushfinch there. Much later, while examining specimens for a study of the genus, J. V. Remsen realised that the 1967 specimens were very distinctive in plumage, and described them as *A. rufinucha terborghi*; that is, a new race and substantial range extension of Bolivian (Rufous-naped) Brushfinch *A. rufinucha* (Remsen 1993). However, in a comparative mitochondrial DNA study of the genus, Garcia-Moreno & Fjeldså (1999) concluded that *terborghi* was a valid full species, with its closest relatives the Black-spectacled, Cuzco and Grey-eared Brushfinches *A. melanopsis*, *A. canigenis* and *A. melanolaemus*.

Description 17 cm. Monotypic. In form a typical *Atlapetes* brushfinch, quite stout-bodied, with a fairly long tail and sturdy legs and feet. Crown rufous, contrasting with blackish cheek-patch and bright yellow throat; most of upperparts blackish, more olive towards the rump; underparts yellow, the breast and flanks washed greenish; iris dark brown, bill black, tarsi blackish.

Habitat Humid montane cloud-forest and elfin forest, 2,520–3,520 m.

Food and feeding No information.

Breeding Nest and eggs undescribed.

Voice Undescribed.

Movements Presumably sedentary.

Range Endemic. Confined to the Cordillera de Vilcabamba, east of the Río Apurímac, Cuzco, Peru.

Conservation status Classified by BirdLife International as Near Threatened, mainly because of the small range. The human population in the area is quite small and much habitat is relatively pristine.

Etymology Named in honour of J. W. Terborgh (b. 1936), who collected the original specimens in 1967. Dr Terborgh is Professor of Environmental Science at Duke University, Durham, NC, and has been enormously influential in the study and conservation of birds and mammals in the Andean region.

PARODI'S HEMISPINGUS
Hemispingus parodii

Although differing markedly in plumage, may be closest to, and a sister taxon of Orange-browed Hemispingus *H. calophrys*. Appears to be a higher-altitude replacement of Black-capped Hemispingus *H. atropileus* (Hilty 2011 HBW 16).

Type information *Hemispingus parodii* Weske & Terborgh 1974, Río Mapitunari, Cordillera Vilcabamba, Dpto. Cuzco, Peru.

Alternative names Parodi's Tanager. Spanish: Hemispingo de Parodi.

Discovery Description was based on eight specimens mist-netted during fieldwork on the ecology and distribution of birds in the eastern Andes of Peru.

Description 14 cm. Monotypic. A relatively large hemispingus; top of head dusky-olive, rather variable, sometimes darker; back and upperwing-coverts dark citrine; flight feathers dusky, the primaries edged yellow, the secondaries and tertials dark citrine; tail dusky with dark citrine outer webs; supercilium yellowish; lores and area around eye olive; ear-coverts dull yellowish; chin, throat and centre of belly bright yellow, flanks duller; vent paler yellow; iris dark brown; bill grey, duskier above; tarsi pale grey.

Habitat Humid upper montane forest and elfin forest with patches of *Chusquea* bamboo at or near treeline; 2,750–3,520 m.

Food and feeding Forages low or near ground in small parties and mixed flocks, especially with Citrine Warblers *Basileuterus luteoviridis*. No data on diet.

Breeding Nest and eggs undescribed. Based on gonadal data of specimens, breeding probably begins in late July.

Voice Day song is a string of somewhat laboured, rapid, chittering 1–2 second phrases, e.g. *p-p-p-psit-sit-sit* etc., moderate to high-pitched. Dawn song, alternates two notes, the first higher than the second *pit-zzre* etc (Isler & Isler 1987). Calls a high lisping descending *tseew* and a sharp *ti* (Schulenberg *et al.* 2007).

Movements Presumably sedentary.

Range Endemic. South-central Peru on the east slope of the Andes (Dptos Cuzco, extreme western Madre de Dios).

Conservation status Recently moved by BirdLife International from Least Concern to Near Threatened, based on estimates of deforestation for agriculture. Restricted range; stated to be "locally common but patchily distributed" (Schulenberg *et al.* 2007) and "uncommon" (Fjeldså & Krabbe 1990).

Etymology Named "in honor of our friend José Parodi Vargas (1930-2012), who.... generously aided us in countless....ways" (Weske & Terborgh 1974).

► Parodi's Hemispingus. Abra Malaga, Peru, December 2008 (*Fabrice Schmitt*).

RUFOUS-BROWED HEMISPINGUS
Hemispingus rufosuperciliaris

A large and very distinctive hemispingus, quite unlike any other members of the genus.

Type information *Hemispingus rufosuperciliaris* Blake & Hocking 1974, Bosque Huaylaspampa, Dpto. Huánuco, Peru.
Alternative names Rufous-browed Tanager. Spanish: Hemispingo cejirrufo.
Discovery Described from a total of 13 specimens taken under the auspices of the Louisiana State University Museum of Zoology, the type specimen having been acquired in July 1973.
Description 15–16 cm. Monotypic. Essentially totally different from any other *Hemispingus* species in plumage, more resembling Rufous-breasted Warbling-finch *Poospiza rubecula* than any other member of its genus. Crown, nape, lores, ear-coverts and side of neck black; upperparts, wings and tail blackish-slate; striking broad cinnamon-rufous supercilium; underparts entirely cinnamon-rufous, except for some whitish in centre of belly and dark grey flanks and vent; iris brown; bill black above, blue-grey below; tarsi brown.
Habitat Upper limits of elfin forest, especially places with thickets of ferns mixed with *Chusquea* bamboo; 2,550–3,500 m (Schulenberg *et al.* 2007).
Food and feeding Few data. Forages low down, below 2 m, in pairs or small (family party?) groups, sometimes in mixed-species flocks. Food items are beetles, caterpillars and seeds (Isler & Isler 1987), and berries (Fjeldså & Krabbe 1990).
Breeding Nest and eggs undescribed.
Voice Song is a loud series of jerky, wiry squeaks and chips; call a dry *tik* or *tsip* (Schulenberg *et al.* 2007).

▲ Rufous-browed Hemispingus. Bosque Unchog, Peru, December 2009 (*Christian Nunes*).

Movements Presumably sedentary.
Range Endemic. Eastern slope of the Peruvian Andes; recorded in three areas, from the Cordillera Colán in the north to the Cordillera Carpish, Dptos Amazonas, San Martín, extreme eastern La Libertad and Huánuco. Might well occur in intervening unexplored regions.
Conservation status Currently classified by BirdLife International as Vulnerable. Apparently "rare and local" (Schulenberg *et al.* 2007). Much of the range has some protection simply because of its inaccessibility; however, conversion of some habitat to agriculture is apparently occurring.
Etymology *Rufosuperciliaris*, Latin, 'rufous-browed'.

PARDUSCO
Nephelornis oneilli

An enigmatic little bird, whose taxonomic relationships were very unclear even to its discoverers, who hesitated to assign it, other than tentatively, to a specific family; they pointed out a series of puzzling anatomical features, such as a frilled tongue (with some analogies to the honeycreepers *Cyanerpes*) and some unique musculature of the tongue. In subsequent years a number of studies based on anatomy and genetic data have been published (Bledsoe 1987, 1988), Bledsoe & Sheldon (1989), Webster (1988, 1992), Burns (1997), Jonsson & Fjeldså (2006), the details of which are outwith the scope of this book. Currently accepted as an odd, atypical tanager, within the clade including *Hemispingus* and relatives (Burns 1997). Its discovery necessitated the erection of a new, monotypic genus.

Type information *Nephelornis oneilli* Lowery & Tallman 1976, Bosque Unchog, on pass between Churubamba and Hacienda Paty above Acomayo, Dpto. Huánuco, Peru.

Alternative names O'Neill's Pardusco. Spanish: Tangara Pardusca.
Discovery The initial specimens were taken in 1973 and 1974 during fieldwork conducted by staff of the Louisiana State University Museum of Zoology under the direction of George H. Lowery (q.v. under *Xenoglaux loweryi*). A large series of specimens was taken by field parties from January to July 1975, allowing the description of a new species and studies of its anatomy.
Description 12.5 cm. Monotypic. Overall brownish; upperparts plain brown, very faint paler eye-ring; upperwing-coverts dull brown with very obscure wing-bar; flight feathers somewhat darker brown; ochraceous below, the throat slightly paler, becoming light buffy on flanks and vent; iris warm brown; bill fine and narrow, dark brown above, flesh-coloured below; tarsi light brown.
Habitat Elfin forest and scrub at the interface of temperate cloud-forest and alpine grassland; always associated with treeline ecotone. 3,000–3,800 m (Capparella & Witt 2009).
Food and feeding Forages in mid-levels of forest, occasionally in canopy or on the ground, in small flocks (5–15 birds), sometimes in mixed flocks with tanagers, conebills and flowerpiercers. Stomach contents of specimens comprised arthropods (beetles, flies, plant bugs and Lepidoptera), along with a little plant matter.
Breeding Nest and eggs undescribed. Two male specimens taken in July were in breeding condition.
Voice Contact call, a high *seep* or *zee*. A series of high, weak, jerky warbled phrases may be the song (Schulenberg *et al.* 2007).
Range Endemic. Eastern slope of the Cordillera Central of Peru, from Dpto. San Martín, through Dptos Huánuco and Pasco to Junín.

▲ Pardusco. Bosque Unchog, Peru, August 2016 (*Miguel Lezama*).

Conservation status Currently classified by BirdLife International as Least Concern. Although the overall range is rather restricted, it appears to be common in some areas of suitable habitat. Some habitat destruction has occurred.
Etymology The generic name, *Nephelornis*, is from the Greek, 'bird of the mists'. The specific epithet honours Dr J. P. O'Neill (b. 1942), Research Associate at the LSUMZ, who, in more than 35 years of fieldwork in South America, has described more new species (including three new genera) than any other living ornithologist. The vernacular name Pardusco is coined from the Spanish, meaning 'brownish'.

ORANGE-THROATED TANAGER
Wetmorethraupis sterrhopteron

A spectacular and unusual tanager, one of the first of the many new species described from South America as a result of the indefatigable efforts, over many years, of the staff of the Louisiana State University Museum of Zoology. A monotypic genus, placed by molecular studies in closest relationship to the genus *Bangsia*, a small group of tanagers endemic to southern Central America and NE South America.

Type information *Wetmorethraupis sterrhopteron* Lowery & O'Neill 1964, Chávez Valdivia, near the confluence of the Río Comania and Río Cenepa, Dpto. Amazonas, Peru.

Alternative names Wetmore's Tanager. Spanish: Tangara Golinaranja.
Discovery The single specimen (probably a male) on which the original description of the species rests was obtained by an Aguaruna Indian sometime in July 1963 near the confluence of the Comania and Cenepa, in lowland Amazonas, Peru. The Aguaruna (or Awajíu, their preferred name) are indigenous to lowland areas of several Peruvian departments and have resisted assimilation until recently; the single specimen was given to Dr J. P. O'Neill by Miss Mildred Larson, a missionary who published the first Aguaruna dictionary. Despite the limited material, several unique features of

▲ Orange-throated Tanager. Yankuam, Zamora-Chinchipe, Ecuador, December 2011 (*Dušan Brinkhuizen*).

the specimen made the erection of a new genus necessary.

Description 17 cm. Monotypic. A thick-billed, long-winged tanager with a relatively short tail. Crown, lores, chin, ear-coverts, upper back and uppertail-coverts velvety black; tail more matt black; marginal wing-coverts and lesser and median secondary coverts deep blue; greater coverts blackish, edged with violet-blue; primaries, secondaries and tertials black, the last two edged pale violet-blue; throat and upper chest deep cadmium-yellow (feathers of lower chin, malar area and throat uniquely stiff and bristly); remainder of underparts yellow-buff, the thighs black; iris dark brown; bill and tarsi black. Sexes similar.

Habitat Mature, humid *terra firme* forest and foothill forest in the upper tropical zone, mostly 600–1,000 m, but also down to 450 m in *Cecropia* forest and once down to 317 m in forest-edge (van Oosten *et al.* 2007). Sometimes in disturbed mature forest (Marin *et al.* 1992).

Food and feeding Forages in pairs or small groups in middle levels of forest, sometimes in mixed-species flocks. Food items insects, including larvae, seeds and fruit pulp (Marin *et al.* 1992).

Breeding One nest so far described (Morrison *et al.* 2014). The nest, which was inaccessible, appeared to be an open cup, about 10 m up in the uppermost fronds of a Walking Palm *Socratia exhorrhiza*, made of twigs, probably lined with grass. Nest construction in late January, incubation probably mid-February. Eggs not observed; this nest was probably predated. The presence of three or more apparently adult tanagers at the nest site suggests cooperative breeding. The form of nest appears to be quite different from the dome-shaped constructions of the *Bangsia* tanagers, the putative closest relatives of *Wetmorethraupis*.

Voice Dawn song is a thick, heavy *jasuu-jasuu-jasuu*, repeated without a pause; also a squeaky, high-pitched slurred series of three-note phrases *we-tsi-tsoo* etc. Call a soft penetrating *seet* (Hilty 2011 HBW 16).

Movements Presumably sedentary.

Range Northern Amazonian Peru (Amazonas) and SE Ecuador (Zamora-Chinchipe) (Marin *et al.* 1992; Balchin & Toyne 1998). Apparently quite easily observed in parts of the Cordillera del Cóndor in Ecuador (Capper & Pereira 2007).

Conservation status Currently listed by BirdLife International as Vulnerable. The main threat is habitat destruction by incoming settlers. Fortunately, the local Aguaruna Indian population, which itself is of fairly low density and has a relatively mild impact on the habitat, is well organised and hostile to outside colonists; the species may thus enjoy unofficial protection from habitat loss (Begazo *et al.* 2001). Road construction may change the situation, however.

Etymology The new genus *Wetmorethraupis* honours Alexander Wetmore (1886–1978), one of the most influential American ornithologists of the 20th century. Dr Wetmore made many contributions to the taxonomy of Neotropical birds and was the author of the highly influential *A Systematic Classification for the Birds of the World* (1930), which gave rise to the 'Wetmore order' of bird families, in use until recently. The specific epithet *sterrhopteron* is derived from the Greek, 'stiff feather', referring to the unique bristly throat feathers.

GOLDEN-BACKED MOUNTAIN TANAGER
Buthraupis aureodorsalis

Probably the most spectacular species in a spectacular genus. May form a superspecies with Black-chested Mountain Tanager *B. eximia*, but very different in plumage.

Type information *Buthraupis aureodorsalis* Blake & Hocking 1974, Quilluacocha, Dpto. Huánuco, Peru.

Alternative names Spanish: Tangara Dorsidorado.

Discovery The type specimen was collected by P. Hocking and M. Villar in early October 1973; a total of seven other specimens were taken and are now in the Louisiana State University Museum of Zoology and the Field Museum, Chicago.

Description 23 cm. Monotypic. A large stout tanager. Crown and nape deep purplish-blue; lores, face, upper back, throat and upper breast black; central back, scapulars, rump and uppertail-coverts bright orange-yellow; lesser wing-coverts blue, remainder of wings, and tail, black; abdomen, sides and flanks bright orange-yellow, with bright chestnut spots and streaks on abdomen; undertail-coverts, lower belly and thighs chestnut; iris dark brown; bill blackish, paler below; tarsi dark grey. Sexes similar.

Habitat Small scattered islands of elfin forest at the treeline, especially in large islands and patches of *Escallonia* and *Clusia*; also in scrub in more open areas near elfin forests; 3,150–3,500 m.

Food and feeding Forages at all levels but mostly at middle heights, in pairs or small groups. Food items include fruits and seeds, and some insects.

Breeding Nest and eggs undescribed. Females in breeding condition in September; immature birds from July to November.

Voice Song is a series of wheedles, squeals and churrs; typically repeats one note 4–6 times and then switches to another in a rather staccato series given hurriedly and rhythmically. Calls include a variety of sharp, metallic and sometimes squeaky notes (Hilty 2011 HBW 16).

Movements Presumably sedentary.

Range Endemic. North-central Peru on eastern slopes of the Andes; currently known from five areas in San Martín, eastern La Libertad and Huánuco.

▲ Golden-backed Mountain Tanager. Bosque Unchoq, Peru, July 2015 (*Miguel Lezama*).

Conservation status Currently classified by BirdLife International as Endangered. It appears to be generally uncommon. Destruction of elfin forest habitat is probably a more serious issue than previously thought. Population estimates are in the range 250–2,500 mature individuals. One area of occurrence, Río Abiseo National Park, San Martín, is protected.

Etymology *Aureodorsalis*, Latin, 'golden-backed'.

GREEN-CAPPED TANAGER
Tangara meyerdeschauenseei

Forms a superspecies with several other generally more or less green-backed tanagers, namely Burnished-buff, Lesser Antillean, Black-backed, Chestnut-backed and Scrub Tanagers *T. cayana, T. cucullata, T. peruviana, T. preciosa* and *T. vitriolina* (Burns & Naoki 2004).

Type information *Tangara meyerdeschauenseei* Schulenberg & Binford 1985, 2 km NE of Sandia, Dpto. Puno, Peru.

Alternative names Spanish: Tangara coroniverde.

Discovery While working for the Louisiana State University Museum of Zoology in southern Peru in 1980,

► Green-capped Tanager, male. Sandia, Puno, Peru, July 2015 (*Miguel Lezama*).

▲ Green-capped Tanager, female. Sandia, Puno, Peru, June 2007 (*Daniel Lane*).

at an altitude of 2,175 m, L. C. Binford collected two specimens of a tanager with an obvious resemblance to the widespread, but generally lowland, Burnished-buff Tanager *T. cayana*. It was later learned that J. Dorst of the Muséum National d'Histoire Naturelle, Paris, had taken two female specimens in the same area in 1960, which he had identified as juvenile examples of *T. cayana* (Dorst 1961) (the female of *meyerdeschauenseei* does indeed have a close resemblance to that species). Based on their two specimens (the holotype is an adult male), and on material loaned by Prof. Dorst, Schulenberg & Binford were able to describe a new species.

Description 14 cm. Monotypic. A rather dull tanager. Crown opalescent greenish-straw, forehead dull bluish-green; ill-defined dark grey mask from lores to ear-coverts; mantle and back pale mealy-green, tinged buff, bluer at sides than in centre; scapulars and wings greenish; rump greenish-straw; underparts largely shining blue-green, the centre of the belly pale yellowish-opal; vent rufous; iris brown; bill blackish above, grey below with black tip; tarsi lead-grey. Sexes similar, but female substantially duller, the crown tinged tawny, underparts pale greenish-straw.

Habitat Open scrub, wooded gardens and forest edge, in Peru mostly 1,750–2,200 m. In Bolivia mostly in dry, open scrub land and forest borders, 1,450–1,700 m, but also in humid yungas forest (Hennessy & Gómez 2003).

Food and feeding Forages in bushes and low trees, in small groups, frequently in mixed-species flocks. Food items are small fruits, sometimes picked up from the ground, nectar, flower petals and arthropods (Naoki 2003).

Breeding Nest and eggs undescribed. Seen collecting nest material in dry scrub habitat in mid-December (Berg *et al.* 2014); in mid-November male and female specimens had quite large gonads (Schulenberg & Binford 1985).

Voice Call is a rather dull, heavy *chup* or *cheeup*. Song consists of a series of high, tinkling notes, mainly on one pitch (A. Spencer, XC 40671).

Movements There may be seasonal movements between different habitat types, from arid scrub to more humid areas (Hennessy & Gómez 2003).

Range Confined to southern Peru (eastern Puno) and western Bolivia (La Paz) (Hennessey & Gómez 2003). May be more widespread than previously believed.

Conservation status Currently listed by BirdLife International as Near Threatened, downgraded from Vulnerable because recent observations show that it is more widely distributed than previously believed. In fact, it may benefit from some man-made habitat changes. The overall population, however, remains small, with a population in the range of 1,000–2,500 mature individuals (BLI SF 2015).

Etymology Named in honour of Rodolphe Meyer de Schauensee (1901–1984), one of the most influential ornithologists of his time. Scion of an aristocratic Swiss family (his father owned a castle, the 'Schloss Schauensee', near Lucerne), he was born in Rome and emigrated to the United States in 1913. He was, for 50 years, curator of ornithology at the Academy of Natural Sciences in Philadelphia, and was responsible for major expansions in the Academy's collections. He was the author of several important books, perhaps the most significant being the groundbreaking *A Guide to the Birds of South America* (1970), although his last work, published just two weeks before his death, was on the birds of China (Ripley 1986).

SIRA TANAGER
Tangara phillipsi

Forms a superspecies with Black-capped Tanager *T. heinei*, and regarded by some authorities as conspecific; however, the major plumage differences and wide geographic separation argue against this. Forms a species group with *T. heinei* and Silver-backed and Straw-backed Tanagers *T. viridicollis* and *T. argyrofenges*.

Type information *Tangara phillipsi* Graves & Weske 1987, Cerros del Sirá, Dpto. Huánuco, Peru.

Alternative names Spanish: Tangara del Sira.

Discovery During 1969 and 1972 J. Terborgh, M. Koepcke and J. Weske made bird collections from the Cerros del Sirá, a largely isolated mountain offshoot of the Cordillera Occidental of central Peru. In late July 1969 a small series of a spectacular new tanager was collected; the species was not, however, formally described until 1987.

Description 13 cm. Monotypic. Sexually dimorphic. Male has a black crown, hindneck, chest and centre belly; throat blue-green; back from nape to rump shining opalescent greenish-blue; upperwing-coverts and flight feathers black, edged slate-blue; flanks opalescent blue; vent bluish-grey; iris dark brown; bill and tarsi black. Female has top of head dusky-green; upperparts shining pale green; tail dull green; throat, sides of neck and chest dull glaucous-green; centre of belly and vent dull greenish-grey to grey.

Habitat Humid montane scrub and edges of montane forest, semi-open, including successional areas following landslides and open areas near streams rather than closed-canopy forest (Mee *et al.* 2002), 1,300–2,220 m (Harvey *et al.* 2011).

Food and feeding Few data. Forages in small parties, but occasionally up to 20 birds, often in mixed flocks with other tanager species, honeycreepers, flowerpiercers and other families, in forest up to 25 m, in bushes to 10 m. Diet is fruit and insects.

Breeding Nest and eggs undescribed. Four specimens collected in July were not in breeding condition.

Movements Presumably sedentary.

Range Endemic. Apparently confined entirely to the Cerros del Sirá in Dpto. Huánuaco, Peru. The northern part of this range has been investigated by several parties of ornithologists (e.g. Graves & Weske 1987; Socolar *et al.* 2013), but the southern section, possibly ecologically separated by intervening lower ground, was not explored ornithologically until much later (Harvey *et al.* 2011). The species occurs in both sections.

Conservation status Classified by BirdLife International as Near Threatened, mainly because of the restricted geographic range. Threats include gold mining and logging. The Cerros were declared a nature reserve in 2001 (Mee *et al.* 2002). The local Ashaninká people are anxious to prevent logging but tend to lack formal title to the land (BLI SF 2014).

Etymology Named in honour of Allan R. Phillips (1914–1996), one of the most influential authorities of his era on the taxonomy of Mexican and Central American birds. A colourful and outspoken character, Dr Phillips took no prisoners among any who disagreed with him on matters ornithological; one of his papers (in *Ser. Zool.* 1: 87–98) "set a record for unwarranted abuse of well-known ornithologists in an obscure journal" (*Auk* 91: 474). His contributions to the ornithology of the American South-west and Central America were huge (see Dickerman *et al.* 1997); among his publications were *The Birds of Arizona* (1964) and the monumental *The Known Birds of North and Middle America* (1986) (published privately in two volumes, which allowed the author to express uninhibited opinions without the interventions of meddlesome editors); nevertheless this work remains fundamental to the present taxonomy of the area.

▼ Sira Tanager, male (left) and female (right). Río Shaani Valley, southern Cerros del Sirá, Ucayali, Peru, September 2008 (*Michael G. Harvey*).

TAMARUGO CONEBILL
Conirostrum tamarugense

A diagnostically plumaged conebill, probably the only member of its genus to show true annual migration as opposed to short-term opportunistic wandering.

Type information *Conirostrum tamarugensis* (sic) Johnson & Millie 1972, Tarapacá, Chile.
Alternative names Spanish: Conirostro de Tamarugal, Comesebo de los Tamarugos.
Discovery Described by A. W. Johnson and W. R. Millie in the Supplement to the first author's *The Birds of Chile*, based on a specimen taken by Dr J. Rottmann in December 1970. Four further specimens were obtained by W. R. Millie the following year.
Description 12.5 cm. Monotypic. In form a typical conebill, with a fine pointed bill. Crown and upperparts dark grey, the wings darker with two narrow wing-bars formed by fine edgings to the greater and lesser coverts and a prominent white flash involving the bases of some of the primaries; lores, supercilium, chin, throat and upper chest rufous; remainder of underparts plain grey except for buffy-white centre of belly and rufous vent; iris dark brown; bill black; tarsi dark grey. Female similar but generally duller and more brown; juveniles (which, apparently uniquely in the genus, are streaked (Valloton 2016)) largely lack the rufous throat.
Habitat Breeding habitat appears to require the presence of Tamarugo *Prosopis tamarugo* forest, with an apparent preference for more mature trees (Estades & López-Calleja 1995), although the species is also found (possibly non-breeding) in riverine scrub, agricultural land and citrus groves; on the wintering area in Peru, mainly in *Polylepis* forest but also in scrub with other species. In Chile, the habitat is extremely arid, with an average annual precipitation of 0.03 mm, and a temperature range of -12° to +35° C (Estades 1996). Ocurs from sea level to ca 3,850 m in Chile, apparently moving to higher altitudes in the pre-puna shrub zone after breeding (Sallabery *et al.* 2010); in Peru, mostly above 3,000 m and up to 4,050 m, but also (possibly while on migration) at 1,300 m (Høgsås *et al.* 2002). Some individuals are present year-round in oases in northern Chile (Schmitt 2015, quoted in Valloton 2016), suggesting breeding.
Food and feeding During the breeding season (when territorial), forages mostly in upper and outer parts of Tamarugo trees; prey items include caterpillars, especially of the butterfly *Leptotes trigemmatus.* On wintering grounds often associates with Cinereous Conebill *C. cinereum.*
Breeding Almost all information is from Estades & López-Callejo (1995). Nest is a hemispheric structure, 6.5–9 cm in diameter and 7–10 cm in height, made of small twigs, feathers, sheep wool and leaf veins, mostly in the central area of Tamarugo trees, and at a mean height of about 4 m. Eggs three, pale grey with irregular brown spots. Breeds late October to early November. No data on incubation or fledging periods.
Voice Song is a fast, buzzy jumble of squeaks lasting only one second. Call a high-pitched *tsee* (Hilty 2011 HBW 16).

◄ Tamarugo Conebill. Camarones, Chile, September 2015 (*Fabrice Schmitt*).

Movements Short-distance migrant. After breeding season may move to higher altitudes, then winters in southern Peru.

Range Endemic breeder; breeding range is in arid areas of NW Chile (mostly Tarapacá, but recent breeding season observations in east-central Antofagasta (Sallaberry *et al.* 2010)). Winters in Peru in Tacna and Moquegua, just into Arequipa; probably also adjacent Chile.

Conservation status Currently listed by BirdLife International as Vulnerable, mainly on the basis of a very restricted breeding range of not much more than 100 km². As a breeding bird, appears to be totally dependent on Tamarugo *Prosopis tamarugo* forest. This is a remarkable tree of the pea (Fabaceae) family which occurs in areas of almost zero rainfall, depending on an enormously deep root system. It is highly prized for wood and charcoal, and was almost extirpated (along, doubtless, with the conebill population) until the Chilean government started a reforestation programme in the 1930s (Briones 1985; Aguirre & Wrann 1985). Although Tamarugo forest is more extensive than previously, pruning for firewood reduces its suitability as a habitat. Further threats include measures taken to control larvae of the butterfly *Leptotes trigemmatus* which would impact food supply, and taking of groundwater which would affect the trees themselves (Estades 1996).

Etymology Both the scientific and common names refer to the obligate habitat, Tamarugo woodland; the scientific name has been amended from *tamarugensis* to *tamarugense* (Mayr & Vuilleumier 1983).

CARRIZAL SEEDEATER
Amaurospiza carrizalensis

Genetic studies (Klicka *et al.* 2007) suggest that the genus *Amaurospiza* may belong in the Cardinalidae (cardinals and grosbeaks) rather than the Emberizidae (New World sparrows and buntings).

▼ Carrizal Seedeater, male. Lower Caroní River, Venezuela, December 2014 (*Igor Castillo/Proyecto AmaurospizaVE*).

◄ Carrizal Seedeater, female. Lower Caroní River, Venezuela, March 2016 (*Igor Castillo/Proyecto AmaurospizaVE*).

Type information *Amaurospiza carrizalensis* Lentino & Restall 2003, Isla Carrizal, Río Caroní, Bolívar, Venezuela.

Alternative names Carrizal Blue Seedeater. Spanish: Semillero de Carrizal.

Discovery In 2001 a unit from the Phelps Ornithological Collection in Caracas, Venezuela, was assisting in a biological survey along the Río Caroní, Bolívar, Venezuela, in an area which would be impacted by the construction of a dam at Guri. In July and August three specimens of a blue seedeater were taken on Carrizal Island, an uninhabited island covered with tropical deciduous bamboo forest. Since no comparable seedeater was known from Venezuela or indeed any closer than central Panama or western Colombia, this was obviously of great interest. Comparison of the three specimens with the other two species of the genus established that a new species was involved.

Description 12 cm. Monotypic. Sexually dimorphic. Male essentially dark slaty-blue all over, blacker on face and underparts, sooty-indigo on lower belly and flanks, with white underwing-coverts and axiliaries; iris dark brown; bill black; tarsi black. Female warm brown above, below paler ochraceous-buff, the wing linings creamy-buff.

Habitat Tropical deciduous forest with bamboo (*Guadua* and *Ripidocladis*); type locality is at 95 m elevation.

Food and feeding From stomach contents of specimens, vegetable matter and insects including weevils.

Breeding Nest and eggs undescribed.

Voice Song is a pleasant, whistled warbled *sweet-sweet pit-swee pit-swoo* (Jaramillo 2011 HBW 16).

Movements Presumably sedentary.

Range Endemic. Presently known from some 14 sites in the lower Caroní Basin, Bolívar state, Venezuela.

Conservation status Currently classified by BirdLife International as Critically Endangered. The type locality, Isla Carrizal, no longer exists, having been flooded by the Guri Dam; several other potential locations were also inundated before they could be investigated. The total global population may be as few as 50 individuals. Currently the Neotropical Bird Club is funding a study of the ecology and conservation of the species (Anon. 2015c).

Etymology Both the vernacular and scientific names are derived from *carrizo*, a word in Spanish meaning reed-grass, but locally used to refer to spiny bamboo. The English name Carrizal Seedeater is preferred over Carrizal Blue Seedeater, to avoid confusion with Blue Seedeater *A. concolor* (SACC, Proposal 74).

FUTURE NEW SPECIES

Species in the Process of Description

Describing a new species is frequently a prolonged process. Even after the initial detection of what appears to be a new taxon, the accumulation of supporting evidence, which may involve extensive field studies, acquisition of vocal and molecular evidence, and lengthy comparison of material with, quite often, specimens from numerous museums or institutions, can be an exhaustive and time-consuming endeavour. Which is exactly as it should be; the Invalid Species chapter illustrates some of the errors and confusions which can arise from over-hasty recourse to the published literature. Consequently, at any one time there will be numerous putative species 'in the pipeline', that is, in the process of being described. This section lists some of these, all of which have been announced to the world in a publication (often with a photo or an illustration) but have not yet been formally described.

It should be noted that this list is very far from complete. Out of respect for the process, and for the ornithologists who are in the middle of writing or working on type descriptions, we have only included species for which some type of preliminary information has already appeared in readily available literature. There are a number of proposed new species which are known to this author (and doubtless an even greater number which are not) which we feel cannot be ethically disclosed prior to the proper, full type description.

'Great Nicobar Rail' *Rallina* sp.

In November 2011 a single crake was observed and photographed at short range at Govind Nagar on the east coast of Great Nicobar Island, Bay of Bengal, by S. Rajeshkumar. It was initially believed to be an abnormal plumage of Band-bellied Crake *Porzana paykulli*, but consultation with standard reference works showed that it was clearly distinct from all known rallids, and probably belonged to the genus *Rallina* (Rajeshkumar *et al.* 2012). There has been only one further observation of the species since (Bharadwaj 2015), so the species is presumably uncommon.

Unidentified turaco from Ethiopia *Tauraco* sp.

A turaco, probably of the Green Turaco *T. persa* complex, observed by D. Robel near Sodere, in the Bale Mountains of Ethiopia in November 1996 may represent a new species. Neither Green Turaco nor Black-billed Turaco *T. schuetti* have been recorded in Ethiopia (Robel 2000). If new, it would be the first new turaco species to be described since 1923. There have been no further observations.

'Eucalypt Cuckoo Dove' *Macropygia* sp.

Eaton *et al.* (2016b) give details of a new cuckoo dove from the islands of Sumba, Alor and Timor, and presumably also Wetar and Pantar, in the Lesser Sunda Islands in the southern Indonesian archipelago. It is similar to Little Cuckoo Dove *M. ruficeps*, but slightly larger, with deeper rufous underparts and lacking a paler-contrasting throat. The voice is distinctive, a slightly inflected (Sumba) or distinctly inflected (Timor, Alor) *whooh whooh whooh*, about six notes in ten seconds.

Unnamed imperial pigeon from New Guinea *Ducula* sp.

Beehler *et al.* (2012) report photographing a new species of imperial pigeon in the Foja Mountains of western New Guinea in 2008. It is clearly related to Shiny (or Rufescent) Imperial Pigeon *D. chalconota*, a wide- ranging species in New Guinea, but is considered by its finders to be a new species based on several plumage features.

► 'Great Nicobar Rail' *Rallina* sp. Govind Nagar tsunami shelter, Great Nicobar, India, November 2011 (*S. Rajeshkumar*).

▲ 'Santa Marta Screech Owl' *Megascops* sp. (grey morph). Cuchilla San Lorenzo, Sierra Nevada de Santa Marta, Dpto. Magdalena, Colombia, October 2008 (*Christian Artuso*).

▲ Unidentified owl *Ciccaba* sp. Cabañas San Isidro, province of Napo, Ecuador, October 2011 (*Dušan Brinkhuizen*).

'Santa Marta Screech Owl' *Megascops* sp.

The existence of a distinctive *Megascops* owl in the Santa Marta massif of northern Colombia has been known for many years; a formal type description is in preparation (N. Krabbe, pers. comm., April 2016). This bird has been seen by large numbers of birdwatchers and ornithologists, with numerous photographs published in a number of journals. An unofficial specific epithet has apparently been suggested (Mikkola 2014).

Unidentified owls from Peru, Ecuador and Colombia *Ciccaba* sp.

Rodríguez *et al.* (2012) published photographs and details of a puzzling owl, obviously of the genus *Ciccaba*, taken at an altitude of 1,435 m on the Manu road, on the eastern slope of the Peruvian Andes. Several possibilities were hypothesised, including a hybrid between Rufous-banded Owl *C. albitarsis* and Black-banded Owl *C. huhula*. Subsequently, another *Ciccaba* owl has been seen numerous times, and photographed, at San Isidro, Ecuador, with recent records in Colombia. The identity of this bird is currently unclear, possibilities being a hybrid between Black-banded Owl and Black-and-white Owl *C. nigrolineata*, an aberrant individual, or a new taxon. No specimen material is available at the time of writing.

'White-spotted Hawk Owl' *Ninox* sp.

Madika *et al.* (2011) speculate that an undescribed hawk owl occurs in the highlands of central Sulawesi, Indonesia. Their hypothesis is based on good quality photographs taken at an altitude of 2,250 m of an obliging individual in July 1999. From these it appears that the bird shows characteristic plumage features, including unique bold white spots on the underparts, as well as other apparent morphological characteristics. The resolution of the status of *Ninox* owls in Indonesia is somewhat confused and would benefit from

◀ 'White-spotted Hawk Owl' *Ninox* sp. Puncak Dingin, Sulawesi, Indonesia, 1999 (*Buttu Madika*).

further fieldwork (Madika *et al.* 2011). What is probably the same species was also photographed at an altitude of 1,700 m on Rorekatimbu (Tebb *et al.* 2008).

Unidentified frogmouth from Sumatra *Batrachostomus* sp.

Verbelen & Demeulemeester (2012) reported a puzzling frogmouth on the island of Siberut, in the Mentawai group of islands, west of Sumatra, based on observations made in 2009. Photographs have been taken and recordings of vocalisations were made. Clearly, further investigation is warranted.

▲ Unidentified frogmouth *Batrachostomus* sp. Pokai village, Siberut, Mentawai Islands, Indonesia, August 2009 (*Bram Demeulemeester*).

Unidentified nightjar from Cameroon *Caprimulgus* sp.

A small nightjar was picked up by N. Borrow and friends in a waterlogged condition on a track in the Bafut-Nguemba Forest Reserve in Cameroon in April 1990. Despite the taking of measurements and photographs it has not been ascribed to any known species; the finders speculated that an undescribed species might be involved (Robertson 1992).

'Timor Nightjar' *Caprimulgus* sp.

Eaton *et al.* (2016b) give details of a new nightjar from Timor, Indonesia. It closely resembles the race *schlegelii* of the widely distributed Large-tailed Nightjar *C. macrurus*, but is vocally distinctive. The song is a rapid series of 10-50 *chok* notes, at the rate of about 10 notes per 2.5 seconds. The species is uncommon, in forest and forest-edge.

Unidentified swift from Honduras *Streptoprocne* sp.

Between 1950 and 1960, numerous observers (including a number of highly competent ornithologists) saw flocks, quite often large, of a big swift over the Caribbean lowlands of northern Honduras. The birds were described as resembling White-collared Swift *Streptoprocne zonaris*, but lacking the full white collar of that species; only the nape was white. In this they resembled White-naped Swift *S. semicollaris* of highland western Mexico. However, *semicollaris* is largely sedentary, moving only small distances altitudinally; the Honduras birds were observed in all months of the year, including during the breeding season of *semicollaris*. According to Monroe (1968), this "must be regarded as the biggest ornithological mystery in (Honduras)"; we are not aware of more recent reports, although Honduras is now much more frequently visited than 50 years ago.

Other unidentified swifts

Apart from the unidentified swift from Honduras mentioned above, several observers have seen unidentified or possibly unknown swifts in various parts of the world. These reports are ably summarised by Chantler & Driessens (1995). These include the 'Beidahe Swift' seen in Hebei province, northern China (Williams *et al.* 1986), also possibly seen in Hong Kong; a swiftlet from Bhutan, resembling but apparently distinct from Himalayan Swiftlet *Aerodramus brevirostris* (Inskipp & Inskipp 1994); a very large, all-black swift seen on Marsabit Mountain, Kenya (Williams 1980); and a very large black spinetail reported from Democratic Republic of the Congo.

Unidentified kingfisher from Sulawesi *Actenoides* sp.

Rheindt *et al.* (2014) report one observation, not presently substantiated by either photographic or specimen evidence, of a distinctive kingfisher seen at close range at Gunung Tumpu, on the eastern peninsula of Sulawesi, in January 2014. The plumage was very distinctive, with some similarities to the lowland Green-backed Kingfisher *A. monachus*, but with scaling on the back, in the fashion of the Scaly-breasted Kingfisher *A. princeps*. In the absence of further evidence the affinities of this bird remain unresolved.

▲ Undescribed antshrike *Thamnophilus* sp. Inírida, Dpto. Guainía, Colombia, August 2016 (*Pablo Florez*).

Undescribed antshrike from Colombia *Thamnophilus* sp.

In August and September 2016 P. Florez was able to make detailed observations and take some excellent photographs of 'a mysterious antshrike in eastern Colombia', in fact in the region of Inírida, Dpto. Guainía, near the Venezuelan border. The bird resembled Chestnut-backed Antshrike *T. palliatus*, a species whose closest occurrence is some 1,600 km from Inírida, but differed in a number of significant features, including (in the male) extensive black from crown to lower breast with little white barring. The song is similar to that of *palliatus* but is slower and softer. The new bird was observed in semi-open *várzea* forest at an elevation of 100 m; comparative photographs are provided in the paper announcing its discovery. Further fieldwork to investigate what is obviously a new taxon is currently being planned (Florez 2017).

'Millpo Tapaculo' *Scytalopus* sp.

A very well-known bird, whose existence has been known for over 25 years and which has been seen by, literally, hundreds of observers including this author; numerous good-quality photographs have been published, including in HBW Alive, and there exist a large number of good recordings (e.g. XC229609 and numerous other examples). Notwithstanding all of this it remains, formally, undescribed. A typical *Scytalopus* tapaculo, the male is slate-grey with heavily barred brown flanks; the song is a rapid series of sharp, single *chip* notes, all on one pitch (see HBW Alive for sonograms). For a tapaculo, a relatively confiding, easily observed species, frequently sitting in plain view on tops of boulders. Occurs in the páramo zone, from 3,600–4,200 m, at least in Dptos Junín and Pasco and possibly elsewhere.

▼ 'Millpo Tapaculo' *Scytalopus* sp. Comas, Dpto. Junín, Peru, June 2015 (*Miguel Lezama*).

'Ampay Tapaculo' *Scytalopus* sp.

Discovered by J. Fjeldså and N. Krabbe in 1987 in Bosque Ampay, near Abancay, Apurímac, Peru (Fjeldså & Krabbe 1990): a description is apparently in preparation (Fjeldså & Schulenberg, in press). A typical *Scytalopus* tapaculo, slate-grey with a hint of a paler brow-patch, the lower flanks brown with fine black bars. Song is a characteristic series of *tcheuw* notes, all on one pitch, about two per second, in a series lasting at least ten seconds. One nest has been described (Baldwin & Drucker 2016): a simple cup, made of dry grasses and plant material, located deep in a crevice in a rocky boulder field, at an altitude of 4,500 m. On 21 December it held two half-grown nestlings. Although most members of the genus build domed nests, open cups have been described for two other species (Greeney & Gelis 2005b).

Undescribed tapaculo from Peru *Scytalopus* sp.

Hosner *et al.* (2015) discovered populations of a *Scytalopus* tapaculo, which they called the 'Above Treeline Tapaculo' (*Scytalopus* cf *simonsi/altirostris*) in the Apurimac Valley of Peru. The birds, which occur in shrubby puna grasslands above the treeline, at altitudes of 3,500–4,200 m, appear to be vocally distinct from 'Ampay Tapaculo' (q.v.) and were the first tapaculos to be observed in such habitat in the valley; the authors comment that an undescribed taxon is probably involved.

'Turimiquire Tapaculo' *Scytalopus* sp.

Ascanio *et al.* (2017) disclose a new *Scytalopus* tapaculo from the Turimiquire region, Sucre, north-eastern Venezuela. In plumage it resembles the Merida and Caracas Tapaculos (*S. meridanus* and *S. caracae*) but with a richer rufous-brown belly, lacking barring. The song consists of 5–6 ascending notes ending with a descending one.

Unidentified spinetail from Peru *Cranioleuca* sp.

In 2006 G. Engblom and colleagues saw a *Cranioleuca* spinetail in the Andamarca area of central Peru. In plumage this resembled the *weskei* race of Marcapta Spinetail *C. marcaptae*; however, vocally it seemed to be very distinct, leading to suggestions that a new taxon might be involved (Birdingblogs.com/2010/Gunnarundescribed-and-new-birds-on the-satipo-road-Peru). Other more recent opinions incline towards it being a range extension of *C. m. weskei* (R. Boegh, Birdforum 21 August 2008).

'Serra do Lontras Treehunter' *Heliobletus* sp.

Silveira *et al.* (2005b), in a survey of the Serra das Lontras-Javi mountain complex of Bahia state, Brazil, disclosed the existence of a new treehunter, of the (formerly monotypic) genus *Heliobletus* (the other species, Sharp-billed Treehunter *H. contaminatus* is widely allopatric in SE Brazil). The call is a characteristic furnariid vocalisation, several sharp *tic* notes followed by a rapid trill, or a trill followed by a series of sharp squeaks. More recently, the species has been seen and photographed in the Serra Bonita range, near Camacan, about 30 km south-west of the original sightings (C. Albano, pers. comm.).

Unknown bowerbird from New Guinea *Amblyornis* sp.

The Fakfak Mountains of Western New Guinea are some of the least accessible and explored regions of the whole island of New Guinea, being extremely inhospitable and deeply broken-up with vertical cliffs and steep-walled fissures. Even local tribesmen rarely penetrate more than 4–5 km into them. However, in 1991 oil exploration teams cut trails into the interior, building helipads at regular intervals. In August and September 1992 D. Gibbs used these access points to study the avifauna at higher altitudes in these mountains, which had previously been inaccessible. He frequently saw bowerbirds on this expedition (Gibbs 1994); previously these had been assigned to the Vogelkop Bowerbird *A. inornatus* but they were darker above and more fulvous below. Most significantly, the construction of the bowers of birds in the Fakfak and Kumawa Mountains differed markedly from that by *inornatus*; since this may be an isolating mechanism in speciation, the population observed by Gibbs might well be a new species. Against this, Uy & Borgia (2000) report that the two populations differ genetically by only 0.5%, while Diamond (1986) points out that bower construction appears to be cultural, involving learning by both sexes. More study is required.

Undescribed honeyeaters from Indonesia and Papua New Guinea

The genus *Myzomela* (Myzomelas or Honeyeaters) comprises some 26 species of the family Meliphagidae, ranging from eastern Indonesia through Melanesia to Fiji, and south into Australia. Typically they are small, sexually dimorphic birds, with fine, decurved bills and, often in the males, bright red plumage. Eaton *et al.* (2016b) disclosed three as yet undescribed species, all endemic to relatively small islands east and south of Sulawesi, Indonesia. These are detailed overleaf.

'Alor Myzomela' *Myzomela* sp.

From the island of Alor, in the Lesser Sundas. In plumage it resembles myzomelas from the adjacent islands of Wetar *M. kuehni* and Sumba *M. (erythrocephala) dammermani*, but differs from the former in having a less extensive red front in the male and from the latter in having a white, not dark, chest. Occurs in montane eucalypt forest in central Alor, above 1,000 m. Song is a series of high-pitched notes, starting *sip-sip*, then rising and speeding up *tit-tit-tit*, about nine notes over 1–2.5 seconds.

'Rote Myzomela' *Myzomela* sp.

From the island of Rote (Roti), off the SW tip of Timor. Very similar to Sumba Myzomela *M. (erythrocephala) dammermani*, but very distinctive vocally; song is a series of hard, high-pitched downslurred *zip* notes, about six notes per second. In plumage differs from Sumba Myzomela in having a narrower black breast-band. Uncommon in woodland and forest-edge, favouring the canopy. An excellent photograph of a singing male is available on the web (M. Nelson, Birdtour Asia, September 2016).

'Taliabu Myzomela' *Myzomela* sp.

From the island of Taliabu, east of Sulawesi. In plumage the male resembles Seram Myzomela *M. (boiei) elisabethae* in having the underparts largely red, but mottled black. Song is a level series of thin, high-pitched *sip* notes, increasing in volume, about seven notes per second.

Unidentified honeyeater from New Guinea *Ptiloprora* sp.

Gibbs (1994) frequently saw examples of a *Ptiloprora* honeyeater during the 1992 penetration of the Fakfak Mountains mentioned above. These differed from the only other member of the genus in the area, Rufous-sided Honeyeater *P. erythropleura*, in the lack of rufous on the flanks, larger size and streaked underparts. In the absence of specimen material the taxonomic status is unclear, whether a new race of *erythropleura* or an undescribed species (Pratt & Beehler 2014). Similar birds occur disjunctly in the Kumawa Mountains to the east.

'Fakfak Honeyeater' *Melipotes* sp.

During his expedition to the Fakfak Mountains mentioned above, D. Gibbs was also able to make observations of a probable new honeyeater of the genus *Melipotes*, which differed from the other then-known *Melipotes* species of Western New Guinea in having a striking elongation of the eye-wattle (Gibbs 1994). What appeared to be the same species was observed briefly by F. Rheindt in June 2009. In the absence of specimen material the taxonomic position of the Fakfak population is uncertain, although it appears likely that an undescribed species is involved. Diamond & Bishop (2015) comment that the species limits within the genus are unclear and require further specimen material for clarification.

'Peleng Fantail' *Rhipidura* sp.

Eaton *et al.* (2016b) give details of a new fantail from the island of Peleng, narrowly separated from the mainland of Sulawesi. A typical fantail in form, with a long, graduated tail, constantly flirted, and generally brown plumage with contrasting white throat. Closely resembles Sulawesi Fantail *R. teysmanni* in plumage; crown and upper back warmer brown, cleaner white throat and more black on the tail. Vocally distinctive, the song is a slow, deliberate series of level metallic notes, first rising then descending in pitch, quite unlike the tinkling songs of neighbouring species. Fairly common in submontane forest above 800 m, sometimes lower.

'Bismarck Flyrobin' *Microeca* sp.

An undescribed *Microeca* flyrobin has been seen on numerous occasions on the islands of New Britain and New Ireland, Papua New Guinea (Dutson 2011). The birds would appear to be most closely related to Lemon-bellied Flyrobin *M. flavigaster*, a widespread Australasian species. It is possible that more than one species is involved; on New Britain, birds are found in lowland forest and forest-edge, whereas the birds on New Ireland seem to occur from 700–1,400 m. The birds are generally rare, with sightings from a limited number of locations on each island; there is no specimen material available as yet.

Unidentified cliff swallow from Ethiopia *Hirundo* or *Petrochelidon* sp.

Several observers have seen an unknown swallow in Ethiopia (Madge & Redman 1989; Atkins & Harvey 1994). It was described as dark blue-grey above, with a square paler rump, buffy-cream to pale rufous; tail slightly notched; darker cheeks; throat white, the remainder of the underparts dirty white or buffy. No specimens were taken and we are not aware of more recent observations, despite the increasing popularity of Ethiopia as a birdwatching destination.

'White-tailed and Kilombero Cisticolas' *Cisticola* spp.

Two presently undescribed cisticolas occur in the Kilombero Swamp, near Ifakara, south-central Tanzania. Both are extremely well known, to the extent of being illustrated in one of the standard East African field guides (Stevenson & Fanshawe 2002) and their continuing lack of formal description is puzzling. The White-tailed Cisticola occurs in sedges and drier ground adjacent to the main Kilombero Swamp; in plumage it resembles the allopatric Carruthers's Cisticola *C. carruthersi*, with a heavily marked back and an obvious white-tipped tail. The Kilombero Cisticola, which lives in flooded reedbeds in the swamp itself, has essentially unmarked warm brown upperparts and a broadly white-tipped tail; the vocalisations of each are diagnostic. Both species were first detected in 1986.

▲ 'Taliabu Bush Warbler' *Locustella* sp. Taliabu, Indonesia, December 2012 (*Robert Hutchinson*).

'Taliabu Bush Warbler' *Locustella* sp.

The island of Taliabu, lying between Sulawesi and the Moluccas in Indonesia, an area of high endemism, remained essentially unexplored ornithologically until recently. In April 2009 F. Rheindt visited the island, concentrating on the especially neglected higher altitudes. In a remnant patch of stunted forest at 1,050–1,100 m he briefly heard the song of a bird reminiscent of Russet Bush Warbler *Locustella mandelli*. After substantial effort he was able to return and on 15 April made some poor-quality recordings and obtained a brief view, confirming that the singer was a *Locustella* warbler. Subsequently B. Demeulemeester and P. Verbelen made good recordings in November 2009, the use of which allowed R. Hutchinson to take good photographs in December 2012. A type description is in the process of preparation (Rheindt 2010a; Rheindt & Hutchinson 2013).

'Peleng Leaf Warbler' *Phylloscopus* sp.

The Banggai Archipelago consists of a group of islands, the largest being Peleng, stretching eastwards from the mainland of Sulawesi, Indonesia. Despite their close proximity to Sulawesi (at its narrowest the channel separating it is only 14 km wide), the islands harbour a number of endemic taxa, including an endemic *Phylloscopus* warbler first observed by F. Rheindt and P. Verbelen on Peleng Island in March 2009. It was very distinct from wintering Arctic Warblers *P. borealis*, with which it sometimes associated, in being smaller, with a yellowish lower breast contrasting with an off-white throat. It also differed vocally and in feeding ecology, being more inclined to forage on larger branches in the manner of a treecreeper *Certhia*. Observations were made in primary and secondary forest at altitudes of 700–900 m (Rheindt *et al.* 2010b).

'Taliabu Leaf Warbler' *Phylloscopus* sp.

A second new leaf warbler occurs on the island of Taliabu, east of the Banggai Archipelago. It was first observed by P. Davidson *et al.* in 1991, who found it to be common on Taliabu; they reported their birds as an undescribed race of the highly polytypic Island Leaf Warbler *P. poliocephalus*. Subsequent work by F. Rheindt and colleagues suggests that a new species may be involved, but while awaiting further comparative evidence – acoustic, morphological and genetic – the taxon remains undescribed (Eaton *et al.* 2016b).

'Rote Leaf Warbler' *Seicercus* sp.

Eaton *et al.* (2016b) give details of a new warbler from the island of Rote (Roti), off the SW tip of Timor, Indonesia. It is a typical *Seicercus* warbler with a single indistinct wing-bar, lemon underparts, broad yellow supercilium and rather large bill. It has two songs, an ascending warble of notes and a more level pitched thin series of notes. A recent photograph is available on the web (M. Nelson, Birdtour Asia, September 2016).

Unidentified thrush-babbler from Cameroon *Ptyrticus* sp.

In October and November 1998, while conducting a study of primates in the Bakossi Mountains, Cameroon, I. Faucher twice observed an unknown bird which appeared to be a *Ptyrticus* thrush-babbler, but which differed in several respects from the only member of this genus, Spotted Thrush-babbler *P. turdinus*, notably in the more extensive, darker and heavier spotting on the underparts (Faucher & Dowsett-Lemaire 2010). The bird would appear to be rare; it was not encountered during extensive fieldwork in the same area in early 1998, 1999 and 2000.

Undescribed white-eyes from Indonesia *Zosterops* spp.

Eaton *et al.* (2016b) disclose three undescribed white-eyes *Zosterops* from Indonesia, two from small islands and the third from mainland Borneo.

'Wangi-Wangi White-eye' *Zosterops* sp.

Occurs on the small island of Wangi-Wangi, off the SE tip of Sulawesi, where it is sympatric and syntopic with the widely distributed Lemon-bellied White-eye *Z. chloris*. Olive-green above, throat yellow, breast and belly white, with a prominent white eye-ring. Habitat is tall forest and forest-edge. Song is a pleasant warble of high- and low-pitched notes, lasting 1–1.5 seconds.

'Obi White-eye' *Zosterops* sp.

Confined to the island of Obi, south of Halmahera, where it occurs in forest, forest-edge and cultivated areas. A rather dark white-eye, lacking yellow in its plumage apart from the vent. Underparts otherwise greyish, olive-green above with darker forehead and narrow whitish eye-ring. Song is a long, winding series of sweet, musical downslurred notes, lasting 4–10 seconds.

'Meratus White-eye' *Zosterops* sp.

In contrast to upland areas of Malaysian Borneo, the highlands of the Indonesian portion of the island have been relatively little explored ornithologically. In July 2016 a short but extraordinarily productive visit was made to the Meratus range in South Kalimantan by J. A. Eaton and colleagues (Eaton *et al.* 2016a). Despite spending only four days in suitable habitat, the expedition discovered two apparently undescribed species, one new race (of Chestnut-hooded Laughingthrush *Garrulax treacheri*) as well as observing eight species not previously

▼ 'Meratus White-eye' *Zosterops* sp. South Kalimantan province, Borneo, Indonesia, July 2016 (*James Eaton*).

recorded from the area. On the first day in the field, a new white-eye *Zosterops* was seen in a mixed-species flock. Subsequently the bird was seen in single-species flocks of up to 40 individuals as well as, regularly, in mixed flocks. It appears to be common above 1,300 m. Detailed notes were made of its plumage and a number of high-quality photographs taken (Eaton *et al.* 2016a). The species differs from other Bornean white-eyes in being uniform yellowish-olive all over, distinguishing it from Hume's White-eye *Z. auriventer* (for taxonomy, see Eaton *et al.* 2016b), which occurs at lower altitudes in the same range. Song is a warbling series of high-pitched short notes, ending with faster lower notes, lasting 1–3 seconds. Call is a high-pitched buzzy *zip*.

▲ 'Meratus Jungle Flycatcher' *Cyornis* sp. South Kalimantan province, Borneo, Indonesia, July 2016 (*James Eaton*).

'Mantaro Wren' *Thryothorus* sp.

In February 2001 G. Engblom heard a distinctive duetting wren song near Otuto, in the Mantaro drainage of Peru. The location is about midway between the ranges of Plain-tailed Wren *T. euophrys* and Inca Wren *T. eisenmanni*, to both of which species the Mantaro birds had obvious resemblance. They did, however, differ from the latter in lacking conspicuous blackish breast spotting. They most closely resemble the race *schulenbergi* of Plain-tailed Wren, which is the largest and dullest of the four known races and the one most close geographically (Brewer 2001); however, unlike all races of that species, it has obvious darker transverse pale bars.

'Meratus Jungle Flycatcher' *Cyornis* sp.

In the same mixed flock observed by J. A. Eaton and colleagues that contained the first observed 'Meratus White-eye' *Zosterops* sp. (q.v.) was a female *Cyornis* flycatcher which puzzled the group, who were very familiar with other members of this genus. The following day, on 8 July 2016, a typical *Cyornis* song was heard; using playback, a male bird, which clearly was an undescribed species, was lured into view. Detailed notes were made of the plumage of both male and female, which differ markedly from the other four *Cyornis* species found on Borneo. The new taxon was observed between 900–1,300 m (Eaton *et al.* 2016a). Song is a series of 3–7 glissading, prolonged, deliberate notes, occasionally with a high-pitched final note, lasting 1–2 seconds, regularly repeated. Calls include a low-pitched hollow *teow* and a high-pitched *sit-sit*.

'Spectacled Flowerpecker' *Dicaeum* sp.

In June 2009, in the Danum Valley conservation area, Sabah, Borneo, Malaysia, R. Webster noticed an unfamiliar flowerpecker feeding on a fruiting mistletoe some 35 m above the ground. The bird was clearly distinct in plumage from all the known flowerpeckers of the area, being medium slate-grey above, the underparts grey with a white central area and, most strikingly, a conspicuous broken white eye-ring, with a white area above the dark lores, and a white throat with a black malar stripe. There were pure white pectoral tufts emerging from the carpal joint; eye, bill and legs were black. At least two individuals were seen; the putative female had yellowish-tinged underparts. A series of good-quality photographs was obtained. There have also been recent sight records in Brunei (R. Orenstein, pers. comm.). It appears that the bird may be a canopy (Edwards *et al.* 2009) or mid-canopy specialist, feeding on mistletoe, possibly particularly found in botanically poor forest types (Eaton *et al.* 2016b). Call is a high-pitched *zic*.

'Toha Sunbird' *Chalcomitra* sp.

In November 1985 G. & H. Welch observed three sunbirds, presumed to be a male, a female and possible immature, in Wadi Toha, Djibouti; the habitat was secondary, mostly *Acacia* forest at an altitude of about 180 m. The male, in particular, was very distinctive and had no resemblance to any sunbird species in the region, having a characteristic bright metallic yellow crown, grey-brown ear-coverts, bright metallic green chin, throat and upper chest separated from whitish underparts by a narrow black band, undertail-coverts pale with darker barring, black tail, two pale wing-bars, long, strongly decurved black bill and black legs (Welch & Welch 1998). The observers surmised that the bird might be most closely related to Marico and Shining Sunbirds *Cinnyris mariquensis* or *C. habessinicus*; however, Redman *et al.* (2011) place it in the genus *Chalcomitra*. There have not, apparently, been any further records.

'Mount Mutis Parrotfinch' *Erythrura* sp.

Eaton *et al.* (2016b) give details of, and illustrate, a new parrotfinch (*Erythrura*) species from Mount Mutis, West Timor, Indonesia. Plumage of the male is basically green with bright red ear-coverts, sky-blue throat and fore-crown, darker on the lores; tail bright red, bill black. Female is duller with indistinct red ear-covert patch. The species occurs in eucalypt forest with dense understorey at 1,400–1,850 m, feeding in low branches or on ground, occasionally in the canopy. Call is a soft, thin, high-pitched *sit*. Several recent photographs, including a female, are available on the web (R. Hutchinson & M. Nelson, Birdtour Asia 2015).

'Pujyani Flowerpiercer' *Diglossa* sp.

Herzog *et al.* (1999) reported several sightings of a new, all-black flowerpiercer in the humid montane forest between 2,700–3,100 m near Pujyani, Dpto. Cochabamba, Bolivia, all made during May 1997. It is apparently most closely related to Moustached Flowerpiercer *D. mysticalis*. There appear to be no further observations since that time (S. Herzog, pers. comm.).

INVALID SPECIES

It is a not uncommon occurrence for the validity of a newly described species to be later called into question. In a substantial proportion of these cases, new evidence, or a different interpretation of observed facts, leads to the 'delisting' of a putative new species or, at times, its relegation into a type of taxonomic limbo while awaiting further evidence. It is more than likely that this fate awaits several of the species given full specific status in this work, as new data from the field and laboratory become available. The purpose of this appendix is to list those forms, claimed as new species since 1960, which have now been shown to be invalid for one reason or another.

How is it possible to describe an invalid species? There are a number of well-established routes. In some cases, a species may be described on the basis of a single specimen. It is a frequent occurrence among birds, of many different orders, to find abnormal individuals. The commonest abnormality is probably albinism, total or partial; this is, however, so well recognised and obvious that such individuals are rarely mistaken for a novel species. In fact, in the list below, there is only one invalidity arising from this cause. However, more subtle abnormalities, such as slight melanism, are not so blatant, and species have been described and retracted due to this cause. For this reason, many taxonomists prefer to have the security of a series of specimens before claiming a new species.

A fertile source of invalid species lies in avian hybrids. Classical evolution theory posits that hybridisation between distinct 'good' species should be selected against and progeny from such genetic mixing should ultimately disappear. Nevertheless, wild-bred hybrids are frequently found (captive hybridisation is, of course, much more extensive); some orders of birds seem particularly prone to hybridisation, notably ducks and geese (Anatidae), gulls and terns (Laridae) and, of particular importance to this work, hummingbirds (Trochilidae). Hybridisation in hummingbirds quite frequently involves individuals not merely of two species, but also of two genera. Under these circumstances it is not surprising that ornithologists relying on one or a few specimens have frequently been misled, and the number of invalid hummingbird species over the history of New World ornithology is legion. In fact, the South American Classification Committee of the American Ornithologists' Union lists (in 2016) no fewer than 63 invalid hummingbird species, their descriptions dating as far back as 1817, in South America alone. Most of these arose from hybrids or aberrantly plumaged individuals. Fortunately for us, since 1960 invalid hybrids have been relatively few in number.

Another cause of improper descriptions of 'new' species is over-hasty publication, before full supporting data have been obtained. There is, understandably, a great pressure to announce a new species as fast as possible, since the 'kudos' of describing a new species is considerable. In some cases it might, perhaps, have been wiser to have taken a more circumspect pathway; Vuilleumier *et al.* (1992) rather caustically observe that they "have noticed that there is an unhealthy disease striking some ornithologists at present, aptly named "new-species fever" by Remsen. They seem impelled to describe as new the bird...without collecting the data and making the careful studies that are necessary". In fairness, we feel that things have improved; in the early years of the 19th century there was a positive horse-race attitude as, particularly, American ornithologists raced to get priority for the naming of new species, to the extent of sometimes publishing type descriptions in newspapers. In fact, the German ornithologist Wilhelm Blasius (1845–1912) routinely used the daily newspaper *Braunschweigisches Anzeigen*, not perhaps the most common publication on the shelves of ornithological libraries, for rapid, priority-gaining descriptions. It was partially in response to such excesses that led to the founding of such publications as the *Bulletin of the British Ornithologists' Club* to provide some discipline. It should be noted that on several recent occasions, ornithologists have, based on field observations, made public the existence of obvious new species but have refrained from naming them because they clearly recognised the need for more data.

The question as to whether a new taxon is indeed a full, valid species or merely a new race of a known one is always a fluid issue. A good number of taxa described as species have since been classified as subspecies; the converse is also frequent. The number of new species resulting from 'splits' of existing species greatly outnumbers the new species *per se* described. In many cases, genetic analysis provides the final arbiter.

Finally, there are those species (mercifully few in number) that have made it into some form of literature by processes comic, curious or bizarre. Perhaps the most extreme of these involves a parrotlet described new to science, in (so far as we can tell) a journal not peer-reviewed by any competent ornithologists, and involving a poor-quality photograph of a cage full of captive birds whose plumage had been dyed by their captors to improve their saleability.

Below we give a systematic list of species that have been described since 1960 and which have since been deemed to be invalid by most competent authorities, with a brief description of the circumstances. Of course, we also have to recognise that, with further evidence, the invalidity of some of these forms may itself prove to be invalid, leading to their resurrection.

Estudillo's Curassow *Crax estudilloi*

This taxon was announced in an editorial note by J. Estudillo López preceding an article on a putative new species of curassow in a gamebird breeders' journal (Allen *et al.* 1977); in fact, there may have been no actual intent to name a new species, but under the rules of international nomenclature that is exactly what was done. The sole individual was acquired as a chick from an Indian who spoke no Spanish, at a location on the eastern slopes of the Andes in Bolivia. It was kept to adulthood in Dr Estudillo's aviary in Mexico; on dying it was, regrettably, only partially preserved. Subsequent DNA work (Joseph *et al.* 1999) established that the bird had, most likely, a hybrid origin, sharing identical sequences with Blue-billed Curassow *C. alberti*.

Vietnamese Pheasant (or Vo Quy's Pheasant) *Lophura hatinensis*

Originally described by Vo Quy in a rather inaccessible publication in Vietnamese (*Chim Viet Nam* [*Birds Vietnam*], 1973). Described from one male specimen, it closely resembles Edwards's Pheasant *L. edwardsi* but differs mostly by having four white tail feathers. DNA investigations by Hennache *et al.* show *hatinensis* to be identical to *edwardsi* and probably a result of inbreeding; these authors point out examples of *edwardsi* with a long pedigree of captivity developing similar plumage characteristics.

Mascarene Shearwater *Puffinus atrodorsalis*

Shirihai *et al.* (1995) described a new species of shearwater, based on a moribund specimen which died subsequently, found on a beach near Durban, South Africa in January 1987. However, Bretagnolle *et al.* (2000) point out that the biometrics of *atrodorsalis* fall within the range of populations of Audubon's Shearwater *P. lherminieri* from the Seychelles and Réunion Island; consequently *atrodorsalis* belongs to that species and is regarded as a synonym.

Kenyon's Shag *Stictocarbo* (*Phalacrocorax*) *kenyoni*

Described by Siegel-Causey (1991) from skeletal measurements, the holotype obtained from a fishing net in February 1959 on Amchitka Island, Alaska; paratypes were collected by K. W. Kenyon in 1957 at the same location and other material obtained from archaeological middens on Amchitka (Siegel-Causey *et al.* 1991). However, Rohwer *et al.* (2000) made a morphometric analysis of 224 specimens of North Pacific shags and concluded that the features claimed to distinguish *kenyoni* were also found in Pelagic and Red-faced Cormorants *P. pelagicus* and *P. urile*; Siegel-Causey's specimens were probably referable to the former.

Cox's Sandpiper *Calidris paramelanotos*

The brief description (Parker 1982) of a new calidrine sandpiper, based on two specimens taken in 1975 and 1977 by J. B. Cox on the presumed wintering grounds in South Australia, caused immediate sensation and immediate controversy in worldwide ornithological circles. It was assumed that, along with other members of the genus *Calidris* wintering in Australia, the breeding grounds were at some unknown location in NE Asia, probably extreme eastern Russia. Russian ornithologists (e.g. Stepanyan 1990) expressed scepticism that a new species could be breeding, undetected, in Siberia, and postulated a hybrid origin, probably between Ruff *Philomachus pugnax* and either Pectoral or Sharp-tailed Sandpipers *C. acuminata* or *C. melanotos*. Meanwhile observations of the mystery bird began to appear worldwide; sightings in Hong Kong, numerous more from Australia (Smith 1984), one from Japan (Ujihara 2002) and, notably, a supposed juvenile trapped and banded in Massachusetts in September 1987 (Buckley 1988; Vickery *et al.* 1988); the matter was finally laid to rest when Christidis *et al.* (1996) conducted a biochemical investigation of three specimens collected in South Australia between 1989 and 1992 and showed, unequivocally, that they were the result of the crossing of a male Pectoral Sandpiper and a female Curlew Sandpiper *C. ferruginea*.

A second 'mystery' calidrid, the so-called 'Cooper's Sandpiper' *C. cooperi*, known from a unique 1833 specimen collected on Long Island (with recent probable Australian records – Cox 1990) has proven to be a hybrid between Curlew Sandpiper and Sharp-tailed Sandpiper.

Kimberley Imperial Pigeon *Ducula constans*

Bruce (1989) conducted a reappraisal of the Pied Imperial Pigeon *D. bicolor* superspecies, and concluded that the population endemic to the Kimberley area of NW Western Australia was consistently distinguishable in plumage and measurements and named it *Ducula constans*, the Kimberley Imperial Pigeon. However, most authorities (e.g. Schodde & Mason 1999) subsume it into the race *D. b. spilorrhoa*, found in northern Australia and parts of New Guinea.

Hocking's Parakeet *Aratinga hockingi*

Described by Arndt (2006b), based on a comprehensive examination of material in seven different museums in Peru, England, Switzerland, Germany and USA. The type specimen was an example taken by J. T. Zimmer in Huánuco, Peru, in 1922. Notwithstanding Dr Arndt's study, and statements that the form is vocally distinct from other *Aratinga* species, the SACC was unconvinced as to the validity of *hockingi* as a good species, preferring to maintain it within the Mitred Parakeet *A. mitratus* group (Proposal 473). The IOC and others also treat *hockingi* as a junior synonym of *A. mitratus*.

It should be noted that the taxonomies of several *Aratinga* and *Pyrrhura* species are currently in a rather fluid situation, with further changes likely as more data become available (J. Bates, pers. comm.).

Yellow-necked Parrotlet *Forpus flavicollis*

The process whereby this species was described has caused considerable hilarity and derision in the ornithological community. Briefly, P. Bertagnolio and L. Racheli "came upon a colour photograph of some parrotlets", in January 2005, taken by an unidentified photographer at an unknown location in Colombia (obtained via an Internet site which subsequently disappeared). The poor-quality picture showed 32 parrotlets in a small cage, but further details are unavailable. "Although the birds are crowded together and are partly obscured by the wire (mesh), on the basis of what parts of the plumage are visible, note especially the broad yellow collar and distinctive.... yellow or orange frontal band". This type description appeared in an avicultural magazine and does not appear to have undergone any peer-review of the rigour normally considered essential for publication of a new species. Indeed, it fell so far short of the standard norms that it was brusquely and unanimously rejected by all members of the South American Classification Committee, who pointed out that the artificial dyeing of captive birds to increase their saleability to gullible purchasers is common practice in South America; in fact one member, G. Stiles, spoke of "a rash of rose-breasted female (Shiny Cowbirds)" in his local botanical garden, as well as "a very striking orange canary sporting a cute green crest of parrot feathers"!

Omani Owl *Strix omanensis*

In March 2013 M. Robb and R. Pop heard unfamiliar owl calls, whose nature suggested that the caller was a member of the genus *Strix*, in a wadi in the Al Hajar Mountains, Oman. Over the next few months, with great effort, good-quality photographs and sound recordings were obtained. Since the known populations (the bird was also detected at a second site some 23 km from the first – van Eijk 2013) were tiny, it was decided that the taking of even one specimen would be unethical, and the species was described (Robb *et al.* 2013) on the basis of the photographs and recordings, a procedure which attracted some criticism (see Kirwan *et al.* 2015). Further molecular work from blood and feather samples from a captured and released specimen (Robb *et al.* 2015) showed that *omanensis* was in fact identical with the type specimen of Hume's Owl *S. butleri*, becoming a junior synonym, while the widely distributed bird previously known as Hume's Owl was in fact an undescribed species, now named Desert Owl *S. hadorami* (q.v.) (Kirwan *et al.* 2015).

Etchecopar's Owlet *Glaucidium etchecopari*

This form was described in 1983 by Érard & Roux, based on material from the Ivory Coast and Liberia, as a race of Chestnut Owlet *G. capense*, a species with a wide distribution from Uganda to NE South Africa. Citing the widely disjunct populations (a gap of some 3,500 km) as well as apparent differences in vocalisations, König *et al.* (2008) elevated the West African form to a full species. However, the IOC and other recent world checklists continue to treat it as a race.

Nyctisirigmus kwalensis (apparently no common name suggested)

This was named by L. I. Davis on the basis of a single "badly faded" recording made by Stuart Keith at Kwale, SE Kenya, and described in 1978 in an obscure privately published journal, *Pan American Studies*. Whether or not this publication was peer-reviewed is uncertain but Vuilleumier *et al.* (1992) recommend treating *kwalensis* as a "*nomen nudum*" (i.e. a description lacking enough detail to be valid).

Allasma northi (apparently no common name suggested)

A second caprimulgid described, without details or type location, by L. I. Davis, based on a sonogram. Rejected by Vuilleumier *et al.* (1992).

Ruschi's Hummingbirds

The Brazilian ornithologist Augusto Ruschi (1915–1986) was a man of wide and varied interests, devoting his life to studying the flora and fauna of Brazil, Ecuador and Peru. He was "celebratedly eccentric" (*Ibis* 141: 358–367). During the 1960s and 1970s he described several new species of hummingbirds, all of which are now believed to be invalid. The invalid species are:

Threnetes cristina Ruschi 1975, probably an immature of *T. loehkeni,* a species described by Grantsau in 1969, itself probably invalid, being a subspecies of Pale-tailed Barbthroat *T. niger leucurus* (Mayr & Vuilleumier 1983).

Threnetes grzimeki Ruschi 1973b. Apparently an immature Rufous-breasted Hermit *Glaucis hirsuta* (Mayr & Vuilleumier 1983).

Phaethornis margarettae Ruschi 1972, 1973c. Usually treated as a race of Great-billed Hermit *P. malaris* (Hinkelmann 1988b).

Phaethornis nigrirostris Ruschi 1973a. Based on a single specimen, "alive" at the time of description. Probably an immature of Scale-throated Hermit *P. euronyme* (Mayr & Vuilleumier 1983.

Phaethornis maranhoensis (Grantsau 1968). A further form, described by R. Grantsau, is, according to Hinkelmann (1988), the previously unknown male plumage of Cinnamon-throated Hermit *P. natteri.*

Black-capped Woodnymph *Thalurania nigricapilla*

Described by Valdés-Velásquez & Schuchmann in 2009 based on specimens taken some 50 km north of Cali, Colombia. Stated to differ from Purple-crowned and Green-crowned Woodnymphs *T. colombica* and *T. fannyi* (which themselves are treated as conspecific by the SACC, Proposal 558) in having a black crown without iridescence; however, doubts have been raised as to whether the holotype and paratype (taken in 1978) were fully adult, which would explain a lack of iridescence (2013 HBW SV). Not accepted by the SACC or the IOC.

Berlioz's Woodnymph *Augasma cyaneoberyllina*

Described by J. Berlioz in 1965, based on two specimens from "Bahia" (Brazil, but no other data). Probably a hybrid of some type, but true status uncertain (Sibley & Monroe 1990).

Handley's Hummingbird *Amazilia handleyi*

Described by A. Wetmore, based on material collected by Dr C. O. Handley on Isla Escudo de Veraguas, off the coast of Bocas del Toro, Panama. Usually treated as a race of Rufous-tailed Hummingbird *A. tzacatl.*

Blue-green Emerald *Agyrtria rondoniae*

Described in 1982 as a species by A. Ruschi, based on material collected in Rondônia, Brazil. Although Weller (1999, HBW 5) states that it is sympatric with Versicolored Emerald *A. versicolor,* the SACC (Proposal 188) prefers to treat it as a race of *versicolor,* citing, among other things, doubts as to the reliability of the field data provided by Ruschi. The IOC and other world checklists also treat it as a race.

Chisos Hummingbird *Phasmornis mystica*

A very appropriately named form; the generic name is derived from the Greek 'ghostly bird', the specific from the Latin 'mysterious'. It was described by H. C. Oberholser in his posthumously published *The Birdlife of Texas* (1974), based on a male specimen shot at Boot Spring in the Chisos Mountains of Texas by C. H. Mueller in 1932. This specimen is apparently lost. Generally believed to be a hybrid of some kind.

Brilliant Sunangel *Heliangelus splendidus*

Described by Weller in 2011 as a new species with two races, based on a study of more than 100 specimens; the nominate and *H. s. pyropus.* Stated to occur from north-central Ecuador to NW Peru (Piura). After spirited debate the SACC unanimously rejected *splendidus* as a new species, preferring that it be kept as a race of Purple-throated Sunangel *H. viola,* or even allowed to go into abeyance totally, subject to the provision of more data (Proposal 544); the IOC and other world checklists agree with this treatment.

Olrog's Tyrannulet *Tyranniscus australis*

Described from one specimen collected in Jujuy, Argentina (Olrog & Contino 1966); later shown (Traylor 1982) to be Sclater's Tyrannulet *Phyllomyias sclateri.*

Chapada Flycatcher *Suiriri islerorum*

Described by Zimmer *et al.* (2001) as a cryptic species, extremely similar to the syntopic and much more widespread Suiriri Flycatcher *S. suiriri*, the taxonomy of which has been fluid and confusing. The new form has been stated to be distinct from *S. suiriri* in a number of features: morphology and bill measurements (Zimmer *et al.* 2001), habitat preferences (Lopes & Marini 2006), egg colour and pattern (Lopes & Marini 2005), feeding techniques (Lopes 2005), display (Zimmer *et al.* 2001) and vocalisations (Robbins 2004 HBW 9).

Notwithstanding all of the above, Kirwan *et al.* (2014) published a detailed analysis of materials in several collections, including some specimens from the mid-19th century, that had not been included in earlier studies, and concluded that short-billed '*islerorum*' was not distinguishable from the race *S. s. affinis* of Suiriri Flycatcher, which is now treated as a full species by the IOC (Chapada Flycatcher *S. affinis*). They also described a new race of Suiriri Flycatcher, *S. s. burmeisteri*, based on material collected by J. Natterer in 1823. This viewpoint is currently accepted by the IOC and others.

Sucunduri Flatbill *Tolmomyias sucunduri*

In 1995 B. M. Whitney and S. Conyne, while making recordings of birds in the Sucunduri–Tapajós interfluvium, encountered and recorded a *Tolmomyias* flycatcher with a distinctive song. Subsequently further recordings were made, and specimens taken, in 2009, leading to the description of a new species (Whitney *et al.* 2013). Notwithstanding the evidence accumulated by Whitney and co-workers, the SACC has not accepted *sucunduri* as a valid species (Proposal 646), a view accepted by the IOC and others.

Chubb's Kingbird *Tyrannus chubbii*

Described by L. I. Davis (1979) without giving either a type location, a type specimen or designating any diagnostic features, in a privately published and little-known journal. Rejected by Vuilleumier & Mayr (1987) who took the opportunity to inject some very caustic comments about the failure to maintain proper standards of zoological nomenclature and the use of obscure and inaccessible publications as a vehicle for such disclosures.

Roosevelt Stipple-throated Antwren *Epinecrophylla dentei*

Described by Whitney *et al.* (2013) from the left bank of the Rio Roosevelt, municipality of Novo Aripuanã, Amazonas, Brazil. Stipple-throated Antwren *E. haematonota* is a widespread species in western Amazonia. In a study published in 2013, Whitney *et al.* proposed elevating the three currently recognised subspecies to full specific rank, while describing, as a new species, the population of the left bank of the Rio Roosevelt. Although this treatment was initially accepted, on a split vote, by the SACC, subsequently it was deemed invalid as a full species and does not appear as such in any of the current world checklists.

Bamboo Antwren *Myrmotherula oreni*

In 1996, during a Rapid Ecological Evaluation of the Serra do Divisor National Park in Acre, Brazil, B. Whitney collected the first recent specimen of the race *heteroptera* of Ihering's Antwren *M. iheringi*. Subsequent fieldwork showed significant vocal differences between different populations of *iheringi*, prompting a major review of the species complex. In a 2013 publication, Miranda *et al.* suggested that *M. (iheringi) heteroptera* and *M. i. iheringi* should be treated as two good species, while a third population, native to eastern Acre, Brazil, northern Bolivia and SE Peru should be described as a new species. The earliest (unrecognised) paratype was a male mist-netted in Dpto. Madre de Dios, Peru, in 1976. These conclusions were based on morphological examination of a large number of specimens, vocal analyses and genetic data. However, the SACC (Proposal 618) has so far unanimously recommended against acceptance of this new taxonomy. The IOC and others treat *heteroptera* and *oreni* as races of *M. iheringi.*

Restinga Antwren *Formicivora littoralis*

Serra Antwren *F. serrana* was described in 1929 as a monotypic species. In 1990, L. Gonzaga and J. Pacheco, basing their study on historic specimens and on material that they themselves had collected, described two new subspecies, *F. s. interposita* and *F. s. littoralis.* Two years after its description, *littoralis* was elevated to full species rank, under the common name of Restinga Antwren (Collar *et al.* 1992; Remsen *et al.* 2009), a treatment followed by Zimmer & Isler in the *Handbook of the Birds of the World* (Vol. 8, 2003). However in a detailed anaylsis, based on morphometric, vocal and plumage characteristics, Firme & Raposo (2011) concluded that *littoralis* was not a valid species, under either the phylogenetic or biological species concepts, and that it and *interposita* should be treated as races of *serrana*, a view accepted by the IOC and others.

Berlioz's Xenops *Megaxenops ferrugineus*

Described by J. Berlioz, on the basis of two specimens collected in Manu, Madre de Dios, Peru, in February 1964; however, Meyer de Schauensee (1966) and Vaurie (1980) have shown that these specimens are in fact the Peruvian Recurvebill *Simoxenops ucayalae*.

Percnostola macrolopha (apparently no common name suggested)

Described as a new species by J. Berlioz, based on a single male specimen taken in March 1964 at Manu, Madre de Dios, Peru. However, Parker (1982) shows this to be the previously unknown male plumage of White-lined Antbird *P. lophotes*.

Cipo Cinclodes *Cinclodes espinhacensis*

In December 2006 a bird obviously closely resembling Long-tailed Cinclodes *Cinclodes pabsti* was observed foraging in the Serra do Cipo National Park, Minas Gerais, Brazil. Several further observations were made in 2007 and a specimen was taken in February 2008. The locations of these records were highly disjunct from the only known locations of *C. pabsti* – more than 1,000 km to the north – prompting obvious speculation as to the taxonomic status of the newly discovered population. Nevertheless it was published as a range extension of *pabsti* (de Freitas *et al.* 2008). Consequently, the same investigator returned to the location on ten occasions to collect more observational data, as well as a series of specimens; as a result, a further paper was published in 2012 naming the northern population as a full species, based on plumage and vocal differences, backed up by molecular data, which it was estimated indicated a divergence from *pabsti* of about 220,000 years standing (de Freitas *et al.* 2012). However, several other ornithologists have expressed the viewpoint that the differences between the two populations deserve no more than subspecific separation (SACC, Proposal 548); in particular, K. Zimmer reported his own observations of '*espinhacensis*' responding very vigorously to recordings of both the song and the call notes of *pabsti*. The IOC and others regard *espinhacensis* as a subspecies of *C. pabsti*.

Brigida's Woodcreeper *Hylexetastes brigidai*

Described as a new species from northern Brazil by da Silva *et al.* (1995); treated as a race of Red-billed Woodcreeper *H. perrotii* by several authorities (e.g. Marantz *et al.* 2003 HBW 8; Ridgely & Tudor 2009).

Carajas Woodcreeper *Xiphocolaptes carajensis*

Described as a full species by da Silva *et al.* (2002b); treated as a race of Strong-billed Woodcreeper *X. promeropirhynchus* by Ridgely & Tudor (2009) and Marantz *et al.* (2003 HBW 8).

Xingu Woodcreeper *Dendrocolaptes retentus*

The Amazonian Barred Woodcreeper *D. certhia*, itself recently split from Northern Barred Woodcreeper (now *D. sanctithomae*), is a wide-ranging species, occurring over much of eastern South America east of the Andes from Venezuela and the Guianas south to Bolivia. In the type description paper for *D. retentus*, Batista *et al.* (2013) suggest raising six races of *D. certhia* to full species status, while a seventh, previously undescribed, becomes a new species, *D. retentus*. Although several authorities (e.g. Clements, 6th edition 2013) and del Hoyo *et al.* (2013) accept the validity of *retentus*, the SACC unanimously rejected it, arguing that the lack of clearly differentiated vocalisations, and possibly weak genetic data, suggest subspecific status. The IOC also retains *retentus* as a subspecies.

Tupana Scythebill *Campylorhynchus gyldenstolpei* and Tapajos Scythebill *C. cardosoi*

The Curve-billed Scythebill *C. procurvoides* is a widespread South American species, inhabiting lowland forest in a broad swathe from eastern Venezuela, the Guianas, NE Brazil and across the Amazon Basin to NE Peru, eastern Ecuador and southern Colombia. In 2013 Aleixo *et al.* published one paper, and Portes *et al.* another, proposing the description of two new species, Tupana Scythebill *C. gyldenstolpei* and Tapajós Scythebill *C. cardosoi*, as well as the elevation of the four previously described races of *procurvoides* to full species; the evidence for these opinions was biochemical and vocal. Presently the SACC (Proposal 623) has not accepted these arguments, while clearly recognising the necessity of reclassifying the whole complex and the desirability of amassing further evidence. The IOC and others treat both new taxa as races of *C. procurvoides*.

Black-and-green Sericornis *Sericornis nigroviridis*

In October 1962 A. H. Miller, working in the Watut River drainage of eastern New Guinea, collected an unknown Australasian tree warbler which he described as a new species. Notwithstanding the fact that the type locality was very well collected, no further specimens was seen or obtained. In a study of the unique specimen and examples of the widespread Buff-faced Sericornis *S. perspicillatus* Beehler (1978) concluded that Dr Miller's specimen was in fact a melanistic example of that species.

Rusty-tailed Gerygone *Gerygone ruficauda*

In a 1983 publication, J. R. Ford and R. E. Johnstone described a new species of flyeater (*Gerygone*), based upon three old specimens, all dating from the late 19th century, two in the Australian Museum, Sydney, the third in the American Museum of Natural History, New York. The two Sydney specimens had labels suggesting that they had been taken in Queensland, the AMNH specimen being without a location. In 1985 R. Schodde published a detailed analysis of this material, including some intricate detective work on the provenance of the specimens, reaching two conclusions. Firstly, the specimens probably do not represent a new species, but were most likely examples of Yellow-bellied Gerygone, a New Guinea species, the plumage colours having been altered by soiling and by having been prepared from alcohol-preserved specimens; and secondly, the Australian locations were almost certainly in error, the specimens having most likely originated in New Guinea. This species is therefore almost certainly invalid.

Bulo Burti Boubou *Laniarius liberatus*

A euphoniously-named bird which in its short lifetime has generated far more than its fair share of controversy on several fronts. In August 1988 E. F. G. Smith briefly glimpsed an unidentified bush-shrike in Bulo Burti, Somalia. The bird was seen several times subsequently, when the possibility of a new species was considered. It was captured in January 1989, and blood and feather samples for DNA analysis were taken; unfortunately these were lost when the airline mislaid the package. Fortuitously, some feathers and traces of dried blood were salvaged, and sent to Copenhagen for analysis. In the meantime the bird itself was taken to Germany, where it was kept alive until being returned to Somalia in March 1990. It could not be returned to the point of capture due to civil war (in any case, most of the habitat in that area had been destroyed) so it was, finally, released in the Balcad Nature Reserve (hence the specific epithet *liberatus*). On the basis of the plumage description and biochemical evidence a new species was described (Smith *et al.* 1991).

The process whereby this form was described, on the basis of a single individual, not collected, and released a substantial distance from the point of capture, has proven controversial (see, for example, Le Croy & Vuilleumier 1992; Collar 1999b) and has given rise to much frank and comradely discussion. Notwithstanding all of the above, a subsequent 2008 publication (Nguembock *et al.* 2008) provides evidence, based on molecular data, that *liberatus* is nothing more than an unusual colour morph of Black Boubou *L. nigerrimus,* which itself is a rare and localised species of coastal NE Kenya and southern Somalia.

Bluntschli's Vanga (or Short-tailed Nuthatch Vanga) *Hypositta perdita*

While examining boxes of unidentified bird skins held in the collections of the Forschungsinstitut Senckenberg, Frankfurt, in 1996, S. Peters discovered two specimens of recently fledged birds which resembled female and juvenile Nuthatch Vanga *H. corallirostris* which he named, in the same genus, as *H. perdita* Bluntschli's Vanga, in recognition of the collector Hans Bluntschli (1877–1962) who had collected them in 1931; the scientific specific name means 'lost'. This treatment was not, however, universally accepted (see Yamagishi & Nakamura 2009 HBW 14). To resolve this issue Fjeldså *et al.* (2013) conducted a genetic analysis that revealed that the Bluntschli's specimens were in fact a rare brown colour morph of White-throated Oxylabes *Oxylabes madagascariensis.*

Mees's Greenlet *Hylophilus puellus*

In 1974 G. F. Mees published a description of a new greenlet based on a single specimen from Tafelberg Mountain, Suriname. Subsequent work by Meyer de Schauensee (1982) showed that this was not a vireonid at all, but an example of a Rufous-rumped Antwren *Terenura callinota.*

Andrew's Swallow *Hirundo andrewi*

In early April 1965, a large concentration of several species of migrating swallows and martins was found on the eastern shore of Lake Naivasha, Kenya. Among these birds, which included both African species and Palaearctic migrants, were several examples of a grey-rumped swallow; with some difficulty one specimen, which proved to be an adult female, was collected. Two days later the migrant concentration had disappeared and the unusual birds were not seen again. On the basis of the single specimen a new species was described by the eminent Kenyan ornithologist J. G. Williams, who named the new bird after his son. Most authorities believe that the Naivasha specimen was a colour morph of Grey-rumped Swallow *Pseudohirundo griseopyga*, although others (e.g. Turner 1989) classify it as a subspecies *P. g. andrewi*.

Degodi Lark *Mirafra degodiensis*

Described by C. Érard in 1975, based on material collected at Bogol Manyo, Degodi, Sidamo province, Ethiopia. Formerly regarded as forming a superspecies with Gillett's Lark *M. gilletti* and treated as such in the *Handbook of the Birds of the World* (Vol. 9, 2004). However, a more recent detailed study by Collar *et al.* (2009) strongly suggests that it is indeed conspecific with *gilletti,* and it is treated as such by all recent authorities.

Sidamo Lark *Heteromirafra sidamoensis*

This form was described in 1975 by C. Érard, based on one specimen taken in 1968 on the Liben Plain near Negelle, Sidamo province, Ethiopia; another specimen was taken in 1974. *H. sidamoensis* is now regarded as a junior synonym of the monotypic *H. archeri* (Archer's Lark), a species described in 1922 from Somaliland and not seen there since (Spottiswoode *et al.* 2013). This viewpoint is accepted by the IOC and others.

Archer's Lark is currently only known from two small sites in Ethiopia, with the bulk of the population on the Liben Plain. It is classified as Critically Endangered and considered to be the most threatened species on the African continent (Spottiswoode *et al.* 2009).

Dorst's Cisticola *Cisticola dorsti*

Described by C. Chappuis and C. Érard in 1991 from three specimens collected in Cameroon and Chad, with accompanying tape recordings of songs, which were distinct from those of the nominate race of the Red-pate Cisticola *C. r. ruficeps*. However, Dowsett-Lemaire & Dowsett (2005) showed that the song was in fact identical to that of the race *guinea* of *C. ruficeps*, and consequently *C. dorsti* is a junior synonym of that subspecies.

Hall's Bulbul *Andropadus hallae*

Described by A. Prigogine in 1972, based on a single specimen taken in September 1970 at Nyamupe, eastern Zaire (Democratic Republic of Congo). Now regarded as a melanistic example of Green Bulbul *A. virens* (Fishpool & Tobias 2005 HBW 10).

Chinese Leaf Warbler *Phylloscopus sichaunensis*

Between April 1986 and June 1989, P. Alström and colleagues observed numerous examples of an unfamiliar *Phylloscopus* warbler in central China, mostly in Sichuan province. No individual could be trapped, but a tentative announcement of a "new species of warbler of the genus *Phylloscopus*" was published (Alström *et al.* 1990). In June 1990, P. Alström and U. Olsson found 20 singing males of the same species in Wolong Nature Reserve, Sichuan, including some paired birds whose nests were found. Three individuals (one male and two females) were trapped, measured and photographed, and blood samples were taken. No specimen was collected; the species was described from a 1962 specimen from Shanxi province, found under the label of *P. proregulus chloronotus* but regarded as a representative of the Wolong birds (Alström *et al.* 1992). However, in a paper published in 2004, Martens *et al.* pointed out that Dr Alström's birds were in fact synonymous with *P. yunnanensis,* a species which had been (briefly) described by La Touche in 1922 based on a few specimens collected on migration in Yunnan province, China. Nevertheless, Alstrom *et al.*'s 1992 paper described the species' song, calls, habitat, breeding biology and interspecific behaviour for the first time, as well as its breeding range from Sichuan to Shanxi and Hebei.

Northern Parrotbill *Paradoxornis polivanovi*

Described by L. S. Stepanyan as a race of the Reed Parrotbill *P. heudei* (Stepanyan 1974). In a later publication (Stepanyan 1998) he argued for full specific status, citing morphological and taxonomic arguments and, among other factors, the very disjunct ranges of *P. (h). polivanovi* and nominate *P. h. heudei*. However, Robson (2007 HBW 12) shows that this gap is less than stated, with previously unknown populations being found, and with further work might be reduced still more; on this basis Penhallurick & Robson (2009) suggest returning to Stepanyan's original assessment, i.e. treating *polivanovi* as a race of *heudei*.

Kibale Ground Thrush *Zoothera kibalensis*

In December 1966 R. Glen and A. Williams collected two ground thrushes in the Kibale Forest, western Uganda. These were identified as Black-eared Thrushes *Z. camaronensis*. Later, in a paper published in 1978, A. Prigogine conducted a comparative examination of the two Kibale specimens and described them as a new species, the Kibale Ground Thrush *Z. kibalensis*. However, several authors, e.g. Traylor (in litt., quoted in Vuilleumier & Mayr 1987) classify it as a race of *camaronensis*, a treatment followed by the *Handbook of the Birds of the World* (Vol. 10, 2005) and the current IOC World Bird List; these taxa are also now placed in the genus *Geokichla*.

Sangha Robin *Stiphrornis sanghensis*

In 1996 P. Beresford and J. Cracraft conducted an expedition to the SW Central African Republic to collect material for the American Museum of Natural History. Several specimens of what was believed to be Forest Robin *Stiphrornis erythrothorax xanthogaster* were obtained; however, on detailed study of the material, it became apparent that specimens from the Dzanga-Sangha Dense Forest Reserve represented an undescribed population. Further material was obtained in 1998. Based on plumage differences from other populations (a deep orange-yellow chin, throat and upper breast, and a yellow wash to the belly feathers), as well as DNA studies, a new species *S. sanghensis* was described (Beresford & Cracraft 1999). This treatment was accepted by the then current IOC World Bird List and by the 3rd edition of Howard & Moore (2003). However, more recently this population has been reclassified as a race of *S. erythrothorax* (Howard & Moore, 4th edition, and the current IOC World Bird List).

Enigmatic Shortwing *Brachypteryx cryptica*

In March 1979 S. Dillon Ripley collected five specimens of a presumed turdine bird in Arunachal Pradesh, NE India, which he described as a new shortwing, closest in size and coloration to Rusty-bellied Shortwing *B. hyperythra*, (Ripley 1980). Shortly afterwards, however, the same author, having access to further specimen material, realised that the birds were not thrushes at all, but instead belonged to the large assemblage of timaline babblers (Ripley 1984). This bird is now subsumed into the race *assamense* of Buff-breasted Babbler *Pellorneum tickelli*.

Benson's Rock Thrush *Monticola bensoni*

Described as a species by Farkas (1971), based on two 19th-century specimens of obscure provenance (this was ultimately clarified, by some remarkable detective work, by Collar & Tattersall in 1987). This form is found in dry or semi-arid rocky scrub in SW Madagascar. Accepted by Collar (2005 HBW 10) who states that it is vocally distinct from Forest Rock Thrush *M. sharpei*, a species found in wetter locations in eastern Madagascar. However, treated as conspecific with *sharpei* by almost all authorities (e.g. Cruaud *et al.* 2011, Safford & Hawkins 2013, Sinclair & Langrand 2013).

Beijing Flycatcher *Ficedula beijingensis*

In 2000 Zheng *et al.* described a new species of flycatcher from the Beijing region of China, based on specimens of breeding birds. However, these authors did not realise that in some species of passerine the fully adult plumage is not acquired until the third year, and birds will breed in apparently immature plumage; the birds described by Zheng *et al.* were in fact immature males of the *elisae* race of Narcissus Flycatcher *F. narcissina* (Töpfer 2006).

Nectarinia sororia (apparently no common name suggested)

Described as a new species in 1960 by S. D. Ripley, based on a single specimen collected in 1957 in the Malange district of Angola. This is apparently a female of Bannerman's Sunbird *Cyanomitra bannermani* (Cheke & Mann 2008 HBW 13).

Lake Victoria Weaver *Ploceus victoriae*

Until fairly recently the status of several *Ploceus* weavers in East Africa was somewhat confused. Consequently a male specimen of a bird, somewhat resembling Northern Masked Weaver *P. taeniopterus*, was taken from a breeding colony near Entebbe, Uganda, in 1983. It was not conclusively identified at the time, but in 1985 it was compared to *Ploceus* specimens in the Natural History Museum, Tring, by J. S. Ash, who considered it to be sufficiently distinct from *taeniopterus* to warrant description as a new species (Ash 1986). This prompted M. Louette (1987) to suggest a hybrid origin, an opinion rebutted by Ash (1987). Currently treated by most authorities as a hybrid between *taeniopterus* and Black-headed Weaver *P. melanocephalus*.

Cream-bellied Munia *Lonchura pallidiventer*

Described by R. L. Restall in 1996, based on a holotype and syntype acquired from bird markets in Jakarta, Indonesia, in 1992. The birds came from a bird-trapper, from SE Kalimantan in Borneo. In all, Restall saw a total of 13 individual birds, of which he acquired three (these went to the Zoological Society of San Diego). In his type description paper Restall gives details of calls, display and courtship of captive birds. However, in a 1998 paper, S. van Balen suggests a hybrid origin, noting the great variability of the specimens, their apparent sterility and inability to breed in captivity, and the fact that no wild observations have ever been made (Smythies 2000); it is distinctly possible that the individuals seen by Restall may have been the product of deliberate aviary hybridisation, a practice frequently used to produce unusual birds of presumably higher resale value. The probable parents are Scaly-breasted and White-bellied Munias *L. punctulata* and *L. leucogaster.* Restall has accepted these arguments (Howard & Moore 2003).

Hypochera (*Vidua*) *lorenzi* (apparently no common name suggested)
Hypochera (*Vidua*) *incognita* (apparently no common name suggested)

In 1972 J. Nicolai described two new species of indigobird, *Hypochera* (*Vidua*) *lorenzi* and *H.* (*V.*) *incognita*, the latter based on two cage birds of unknown provenance acquired in Angola and Zaire (DRC). Both are obligate brood parasites, *lorenzi* on Black-bellied Firefinch *Lagonosticta rara*, and *incognita* on Bar-breasted Firefinch *L. rufopicta.* Both species as described were essentially inseparable from other *Vidua* species in the field, complicated by the fact that each had differing tones of gloss in the male plumage, blue or violet in *lorenzi* , green or blue in *incognita*, according to geographic location. Important distinctions involved the pattern of the inside of the mouth of nestlings, mimicking those of the host species. Neither species is currently recognised, following the work of Payne (1982). *Lorenzi* is treated as conspecific with Wilson's Indigobird *V. wilsoni*, and *incognita* as conspecific with Village Indigobird *V. chalybeata* (Payne 2010 HBW 15).

Long-tailed Pipit *Anthus longicaudatus*

Described by R. Liversidge (1996), who had noticed that during the austral winter, flocks of an unusual pipit were seen around Kimberley, South Africa. Consequently a number of specimens were taken and a new species described. It was also claimed (Liversidge 1998) that the new species could be picked out in the field by posture and mannerisms. However, Davies & Peacock (2014) examined the specimen material and concluded that four, including the male paratype of *A. longicaudatus*, were in fact Buffy Pipits *A. vaalensis*, while the fifth, a female paratype, was actually a Long-billed Pipit *A. similis.* This viewpoint is currently accepted by the IOC and others.

Kimberley Pipit *Anthus pseudosimilis*

The history of this form has been complex and confusing. DNA evidence of specimens taken in the region of Kimberley, South Africa, suggested that a new species, not close biochemically to the morphologically-similar Long-billed Pipit *A. similis*, was involved (Voelker 1999; Liversidge & Voelker 2002). Subsequently, Peacock (2006) stated that the new form differed from *similis* in plumage and wing-formula, as well as appearing to have a different nest construction and egg colour. Notwithstanding all of this, a recent paper (Davies & Peacock 2014) claimed that *pseudosimilis* is an invalid species, the specimens putatively ascribed to it being one African Pipit *A. cinnamomeus*, three indistinguishable from *A. similis*, while the last was a Mountain Pipit *A. hoeschi.* This viewpoint is currently accepted by the IOC and others.

South Hills Crossbill *Loxia sinesciurus*

The taxonomy of different populations of the Red Crossbill *Loxia curvirostra* across temperate coniferous forests in both in Eurasia and North America is complex, with differing vocalisations and subtle differences in bill morphology in different groups. C. W. Benkman and his colleagues have performed long-term fieldwork and have published numerous studies of crossbill populations in western North America (e.g. Benkman 1993, 1999, 2003; Benkman *et al.* 2003). In 2009 a paper was published (Benkman *et al.* 2009) describing a new species, the South Hills Crossbill *L. sinesciurus*, citing as evidence for specific status vocalisations, geographic isolation and apparent reproductive isolation from sympatric crossbills. It was suggested that the new species has evolved specifically in an environment lacking the food-competitive American Red Squirrel *Tamiasciurus hudsonicus* (hence the given name *Loxia sinesciurus*, the 'squirrel-less crossbill'). This form is currently classified by the IOC World Bird List as a race of the Red Crossbill *L. curvirostra*, although undoubtedly further evidence might change this (White 2010).

Narosky's Seedeater (or Entre Rios Seedeater) *Sporophila zelichi*

In 1969 S. Narosky was sent several specimens of seedeaters *Sporophila* by Dr M. R. Zelich, originating in Entre Rios, Argentina. Two skins (and two further specimens in captivity) could not be assigned to any known species, notably having a white collar. Consequently Narosky described them as a new species, *S. zelichi*. Subsequent observations were made in Uruguay (Azpiroz 2003), Brazil (Bencke 2004) and Paraguay (Clay & Field 2005). Nevertheless, doubts as to the validity of the species were raised, postulating among other possibilities a colour morph or hybrid origin (Ridgely & Tudor 1989). Finally, Areta (2006) showed that the most likely explanation is that *zelichi* is a colour morph of Marsh Seedeater *S. palustris*.

Guerrero Brushfinch *Arremon kuehneri*

Described by Navarro-Siguënza *et al.* in 2013, based on comparative studies of mitochondrial DNA. The authors stated that birds from the state of Guerrero, SW Mexico, which are indistinguishable from examples of the race *suttoni* of Chestnut-capped Brushfinch *A. brunneiceps* in plumage and measurements, were nevertheless a discrete species, counter-intuitively most closely related to Green-striped Brushfinch *A. virenticeps*, a species of very different plumage. The validity of this 'super-cryptic' species has not, currently, been accepted by any of the major world checklists.

Frisch's Tanager *Rhamphocelus ciroalbicaudatus*

In 2007 J. D. Frisch named a new species of *Rhamphocelus* tanager that he had photographed in São Paulo, Brazil, using as a vehicle a publication of somewhat less than global circulation, namely *Nature Society News, Griggsville* (Griggsville is a town of some 1,200 population in Illinois). This fell rather short of acceptable documentation for the description of a new species; the bird in question was almost certainly a partially albino Silver-beaked Tanager *R. carbo*.

FUTURE DISCOVERIES

How many new species are there still to come?

The ultimate question is, of course, how many species have yet escaped detection and remain still to be discovered? Only a reckless and courageous person would come up with an estimate. In a celebrated paper published in 1946, no less an authority than Ernst Mayr, one of the 20th century's most distinguished ornithologists, opined that "The period of new discoveries is practically at its end. I doubt that in the entire world even as many as 100 new species remain to be discovered". One is, perhaps, reminded of the situation at the end of the 19th century when some of the pundits of physics stated that their science was "pretty well settled" with only a few minor loose ends to be tied up. Then along came Planck, and Einstein, and Bohr, and Rutherford, and Heisenberg, and... Nobody, I think, would regard present-day physics as "pretty well settled". In the 14 years between Mayr's paper and the period covered by this book, some 57 species were described; a further 288 are covered by this work, while to this author's certain knowledge, at least another 40 are 'in the process of being described'. And this does not include new species arising from 'splits'.

Where are the as-yet-undescribed species going to come from? In the past, considerable numbers of species have been described from geographic locations that were simply too remote or inaccessible to have been adequately explored previously. Locations like the Foja and Fakfak ranges of New Guinea, essentially unknown even to local people until accessed by helicopter, the Chiribiquete Mountains of Colombia or 'Hill 1538' in Loreto, Peru, when finally accessed, immediately produced spectacular new species. Similarly, islands not previously investigated – notably in Indonesia – have provided a trove of new species. However, most potential new sites have now been explored (with some possible exceptions in New Guinea and Indonesia), so the scope for spectacular new discoveries is becoming more limited. (Although, it must be said that some extraordinary discoveries have been made in locations so well-known that one is left with a sense of wonder as to how the species concerned could possibly have been undetected for so long; the Elfin Woods Warbler type specimen was taken in a US national park (in

▼ Elfin Woods Warbler, male. El Yunque, Río Grande, Puerto Rico, USA, February 2015 (*Dubi Shapiro*).

▲ Mayotte Scops Owl. Pic Combani, Mayotte, Comoro Islands, December 2015 (*Ken Behrens*).

▲ Pacific Pygmy Owl. Balsas, Dpto. Cajamarca, Peru, October 2015 (*David Beadle*).

Puerto Rico), a few minutes' hike from the park headquarters which is staffed by government employees, while this author located two singing males within five minutes of getting out of his car. The Cambodian Tailorbird was described as recently as 2013 from the suburbs of a capital city, equivalent to finding a novel species in Kingston-upon-Thames or Mississauga.) However, one can surmise that the majority of new species in the future are going to be 'cryptic' species, that is, species closely resembling others, the discoverer often having been alerted by a distinctive vocalisation. Subsequent collection and careful examination of specimen material, usually in comparison with existing museum specimens and generally nowadays backed up by molecular studies, then reveals a distinctive form. Often, retrospective examination of museum material reveals that examples of a new species were already in collections, sometimes for many years, but had simply been overlooked. One might even surmise that the world's museums may at this very moment contain several undescribed species, waiting for an alert, perceptive and painstaking ornithologist to uncover them.

What are the future's undescribed species likely to be? There is a clear bias against obvious, large, diurnal species. Thus this book contains no gulls, terns, penguins, storks or herons, while the only shorebird is a semi-nocturnal woodcock. Instead, we list some 30 new owls (about 15% of all known owls, and more than 10% of all new species covered in this book), and numerous nightjars and rails. For obvious reasons, being nocturnal or extremely secretive delays detection. Other groups

▲ Cambodian Tailorbird. Udong Mountain Temple, Cambodia, February 2016 (*David Beadle*).

▲ Atiu Swiftlet in front of coconut palm. Atiu, Cook Islands, September 2009 (*Gerald McCormack*).

▲ Alta Floresta Antpitta. Alta Floresta, Mato Grosso, Brazil, August 2016 (*João Quental*).

well represented are swifts – often very difficult to identify – and hummingbirds, with frequently a very limited geographic distribution. In the passerines, again there is obvious bias towards some groups, often secretive or confusingly plumaged; antbirds, antpittas, tyrannid flycatchers and, above all else, tapaculos, where there has been a positive epidemic of new species in the last 20 years, driven by careful analysis of vocalisations.

So, we feel emboldened to make two predictions; that genuine new species, as opposed to 'splits', will come from the ranks of small, inconspicuous, furtive or nocturnal birds, with relatively few spectacular, conspicuous examples such as the Araripe Manakin or the Scarlet-banded Barbet; and that whatever predictions we make, we are almost certain to be proven wrong in the near future. This is, quite certainly, one of the fascinations of ornithology and why for so many of us it is such an obsession.

► Male Scarlet-banded Barbet. Plataforma, Dpto. San Martín, Peru, July 2014 (*Dubi Shapiro*).

▼ Magdalena Tapaculo. Reinita, Colombia, January 2010 (*Nigel Voaden*).

THE ETHICS OF COLLECTING

To Collect or not to Collect?

For most of the history of taxonomic ornithology, the question of the ethics and necessity of collecting specimens was rarely, if ever, raised. The acquisition of a specimen, preferably "a good long series" of specimens, was regarded as a necessity for the description of a new species. Virtually all of 19th century taxonomic ornithology was based on specimens; in many cases the type description of a form was written by a museum-based ornithologist, based on specimen material sent by workers in the field. Thus, the specific or subspecific epithet '*sharpei*', based on the indefatigable British ornithologist Richard Bowdler Sharpe (1847–1909), occurs 28 times in the Howard & Moore Checklist, even though Sharpe himself did not collect them. (Almost 200 species and subspecies carry the name 'Sharpe' as the naming authority.) J. J. Audubon (1785–1851), whose work was hugely influential in the early years of the study of American birds, is known to have shot massive numbers of specimens; Alexander Wilson (1766–1813), frequently referred to as the "father of American ornithology", is commemorated in his native Paisley, in Scotland, with a statue on the Abbey Green, the inscription on the plinth describing him as "ornithologist and poet". In one hand he holds a pen; in the other, a dead bird, looking suspiciously like a Passenger Pigeon.

Thus, for much of the time since Linnaeus, the acquisition of specimens was simply a non-issue. It was deemed an essential part of the process. Questions of conservation, i.e. whether the taking of specimens had a significant deleterious effect on the viability of the species seem rarely, if ever, to have been raised. When the veteran American collector Rollo Beck (1870–1950) visited the island of Guadelupe, off Baja California in 1900 he took nine specimens of the endemic caracara (*Caracara lutosa*), that is, all the ones that he saw (he shot at and missed, or merely wounded, two others). He may well have been the last ornithologist to see the species alive, although in fairness the final extinction of the species can probably be blamed on sheep- and goat-herders. Earlier, the distinguished American ornithologist Dr Edward Palmer took about a dozen specimens of the "Guadalupe eagle" (sic), which "he knew to be a rare bird in process of extinction", a circumstance that does not appear to have inhibited his collecting activities one whit (Abbott 1933).

▼ Varzea Thrush. Rio Ariaú, Iranduba, Brazil, September 2016 (*Thiago Laranjeiras*).

In recent years, however, the ethics and necessity of collecting specimens has been the subject of vigorous debate, with widely divergent views expressed in the literature and in various public forums. At one end, there are naturalists who believe that any collection, for any purpose, is unjustifiable and hence morally repugnant. These views command widespread respect but, I think it is safe to say, do not reflect the opinions of the majority of taxonomic ornithologists, who are prepared to countenance further and continued collection, albeit with restrictions. The opposing viewpoints have been aired quite extensively in the literature; Vuilleumier (1998) gives a good overview of, in his opinion, the necessity of collection, while Donegan (2000, 2008) provides a thoughtful counterbalance. Numerous other published opinions occur in the literature; Peterson (2014), for example, maintains adamantly that type specimens are essential, and suggests that the formal naming of new species be delayed until such material is available.

Part of the controversy arises from different interpretations of the International Code for Zoological Nomenclature, boiling down to whether an actual specimen in the hand is required to describe a new taxon, or whether other convincing evidence may suffice in certain circumstances. Dubois and Nemésio (2007) argue for the first viewpoint, Donegan (2008) for the latter.

The arguments for having an actual specimen in the hand are several. The specimen allows direct comparison with other, possibly related or potentially conspecific forms, in a way that is difficult from photographs. Especially nowadays, one specimen will yield multiple layers of information; not merely plumage and measurements, but diet (based on stomach contents), moult, internal and external parasites, etc, and even blood or tissue material for genetic analysis. In older collections, much of these data were not recorded, and hence the specimens are of lesser value than ones taken recently – a fact that Vuilleumier (1998) makes in support of continued collection. In their type description of Varzea Thrush *Turdus sanchezorum*, O'Neill *et al.* (2011) make a strong argument for the acquisition of recent material with careful recording of data, pointing out that the new species was only discovered and documented with modern specimen material. One unarguable consideration for the taking of specimens involves detailed anatomical studies. When a form of debatable affinities is discovered – notable examples in

▼ Pardusco. Bosque Unchoq, Peru, September 2015 (*Miguel Lezama*).

this book being the Pardusco *Nephelornis oneilli* and the Pink-legged Gravateiro *Acrobatornis fonsecai* – detailed anatomical studies of features such as skull structure, tongue morphology and musculature are required to elucidate its taxonomic relationships. Clearly, these cannot be done without sacrificing some birds as specimens.

It has also been argued that the new knowledge arising from collection actually aids conservation, in that by better defining habitat requirements the overall survival of the species is aided; it should be noted that, above all other causes, extinctions have been caused by habitat destruction. Although, especially in the earlier years of the 19th century, unseemly competition between museums and similar institutions led to uncontrolled collection of shrinking populations (the extreme example being the race to get skins and eggs of Great Auk *Pinguinus impennis*, which undoubtedly contributed to the demise of the species), in recent years it can be argued that in no case has collecting been shown to have significant effects on the populations of any species. Other, more inexorable, forces have taken their toll.

▲ Pink-legged Graveteiro, immature. Serra Bonita Reserve, Brazil, June 2017 (*Ciro Albano*).

The opposing view, ably stated by Donegan (2000, 2008), denies most of these points. (Donegan does not state that *all* collecting should cease, merely that very clear justification is required.) In fact, several species in this book were described without a full, formal specimen being taken; but in all these cases the population appeared to be very small and, tellingly, all were in recent years. Thus, the Bugun Liocichla *Liocichla bugunorum* was described from photographs of a captured specimen and feather samples; the Omani Owl *Strix omanensis* from photographs alone, along with recordings of vocalisations (although this species has turned out to be invalid – see below); and, perhaps most controversially, the Bulo Burti Boubou *Laniarius liberatus* (also now considered invalid – see chapter on Invalid Species), the lone example of which was captured in Somalia, transported to Germany, returned to Somalia and released in a different location from its capture. This sequence of events gave rise to much spirited discussion (Le Croy & Vuilleumier 1992; Collar 1999). In the case of the Omani Owl, the situation became yet more complex with the subsequent description of Desert Owl *Strix hadorami* by Kirwan *et al.* (2015). The discov-

► Bugun Liocichla. Eaglenest Wildlife Sanctuary, Arunachal Pradesh, India, December 2015 (*Jainy Kuriakose*).

erers of *omanensis* (q.v.) gave a very convincing argument as to why the collection of even one specimen would have been unethical (Robb *et al.* 2013). Kirwan *et al.* disputed these opinions, stating that it would have been preferable to delay publication of *omanensis* until more data became available (which, in fact, has now been done, and the name *omanensis* has become a junior synonym of a bird that had already been described).

▲ Desert Owl, adult, with Scrub Warbler. Dead Sea, Israel, June 2016 (*Amir Ben Dov*).

However, the absence of a formal description of a species does tend to inhibit conservation efforts – it was for this reason that, of all things, the Loch Ness monster was formally named (by Sir Peter Scott) in 1975, based on a spectacularly implausible and fuzzy image of an obscure object in the depths of the loch. Having been named, in the inherently unlikely event of its existence being proven, it was claimed that legal protection would have been made easier.

It seems probable that scientific collection will continue to be a mainstay of the description of new taxa, albeit, one hopes, with more sensitivity than was the case in the early days. The case of the Serendib Scops Owl *Otus thilohoffmanni* is a case in point; only after the discoverers had ascertained that the population was of a significant size was it considered ethical to take the one unique museum specimen. It is to be hoped that similar restraint will be shown in future discoveries.

► Serendip Scops Owl, female. Kitulgala Forest, Sri Lanka, June 2006 (*Uditha Hettige*).

CONSERVATION ISSUES

According to the 2015 Red List of birds, of the 10,279 extant species of birds in the world, no fewer than 1,375, or 13%, are under some type of threat, varying from imminent danger of worldwide extinction (Critically Endangered) to Vulnerable. A further 971 species (9%) are classified as Near Threatened, giving a total of more than one in five species which are the cause of some degree of concern (A. Symes, pers. comm.).

The categories used by BirdLife International (Cambridge, UK) are, in descending order of seriousness, Extinct Globally, Extinct in the Wild, Critically Endangered, Endangered, Vulnerable, Near Threatened, Least Concern and Data Deficient.

This situation is alarming enough on its own. However, for the species covered by this book, the situation is substantially *more* dire. This book deals with 288 species; 37 of these do not currently have a BirdLife International conservation assessment (in some cases this is because their very recent discovery has left insufficient time to develop a meaningful assessment, in others because we have given them the benefit of the doubt as to whether a species is valid or a subspecies, while BirdLife has excluded them for that reason. In addition, some species have not been assessed by BirdLife (Data Deficient category) because of the lack of recent or reliable information). Subtracting these 37, we have a total of 251 species. Of these, a mere 80 (32%) are in the category of Least Concern. (It should be noted, in addition, that even species so classified may well be declining in population; it is simply that these declines have not – as yet – reached a level that would trigger a movement into the Near Threatened category.) It should also be noted that some Data Deficient species are in that category because the number of authenticated records of them is tiny – for several species, one specimen in each case, with no recent observations in the wild – so that in reality a good proportion of the Data Deficient species probably belong in the Critically Endangered or even Extinct categories; adding in such species would make the overall situation look even worse.

The remaining 68% of species in this book all fall into one of the categories of conservation concern, as defined by BirdLife International. Two species have definitely become extinct since their discovery, i.e. since 1960. About 24 species are Critically Endangered, although some of these may indeed be already extinct (there is a certain reluctance to declare, finally, that a species has been irrevocably lost; in some cases, species do, miraculously, reappear after years of absence, while proving a negative is always difficult). Nevertheless, there is almost a sad certainty that several of the species in this book are indeed on their way to oblivion, if not already there.

A further 50 species are Endangered, 48 are Vulnerable and 47 are Near Threatened (these figures are very much moving targets, and will doubtless have changed in the gap between this chapter having been written and being published).

Why is the situation so much more desperate for newly described species than birds as a whole? There are a number of reasons. Firstly, bird species may have escaped earlier detection because they were sparse or uncommon to begin with. Obviously a species with a low population density or small total numbers is more vulnerable than an abundant one. More frequently, however, a newly described species has a very limited geographic distribution, often a limited altitude range in a mountainous area that may itself be limited in extent, or a small island. Under these circumstances, there is frequently very little leeway when habitat destruction or other negative influences occur. Most newly described species are found in tropical areas, frequently in Third World countries; and it is here that the effects on habitat of uncontrolled human population growth are at their most devastating.

What are the most serious causes of extinction, both avian and for all other biological species? Above all other factors, it is habitat destruction. Primarily habitat destruction by logging and by conversion to agricultural land, whether for slash-and-burn, largely unsustainable, farming as in much of the Third World, or large-scale conversion of habitat for such uses as palm oil production in South-East Asia or soya bean farming and ranching in Brazil. It is difficult to be optimistic about wildlife conservation in, especially, tropical areas when faced with inexorable population growth (to give two examples, the human populations of the Philippines and Ethiopia, two areas where there are major concerns about the survival of several bird species, have, in the period covered by this book (1960 to 2015), increased from 26 million to 101 million and 22 million to 94 million respectively; Ethiopia's population is projected to reach 243 million by the year 2100).

A more subtle and indirect, but potentially very serious, form of habitat destruction is that caused by climate change. In the coming decades, changes in temperatures and rainfall patterns are likely to cause major habitat shifts. Not all of these have to be, by necessity, negative in impact; in some cases a warming trend may have positive effects on some species. However, in the coming years, increased desertification, driven by rising temperatures, overgrazing and destruction of trees for firewood and other purposes, can be expected to result in overall habitat loss, not only for breeding species but also, as in the Sahel region of Africa, for migratory species on their wintering grounds.

In comparison to habitat destruction, other causes of extinction, though devastating in certain cases, are of lesser importance. A perennial cause, particularly damaging on oceanic islands, is the introduction by human agency, deliberately or accidentally, of alien organisms. These can be overt, involving rats, pigs, goats, cats or (in one particularly galling and unnecessary recent instance) mink; or more subtle, but sometimes equally destructive, such as the decimation currently occurring in the Hawaiian archipelago by avian malaria.

Although, in the past, numerous species have been hunted to extinction – some of the most famous examples including the Eskimo Curlew, the Great Auk, the New Zealand moas and the Passenger Pigeon – today the greatest threat to birds by direct human action is probably trapping for the cage bird industry: this has a disproportionate influence on members of the parrot family, and to a slightly lesser extent on small finches, although, especially in South-East Asia, other species such as green magpies and bulbuls are also severely impacted. In a few instances, however, direct hunting for food is still significant.

What can be done to slow or even reverse this dreadful progression? In many cases, we are looking at a rearguard action, often too little and too late. One of the major centres of extinction, in progress or potential, is the Atlantic coastal rainforest of Brazil, home to a large number of endemic taxa; what little remains is a pitiful remnant of its former glory of even 50 years ago, and with no resemblance at all to the spectacular forests that were so poetically described by Charles Darwin on his first major landfall from HMS *Beagle* in February 1832. Similar stories abound in other areas of Latin America, in tropical Africa, Madagascar, South-East Asia and the islands of Indonesia and the Philippines. The key to much wildlife conservation in these areas is, and will be, the acquisition of habitat and its maintenance through proper protection and wardening. This takes money. Fortunately, in almost all areas of the world, there exist organisations dedicated to just these goals, but there are still major regions that are in desperate need of effective conservation organisations.

It is to be hoped that this author will not be accused of excessive partiality if he makes particular mention of Fundación Jocotoco. This organisation was founded at very short notice after the expedition of January 1998, which trapped and photographed the first live specimen of Jocotoco Antpitta *Grallaria ridgelyi* (q.v.). It became very obvious that the tiny habitat of this species was being very rapidly destroyed; as a result the Fundación was very quickly created, resulting in immediate land purchases to protect this and other species: in the years since, the Fundación has expanded to 11 reserves (with two more in the pipeline) in Ecuador, providing, in some cases, the only significant protection for a number of Critically Endangered or very rare species such as the El Oro Parakeet, the Pale-headed Brushfinch, the Black-breasted Puffleg and, of course, the Jocotoco Antpitta.

If this book achieves nothing else, if it spreads the realisation that we are living at a critical time for the world's wildlife, in all its forms, and that we still have chances, rapidly disappearing, to retrieve some of the damage done by our species, then perhaps it will have achieved something worthwhile.

◀ Jocotoco Antpitta. Tapichalaca, southern Ecuador, February 2009 (*Jon Hornbuckle*).

▲ The team that captured and photographed the first live example of Jocotoco Antpitta *Grallaria ridgelyi*, a discovery that led to the creation of the Tapichulaca Reserve and the Fundación Jocotoco. Left to right: Doug Wechsler, David Agro, Niels Krabbe, Xavier Muñoz, Nigel Simpson, Francisco Sornoza, Bob Ridgely, David Brewer, Lelis Navarrete. Quebrada Honda, southern Ecuador, January 1998 (*Doug Wechsler*).

BIBLIOGRAPHY

Abbreviated references

To save space, some repetitive and lengthy references have been abbreviated. These include all references to the *Handbook of the Birds of the World*. Thus, for example:

Kroodsma, D. E. & Brewer, D. (2005). Family Troglodytidae (Wrens). Pp 356–447 in: del Hoyo, J., Elliott, A. & Christie, D.A. eds. (2005). *Handbook of the Birds of the World*. Vol. 10. Cuckoo-shrikes to Thrushes. Lynx Edicions, Barcelona

is abbreviated to: **Kroodsma & Brewer 2005 HBW 10**.

References to the supplemental volume of the Handbook, the full reference for which is:

del Hoyo, J., Elliott, A., Sargatal, J. & Christie, D. A. (eds) (2013). *Handbook of the Birds of the World*. Special Volume: New Species and Global Index. Lynx Edicions, Barcelona

are abbreviated to: **2013 HBW SV**.

Online HBW references (*HBW Alive*), the full reference for which is:

del Hoyo, J., Elliott, A., Sargatal, J., Christie, D. A. & de Juana, E. (eds). *Handbook of the Birds of the World Alive*. Lynx Edicions, Barcelona

would be abbreviated to: **del Hoyo *et al.* 2016 HBWAlive**.

BirdLife International species factsheets are available online at www.datazone.birdlife.org/species/search. References to these are abbreviated as: **BLI SF (year)**.

The notation **TD** in front of a reference means that it is the type description for the species; **TD(I)** refers to a type description of an invalid species.

Recordings available from Xeno-canto are referenced as **XC**, then the Xeno-canto accession number.

Abalaka, J.I., Ottosson, U., Tende, T. & Larson, K., 2010. The Rock Firefinch *Lagonosticta sanguinodorsalis* in the Mandare Mountains, north-east Nigeria: a new subspecies? *Bull. African Bird Club* 17(2): 210–211.

Abbot, C.A., 1933. Closing history of the Guadeloupe Caracara. *Condor* 35: 10–14.

ACAP, 2009. ACAP Species Assessment: Antipodean Albatross *Diomedea antipodensis*. Available at: http://www.acap.aq/acap-species/download-document/1210–antipodean-albatross#.

Acevedo-Charry, O., Cárdenas, A., Coral-Jaramillo, B., Daza Díaz, W., Jaramillo, J. & Freile, J.F., 2015. First record of Subtropical Pygmy Owl *Glaucidium parkeri* in the Colombian Andes. *Bull. Brit. Orn. Club* 135(1): 77–79.

Ágreda, A., Nilsson, J., Tonato, L. & Román, H., 2005. A new population of Cinnamon-breasted Tody-tyrant *Hemitriccus cinnamomeipectus* in Ecuador. *Cotinga* 24: 16–18.

Agüero, M.L., García Borboroglu, J.P. & Esler, D., 2010. Nesting habitat of Chubut Steamer Ducks in Patagonia, Argentina. *Emu* 110: 302–306.

Aguero, M.L., García Borboroglu, P. & Esler, D., 2011. Distribution and abundance of Chubut Steamerducks: an endemic species to central Patagonia, Argentina. *Bird Cons. Int.* 22: 307–315.

Aguirre, J.J. & Wrann, J., 1985. Especies del género *Prosopis* y su manejo en la Pampa del Tamarugal. Pp 3–33, *in* Estado actual del conocimiento sobre *Prosopis tamarugo* (M.A. Habit, Ed.). United Nations Food and Agriculture Organisation, Santiago.

Albano, C., 2009. First record of the Orange-bellied Antwren (*Terenura sicki*) in the lowland Atlantic Forest of Pernambuco, northeastern Brazil. *Revista Brasiliera de Ornitologia* 17(2): 150–151.

Albano, C. & Girão, W., 2009. Araripe Manakin *Antilophia bokermanni*. *Neotropical Birding* 5: 71–74.

TD(I) Aleixo, A., Portes, C.E.B., Whittaker, A., Weckstein, J.D., Gonzaga, L.P., Zimmer, K.J., Ribas, C.M. & Bates, J.B., 2013. Molecular systematics and taxonomic revision of the Curve-billed Scythebill (*Campylorhamphus procurvoides*: Dendrocolaptidae), with description of a new species from western Amazonian Brazil. *Handbook of the Birds of the World* Special Volume, pp 253–257. Lynx Edicions, Barcelona.

Allen, D., Espanola, C., Broad, G., Oliveros, C. & Gonzalez, J.C.T., 2006. New bird records for the Babuyan Islands, Philippines, including two first records for the Philippines. *Forktail* 22: 57–70.

TD Allen, D., Oliveros, C., Española, C., Broad, G. & Gonzalez, J.C.T., 2004. A new species of *Gallirallus* from Calayan Island, Philippines. *Forktail* 20: 1–7.

TD(I) Allen, G.A. Jr, Allen, G.A. III & Allen, L., 1977. A new species of curassow discovered. *Game Breeders Cons. Gaz.* 26(6): 6–7.

Allen, S. & Catsis, M., 2007. On the trail of the Bugun Liocichla *Liocichla bugunorum* in north-east India. *BirdingASIA* 7: 75–80.

Allport, G.A., Ausden, M., Hayman, P.V., Robertson, P. & Wood, P., 1989. *The Conservation of the Birds of the Gola Forest, Sierra Leone*. International Council for Bird Preservation, Cambridge, UK.

TD Alonso, J.A. & Whitney, B.M., 2001. A new *Zimmerius* Tyrannulet (Aves: Tyrannidae) from white sand forests of northern Amazonian Peru. *Wilson Bull.* 113(1): 1–9.

Alström, P., Colston, P.R. & Olsson, U., 1990. Description of a possible new species of leaf warbler of the genus *Phylloscopus* from China. *Bull. Brit. Orn. Club* 110(1): 43–47.

TD Alström, P., Davidson, P., Duckworth, J.W., Eames, J.C., Lê Trong Trai, Nguyên Cu., Olsson, U., Robson, C. & Timmins, R., 2010. Description of a new species of *Phylloscopus* warbler from Vietnam and Laos. *Ibis* 152: 145–168.

Alström, P., Ericson, P.G.P., Olsson, U. & Sundberg, P., 2006. Phylogeny and classification of the avian superfamily Sylvioidea. *Mol. Phylogenet. Evol.* 38(2): 381–397.

Alström, P., Höhna, S. & Gelang, M., 2011. Non-monophyly and intricate morphological evolution within the avian family

Cettiidae revealed by multilocus analysis of a taxonomically densely sampled dataset. *Evolutionary Biology* 11: Suppl. art. 352: pp 16.

Alström, P. & Mild, K., 2003. *Pipits and Wagtails of Europe, Asia and North America*. Christopher Helm, London.

Alström, P. & Olsson, U., 1990. Taxonomy of the *Phylloscopus proregulus* complex. *Bull. Brit. Orn. Club* 110(1): 38–43.

TD Alström, P. & Olsson, U., 1995. A new species of *Phylloscopus* warbler from Sichuan Province, China. *Ibis* 137: 459–468.

TD Alström, P. & Olsson, U., 1999. The Golden-spectacled Warbler: a complex of sibling species, including a previously undescribed species. *Ibis* 141: 545–568.

Alström, P. & Olsson, U., 2000. Golden-spectacled Warbler systematics. *Ibis* 142: 495–500.

TD(I) Alström, P., Olsson, U. & Colston, P.R., 1992. A new species of *Phylloscopus* warbler from central China. *Ibis* 134: 329–334.

TD Alström, P., Xia, C-W., Rasmussen, P.C., Olsson, U., Dai, B., Zhao, J., Song, G., Liu, Y., Zhang, Y-Y. & Lei, F-M., 2015. Integrative taxonomy of the Russet Bush Warbler *Locustella mandelli* complex reveals a new species from central China. *Avian Research* 6(9): doi:10.1186/s40657-015-0016-z(OA)

Alvarenga, H.M.F., 1995. *In memoriam*: Werner C.A. Bokermann. *Ararajuba*: 3: 101–102.

TD Álvarez, A.J. & Whitney, B.M., 2001. A new *Zimmerius* tyrannulet (Aves: Tyrannidae) from white sand forests of northern Amazonian Peru. *Wilson Bull.* 113(1): 1–9.

Alves, M.A.S., Pimm, S.L., Storni, A., Raposo, M.A., Brooke, M de L., Harris, G., Foster, A. & Jenkins, C.N., 2008. Mapping and exploring the distribution of the Vulnerable Grey-winged Cotinga *Tijuca condita*. *Oryx* 42(4): 562–566.

TD Amadon, D., 1962. A new genus and species of Philippine bird. *Condor* 64: 3.

TD Amadon, D. & duPont, J.E., 1970. Notes on Philippine birds. *Occ. Pap. Delaware Mus. Nat. Hist.* 1: 1–14.

American Bird Conservancy, 2013. https://abcbirds.org/article/new-population-of-very-rare-long-whiskered-owlet-discovered-in-peru/

Andrew, D. & Rogers, D., 1993. Australian Babblers. *Wingspan* September 1993, 15–16.

Anon. 1993. (Chinese ornithologists claim to have collected specimens of Vaurie's Nightjar some 250 km from type-location). *Wingspan* September 1993, 11: 32.

Anon. 1994. (Vaurie's Nightjar, invalidity of 1993 reports). *Wingspan* June 1994,12.

Anon. 2000. Bárbara D'Achille: Un ejemplo de amor por la naturaleza y el medio ambiente. *Perú Héroes Civiles* PDF Scribd.

Anon. 2001. Murici protected. *Cotinga* 16: 9.

Anon. 2002a. A significant new population of the Flores Monarch *Monarcha sacerdotum*. *World Birdwatch* 24(1): 3.

Anon. 2002b. More Flores Monarchs. *Bull. Oriental Bird Club* 35: 57.

Anon. 2003. Neotropical Bird Club Conservation Awards. *Cotinga* 19: 10.

Anon. 2006. Bugun Liocichla: a sensational discovery in northeast India. *Birdlife International* 2006-09-12. http://www.birdguides.com/webzine/article.asp?a=756

Anon. 2009. Rarest bird on Earth: spotted at last! *Africa: Birds and Birding*. 14(4): 58.

Anon. 2012a. Conserving forests for birds and people. *World Birdwatch* December 2012.

Anon. 2012b. Araripe Manakin *Antilophia bokermanni*. *World Birdwatch* March 2012.

Anon. 2012c. Rediscovery of Sillem's Mountain-finch. *Oriental Bird Images: a Database of the Oriental Bird Club.*

Anon. 2013a. Hooded Grebe *Podiceps gallardoi*. *World Birdwatch* September 2013, pp 20–21.

Anon. 2013b. Saving the threatened endemic birds of southern Ethiopia. *World Birdwatch* March 2103, pp 22–24.

Anon. 2013c. Hiding in plain sight: new bird species found in Cambodia's capital. *World Birdwatch* September 2013, pp 2–3.

Anon. 2013d. *Birding* May/June 2013, 47–52.

Anon. 2013e. First nest discovered of world's most endangered birds. *Avian Biology Research* 6(1): 79–80.

Anon. 2015a. International Conservation Fund of Canada, 2014 Annual Report. International Conservation Fund of Canada, Box 40, Chester, NS B0J 1J0.

Anon. 2015b. Focusing on Wildlife, 30 March 2015. Contact: info@focusingonwildlife.com.

Anon. 2015c. NBC Conservation Awards Update. *Neotropical Birding* 16: 63.

Anon. 2017. Nuevo Tapaculo en la Reserva Natural de las Aves Reinita Cerúlea. Noticias Caracol, www.proaves.org, 20 February 2017.

Apa, A.D. & Wiechman, L.A., 2015. Captive-rearing of Gunnison Sage-grouse from egg collection to adulthood to foster proactive conservation and recovery of a conservation-reliant species. *Zoo Biology* 34(5): 438–452.

Araujo Gussoni, C.O., 2010. Novas informaçãos sobre a história natural da maria-de-restinga *Phylloscartes kronei* (Aves: Tyrannidae). Instituto de Biociências do Campus de Rio Claro, Universidade Estadual Paulista.

Araujo Gussoni, C.O. & de Oliveira Santos, M.C., 2011. Foraging behavior of the Restinga Tyrannulet (*Phylloscartes kronei*) (Aves, Tyrannidae). *Ornitología Neotropical* 22(4): 495–504.

Arctander, P. & Fjeldså, J., 1994. Andean tapaculos of the genus *Scytalopus* (Aves: Rhinocryptidae): a study of modes of differentiation, using DNA sequence data. Pp 205–225 in *Conservation Genetics* (V. Loeschcke, J. Tomiuk & S.K.Kain, Eds). Birkhauser Verlag, Basel, Switzerland.

Arendt, W.J., Qian, S.S. & Mineard, K.A., 2013. Population decline of the Elfin-woods Warbler *Setophaga angelae* in eastern Puerto Rico. *Bird Cons. Int.* 23: 136–146.

Areta, J.I., 2008. The Entre Ríos Seedeater (*Sporophila zelichi*): a species that never was. *J. Field Ornithol.* 79: 352–363.

TD Arndt, T., 2006a. A new hanging-parrot from Caminguin Island, Philippines. *BirdingASIA* 5: 55–58.

TD(I) Arndt, T., 2006b. A revision of the *Aratinga mitrata* complex, with the description of one new species, two new subspecies and species-level status of *Aratinga alticola*. *J. Orn.* 147: 73–86.

TD Arndt, T., 2008. Anmerkungen zu einigen *Pyrrhura*-formen mit der Beschreibung einer neuen Art und zweier neuer Untarten. *Papageien* 21: 278–286.

Arroyo-Vásquez, B., 1992. Observations of the breeding biology of the Elfin-woods Warbler. *Wilson Bull.* 104(2): 362–365.

Ascanio, D., Rodríguez, G. & Restall, R., 2017. *Birds of Venezuela (Helm Field Guide)*. Bloomsbury, London

TD Ash, J.S., 1979. A new species of serin from Ethiopia. *Ibis* 121: 1–7.

TD(I) Ash, J.S., 1986. A *Ploceus sp. nov.* from Uganda. *Ibis* 128(3): 330–336.

Ash, J.S., 1987. *Ploceus victoriae*. *Ibis* 129(3): 406–407.

Ash, J.S. & Atkins, J., 2009. *Birds of Ethiopia and Eritrea: an Atlas of Distribution*. Christopher Helm, London.

Ash, J.S. & Miskell, J.E., 1998. *Birds of Somalia*. Pica Press, Robertsbridge, UK.

Ash, J.S. & Olson, S.L., 1985. A second specimen of *Mirafra (Heteromirafra) sidamoensis* Érard. *Bull. Brit. Orn. Club* 105(4): 141–143.

TD Athreya, R., 2006. A new species of Liocichla (Aves: Timaliidae) from Eaglenest Wildlife Sanctuary, Arunachal Pradesh, India. *Indian Birds* 2(4): 82–94.

Atkins, J.D., 1992. A new location for the Ankober Serin *Serinus ankoberensis* near Debra Sine, Ethiopia. *Scopus* 16: 105–107.

Atkins, J.D. & Harvey, W.G., 1994. Further sightings of an unnamed Cliff Swallow *Hirundo* sp. in Ethiopia. *Scopus* 18: 52–54.

Atkinson, P.W., Whittingham, M.J., Gómez de Silva, H., Kent, A.M. & Maier, R.T., 1993. Notes on the ecology, conservation and taxonomic status of *Hylorchilus* wrens. *Bird Cons. Int.* 3: 75–85.

TD Avendaño, J.E., Cuervo, A.M., López-O., J.P., Gutiérrez-Pinto, N., Cortes-Diago, A. & Cadena, C.D., 2015. A new species of tapaculo (Rhinocryptidae: *Scytalopus*) from the Serranía de Perijá of Colombia and Venezuela. *Auk* 132: 450–466.

Azpiroz, A.B., 2003. Primeros registros del Capuchino de Collar (*Sporophila zelichi*) en Uruguay. *Ornitología Neotropical* 14: 117–119.

Baird, S.F., Brewer, T.M. & Ridgway, R., 1874. *A History of North American Birds*. Little, Brown, Boston, MA.

Baker, C., Bottomley, C., Kelly, L., Payne, L. & Whittle, M., 1998. Mangaia '96. Final report of the Oxford University expedition to the Cook Islands to study the Mangaia Kingfisher (Tanga'eo), 21 June-21 August 1996. pp 1–61, published privately, Oxford University.

TD Baker, N.E. & Baker, E.M., 1990. A new species of weaver from Tanzania. *Bull. Brit. Orn. Club* 110(1): 51–58.

Baker, P.E., 1998. A description of the first live Po'o-uli captured. *Wilson Bull.* 110(3): 307–310.

Baker, P.E., 2001. Status and distribution of the Po'o-uli in Hanawi Natural Area Reserve between December 1995 and June 1997. *Studies in Avian Biology* 22: 144–150.

Balchin, C., 2007. Back from the dead! A potpourri of recent rediscoveries in the Neotropics. *Neotropical Birding* 2: 4–11.

Balchin, C.S. & Toyne, E.P., 1998. The avifauna and conservation status of the Río Nangaritza valley, southern Ecuador. *Bird Cons. Int.* 8: 237–253.

Baldwin, J.W. & Drucker, J.R., 2016. Nest and nestling data for the unnamed "Ampay" Tapaculo *Scytalopus* sp. from Apurímac, Peru. *Cotinga* 38: 10–14.

Baldwin, P.H. & Casey, T.L.C., 1983. A preliminary list of foods of the Poo-uli. *Elepaio* 43: 53–56.

van Balen, S., 1998. A hybrid munia? *Bull. Brit. Orn. Club* 118(2): 118–119.

Banks, R.C., 1964. Geographic variation in the White-crowned Sparrow *Zonotrichia leucophrys*. *Univ. Calif. Publ. Zool.* 70: 1–122.

Bannerman, D.A., 1939. *The Birds of Tropical West Africa*, Vol 5. Oliver & Boyd, London.

Barlow, C., Wacher, T. & Disley, T., 1997. *A Field Guide to the Birds of The Gambia and Senegal*. Pica Press, Robertsbridge, UK.

Barnes, R., Butchart, S.H.M., Davies, C.W.N., Fernández, M. & Seddon, N., 1997. New distributional information on eight bird species from northern Peru. *Bull. Brit. Orn. Club* 117(1): 69–74.

Barnet, L.K. & Emms, C., 2001. New species and breeding records for The Gambia. *Bull. African Bird Club* 8(1): 44–45.

TD Barrera, L.F., Bartels, A. & Fundación ProAves de Colombia, 2010. A new species of Antpitta (family Grallariidae) from the Colibrí del Sol Bird Reserve, Colombia. *Conservación Colombiana* 13: 8–24.

Barrowclough, G.M., Cracraft, J., Klicka, J. & Zink, R.M., 2016. How many kinds of birds are there and does it matter? PLoS ONE 11(11): e0166307.

TD(I) Batista, R., Aleixo, A., Vallinoto, M., Azevedo, L., Sena do Rêgo, P., Siveira, L.F., Sampaio, I. & Schneider, H., 2013. Molecular systematics and taxonomic revision of the Amazonian Barred Woodcreeper complex (*Dendrocolaptes certhia*: Dendrocolaptidae), with description of a new species from the Xingu-Tocantins interfluve. *Handbook of the Birds of the World* Special Volume pp 245–247. Lynx Edicions, Barcelona.

Bauerfeind, E., Dickinson, E.C. & Steinheimer, F.D., 2014. Contested spinetail systematics: nomenclature and the Code to the rescue. *Bull. Brit. Orn. Club* 134: 70–76.

Beadle, D., Grosset, A., Kirwan, G.M. & Minns, J., 2003. Range extension for the Manu Antbird *Cercomacra manu* in north Brazil. *Bull. Brit. Orn. Club* 123(4): 236–239.

Beehler, B., 1978. Status of *Sericornis nigroviridis*. *Condor* 80(1): 115–116.

Beehler, B., 1983. Thoughts on an ornithological mystery from Bougainville Island, Papua New Guinea. *Emu* 83: 114–115.

Beehler, B.M., Diamond, J.M., Kemp, N., Scholes, E. III, Milensky, C. & Laman, T.G., 2012. Avifauna of the Foja Mountains of western New Guinea. *Bull. Brit. Orn. Club* 132(2): 84–101.

TD Beehler, B.M., Prawiradilaga, D.M., de Fretes, Y. & Kemp, N., 2007. A new species of smoky honeyeater (Meliphagidae: *Melipotes*) from western New Guinea. *Auk* 124(3): 1000–1009.

Begazo, A.J., Valqui, T., Sokol, M. & Langlois, E., 2001. Notes on some birds from central and northern Peru. *Cotinga* 15: 81–87.

Bellatreche, M., 1991. Deux nouvelles localisations de la Sitelle kabyle en Algérie. *Oiseau et RFO* 61(3): 269–272.

Bellatreche, M. & Chalabi, B., 1990. Données nouvelles sur l'aire de distribution de la Sittelle kabyle *Sitta ledanti*. *Alauda* 58: 95–97.

Belton, W., 1974. More new birds for Rio Grande do Sul, Brazil. *Auk* 91: 429–432.

Belton, W., 1985. Birds of Rio Grande do Sul, Brazil. 2: Formicariidae through Corvidae. *Bull. Am. Mus. Nat. Hist.* 178: 369–636.

Bencke, G.A., 2004. O caboclinho *Sporophila zelichi* observando no Rio Grande do Sul, Brasil. *Ararajuba* 12(2): 88–89.

Bencke, G.A., Kindel, A., Mähler, J.K., 2000. Adiçoes à avifauna de Mata Atlântica do Rio Grande do Sul. In: Alves, M.A.S., Van Sluys, M., Bergallo, H.G. & Rocha, C.F.D. (eds). *A ornitologia no Brasil: pesquisa atual, conservação e perspectivas*, pp 317–323. Sociedad Brasiliera de Ornitología, Brasilia.

Bencke, G.A., Suertegaray Fontana, C. & de Mendonça-Lima, A., 2001. Registro de dois novos passeriformes para o Brasil: *Serpophaga griseiceps* (Tyrannidae) e *Asthenes pyrrholeuca* (Furnariidae). *Ararajuba* 10(2): 261–277.

Benham, P.M. Cuervo, A.M., Mcguire, J.A. & Witt, C.C., 2015. Biogeography of Andean metaltail hummingbirds: contrasting evolutionary histories of tree-line and habitat-general clades. *Journal of Biogeography* 42(4): 763–777.

Benkman, C.W., 1993. Adaptation to single resources and the evolution of crossbill (*Loxia*) diversity. *Ecological Monographs* 63: 305–325.

Benkman, C.W., 1999. The selection mosaic and diversifying coevolution between crossbills and lodgepole pine. *American Naturalist* 153 Suppl. S., S75–S91.

Benkman, C.W., 2003. Divergent selection drives the adaptive radiation of crossbills. *Evolution* 57(5): 1176–1181.

Benkman, C.W., Parchman, T.L., Favis, A. & Siepielski, A.M., 2003. Reciprocal Selection Causes a Coevolutionary Arms Race between Crossbills and Lodgepole Pine. *American Naturalist* 162(2): 182–184.

TD(I) Benkman, C.W., Smith, J.W., Keenan P.C., Parchman, T.L. & Santisteban, L., 2009. New species of the Red Crossbill (Fringillidae: *Loxia*) from Idaho. *Condor* 111(1): 169–176.

TD Benson, 1960. The birds of the Comoro Islands: results of the British Ornithologists' Union Centenary Expedition 1958. *Ibis* 103b: 5–106.

TD Benson, C.W. & Irwin, M.P.S., 1965. A new species of tinkerbarbet from Northern Rhodesia. *Bull. Brit. Orn. Club* 85: 5–9.

TD Benson, C.W. & Penny, M.J., 1968. A new species of warbler from the Aldabra Atoll. *Bull. Brit. Orn. Club* 88: 102–108.

Benz, B.W. & Robbins, M.B., 2011. Molecular phylogenetics, vocalisations and species limits in *Celeus* woodpeckers (Aves: Picidae). *Molecular Phylogenetics and Evolution* 61: 29–44.

Beolens, B. & Watkins, M., 2003. *Whose Bird?* Christopher Helm, London.

Beolens, B., Watkins, M. & Grayson, M., 2014. *The Eponym Dictionary of Birds.* Bloomsbury, London.

Beresford, P., Barker, F.K., Ryan, P.G. & Crowe, T.M., 2005. African endemics span the tree of songbirds (Passeri): molecular systematics of several evolutionary "enigmas". *Proc. Roy. Soc. Lond.* B 272 (1565): 849–858.

Beresford, P. & Cracraft, J., 1999. Speciation in African Forest Robins (*Stiphrornis*): species limits, phylogenetic relationships and molecular biogeography. *Amer. Mus. Novit.* 3270: 1–22.

TD Beresford, P., Fjeldså, J. & Kiure, J., 2004. A new species of Akalat (*Sheppardia*) narrowly endemic in the Eastern Arc of Tanzania. *Auk* 121(1): 23–34.

Berg, M., van Kleunen, A. & Hennessey, A.B., 2014. Range and status of Green-capped Tanager *Tangara meyerdeschauenseei* in Bolivia. *Cotinga* 36: 52–55.

TD(I) Berlioz, J., 1959. Description de deux éspeces nouvelles d'oiseaux de Bolivie. *Bull. Mus. Nat. d'Hist. Nat.* 3: 218–219.

TD Berlioz, J., 1962. (*Contopus albigularis*). *Bull. Mus. Hist. Nat. Paris* 34: 138.

TD(I) Berlioz, J., 1965. (*Augasma cyaneo-beryllina*) *L'Oiseau* 35: 7.

TD(I) Berlioz, J., 1966. Descriptions de deux éspeces nouvelles d'oiseaux du Pérou. *L'Oiseau et RFO* 36: 1–3.

Bernardo, C.S.S. & Thorns, D., 2009. First breeding data for Slaty Bristlefront *Merulaxis ater* in Rio de Janeiro, Brazil. *Cotinga* 31: 146–147.

TD(I) Bertagnolio, P. & Racheli, L., 2010. A new parrotlet from Colombia. *Avicultural Magazine* 116: 128–133.

Bertoni, A.W., 1919. Especies de aves nuevas para el Paraguay. *Hornero* 1: 255–258.

Best, B.J., Clarke, C.T., Checker, M., Broom, A.L., Thewlis, R.M., Duckworth, W. & McNab, A., 1993. Distributional records, natural history notes and conservation of some poorly known birds from southwestern Ecuador and northwestern Peru. *Bull. Brit. Orn. Club* 113(2): 108–119.

Bharadwaj, 2015. Recent sightings of Great Nicobar Crake *Rallina* sp. nov. and Nicobar Megapode *Megapodius nicobarensis* at Great Nicobar Island. *Ela Journal of Forestry and Wildlife* 4(2): 16–17. (http://www.elafoundation.org/Ela_Foundation/journals/EJFW-4-2.pdf#page-16)

Bierregard, R.O., Cohn-Haft, M. & Stotz, D.F., 1997. Cryptic biodiversity: an overlooked species and new subspecies of Antbird (Aves: Formicariidae) with a revision of *Cercomacra tyrannina* in northeastern South America. *Orn. Monographs* 48: 111–128.

Binford, L.C., 1989. *A Distributional Survey of the Birds of the Mexican State of Oaxaca.* Ornithological Monographs 43. American Ornithologists' Union, New York.

Birdlife Data Zone (Socotra Buzzard). http://www.birdlife.org/datazone/sowb/casestudy/27, 07–08–2013.

Bishop, D. & Brickle, N.W., 1999. An annotated checklist of the birds of the Tanimbar Islands. *Kukila* 10: 115–150.

Bishop, K.D. & Jones, D.N., 2001. The montane avifauna of West New Britain, with special reference to the Nakanai Mountains. *Emu* 101(3): 205–220.

Black, A., Carpenter, G. & Pedler, R., 2012. Habitats of the Grey Grasswren *Amytornis barbatus diamantina* and a review of the species' distribution. *Corella* 36(2): 29–37.

TD Blake, E.R., 1971. A new species of spinetail (*Synallaxis*) from Peru. *Auk* 88(1): 179.

TD Blake, E.R. & Hocking, P., 1974. Two new species of tanager from Peru. *Wilson Bull.* 86(4): 321–324.

Bledsoe, A.H., 1987. DNA evolutionary rates in nine-primaried passerine birds. *Molecular Biology and Evolution* 4: 559–571.

Bledsoe, A.H., 1988. Nuclear DNA evolution and phylogeny of the New World nine-primaried oscines. *Auk* 105(3): 504–515.

Bledsoe, A.H. & Sheldon, F.H., 1989. The metric properties of DNA-DNA hybridization dissimilarity measures. *Systematic Zoology* 38(2): 93–105.

Boano, G., Vinals, N., Durante, A. & Pavia, M., 2015. Apparent sympatry of *Stiphrornis pyrrholaemus* Schmidt & Angehr, 2008 and *S. xanthogaster* Sharpe, 1903 (Passeriformes: Muscicapidae) in Gabon, and taxonomic implications. *Zootaxa* 4032(1): 127–133.

Bock, W.J., 1978. Tongue morphology and affinities of the Hawaiian honeycreeper *Melamprosops phaeosoma. Ibis* 120: 467–479.

de Boer, M.N., 2015. First record of Barau's Petrel *Pterodroma baraui* for Namibia, South Atlantic. *Bull. African Bird Club* 22(2): 211–213.

Boles, W.E., Bartram, K. & Clancy, G.P., 1985. First Australian specimen of the White-necked Petrel. *Australian Birds* 19: 51–54.

Bolton, M., 2007. Playback experiments indicate absence of vocal recognition among temporally and geographically separated populations of Madeiran Storm-petrels *Oceanodroma castro. Ibis* 149: 255–263.

TD Bolton, M., Smith, A.L., Gómez-Díaz, E., Friesen, V.L., Medeiros, R., Bried, J., Roscales, J.L. & Furness, R.W., 2008. Monteiro's Storm-Petrel *Oceanodroma monteiroi:* a new species from the Azores. *Ibis* 150: 717–727.

Bond, J. & Meyer de Schauensee, R., 1940. Description of a new Puff-bird from the lower Amazon. *Notulae Naturae* 50: 1–2.

TD Bornschein, M.R., Maurício, G.N., Belmonte-Lopes, R., Mata, H. & Bonatto, S.L., 2007. Diamantina Tapaculo, a new *Scytalopus* endemic to the Chapada Diamantina, northeastern Brazil (Passeriformes: Rhinocryptidae). *Rev. Bras. Orn.* 15: 151–174.

Bornschein, M., Pichorim, R. & Reinert, B.L., 2001. Novos registros de *Scytalopus iraiensis. Natteria* 2: 29–33.

Bornschein, M.R., Pizo, M.A., Sobotka, D.D., Belmonte-Lopes, R., Golec, C., Machado-de-Souza, T., Pie, M.R. & Reinert, B.L., 2015. Longevity records and signs of aging in Marsh Antwren *Formicivora acutirostris* (Thamnophilidae). *Wilson J. Orn.* 127(1): 98–102.

TD Bornschein, M., Reinert, B.L. & Pichorim, M., 1998. Descrição, ecologia e conservaça de um novo *Scytalopus* (Rhinocryptidae) do sul do Brasil, com comentários sobre a morfologia da família. *Ararajuba* 6(1): 3–36.

TD Bornschein, M.R., Reinert, B.L. & Teixeira, D.M., 1995. Um novo Formicariidae do sul do Brasil (Aves, Passeriformes). Publicação Técnico-Cientifica do Instituto Iguaçu No 1: 18, Rio de Janeiro.

Borrow, N., 2010. Recent reports. *Bull. African Bird Club* 17(2): 243.

Borrow, N. & Demey, R., 2001. *A Guide to the Birds of Western Africa.* Christopher Helm, London.

Borrow, N. & Demey, R., 2014. *Birds of Western Africa,* 2nd Edition. Christopher Helm, London.

Bory de St-Vincent, J.B.G.M., 1804. Voyage dans les quatre principales îles des mers d'Afrique, etc., Paris, 3 vols.

Bosch, K., 1991. Eine neue Amazonenarte? *Amazilia kawalli. Gefierdete Welt* 115(6): 190–191.

Bourne, W.R.P. & David, A.C.F., 1995. The early history and ornithology of St Paul and Amsterdam Islands, southern Indian Ocean. *Le Gerfaut* 85: 19–36.

Bowden, C.G.R., Hayman, P.V. & Martins, R.P., 1995. The *Melignomon* honeyguides: a review of recent range extensions and some remarks on their identification, with a description of the song of Zenker's Honeyguide. *Bull. African Bird Club* 2(1): 31–38.

TD Bowie, R.C.K. & Fjeldså, J., 2005. Genetic and morphological evidence for two species in the Udzungwa Forest Partridge *Xenoperdix udzungwensis*. *Journal of East African Natural History* 94(1): 191–205.

TD Bowie, R.C.K., Fjeldså, J. & Kiure, J., 2009. Multilocus molecular DNA variation in Winifred's Warbler *Sceptomycter winifredae* suggests cryptic speciation and the existence of a threatened species in the Rubeho-Ukaguru Mountains of Tanzania. *Ibis* 151(4): 709–719.

Brandt, M.J. & Cresswell, W., 2008. Breeding behaviour, home range and habitat selection in Rock Firefinches *Lagonosticta sanguinodorsalis* in the wet and dry seasons in central Nigeria. *Ibis* 150(3): 495–507.

Brandt, M.J. & Cresswell, W., 2009. Diurnal foraging routines in a tropical bird, the Rock Firefinch *Lagonosticta sanguinodorsalis*: how important is predation risk? *J. Avian Biology* 40(1): 90–94.

Braun, C.E., Oyler-McCance, S.J., Nehring, J.A., Commons, M.L., Young, J.A. & Potter, K.M., 2014. The historical distribution of Gunnison Sage-grouse in Colorado. *Wilson J. Orn.* 126(2): 207–217.

Braun, M.T. & Parker, T.A. III, 1985. Molecular, morphological and behavioral evidence concerning the taxonomic relationships of "*Synallaxis*" *gularis* and other Synallaxines. *Orn. Monogr.* 36: 333–346.

Brazil, M.A., 1991. *The Birds of Japan*. Smithsonian Inst. Press, Washington, DC.

Brazil, M., 2009. *Birds of East Asia: Eastern China, Taiwan, Korea, Japan, Eastern Russia*. Christopher Helm, London.

Bretagnolle, V. & Attie, C., 1991. Status of Barau's Petrel (*Pterodroma baraui*) – colony sites, breeding population and taxonomic affinities. *Colonial Waterbirds* 14(1): 25–33.

Brewer, D., 2001. *Wrens, Dippers and Thrashers*. Christopher Helm, London.

Brinkhuizen, D.M., Shackelford, D. & Oriel Altamirano, J., 2012. The Long-whiskered Owlets *Xenoglaux loweryi* of Abra Patricia. *Neotropical Birding* 10: 39–46.

Brinkhuizen, D.M. & Solano-Ugalde, A. Range extension of Choco Vireo *Vireo masteri* in Ecuador and a description of the species' song. In press.

Briones, L., 1985. Visión retrospectiva antropológica del *Prosopis*. Pp 51–53 *in* Estado actual del conocimiento sobre *Prosopis tamarugo* (M.A.Habit, Ed.). United Nations Food and Agriculture Organisation, Santiago.

Brooke, M. de L., 2004. *Albatrosses and Petrels Across the World*. Oxford University Press, Oxford.

Brooke, R.K., 1969. *Apus berliozi*, its races and siblings. *Bull. Brit. Orn. Club* 89(1): 11–16.

Brooke, R.K., 1971a. Geographical variation and distribution of the swift genus *Schoutedenapus*. *Bull. Brit. Orn. Club* 91: 25–28.

Brooke, R.K., 1971b. Taxonomic history of *Schoutedenapus schoutedeni*. *Bull. Brit. Orn. Club* 91: 93–94.

Brooke, R.K., 1972a. On the breeding and migratory status of *Apus berliozi bensoni*. *Bull. Brit. Orn. Club* 92: 114.

Brooke, R.K., 1972b. Generic limits in old world Apodidae and Hirundinidae. *Bull. Brit. Orn. Club* 92: 53–57.

Brooks, T.M., Clay, R.P., Lowen, J.C., Butchart, S.H.M., Barnes, R., Esquivel, E.Z., Etcheverry, N.I. & Vincent, J.P., 1995. New information on nine birds from Paraguay. *Orn. Neotrop.* 6: 129–134.

Brown, J.L. & Balda, R.P., 1977. The relationship of habitat quality to group size in Hall's Babbler (*Pomatostomus halli*). *Condor* 79: 312–320.

Bruce, M.D., 1989. A reappraisal of the species limits in the Pied Imperial Pigeon *Ducula bicolor* (Scopoli, 1786) superspecies. *Riv. Ital. Orn.* 59: 218–222.

Buchanan, G.M., Butchart, S.H.M., Dutson, G., Pilgrim, J.D., Steininger, M.K., Bishop, K.D. & Mayaux, P., 2008. Using remote sensing to inform conservation status assessment: estimates of recent deforestation rates on New Britain and the impacts on endemic birds. *Biological Conservation* 141(1): 56–66.

Buckley, P.A., 1998. The world's first known juvenile Cox's Sandpiper. *British Birds* 81: 253–257.

Bull, J. & Amadon, D., 1983. In Memoriam: Eugene Eisenmann. *Auk* 100(1): 188–191.

Burbridge, M.L., Colbourne, R.M., Robertson, H.A. & Barker, A.J., 2003. Molecular and other biological evidence supports the recognition of at least three species of kiwi. *Conservation Genetics* 4: 167–177.

Burnier, E., 1976. Une nouvelle espèce de l'avifaune paléarctique: la Sitelle kabyle, *Sitta ledanti*. *Nos Oiseaux* 33: 337–340.

Burns, K.J., 1997. Molecular systematics of tanagers (Thraupinae): Evolution and biogeography of a diverse radiation of Neotropical birds. *Molecular Phylogenetics and Evolution* 8: 334–348.

Burns, K.J. & Naoki, K., 2004. Molecular phylogenetics and biogeography of Neotropical tanagers of the genus *Tangara*. *Molecular Phylogenetics and Evolution* 32: 838–854.

Butchart, S.H.M., 2007a. Prigogine's Nightjar *Caprimulgus prigoginei*. *Bull. African Bird Club* 14(2): 145.

Butchart, S.H.M., 2007b. Birds to find: a review of 'lost', obscure and poorly known African bird species. *Bull. African Bird Club* 14(2): 139–157.

Butchart, S.H.M., Brooks, T.M., Davies, C.W.N., Dharmaputra, G., Dutson, G.C.L., Lowen, J.C. & Sahu. A., 1996. The conservation status of forest birds on Flores and Sumbawa, Indonesia. *Bird Cons. Int.* 6: 335–370.

TD Buzzetti, D.R.C., Belmonte-Lopes, R., Reinert, B.L., Silveira, L.F. & Bornschein, M.R., 2013. A new species of *Formicivora* Swainson, 1824 (Thamnophilidae) from the state of São Paulo, Brazil. *Revista Brasiliera de Ornitologia* 21(4): 269–291.

Cadena, C.D. & Stiles, F.G., 2010. The price of priority. *Ornitología Colombiana* 9: 6–10.

Capparella, A.P. & Witt, C.C., 2009. Pardusco (*Nephelornis oneilli*). Neotropical Birds Online (T.S.Schulenberg, ed.). Cornell Laboratory of Ornithology, Ithaca, NY.

Capper, D. & Pereira, P., 2007. Orange-throated Tanager the easy way – the Cordillera del Cóndor in south-eastern Ecuador. *Neotropical Birding* 2: 44–47.

TD Carantón-Ayala, D. & Certuche-Cubillos, K., 2010. A new species of antpitta (Grallariidae: *Grallaria*) from the northern sector of the western Andes of Colombia. *Ornitología Colombiana* 9: 56–70.

TD Carneiro, L.S., Gonzaga, L.P., Rêgo, P.S., Sampaio, I., Schneider, H. & Aleixo, A., 2012. Systematic revision of the Spotted Antpitta (Grallariidae: *Hylopezus macularius*), with description of a cryptic new species from Brazilian Amazonia. *Auk* 129(2): 338–351.

Carpenter, G., 2002. The Grey Grasswren on Cooper Creek, south west Queensland. *Sunbird* 32: 52–53.

Carriker, M.A., Jr, 1954. Additions to the avifauna of Colombia. *Novedades Colombianas* 1: 14–16.

Carruthers, R.K., Horton, W. & Vernon, D.F., 1970. Distribution, habits and sexual dimorphism of the Western Grass-wren *Amytornis textilis ballarae* Condon in north-western Queensland. *Mem. Qld. Mus.* 15: 335–341.

TD Casey, T.L.C. & Jacobi, J.D., 1974. A new genus and species of bird from the Island of Maui, Hawaii. (Passeriformes: Drepanididae). *Occ. Pap. Bernice P. Bishop Mus.* 24: 216–226.

Carter, M., Reid, T. & Langley, P., 1989. Barau's Petrel, a new species for Australia. *Australian Bird Watcher* 13: 39–43.

Chan Bosco Pui Lok, Lee Kwok Shing, Zhang Jian-Feng & Su Wen-Ba, 2005. Notable bird records from Bawangling National Nature Reserve, Hainan Island, China. *Forktail* 21: 33–41.

Chantler, P., 1998. The migratory status of the White-fronted Swift *Cypseloides storeri*. *Cotinga* 9: 40.

Chantler, P. & Driessens, G., 1995. *Swifts: a Guide to the Swifts and Treeswifts of the World.* Pica Press, Sussex, UK.

Chantler, P. & Rye, P., 2008. Nepal Wren-babbler. *Dutch Birding* 30: 104–106.

Chapman, G.S., 1996. The Grasswrens: a brief pictorial. *Wingspan* 6(1): 8–15.

TD Chappuis, C., 1974. Illustration sonore de problèmes bioacoustiques posés par les oiseaux de la zone éthiopienne. *Alauda* 42: 467–500.

TD(I) Chappuis, C. & Érard, C., 1991. A new cisticola from west-central Africa. *Bull. Brit. Orn. Club* 111(2): 59–70.

Chappuis, C., Érard, C. & Morel, G.J., 1989. Type specimens of *Prinia suflava* (Gmelin) and *Prinia fluviatilis* Chappuis. *Bull. Brit. Orn. Club* 109(2): 108–110.

Chappuis, C., Érard, C., & Morel, G.J., 1993. Morphology, habitat, vocalisations and distribution of the River Prinia *Prinia fluviatilis* Chappuis. *Proc. 7 Pan-Afr. Orn. Congr.*, 481–487.

Cheke, A.S.& Diamond, A.W., 1986. Birds on Moheli and Grande Comore (Comoro Islands) in February 1975. *Bull. Brit. Orn. Club* 106(4): 138–148.

Cheke, R.A. & Mann, C.F., 2001. *Sunbirds. A Guide to the Sunbirds, Flowerpeckers, Spiderhunters and Sugarbirds of the World.* Christopher Helm, London.

Chesser, R.T., 2004. Systematics, evolution and biogeography of the South American ovenbirds genus *Cinclodes*. *Auk* 121(3): 752–766.

Chikara, O., 2011. Possible records of the newly described Bryan's Shearwater *Puffinus bryani* in Japan. *BirdingASIA* 16: 86–88.

Chittenden, H., 1998. Nesting behaviour in the Lemon-breasted Canary *Serinus citrinipectus*. *Bird Numbers* 7: 15.

Christian, D.G., 2001. Nests and nesting behavior of some little known Panamanian birds. *Ornitología Neotropical* 12: 327–336.

Christian, E.W., 1920. List of Mangaia Island birds. *J. Polynesian Soc.* 29: 87.

Christidis, L., 1999. Evolution and biogeography of the Australian grasswrens, *Amytornis* (Aves: Maluridae): biochemical perspectives. *Australian Journal of Zoology* 47: 113–124.

Christidis, L. & Boles, W.E., 2008. *Systematics and Taxonomy of Australian Birds.* CSIRO Publishing, Collingwood, Australia.

Christidis, L., Davies, K., Westerman, M., Christian, P.D. & Schodde, R., 1996. Molecular assessment of the taxonomic status of Cox's Sandpiper. *Condor* 98(3): 459–463.

Christidis, L., Rheindt, F.E., Boles, W.E. & Norman, J.A., 2010. Plumage patterns are good indicators of taxonomic diversity, but not phylogenetic affinities, in Australian grasswrens *Amytornis* (Aves: Maluridae). *Molecular Phylogenetics and Evolution* 57: 868–877.

Christidis, L., Rheindt, F.E., Boles, W.E. & Norman, J.A., 2013. A re-appraisal of species diversity within the Australian grasswrens *Amytornis* (Aves: Maluridae). *Australian Zoologist* 36(4): 429–437.

Choudhary, H., 2000. Nepal Wren Babbler *Pnoepyga immaculata* nest-building in the Langtang Valley, Nepal. *Forktail* 16: 170.

Cibois, A., David, N., Gregory, S.M.S. & Pasquet, E., 2010. Bernieridae (Aves: Passeriformes): a family-group name for the Malagasy sylvioid radiation. *Zootaxa* 2554: 65–68.

Cibois, A., Pasquet, E., & Schulenberg, T.S., 1999. Molecular systematics of the Malagasy Babblers (Passeriformes: Timaliidae) and Warblers (Passeriformes: Sylviidae), based on Cytochrome *b* and 16S rRna sequences. *Molecular Phylogenetics and Evolution* 13(3): 581–595.

Cibois, A., Slikas, B., Schulenberg, T.S. & Pasquet, E., 2001. An endemic radiation of Malagasy songbirds is revealed by mitochondrial DNA sequence data. *Evolution* 55(6): 581–595.

Cibois, A., Thibault, J.-C. & Pasquet, E., 2011. Molecular and morphological analysis of Pacific reed-warbler specimens of dubious origins, including *Acrocephalus luscinius astrolabii*. *Bull. Brit. Orn. Club* 131: 32–40.

Cizek, A., 2008. Interesting records from Vilancoulos, Mozambique, with a breeding record of the Lemon-breasted Canary. *Honeyguide* 54(1–2): 63–68.

TD Clancey, P.A. & Lawson, W.J., 1960. A new species of canary from southern Mozambique. *Durban Mus. Novit.* 6(4): 61–64.

Claramunt, S., 2014. Morphometric insights into the existence of a new species of *Cichlocolaptes* in northeastern Brazil. *Revista Brasiliera de Ornitologia* 22(2): 95–101.

Clay, R.P. & Field, B., 2005. First records of Narosky's Seedeater *Sporophila zelichi* in Paraguay. *Cotinga* 23: 84–85.

Clay, R.P., Tobias, J.A., Lowen, J.C. & Beadle, D., 1998. Field identification of *Phylloscartes* and *Phyllomyias* tyrannulets in the Atlantic rain forest region. *Cotinga* 10: 82–95.

Cleere, N., 1998. *Nightjars; A Guide to Nightjars and Related Nightbirds.* Pica Press, Sussex.

Clement, P., 1993. *Finches and Sparrows.* Christopher Helm/A & C Black, London.

Clement, P. & Hathway, R., 2000. *Thrushes.* Christopher Helm/A & C Black, London.

Clemlow, E., 1993. Breeding the Lemon-breasted Canary. *Avic. Mag.* 99(4): 205–207.

Clouet, M., Canu, J.G. & Lipinski, F., 1994. Sur la nidification de la Buse de Socotra *Buteo buteo* spp. *Alauda* 62(2): 144–145.

Clouet, M. & Wink, M., 2000. The Buzzards of Cape Verde *Buteo* (*buteo*) *bannermani* and Socotra *Buteo* (*buteo*) spp: First results of a genetic analysis based on nucleotide sequences of the cytochrome b gene. *Alauda* 68(1): 55–58.

Cobos, V. & Miatello, R., 2001. Descripción del nido, huevo y pichón de la Monjita Salinera (*Neoxolmis salinarum*). *Hornero* 16(1): 47–48.

TD Coelho, G. & Silva, W., 1998. A new species of *Antilopha* (Passeriformes: Pipridae) from Chapada do Araripe, Brazil. *Ararajuba* 6(2): 81–84.

TD Cohn-Haft, M. & Bravo, G., 2013. A new species of *Herpsilochmus* antwren from west of the Rio Madeira in Amazonian Brazil. *Handbook of the Birds of the World* Special Volume, pp 272–276. Lynx Edicions, Barcelona.

TD Cohn-Haft, M., Santos Junior, M.A., Fernandes, A.M. & Ribas, C.A., 2013. A new species of *Cyanocorax* jay from savannas of the central Amazon. *Handbook of the Birds of the World* Special Volume, pp 306–310. Lynx Edicions, Barcelona.

Collar, N.J., 1999a. The type locality and conservation status of *Monticola bensoni*. *Ostrich* 70(2): 151.

Collar, N.J., 1999b. New species, high standards and the case of *Laniarus liberatus*. *Ibis* 141(3): 358–367.

Collar, N.J., 2006. A partial revision of the Asian babblers (Timaliidae). *Forktail* 22: 85–112.

Collar, N.J., 2008. New bird descriptions without proper voucher specimens: further to Kannan. *J. Bombay Nat. Hist. Soc.* 105(2): 222–223.

Collar, N.J., 2009. Conservation-driven changes in English bird names, and the case of the Liben Lark. *Bull. African Bird Club* 16(2): 245.

Collar, N.J., 2011. Taxonomic notes on some Asian babblers (Timaliidae). *Forktail* 27: 100–102.

Collar, N.J., Dellelegn Abebe, Y., Fishpool, L.D.C., Gabremichael, M.N., Spottiswoode, C.N. & Wondafrash, M., 2008. Type locality, habitat, behaviour, voice, nest, eggs and plight of the Sidamo Lark *Heteromirafra sidamoensis*. *Bull. African Bird Club* 15(2): 180–190.

Collar, N.J. Dingle, C., Gabremichael, M.N. & Spottiswoode, C.N., 2009. Taxonomic status of the Degodi Lark *Mirafra degodiensis*, with notes on the voice of Gillett's Lark *M. gilletti*. *Bull. Brit. Orn. Club* 129: 49–62.

Collar, N.J. & Fishpool, L.D.C., 2006. What is *Pogoniulus makawai*? *Bull. African Bird Club* 13(1): 18–27.

Collar, N.J., Gonzaga, L.P., Krabbe, N., Madrono Nieto, A., Naranjo, L.G., Parker, T.A. & Wege, D.C., 1992. *Threatened birds of the Americas: the ICBP/IUCN Red Data Book,* International Council for Bird Preservation, Cambridge, 1150 pp.

Collar, N.J., Mallari, N.A.D. & Tabaranza, B.R.J., 1999. *Threatened Birds of the Philippines*. Bookmark Inc., Makati City, Philippines.

Collar, N.J. & Pilgrim, J.D., 2007. Species-level changes proposed for Asian birds, 2005–2006. *BirdingASIA* 8:14–30.

Collar, N.J. & Pittman. A.J., 1996. *Amazona kawalli* is a valid name for a valid species. *Bull. Brit. Orn. Club* 116(4): 256–265.

Collar, N.J. & Tattersall, I., 1987. J. T. Last and the type-locality of Benson's Rock-thrush *Monticola bensoni*. *Bull. Brit. Orn. Club* 107(2): 55–59.

TD Collins, C.T., 1972. A new species of swift of the genus *Cypseloides* from North-eastern South America. (Aves: Apodidae). *Contrib. Sci. Los Angeles Co. Mus. Nat. Hist.* 229: 1–9.

TD Colston, P.R., 1972. A new bulbul from southwestern Madagascar. *Ibis* 114(1): 89–92.

TD Colston, P.R., 1981. A newly described species of *Melignomon* (Indicatoridae) from Liberia, West Africa. *Bull. Brit. Orn. Club* 101(2): 289–291.

Colston, P.R., 1982. A new species of *Mirafra* (Alaudidae) and new races of the Somali Long-billed Lark *Mirafra somalica*, Thekla Lark *Galerida malabarica* and Malindi Pipit *Anthus melindae* from southern coastal Somalia. *Bull. Brit. Orn. Club* 102(3): 106–114.

Comité Editorial, Conservación Colombiana, 18 Mayo 2010. Editorial sobre la descripción de una nueva especie de Grallaria. *Conservación Colombiana* 13: 4–7.

TD Condon, H.T., 1969. A new subspecies of the Western Grasswren *Amytornis textilis* in northwestern Queensland; *Amytornis textilis ballarae* new subspecies. *Mem. Queensland Mus.* 15(3): 205–206.

Cooney, S.J.N., Watson, D.M. & Young, J., 2006. Mistletoe nesting in Australian birds: a review. *Emu* 106(1): 1–12.

TD Coopmans, P. & Krabbe, N., 2000. A new species of flycatcher (Tyrannidae: *Myiopagis*) from eastern Ecuador and eastern Peru. *Wilson Bull.* 112(3): 305–312.

Corbin, K.W., Livezey, B.C. & Humphrey, P.S., 1988. Genetic differentiation among Steamerducks (*Tachyeres*): an electrophoretic analysis. *Condor* 90: 773–781.

Cordeiro, N.J. & Githiru, M., 2000. Conservation evaluation for birds of *Brachylaena* woodland and mixed dry forest in northeastern Tanzania. *Bird Cons. Int.* 10: 47–65.

Cordeiro, P.H.C., de Melo Junior, T.A. & Ferreira de Vasconcelos, M., 1998. A range extension for Cipó Canastero *Asthenes luizae* in Brazil. *Cotinga* 10: 64–65.

Cortes, O., López, J.P. & Roa, A., undated. *Assessment and conservation of Cundinamarca Antpitta* (*Grallaria kaestneri*) at Farallon de Medina, Cundinamarca.

TD Cortés-Diago, A., Ortega, L.A., Mazariegos-Hurtado, L. & Weller, A-A., 2007. A new species of *Eriocnemus* (Trochilidae). *Ornitología Neotropical* 18(2): 161–169.

da Costa, T.V.V., Andretti, C.B., Olmos, F. & Pacheco, J.F., 2011. New records of Sulphur-breasted Parakeet *Aratinga maculata* in Pará and Amapá states, Brazil. *Cotinga* 33: 136–137.

Cotterill, F.D.P., 2004. Drainage evolution in south-central Africa and vicariant speciation in swamp-dwelling weaver birds and swamp flycatchers. *Honeyguide* 50: 7–25.

TD Cowles, G.S., 1964. A new Australian babbler. *Emu* 64(1): 1–5.

Cox, J.B., 1987. Some notes on the perplexing Cox's Sandpiper. *S. Aust. Ornith.* 30: 85–97.

Cox, J.B., 1989a. Notes on the affinities of Cooper's and Cox's Sandpipers. *S. Aust. Ornith.* 30: 169–181.

Cox, J.B., 1989b. The story behind the naming of Cox's Sandpiper. *Aust. Bird Watcher* 12: 50–57.

Cox, J.B., 1990a. The enigmatic Cooper's and Cox's Sandpipers. *Dutch Birding* 12: 53–64.

Cox, J.B., 1990b. The measurements of Cooper's Sandpiper and the occurrence of a similar bird in Australia. *S. Aust. Ornith.* 31: 38–43.

Cracraft, J., 1983. Species concepts and speciation analysis. *Curr. Ornithol.* 1: 159–187.

Craig, A.J.F.K., Hasson, M., Jordaens, K., Breman, F.C. & Louette, M., 2011. Range extension of the Lufira Masked Weaver *Ploceus ruweti*, endemic to Katanga province, Democratic Republic of Congo. *Ostrich* 82(1): 77–78.

TD Crossin, R.S. & Ely, C.A., 1973. A new race of Sumichrast's Wren from Chiapas, *México. Condor* 75: 137–139.

Cruaud, A., Raherilalao, M.J., Pasquet, E. & Goodman, S.M., 2011. Phylogeny and systematics of the Malagasy rock-thrushes (Muscicapidae, *Monticola*). *Zool. Scripta* 40(6): 554–566.

Cruz, A. & Delannoy, C.A., 1984a. Ecology of the Elfin-woods Warbler (*Dendroica angelae*). I. Distribution, habitat usage and population densities. *Caribbean Journal of Science* 20: 89–96.

Cruz, A. & Delannoy, C.A., 1984b. Ecology of the Elfin-woods Warbler (*Dendroica angelae*). II. Feeding ecology of the Elfin-woods Warbler and associated insectivorous birds in Puerto Rico. *Caribbean Journal of Science* 20: 153–162.

Cuddihey, L.W. & Stone, C.P., 1990. *Alteration of Native Hawaiian Vegetation: Effects of Humans, their Activities and Introductions.* University of Hawaii Press, Honolulu.

Cuervo, A.M., 2014. *Lipaugus weberi*. Pp 267–270 in: Renjifo, L.M., Gómez, M.F., Velásquez-Tibatá, J., Amaya-Villareal, A.M., Kattan, G.H., Amaya-Espinal, J.D. & Burbano-Girón, J., (eds). *Libro rojo de aves de Colombia, Volumen I: bosques húmedos de los Andes y la costa Pacífica.* Bogotá: Editorial Pontificia Universidad Javeriana & Instituto Alexander von Humboldt.

TD Cuervo, A.M., Cadena, C.D., Krabbe, N. & Renjifo, L.M., 2005. *Scytalopus stilesi*, a new species of tapaculo (Rhinocryptidae) from the Cordillera Central of Colombia. *Auk* 122(2): 445–463.

Cuervo, A.M., Pulgarín, P.C. & Calderón, D., 2008. New distributional data from the Cordillera Central of the Colombian Andes, with implications for the Biogeography of northwestern South America. *Condor* 110(3): 526–537.

Cuervo, A.M. & Restrepo, C., 2007. Assemblage and population-level consequences of forest fragmentation on bilateral asymmetry in tropical montane birds. *Biol. J. Linn. Soc.* 92: 119–133.

TD Cuervo, A.M., Salaman, P.G.W., Donegan, T.M. & Ochoa, J.M., 2001. A new species of piha (Cotingidae: *Lipaugus*) from the Cordillera Central of Colombia. *Ibis* 143: 353–368.

Cuervo, A.M., Stiles, F.G., Lentino, M., Brumfield, R.T. & Derryberry, E.P., 2014. Geographic variation and phylogenetic relationships of *Myiopagis olallai* (Aves: Passeriformes:

Tyrannidae), with the description of two new taxa from the Northern Andes. *Zootaxa* 3873: 001–024.

Curry-Lindahl, K., 1961. Obituary: Nils Gyldenstolpe. *Ibis* 103A: 627.

Curson, J. & Lowen, J., 1997. Neotropical Notebook. *Cotinga* 7: 9–10.

D'Angelo Neto, S. & de Vasconcelos, M.F., 2003. Novo registro estende a distribuição conhecida de *Arremon franciscanus* (Passeriformes: Emberizidae) ao sul. *Ararajuba* 11(2): 215.

Dantas Santos, M.P., Silveira, L.F. & Cardoso da Silva, J.M., 2011. Birds of Serra do Cachimbo, Pará State, Brazil. *Revista Brasiliera de Ornitologia* 19(2): 244–259.

Dantas, S.M., Weckstein, J.D., Bates, J.M., Krabbe, N.K., Cadena, C.D., Robbins, M.B., Valderrama, E. & Aleixo, A., 2016. Molecular systematics of the new world screech-owls (*Megascops*: Aves, Strigidae): biogeographic and taxonomic implications. *Molecular Phylogenetics and Evolution* 94: 626–634. Part B.

Darwin, C., 1839. *Journal of Researches into the Geology & Natural History of the Various Countries visited during the Voyage of HMS Beagle round the World.* Reprinted by J.M.Dent & Co., New York, 1908.

Dauphiné, N., Cooper, R.J. & Yagkuag, A.T., 2008. A new location and altitudinal range for the Royal Sunangel *Heliangelus regalis*. *Cotinga* 30: 83–84.

Dauphiné, N.S., Mahoney, A.C., Cooper, R.J. & Brooks, D.M., 2006. Selective logging and bird conservation in the Cordillera de Colán, northern Peru. 24th International Ornithological Congress, Hamburg, Germany, 13–19 August, 2006, p 154.

Davidson, P., Duckworth, W. & Poole, C., 2001. Mekong Wagtail: the great river's only avian endemic. *Bull. Oriental Bird Club* 34: 56–59.

Davidson, P.J., Lucking, R.S., Stones, A.J., Bean, N.J., Raharjaningtrah, W. & Banjaransari, H., 1991. *Report of an ornithological survey of Taliabu, Indonesia.* Univ. of East Anglia, Norwich, UK.

Davies, C.W.N., Barnes, R., Butchart, S.H.M., Fernandez, M. & Seddon, N., 1997. The conservation status of birds on the Cordillera de Colán, Peru. *Bird Cons. Int.*, 7: 181–195.

Davies, C.W.N. & Fernandez, M.,1996. Recent observations on the ecology of the Royal Sunangel *Heliangelus regalis*. *Bull. Brit. Orn. Club* 116 (1): 46–49.

Davies, G.B.P. & Peacock, D.S., 2014. Reassessment of plumage characters and morphometrics of *Anthus longicaudatus* Liversidge, 1996 and *Anthus pseudosimilis* Liversidge & Voelker, 2002 (Aves: Motacillidae). *Annals of the Ditsong National Museum of Natural History* 4: 187–206.

TD(I) Davis, L.I., 1978 (*Nyctisygmus kwalensis*). *Pan American Studies* 1(2): 47.

TD(I) Davis, L.I., 1978. (*Allasma northi*). *Pan American Studies* 1(2): 52.

TD(I) Davis, L.I., 1979. (*Tyrannus chubbi*). *Pan American Studies* 2(1): 39.

Davis, T.J., 1986. Distribution and natural history of some birds from the departments of San Martín and Amazonas, northern Peru. *Condor* 88: 50–56.

TD Davis, T.J. & O'Neill, J.P., 1986. A new species of antwren (Formicariidae: *Herpsilochmus*) from Peru, with comments on the systematics of other members of the genus. *Wilson Bull.* 98(3): 337–352.

Degnan, S.M., 1993. Genetic variability and population differentiation inferred from DNA fingerprinting in silvereyes (Aves: Zosteropidae). *Evolution* 47(4): 1105–1117.

Delannoy, C.A., 2007. Distribution, abundance and description of habitats of the Elfin-woods Warbler, *Dendroica angelae*, in southwestern Puerto Rico. Final Report, US Fish and Wildlife Service grant Agreement No 401814G078, August 1, 2004–December 30, 2006.

Delannoy-Julia, C.A., 2009. Elfin-woods Warbler (*Setophaga angelae*). *Neotropical Birds Online.* (T.S.Schulenberg, Editor). Cornell Lab. of Ornithology, Ithaca, NY.

Delacour, J., 1952. The specific grouping of the bush warblers *Bradypterus luteoventris, Bradypterus montis* and *Bradypterus seebohmi*. *Ibis* 94: 362–363.

TD Delgado B., F.S., 1985. A new subspecies of the Painted Parakeet (*Pyrrhura picta*) from Panama. *Orn. Monographs* 36: 16–20.

del Hoyo, J. & Collar, N.J., 2011.Acrobatic display in the Black-crowned Barwing *Actinodura sodangorum*. *Forktail* 27: 112–113.

del Hoyo, J. & Collar, N.J. 2014. *HBW and BirdLife International Illustrated Checklist of the Birds of the World.* Volume 1: Non-passerines. Lynx Edicions, Barcelona.

del Hoyo, J. & Collar, N.J. 2016. *HBW and BirdLife International Illustrated Checklist of the Birds of the World.* Volume 2: Passerines. Lynx Edicions, Barcelona.

del Hoyo, J., Collar, N.J. & Kirwan, G.M., 2016. Azuero Parakeet. *HBWAlive.*

Del-Rio, G. & Silveira, L.F., 2016. Remarks on the natural history of the São Paulo Marsh Antwren *Formicivora paludicola*. *Wilson Journal of Ornithology* 128: 445–448.

Demey, R. & Hester, A., 2008. First records of Nimba Flycatcher *Melaeornis annamarulae* for Ghana. *Bull. African Bird Club* 15(1): 95–96.

Demey, R., & Ossom, W., 2007. Rapid survey of the birds of the Atewa Range Forest Reserve, Ghana. *RAP Bulletin of Biological Assessment* 47: 84–89.

Diamond, J.M., 1971. Bird records from west New Britain. *Condor* 73: 481–483.

Diamond, J., 1975. Distributional ecology and habits of some Bougainville birds (Solomon Islands). *Condor* 77: 14–23.

Diamond, J.M., 1985. New distributional records and taxa from the outlying mountain ranges of New Guinea. *Emu* 85: 65–91.

Diamond, J., 1986. Animal art: variation in bower style among male bowerbirds *Amblyornis inornatus*. *Proc. Natl. Acad. Sci. USA* 83: 3042–3046.

TD Diamond, J., 1991. A new species of rail from the Solomon Islands and convergent evolution of insular flightlessness. *Auk* 108(3): 461–470.

Diamond, J. & Bishop, D., 2015. Avifauna of the Kumawa and Fakfak Mountains, Indonesian New Guinea. *Bull. Brit. Orn. Club* 135(4): 292–336.

Dickerman, R.W. (Comp.), 1997. *The Era of Allan Phillips: a Festschrift.* Albuquerque, NM.

Dickerman, R.W. & Parkes, K.C., 1997. Taxa described by Allan R. Phillips, 1939–1994: a critical list. *In* Dickerman, R.W. (Comp.), 1997. *The Era of Allan Phillips: a Festschrift.* Albuquerque, NM.

Dickinson, E.C., Kennedy, R.S. & Parkes, K.C., 1991. *The Birds of the Philippines; an Annotated Checklist.* BOU 1991, Tring, UK.

Dickinson, E., 1986. Does the White-eyed River Martin *Pseudochelidon sirintarae* breed in China? *Forktail* 2: 95–96.

Di Giacomo, A.S., 2005. *Áreas importantes para la Conservación de las aves en la Argentina. Sitios Prioritarios para la Conservación de la Biodiversidad.* Temas de Naturaleza y Conservación 5. Aves Argentinas/AOP, Buenos Aires.

TD Di Giacomo, A. & Kopuchian, C., 2016. Una nueva especie de Capuchino (*Sporophila*: Thraupidae) de los Esteros del Iberá, Argentina. *Nuestras Aves* 61: 3–5.

Di Giacomo, A.S., Vickery, P., Casañas, H., Spitznagel, O., Ostrosky, C., Krapovickas, S. & Bosso, A., 2010. Landscape associations of globally threatened grassland birds in the Aguapey River Important Bird Area, Corrientes, Argentina. *Bird Cons. Int.* 20: 62–73.

Diller, J., 2011. *When I Fell from the Sky.* Titletown Publishing, Green Bay, Wis.

TD Dinesen, L., Lehmberg, T., Svendsen, J.O., Hansen, L.A. & Fjeldså, J., 1994. A new genus and species of perdicine bird (Phasianidae: Perdicini) from Tanzania: a relict form with Indo-Malayan affinities. *Ibis* 136: 2–11.

Dinesen, L., Lehmberg, T., Rahner, M.C.& Fjeldså, J. 2001. Conservation priorities for the forests of the Udzungwa Mountains, Tanzania, based on primates, duikers and birds. *Biological Conservation* 99(2): 223–236.

Doggart, N., Perkin, A., Kiure, J., Fjeldså, J., Poynton, J & Burgess, N., 2006. Changing places; how the results of new field work in the Rubeho Mountains influence conservation priorities in the Eastern Arc Mountains of Tanzania. *African J. Ecol.* 44: 134–144.

Donegan, T.M., 2000. Is specimen-taking of birds in the Neotropics really "essential"? Ethical and practical objections to further collection. *Orn. Neotrop.* 11: 263–267.

TD Donegan, T.M., 2007. A new species of brush finch (Emberizidae: *Atlapetes*) from the northern central Andes of Colombia. *Bull. Brit. Orn. Club* 127(4): 255–267.

Donegan, T.M., 2008. New species and subspecies descriptions do not and should not always require a dead type specimen. *Zootaxa* 1761: 37–48.

Donegan, T.M., Avendaño, J.E., Briceño-L., E.R., Luna, J.C., Roa, C., Parra, R., Turner, C., Sharp, M. & Huertas, B., 2010. Aves de la Serranía de los Yariguies y tierras bajas circundantes, Santander, Colombia. *Cotinga* 32: 23–40.

TD Donegan, T.M., Avedaño, J.E. & Lambert, F., 2013. A new tapaculo related to *Scytalopus rodriguezi* from Serranía de los Yarigues, Colombia. *Bull. Brit. Orn. Club* 133(4): 256–271.

Donegan, T.M. & Huertas, B., 2006. A new brush-finch in the *Atlapetes latinuchus* complex from the Yariguíes Mountains and adjacent Eastern Andes of Colombia. *Bull. Brit. Orn. Club* 126(2): 94–115.

Donegan, T., Quevedo, A., McMillan, M. & Salaman, P., 2011. Revision of the status of bird species occurring or reported in Colombia, 2011. *Conservación Colombiana* 15: 4–21.

Donegan, T., Salaman, P., Caro, D. & McMullan, M., 2010. Revision of the status of bird species occurring in Colombia, 2010. *Conservación Colombiana* 13: 25–54.

Dornas, T., Leite, G.A., Pinheiro, R.T. & Crozariol, M.A., 2011. Primeiro registro do criticamente ameaçado pica-pau-do-parnaiba *Celeus obrieni* do Mato Grosso (Brasil) e comentários sobre distribuição geográfica e conservação. *Cotinga* 33: 140–143.

Dorst, J., 1961. Étude d'une collection d'oiseaux rapportée la vallée de Sandia, Pérou meridional. *Bull. Mus. Natl. Hist. nat.*33: 563–570.

Dorst, J. & Roux, F., 1991. Robert Daniel Etchécopar (1905–1990). *L'Oiseau et RFO* 60(4).

Doughty, R.W., 1975. *Feather Fashions and Bird Preservation. A Study in Nature Protection.* University of California Press, Berkeley, CA.

Dow, D.D., 1980. Communally breeding Australian birds with an analysis of distribution and environmental factors. *Emu* 80(3): 121–140.

Dow, D.D., 1982. Communal breeding. *RAOU Newsletter* 54: 6–7.

Downing, C. & Hickman, J., 2002. The first White-chested Swift *Cypseloides lemosi* in Amazonian Colombia. *Cotinga* 18: 102–103.

Dowsett, R.J. & Dowsett-Lemaire, F., 1980. The systematic status of some Zambian birds. *Gerfaut* 70: 151–199.

Dowsett, R.J. & Dowsett-Lemaire, F., 1993. Comments on the taxonomy of some Afrotropical bird species. *Tauraco Res.Rep.* 5: 323–389.

Dowsett, R.J. & Dowsett-Lemaire, F., 2000. New species and amendments to the avifauna of Cameroon. *Bull. Brit. Orn. Club* 120(3): 179–185.

Dowsett-Lemaire, F., 2009. The song of presumed Prigogine's Nightjar *Caprimulgus prigoginei* and its possible occurrence in Lower Guinea. *Bull. African Bird Club* 16(2): 174–179.

Dowsett-Lemaire, F. & Dowsett, R.J., 2000. Birds of the Lobéké Faunal Reserve, Cameroon, and its regional importance for conservation. *Bird Cons. Int.* 10: 67–87.

Dowsett-Lemaire, F. & Dowsett, R.J., 2005. *Cisticola dorsti* (Dorst's Cisticola) and *C. ruficeps guinea* are conspecific. *Bull. Brit. Orn. Club* 125(4): 305–313.

Dowsett-Lemaire, F. & Dowsett, R.J., 2008. Selected notes on birds of Gola Forest and surroundings, Sierra Leone, including three new species for the country. *Bull. African Bird Club* 15(2): 215–227.

Driskell, A.C., Norman, J.A., Pruett-Jones, S., Mangall, E., Sonsthagen, S. & Christidis, L., 2011. A multigene phylogeny examining evolutionary and ecological relationships in the Australo-papuan wrens of the subfamily Malurinae (Aves). *Molecular Phylogenetics and Evolution* 60: 480–485.

Drovetski, S.V., Semenov, G., Red'kin, Ya.A., Sotnikov, V.N., Fadeev, I.V. & Koblik, E.A., 2015. Effects of asymmetric nuclear introgression, introgressive mitochondrial sweep, and purifying selection on phylogenetic reconstruction and divergence estimates in the Pacific clade of *Locustella* warblers. *PLoS ONE* 10(4): e0122590.

Drovetski, S.V., Zink, R.M., Fadeev, I.V., Nesterov, E.V., Koblik, E.A., Red'kin, Ya.A. & Rohwer, S., 2004. Mitochondrial phylogeny of *Locustella* and related genera. *J. Avian Biol.* 35(2): 105–110.

Duarte, J.M.B. & Caparroz, R., 1995. Cytotaxonomic analysis of Brazilian species of the genus *Amazona* (Aves: Psittacidae) and confirmation of the genus *Salvatoria* (Ribeiro, 1920). *Braz. J. Gen.* 18: 623–628.

Dubois, A. & Nemésio, A., 2007. Does nomenclatural availability of nomina of new species or subspecies require the deposition of vouchers in collections? *Zootaxa* 1409: 1–22.

TD Duckworth, J.W., Alström, P., Davidson, P., Enans, T.D., Poole, C.M., Tan Setha & Timmins, R.J., 2001. A new species of wagtail from the lower Mekong basin. *Bull. Brit. Orn. Club* 121: 152–182.

Duckworth, J.W., Tizard, R.J., Timmins, R.J., Thewlis, R.M., Robichaud, W.G. & Evans, T.D., 1998. Bird records from Laos, October 1994–August 1995. *Forktail* 13: 33–68.

Dumont D'Urville, J.S.C., 1841–1846. *Voyage au Pole Sud et dans l'Oceanie sur les corvettes l'*Astrolabe *et la* Zélée*, execute par l'ordre du roi pendant les années 1837–1838–1839–1840, sous le commandement de M.J.Dumont D'Urville.* Paris, 10 vols.

Dunning, J.S., 1982. *South American Landbirds: a Photographic Guide to Identification.* Harrowood Books, Newton Square, PA.

Dutson, G., 1993. A sighting of *Ficedula (crypta) disposita* in Luzon, Philippines. *Forktail* Suppl. 8: 144–147.

TD Dutson, G., 2008. A new species of white-eye *Zosterops* and notes on other birds from Vanikoro, Solomon Islands. *Ibis* 150: 698–706.

Dutson, G., 2011. *Birds of Melanesia: Bismarcks, Solomons, Vanuatu and New Caledonia.* Christopher Helm, London.

Eames, J.C., 2001. On the trail of Vietnam's endemic babblers. *Bull Oriental Bird Club* 33: 2027.

TD Eames, J.C. & Eames, C., 2001. A new species of Laughingthrush (Passeriformes: Garrulacinae) from the Central Highlands of Vietnam. *Bull. Brit. Orn. Club* 121(1): 10–23.

TD Eames, J.C., Lê Trong Trai & Nguyên Cu, 1999a. A new species of Laughingthrush (Passeriformes: Garrulacinae) from the western highlands of Vietnam. *Bull. Brit. Orn. Club* 119: 4–15.

TD Eames, J.C., Lê Trong Trai, Nguyên Cu & Eve, R., 1999b. New species of Barwing *Actinodura* (Passeriformes: Sylviinae: Timaliini) from the western highlands of Vietnam. *Ibis* 141: 1–10.

Eames, J.C. & Nguyen Van Truong, 2016.The discovery of the Nonggang Babbler *Stachyris nonggangensis* in Vietnam. *BirdingASIA* 26: 29–31.

Eaton, J.A., Mitchell, S.L., Navario Gonzalez Bocos, C. & Rheindt, F.E., 2016a. A short survey of the Meratus Mountains, south Kalimantan province, Indonesia: two undescribed avian species discovered. *BirdingASIA* 26: 107–113.

Eaton, J.A., van Balen, B., Brickle, N.W. & Rheindt, F.E. 2016b. *Birds of the Indonesian Archipelago: Greater Sundas and Wallacea.* Lynx Edicions, Barcelona.

Edwards, D.P., Webster, R.E. & Rowlett, R.A., 2009. 'Spectacled Flowerpecker': a species new to science discovered in Borneo? *BirdingASIA* 12: 38–41.

TD Eisenmann, E. & Lehmann, F.C., 1962. A new species of swift of the genus *Cypseloides* from Colombia. *Amer. Mus. Novit.* 2117: 1–16.

Engblom, G., 2010. http://birdingblogs.com/2010/Gunnar/long-whiskered-owlet

Engilis, A., Jr, Pratt, T.K., Kepler, C.B., Ecton, A.M. & Fluetsch, K.M., 1996. Description of adults, eggshells, nestling, fledgling and nest of the Poo-uli. *Wilson Bull.* 108(4): 607–619.

TD(I) Érard, C., 1975. Une nouvelle alouette du sud de l'Ethiopie. *Alauda* 43(2): 115–124.

Érard, C. & Fry, C.H., 1997 in *The Birds of Africa* Vol. 5 (E.K.Urban, C.H.Fry & S.Keith, Eds.). Academic Press, London.

TD(I) Érard, C. & Roux, F., 1983. La Chevêchette du Cap *Glaucidium capense* dans l'ouest africain. Description d'une race géographique nouvelle. *L'Oiseau et RFO* 53(2): 97–104.

Estades, C.F., 1996. Natural history and conservation status of the Tamarugo Conebill in northern Chile. *Wilson Bull.* 108(2): 268–279.

Estades, C.F. & López-Calleja, M.V., 1995. First nesting record of the Tamarugo Conebill (*Conirostrum tamarugense*). *Auk* 112(3): 797–800.

Evans, T.D., 1997a. Records of birds from the forests of the East Usambara lowlands, Tanzania, August 1994–February 1995. *Scopus* 19: 92–108.

Evans, T.D., 1997b. Preliminary estimates of the population density of the Sokoke Scops Owl *Otus ireneae* Ripley in the East Usambara lowlands, Tanzania. *African Journal of Ecology* 35: 303–311.

Evans, T.D. & Anderson, G.Q.A., 1992. Results of an ornithological survey in the Ukaguru and East Usambara Mountains, Tanzania. *Scopus* 19: 95–108.

Falla, R.A., 1976. Notes on the gadfly petrels *Pterodroma externa* and *P. e. cervicalis*. *Notornis* 23: 320–322.

de Faria, I.P., 2007. Globally threatened, rare and endemic bird records to Vicente Pires region, Federal District, Brazil. *Revista Brasiliera de Ornitologia* 15(1): 117–122.

de Farias, F. B., 2016. Primeiro registro de alegrinho-trinador (*Serpophaga griseicapilla*) e gaivota-defranklin (*Leucophaeus pipixcan*) em Santa Catarina, Sul do Brasil. *Ornithologia* 9(2): 32–113.

TD(I) Farkas, T., 1971. *Monticola bensoni*; a new species from southwestern Madagascar. *Ostrich Suppl. Issue* 9: 83–90.

Faucher, I. & Dowsett-Lemaire, F., 2011. A new form of thrush babbler *Ptyrticus* sp. from the Bakossi Mountains, Cameroon? *Bull. African Bird Club* 18(1): 74–75.

Faulquier, L., Fontaine, R., Vidal, E., Salamolard, M. & Le Corre, M., 2009. Feral cats *Felis catus* threaten the endangered endemic Barau's Petrel *Pterodroma baraui* at Reunion Island (Western Indian Ocean). *Waterbirds* 32(2): 330–336.

TD Favoloro, N.J. & McEvey, A., 1968. A new species of Australian grasswren. *Mem. Nat. Mus. Vic.* 28: 1–9.

TD Ferreira de Vasconcelos, M., D'Angelo Neto, S. & Fjeldså, J., 2008. Redescription of Cipó Canastero *Asthenes luizae*, with notes on its systematic relationships. *Bull. Brit. Orn. Club* 128(3): 179–186.

Ferreira de Vasconcelos, M., Ribeiro, F.M.F. & Pardini, H., 2008. Primeiro registro do João-Cipó (*Asthenes luizae*) no Parque Nacional das Sempre-Vivas, Minas Gerais. *Atualidades Ornitológicas* 115: 8.

Ferreira de Vasconcelos, M. & Sa,F., 2002. O João-Cipó (*Asthenes luizae*) no Parque Estadual do Pico Itambe, Minas Gerais, Brasil. *Atualidades Ornitológicas* 107: 10.

Field, G.D., 1979. A new species of *Malimbus* sighted in Sierra Leone and a review of the genus. *Malimbus* 1: 2–13.

Fierro-Calderón, K. & Montealegre, C., 2010. Nuevo registro del Buhito nubícola *Glaucidium nubicola* en la Cordillera Occidentale de Colombia. *Boletin SAO* 20 (1): 29–33.

Finch, B.W., 2005. River Prinia *Prinia fluviatilis* near Lokichokio: a new species for Kenya and East Africa. *Scopus* 25: 55–59.

Firme, D.H. & Raposo, M.A., 2011. Taxonomy and geographic variation of *Formicivora serrana* (Hellmayr, 1929) and *Formicivora littoralis* Gonzaga and Pacheco, 1990 (Aves: Thamnophilidae). *Zootaxa* 2742: 1–33.

Fishpool, L.D.C. & Evans, M.I., 2001. *Important Bird Areas of Africa and Associated Islands: Priority sites for Conservation.* Pisces Publications and BirdLife International, Newbury and Cambridge.

TD Fitzpatrick, J.W. & O'Neill, J.P., 1979. A new tody-tyrant from northern Peru. *Auk* 96: 443–447.

TD Fitzpatrick, J.W. & O'Neill, J.P., 1986. *Otus petersoni*, a new screech-owl from the eastern Andes, with systematic notes on *O. colombianus* and *O. ingens*. *Wilson Bull.* 98(1): 1–14.

TD Fitzpatrick, J.W. & Stotz, D.F., 1997. A new species of tyrannulet (*Phylloscartes*) from the Andean foothills of Peru and Bolivia. Pp 37–44 in J.V.Remsen Jr (ed.), Studies in Neotropical Ornithology honoring Ted Parker. *Ornithological Monographs 48*. American Ornithologists' Union, Washington, DC.

TD Fitzpatrick, J.W., Terborgh, J.W. & Willard, D.E., 1977. A new species of wood-wren from Peru. *Auk* 94(2): 195–201.

TD Fitzpatrick, J.W. & Willard, D.E., 1990. *Cercomacra manu*, a new species of antbird from southwestern Amazonia. *Auk* 107: 239–245.

TD Fitzpatrick, J.W., Willard, D.E. & Terborgh, J.W., 1979. A new species of hummingbird from Peru. *Wilson Bull.* 177–186.

Fjeldså, J., 1986. Feeding ecology and possible life history tactics of the Hooded Grebe *Podiceps gallardoi*. *Ardea* 74: 40–58.

Fjeldså, J., 2004. *The Grebes.* Oxford University Press.

Fjeldså, J., Baiker, J., Engblom, G., Franke, I., Geale, D., Krabbe, N.K., Lane, D.F., Lezama, M., Schmidt, F., Williams, R.S.R., Ugarte-Núñez, J., Yábar, V. & Yábar, R., 2012. Reappraisal of Koepcke's Screech Owl *Megascops koepckeae* and description of a new subspecies. *Bull. Brit. Orn. Club* 132(3): 180–193.

TD Fjeldså, J., Bowie, R.C.K. & Kiure, J., 2006. The forest batis, *Batis mixta*, is two species: description of a new, narrowly distributed *Batis* species in the Eastern Arc biodiversity hotspot. *J. Ornithol.* 147: 578–590.

Fjeldså, J. & Kiure, J., 2003. A new population of the Udzungwa Forest Partridge. *Bull. Brit. Orn. Club* 123(1): 52–57.

Fjeldså, J. & Krabbe, N., 1990. *The Birds of the High Andes.* Zoological Museum, Copenhagen.

Fjeldså, J., Mayr, G., Jønsson, K.A. & Irestedt, M., 2013. On the true identity of Bluntschli's Vanga *Hypositta perdita* Peters, 1996, a presumed extinct species of Vangidae. *Bull. Brit. Orn. Club* 133(1): 72–75.

Fjeldså, J. & Sharpe, C.J., 2016. Pincoya Storm-petrel. *HBWAlive.*

Florez, P. 2017. A mysterious antshrike in eastern Colombia. *Neotropical Birding* 20: 21–24.

TD Forbes-Watson, A.D., 1970. A new species of *Melaeornis* (Muscicapinae) from Liberia. *Bull. Brit. Orn. Club* 90(6); 145–148.

Ford, J., 1977. Sympatry in Hall's and White-browed Babblers in New South Wales. *Emu* 77: 40.

TD Ford, J. & Johnstone, R.E., 1983. The Rusty-tailed Flyeater *Gerygone ruficauda*: New species from Queensland, Australia. *Western Australian Naturalist* 15(6): 133–135.

Ford, J. & Parker, S.A.,1974. Distribution and taxonomy of some birds from south-western Queensland. *Emu* 74: 177–194.

Forero, M.G. & Tella, J.L., 1997. Sexual dimorphism, plumage variability and species determination in nightjars; the need for further examination of the Nechisar Nightjar *Caprimulgus solala*. *Ibis* 139: 410–411.

Forshaw, J.M., 2010. *Parrots of the World*. Princeton University Press, Princeton, NJ.

Fowlie, M., 2012. Red List for Birds. *World Birdwatch* June 2012, 25.

França, L.F. & Marini, M.A., 2009. Low and variable reproductive success of a neotropical tyrant-flycatcher, Chapada Flycatcher (*Suiriri islerorum*). *Emu* 109: 265–269.

Freeman, B.G. & Rojas, C.J., 2010. The nest and egg of the Cinnamon Screech Owl *Megascops petersoni* in central Colombia. *Cotinga* 32: 148–149.

Freile, J.F., Chaves, J.A., Iturralde, G. & Guevara, E., 2003. Notes on the distribution, habitat and conservation of the Cloud-forest Pygmy-owl *Glaucidium nubicola* in Ecuador. *Ornitol. Neotrop.* 14: 275–278.

Freile, J.F., Piedrahita, P., Buitrón-Jurado, G., Rodríguez, C.A., Jadán, O. & Bonaccorso, E., 2011. Observations on the natural history of the Royal Sunangel *Heliangelus regalis* in the Nangaritza Valley, Ecuador. *Wilson J. Orn.*123(1): 85–92.

TD(I) de Freitas, G.H.S, Chaves, A.V., Costa, L.M., Santos, F.R. & Rodrigues, M., 2012. A new species of *Cinclodes* from the Espinhaço Range, southeastern Brazil: insights into the biogeographical history of the South American highlands. *Ibis* 154: 738–755.

de Freitas, G.H.S, Costa, L.M., Ferreira, J.D. & Rodrigues, M., 2008. The range of Long-tailed Cinclodes extends to Minas Gerais. *Bull. Brit. Orn. Club* 128(3): 215–216.

Frisch, J.D., 2007. (Description of *Ramphocelus ciroalbicaudatus* as a new species). *Nature Society News, Griggsville* 5: 13.

Fry, C.H., Fry, K. & Harris, A., 1992. *Kingfishers, Bee-eaters and Rollers*. Christopher Helm/A&C Black, London.

Fry, C.H., Keith, S. & Urban, E.K., 1988. *The Birds of Africa* Vol. 3. Academic Press, London and San Diego.

TD Fry, C.H. & Smith, D.A., 1985. A new swallow from the Red Sea. *Ibis* 127(1): 1–6.

Fuchs, J., Lemoine, D., Parra, J.L., Pons, J-M., Raherilalao, M.J., Prŷs-Jones, R., Thebaud, C., Warren, B.H. & Goodman, S.M., 2016. Long-distance dispersal and inter-island colonization across the western Malagasy Region explain diversification in brush-warblers (Passeriformes: *Nesillas*). *Biol. J. Linn. Soc.* D01: 10.1111/bij.12825.

Fullard, J.H., Barclay, R.M. & Thomas, D.W., 2010. Observations on the behavioural ecology of the Atiu Swiftlet *Aerodromus sawtelli*. *Bird Cons. Int.* 20: 385–391.

Fund, W., 2013. Mount Lofty Woodlands, Australia. http://editors.eol.org/eoearth/wiki/Mount_Lofty_woodlands,_Australia

TD Gaban-Lima, R., Raposo, M.A. & Hofling, E., 2002. Description of a new species of *Pionopsitta* (Aves: Psittacidae) endemic to Brazil. *Auk* 119(3): 815–819.

Gamauf, A., Gjershaug, J-O., Røv, N., Kvaløy, K. & Haring, E., 2005. Species or subspecies? The dilemma of taxonomic ranking of some South-east Asian hawk-eagles (genus *Spizaetus*). *Bird Cons. Int.*15(1): 99–117.

García-Moreno, J. & Fjeldså, J., 1999. Re-evaluation of species limits in the genus *Atlapetes* based on mtDNA sequence data. *Ibis* 141: 199–207.

García R., J.C., Gibb, G.C. & Trewick, S.A., 2014. Deep global evolutionary radiation in birds: diversification and trait evolution in the cosmopolitan bird family Rallidae. *Molecular Phylogenetics and Evolution* 81: 96–108.

Gartshore, M.E., 2005. *Report of Field Trip to Cross River National Park, 2004, Nigerian National Parks Service*. A. P. Leventis Conservation Foundation, Nigerian Conservation Foundation, Wildlife Conservation Society.

TD Gatter, W., 1985. Ein neuer Bülbül aus Westafrika (Aves, Pycnonotidae). *J. Orn.* 126(2): 155–161.

Gatter, W. & Gardner, R., 1993. The biology of the Gola Malimbe *Malimbus ballmanni* Wolters, 1974. *Bird Cons. Int.* 3(2): 87–103.

Gatter, W. & Mattes, H., 1979. Zur Populationsgrösse und Ökologie des neuendeckten Kabylienkleibers *Sitta ledanti* Vielliard 1976. *J. Ornithol.* 120: 390–405.

Gerhart, N.G., 2004. Rediscovery of the Selva Cacique (*Cacicus koepckeae*) in southeastern Peru with notes on habitat, voice and nest. *Wilson Bull.* 116: 74–82.

Gerlang, M., Cibois, A., Pasquet, E., Olsson, U., Alström, P. & Ericson, P.G.P., 2009. Phylogeny of babblers (Aves: Passeriformes): major lineages, family limits and classification. *Zool. Scripta* 38: 225–236.

Giai, A.G., 1951. Notas sobre la avifauna de Salta y Misiones. *Hornero* 9: 247–276.

Gibbs, D., 1994. Undescribed taxa and new records from the Fakfak Mountains, Irian Jaya. *Bull. Brit. Orn. Club* 114(1): 4–12.

Gibbs, D., 1996. Notes on Solomon Island birds. *Bull. Brit. Orn. Club* 116(1): 18–25.

Gill, F. & Donsker, D. (eds), 2015. Finches, euphonias. *World Bird List Version 5.2*. International Ornithologists' Union.

TD Gilliard, E.T., 1960a. Results of the 1958–1959 Gilliard New Britain Expedition. 1. A New Genus of Honeyeater (Aves). *Amer. Mus. Novit.* 2001: 2.

TD Gilliard, E.T., 1960b. Results of the 1958–1959 Gilliard New Britain Expedition. 2. A new species of thicket warbler (Aves, *Cichlornis*) from New Britain. *Amer. Mus. Novit.* 2008: 1–6.

Ginés, H., Aveledo, R., Pons, A.R., Yépez, G. & Múñoz Tebar, R., 1953. Lista y comentario de las aves colectadas en la región. In *La Región de Perijá y sus Habitantes* (Sociedad de Ciencias Naturales La Salle, Editor). Editorial Sucre, Caracas, Venezuela.

Girão, W. & Souto, A., 2005. Breeding period of Araripe Manakin inferred from vocalisation activity. *Cotinga* 24: 35–37.

Gochfield, M., Hill, D.O. & Tudor, G., 1973. A second population of the recently described Elfin-woods Warbler and other bird records from the West Indies. *Caribbean Journal of Science* 13: 231–235.

Gomes, H.B. & Rodrigues, M., 2010. The nest of the Cipo Canastero (*Asthenes luizae*), an endemic furnariid from the Espinhaço Range, southeastern Brazil. *Wilson J. Orn.*122(3): 600–603.

Gómez, L.G., Houston, D.C., Cotton, P. & Tye, A. (1994). The role of Greater Yellow-headed Vultures as scavengers in neotropical forest. *Ibis* 136: 193–196.

Gómez de Silva, H., 1997a. Distribution and conservation status of *Hylorchilus* wrens (Troglodytidae) in Mexico. *Bird Cons. Int.* 7: 409–418.

Gómez de Silva, H., 1997b. Comparative analysis of the vocalisations of *Hylorchilus* wrens. *Condor* 99: 981–984.

Gómez de Silva, H., Marantz, C.A. & Pérez-Villafaña, M., 2004. Song in female *Hylorchilus* wrens. *Wilson Bull.* 116(2): 186–188.

TD Gonzaga, L.P., 1988. A new antwren *Myrmotherula* from southeastern Brazil. *Bull. Brit. Orn. Club* 108(3): 132–135.

TD Gonzaga, L.P., Carvalhaes, A.M.P. & Buzzetti, D.R.C., 2007. A new species of *Formicivora* antwren from the Chapada Diamantina, eastern Brazil (Aves: Thamnophilidae). *Zootaxa* 1473: 25–38.

TD(I) Gonzaga, L.P. & Pacheco, J.F., 1990. Two new subspecies of *Formicivora serrana* (Hellmayr) from southeastern Brazil, and notes on the type locality of *Formicivora deluzae* Ménétries. *Bull. Brit. Orn. Club* 110(4): 187–193.

TD Gonzaga, L.P. & Pacheco, J.F., 1995. A new species of *Phylloscartes* (Tyrannidae) from the mountains of southern Bahia, Brazil. *Bull. Brit. Orn. Club* 115: 88–97.

TD Gonzales, P.C. & Kennedy, R.S., 1990. A new species of *Stachyris* babbler (Aves: Timaliidae) from the island of Panay, Philippines. *Wilson Bull.* 102(3): 367–379.

González, G.M., 2008. Distribución y abundancia de la Reinita de Bosque Enano (*Dendroica angelae*) en el Bosque de Maricao y áreas adyacentes. MS Thesis, Universidad de Puerto Rico, Recinto Universitario de Mayagüez, pp 81.

TD Goodman, S.M., Hawkins, A.F.A. & Domergue, C.A., 1997. A new species of vanga (Vangidae, *Calicalius*) from southwestern Madagascar. *Bull. Brit. Orn. Club* 117(1): 5–10.

TD Goodman, S.M., Langrand, O. & Whitney, B.M., 1996. A new genus and species of passerine from the eastern rain forest of Madagascar. *Ibis* 138(2): 153–159.

TD Goodman, S.M., Raherililao, M.J. & Block, N.L., 2011. Patterns of morphological and genetic variation in the *Mentocrex kioloides* complex (Aves: Gruiformes: Rallidae) from Madagascar, with the description of a new species. *Zootaxa* 2776: 49–60.

Goodman, S.M. & Weigt, L.A., 2002. The generic and species relationships of the reputed endemic Malagasy genus *Pseudocossyphus* (family Turdidae). *Ostrich* 73(1–2): 26–35.

Goodwin, D., 1965. Some remarks on the new barbet. *Bull. Brit. Orn. Club* 85: 9–10.

Goodwin, D., 1982. *Estrildid Finches of the World.* British Museum (Natural History) & Cornell University Press, Ithaca, NY.

Gorman, G., 2014. *Woodpeckers of the World: the Complete Guide.* Christopher Helm, London.

Grantsau, R., 1967. Sôbre o gênero *Augastes*, com a descrição de uma subespécie nova (Aves: Trochilidae). *Papéis Avulsos de Zoologia de São Paulo* 21(3): 21–31.

TD Grantsau, R. & Camargo, H.F.A., 1989. Nova espécie de *Amazona* (Aves: Psittacidae). *Rev. Brasil. Biol.* 49: 1017–1020.

Grantsau, R. & Camargo, H.F.A., 1990. Eine neue Papageienart aus Brasilien, *Amazona kawalli* (Aves: Psittacidae). *Trochilus* 11: 103–108.

TD Graves, G.R., 1980. A new species of Metaltail Hummingbird from Northern Peru. *Wilson Bull.* 92(1): 1–7.

TD Graves, G.R., 1987. A cryptic new species of antpitta (Formicaridae: *Grallaria*) from the Peruvian Andes. *Wilson Bull.* 99: 313–321.

TD Graves, G.R., 1988. *Phylloscartes lanyoni*, a new species of Bristle-tyrant from the lower Cauca Valley of Colombia. *Wilson Bull.* 100(4): 529–534.

Graves, G.R., 1992. Greater Yellow-headed Vulture (*Cathartes melambrotus*) locates food by olfaction. *J. Raptor Research* 26(1): 38–39.

TD Graves, G.R., 1993. Relic of a lost world: a new species of Sunangel (Trochilidae: *Heliangelus*) from "Bogotá". *Auk* 110(1): 1–8.

TD Graves, G.R., 1997. Colorimetric and morphometric gradients in Colombian populations of Dusky Antbirds (*Cercomacra tyrannina*), with a description of a new species, *Cercomacra parkeri*. *Orn. Monographs* 48: 20–35.

Graves, G.R., Lane, D.F., O'Neill, J.P. & Valqui, T., 2011. A distinctive new subspecies of the Royal Sunangel (Aves: Trochiliformes: *Heliangelus regalis*) from the Cordillera Azul, northern Peru. *Zootaxa* 3002: 52–58.

TD Graves, G.R., O'Neill, J.P. & Parker, T.A., 1983. *Grallaricula ochraceifrons*, a new species of antpitta from northern Peru. *Wilson Bull.* 95: 1–6.

TD Graves, G.R. & Weske, J.S., 1987. *Tangara phillipsi*, a new species of tanager from the Cerros del Sira, eastern Peru. *Wilson Bull.* 99: 1–6.

Greeney, H.F., 2008. Additions to our understanding of *Scytalopus* tapaculo reproductive biology. *Ornitología Neotropical* 19: 463–466.

Greeney, H.F., 2012a. Antpittas and worm-feeders; a match made by evolution? Evidence for possible commensal foraging relationships between antpittas (Grallariidae) and mammals. *Neotropical Biology and Conservation* 7: 140–143.

Greeney, H.F., 2012b. The natal plumages of antpittas (Grallariidae). *Ornitología Colombiana* 12: 65–68.

Greeney, H.F., Dobbs, R.C., Martin, P.R. & Gelis, R.A., 2008. The breeding biology of *Grallaria* and *Grallaricula* antpittas. *J. Field Ornith.* 79: 113–129.

Greeney, H.F. & Gelis, R.A., 2005a. Juvenile plumage and vocalization of the Jocotoco Antpitta *Grallaria ridgelyi*. *Cotinga* 23: 79–81.

Greeney, H.F. & Gelis, R.A., 2005b. The nest and nestlings of the Long-tailed Tapaculo (*Scytalopus micropterus*) in Ecuador. *Orn. Colombiana* 3: 88–91.

Greeney, H.F. & Juiña J., M.E., 2010. First description of the nest of the Jocotoco Antpitta (*Grallaria ridgelyi*). *Wilson J. Orn.* 122(2): 392–395.

Greeney, H.F. & Rombough, C.J.F., 2005. First nest of the Chusquea Tapaculo (*Scytalopus parkeri*) in southern Ecuador. *Ornitología Neotropical* 16: 439–440.

Grieve, A. & Kirwan, G.M., 2012. Studies of Socotran birds VII. Forbes-Watson's Swift *Apus berliozi* in Arabia – the answer to the mystery of the 'Dhofar Swift'. *Bull. Brit. Orn. Club* 132(3): 194–206.

Grilli, P., Soave, G., Valqui, T. & Fraga, R.M., 2009. Observations on the distribution, ecology, behavior and nesting of the Selva Cacique (Unpubl., from Fraga, R.M. 2011 HBW 16)

Grilli. P., Soave, G. & Fraga, R., 2012. Natural History and distribution of Selva Cacique (*Cacicus koepckeae*) in the Peruvian Amazon. *Ornitología Neotropical* 23(3): 375–383.

Groombridge, J.J., in Pratt, T.K., Atkinson, C.T., Banko, P.C., Jacobi, J.D. & Woodworth, B.T. (Eds), 2009. *Conservation Biology of Hawaiian Forest Birds.* Yale University Press, New Haven and London.

Groombridge, J.J., Massey, J.G., Bruch, J.C., Malcolm, T., Brosius, C.N., Okada, M.M., Sparklin, B., Fretz, J.S. & VanderWerf, E.A., 2004. An attempt to recover the Po'ouli by translocation and an appraisal of recovery strategy for bird species of extreme rarity. *Biological Conservation* 118: 365–375.

Guilherme, E., 2013. A range extension for Várzea Thrush *Turdus sanchezorum* in south-western Amazonia. *Bull. Brit. Orn. Club* 133(3): 249– 251.

Guilherme, E. & Sousa Santos, G., 2013. A new locality and habitat type for Rondônia Bushbird *Clytoctantes atrogularis*. *Bull. Brit. Orn. Club* 133(1): 68–71.

Gunawadana, J., 2014. (Obituary of Thilo W. Hoffmann). *Ceylon Today*, 2014–05–20. (http://ceylontoday.lk/index.html)

Gussoni, C.O.A., 2010. Novas informações sobre a história natural da Maria-da-Restinga (*Phylloscartes kronei*) (Aves. Tyrannidae). MSc Dissertation, Univ. Estadual Paulista, Rio Claro, Brazil.

Gussoni, C.O.A., 2014. Area de vida e biologia reprodutiva da Maria-da-Restinga (*Phylloscartes kronei*) (Aves, Tyrannidae). PhD Thesis, Univ. Estadual Paulista, Rio Claro, Brazil.

Gussoni, C.O.A. & Santos, M.C.O., 2011. Foraging behavior of Restinga Tyrannulet. *Orn. Neotrop.* 22: 495–504.

Gyldenstolpe, N., 1945a. The bird fauna of Rio Juruá in western Brazil. *Kungl. Sv. Vet. Akad. Handlingar* 22: 1–338.

Gyldenstolpe, N., 1945b. A contribution to the ornithology of northern Bolivia. *Kungl. Sv. Vet. Akad. Handlingar* 23: 1–300.

Gyldenstolpe, N., 1951. The ornithology of the Rio Purús region in western Brazil. *Arkiv för Zoologi Serie* 2 Band 2,1: 1–320.

Gyldenstolpe, N., 1916. Zoological results of the Swedish Zoological Expeditions to Siam 1911–1912 & 1914–1915. *Kungl. Sv. Vet. Akad. Handlingar* 56(2): 3–160.

Hachisuka, M. & Udagawa, T., 1951. Contribution to the ornithology of Formosa. Part II. *Quarterly Journal of the Taiwan Museum* 4: 1–180.

TD Hadden, D., 1983. A new species of thicket warbler *Cichlornis* (Sylviinae) from Bougainville Island, North Solomons Province, Papua New Guinea. *Bull. Brit. Orn. Club* 103(1): 22–25.

Haffer, J., 1969. Speciation in Amazonian forest birds. *Science* 165: 131–137.

Haffer, J., 1982. General aspects of the refuge theory. Pp 6–24 in *Biological Diversification in the Tropics* (G.T. Prance, Ed.). Columbia University Press, New York.

Hall, B.P. & Moreau, R.E., 1962. A Study of the Rare Birds of Africa. *Bull. Brit. Mus. (Nat. Hist.) Zoology* 8(7): 316-378.

Handschuh, M. & Packman, C., 2010. First nest record of Mekong Wagtail *Motacilla samveasnae*. *BirdingASIA* 14: 84.

Hansen, L.A., 2007. Population density estimates and threats evaluation of the highly endangered Udzungwa Forest Partridge in the Udzungwa Mountains of Tanzania. Unpublished report for the Critical Ecosystem Partnership Fund, 18pp

Hansson, S., 2006. In 'Neotropical Notebook', *Cotinga* 26: 92.

Harato, T. & Ozaki, K., 1993. Roosting behavior of the Okinawa Rail. *J. Yamashina Inst. Orn.* 25: 40–53.

Hardy, J.W., 2002. A Banding Study of the Grey Grasswren *Amytornis barbatus barbatus* in the Caryapundy Swamp of South-Western Queensland. *Corella* 26(4): 106–109.

Haring, E., Kvaløy, K., Gjershaug, J-O., Røv, N. & Gamauf, A., 2007. Convergent evolution and paraphyly of the hawk-eagles of the genus *Spizaetus* (Aves: Accipitridae) – phylogenetic analyses based on mitochondrial markers. *J. Zool. Syst. Evol. Research* 45: 353–365.

Harrap, S., 1989. Identification, vocalisations and taxonomy of *Pnoepyga* wren-babblers. *Forktail* 5: 61–70.

Harrap, S., 2011. Nepal Wren-babbler *Pnoepyga immaculata*: 25 years on. *BirdingASIA* 15: 81–83.

Harrap, S. & Fisher, T., 1994. A mystery woodcock in the Philippines. *Oriental Bird Club Bull.* 19: 54–56.

Harrap, S. & Mitchell, K., 1994. More notes on Rabor's Wren-babbler *Napothera rabori*. *Bull. Oriental Bird Club* 20: 50–51.

Harrap, S. & Quinn, D., 1996. *Tits, Nuthatches and Creepers.* A&C Black, London.

TD Harris, J.B.C., Rasmussen, P.C., Yong, D.L., Prawiradilaga, D.M., Putra, D.D., Round, P.D. & Rheindt, F.E., 2014. A new species of *Muscicapa* Flycatcher from Sulawesi, Indonesia. *PLoS ONE.* doi:10.1371/journal.pone.0112657

Harris, T. & Franklin, K., 2000. *Shrikes and Bush-shrikes.* Christopher Helm, London.

TD Harrison, P., Sallaberry, M., Gaskin, C.P., Baird, K.A., Jaramillo, A., Metz, S.M., Pearman, M., O'Keefe, M., Dowdall, J., Enright, S., Fahy, K., Gilligan, J. & Lillie, G., 2013. A new storm-petrel species from Chile. *Auk* 130(1): 180–191.

Hartlaub, G. & Finsch, O., 1872. On a fourth collection of birds from the Pelew and Mackenzie Islands. *Proc. Zool. Soc. London* (1872): 87–114.

Harvey, M.G., Lane, D.F., Hite, J., Terrill, R.S., Figueroa Ramírez, S., Smith, B.T., Klicka, J. & Vargas Campos, W., 2014. Notes on bird species in bamboo in northern Madre de Dios, Peru, including the first Peruvian record of Acre Tody-Tyrant (*Hemitriccus cohnhafti*). *Occ. Pap. Mus. Nat. Sci. Louisiana State Univ.* 81: 1–38.

Harvey, M.G., Winger, B.M., Seeholzer, G.F. & Cáceres A., D., 2011. Avifauna of the Gran Pajonal and southern Cerros del Sirá, Peru. *Wilson J. Orn.*123: 289–315.

Hawkins, F., Rabenandrasana, M., Virginie, M.C., Manese, R.O., Mulder, R., Ellis, E.R. & Robert, R., 1998. Field observations of the Red-shouldered Vanga *Calicalius rufocarpalis*: a newly described Malagasy endemic. *Bull. African Bird Club* 5(1): 30–32.

Hebert, P.D.N., Stoeckle, M.Y., Zemlak, T.S. & Francis, C.M., 2004. Identification of birds through DNA barcodes. *PLoS Biol* 2(10): e312.

Heim de Balsac, H. & Mayaud, N., 1962. *Les Oiseaux du Nord-Ouest de l'Afrique.* Paris.

Heindl, M. & Schuchmann, K-L., 1998. Biogeography, geographical variation and taxonomy of the Andean hummingbird genus *Metallura* Gould 1847. *J. Orn.* 139: 425–473.

Heinz, M., Schmidt, V. & Schaefer, M., 2005. New distributional record for the Jocotoco Antpitta *Grallaria ridgelyi* in south Ecuador. *Cotinga* 23: 24–26.

Hekstra, G.P., 1982. Description of twenty-four new subspecies of American *Otus* (Aves: Strigidae). *Bull. Zool. Mus. Amsterdam* 9(7): 49–63.

Helbig, A.J., Kocum, A., Seibold, I. & Braun, M.J., 2005. A multi-gene phylogeny of aquiline eagles (Aves: Accipitriformes) reveals extensive paraphyly at the genus level. *Molecular Phylogenics and Evolution* 35(1): 147–164.

Hellmayr, C.E., 1910. The birds of the Rio Madeira. *Novit. Zool.* 17(3): 172 pp.

Hellmayr, C.E., 1927. Catalogue of the birds of the Americas and adjacent islands. Part V. Field Museum of Natural History Zoological Series volume 13, part 5; (1934), volume 13, part 7.

Hennache, A., Mahood, S.P. Eames, J.C. & Randi, E., 2012. *Lophura hatinensis* is an invalid taxon. *Forktail* 28: 129–135.

Hennesey, A. B. & Gómez, M. I., 2003. Four bird species new to Bolivia: an ornithological survey of the Yungas Site Tokoaque, Madidi National Park. *Cotinga* 19: 25–33.

Henry, P.-Y., 2008. Aves, Cotingidae, *Doliornis remseni*: filling distribution gap, habitat and conservation, Ecuador. *Check List* 4(1): 1–4.

Herzog, S.K., Ewing, S.R., Evans, K.L., MacCormick, A., Valqui, T., Bryce, R., Kessler, M. & MacLeod, R., 2009. Vocalisations, distribution and ecology of the Cloud-forest Screech-owl *Megascops marshalli*. *Wilson J. Orn.* 121(2): 240–252.

Herzog, S.K., Fjeldså, J., Kessler, M. & Balderrama, J.A., 1999. Ornithological surveys in the Cordillera Cocapata, depto. Cochabamba, Bolivia, a transition zone between humid and dry intermontane Andean habitats. *Bull. Brit. Orn. Club* 119(3): 162–177.

TD Herzog, S.K., Kessler, M. & Balderrama, J.A., 2008. A new species of Tyrannulet (Tyrannidae: *Phyllomyias*) from Andean foothills in northwestern Bolivia and adjacent Peru. *Auk* 125(2): 265–276.

Herzog, S.K. & Mazar Barnett, J., 2004. On the validity and confused identity of *Serpophaga griseiceps* Berlioz 1959 (Tyrannidae). *Auk* 121(2): 415–421.

Hesket Prichard, H., 1902. *Through the Heart of Patagonia.* D. Appleton & Company, New York.

Hidasi, J., Mendonça, L.G.A. & Blamires, D., 2008. Primeiro registro documentado de *Celeus obrieni* (Picidae) para el estado de Goiás, Brasil. *Revista Brasiliera de Ornitologia* 16(4): 373–375.

Higgins, P.J., Peter, J.M. & Steele, W.K. (eds) 2001. *Handbook of Australian, New Zealand and Antarctic Birds. Volume 5: Tyrant-flycatchers to Chats.* Oxford University Press, Melbourne.

Higgins, P.J. & Peter, J.M., (eds) 2002. *Handbook of Australian, New Zealand and Antarctic Birds. Volume 6. Pardalotes to Shrike-thrushes.* Oxford University Press, Melbourne.

Higgins, P.J., Peter, J.M. & Cowling, S.J. (eds) 2006. *Handbook of Australian, New Zealand and Antarctic Birds. Volume 7: Boatbill to Starlings.* Oxford University Press, Melbourne.

TD Hilty, S.L. & Ascanio, D., 2009. A new species of spinetail (Furnariidae: *Synallaxis*) from the Río Orinoco of Venezuela. *Auk* 126(3): 485–492.

TD Hilty, S.L., Ascanio, D. & Whittaker, A., 2013. A new species of softtail (Furnariidae: *Thripophaga*) from the delta of the Orinoco River in Venezuela. *Condor* 115(1): 143–154.

Hilty, S.L. & Brown, W.L., 1983. Range extensions of Colombian birds as indicated by the M.A.Carriker Jr collection at the National Museum of Natural History, Smithsonian Institute. *Bull. Brit. Orn. Club* 103(1): 5–17.

Hilty, S.L. & Brown, W.L., 1986. *A Guide to the Birds of Colombia.* Princeton UP, Princeton, NJ.

Hinkelmann, C., 1988. Comments on recently described new species of hummingbirds. *Bull. Brit. Orn. Club* 108(4): 159–169.

Hinkelmann, C., Nicolai, B. & Dickerman, R.W., 1991. Notes on a hitherto unknown specimen of *Neolesbia nehrkorni* (Berlepsch, 1887: Trochilidae) with a discussion of the hybrid origin of the 'species'. *Bull. Brit. Orn. Club* 111: 190–199.

Hinkelmann, C. & Schuchmann, K.-L., 1997. Phylogeny of the Hermit Hummingbirds (Trochilidae: Phaethorninae). *Stud. Neotrop. Fauna & Environm.* 32: 142–163.

Hiragi, T. & Ikeuchi, E., 2015. An observation of predation on Okinawa Rail (*Gallirallus okinawae*) by Habu (*Protobothrops flavoviridis*). *Biological Magazine Okinawa* 53: 45–47.

Høgsås, T.E., Málaga-Arenas, E. & Neyra, J.P., 2002. Noteworthy bird records from south-west Peru. *Cotinga* 17: 60–61.

Holmgren, J., 1998. A parsimonious phylogenetic tree for the swifts, Apodi, compared with DNA-analysis phylogenetics. *Bull. Brit. Orn. Club* 118(4): 238–248.

Holt, D.W., Berkley, R., Deppe, C., Enríquez Rocha, P., Petersen, J.L., Rangel Salazar, J.L., Segars, K.P., Wood, K.L., Bonan A. & de Juana, E. 2016. Amazonian Pygmy-owl, *HBWAlive.*

Holt, D.W., Berkley, R., Deppe, C., Enríquez Rocha, P., Petersen, J.L., Rangel Salazar, J.L., Segars, K.P. & Wood, K.L., 2016. Yungas Pygmy-owl, *HBWAlive.*

TD Holyoak, D.T., 1974. Undescribed landbirds from the Cook Islands, Pacific Ocean. *Bull. Brit. Orn. Club* 94: 145–150.

Holyoak, D.T., 1976. In brief (b). *Halcyon ruficollaris. Bull. Brit. Orn. Club* 96(1): 46.

Holyoak, D.T. & Thibault, J-C., 1978a. Notes on the biology and systematics of Polynesian swiftlets *Aerodromus. Bull. Brit. Orn. Club* 98(2): 59–65.

TD Holyoak, D.T. & Thibault, J-C., 1978b. Undescribed *Acrocephalus* warblers from Pacific Ocean Islands. *Bull. Brit. Orn. Club* 98(4): 122–127.

Honkala, J. & Niiranen, S., 2010. *A Birdwatching Guide to South-east Brazil.* Portal do Bosque.

Hooper, H.L.N., 1974. Hall's Babbler in New South Wales. *Aust. Bird-Watcher.* 195–197.

Hornbuckle, J., 1999. The birds of Abra Patricia and the upper Río Mayo, north Peru. *Cotinga* 12: 11–28.

Hornbuckle, J., 2001. New to Science. *World Birdwatch* 112: 30–33.

Hornskov, J., 1995. Recent observations of birds in the Philippine Archipelago. *Forktail* 11: 1–10.

Hosner, P.A., Andersen, M.J., Robbins, M.B., Urbay-Tello, A., Cueto-Aparicio, L., Verde-Guerra, K., Sánchez-González, L.A., Navarro-Sigüenza, A.G., Boyd, R.L., Núñez, J., Tiravanti, J., Combe, M., Owens, H.L. & Peterson, A.T., 2015a. Avifaunal surveys of the upper Apurímac River Valley, Ayacucho and Cuzco Departments, Peru: new distributional records and biogeographic, taxonomic and conservation implications. *Wilson J. Orn.* 127(4): 563–581.

TD Hosner, P.A., Boggess, N.C., Alviola, P., Sánchez-González, L.A., Oliveros, C.A., Urriza, R. & Moyle, R.G., 2013. Phylogeography of the *Robsonius* Ground-warblers (Passeriformes: Locustellidae) reveals an undescribed species from north-eastern Luzon, Philippines. *Condor* 115(3): 630–639.

Hosner, P.A., Cueto-Aparicio, L., Ferro-Meza, G., Miranda, D. & Robbins, M.A., 2015b. Vocal and molecular phylogenetic evidence for recognition of a thistletail species (Furnariidae: *Asthenes*) endemic to the elfin forests of Ayacucho, Peru. *Wilson J. Orn.* 127(4): 724–730.

TD Hosner, P.A., Robbins, M.A., Valqui, T. & Peterson, A.T., 2013. A new species of *Scytalopus* tapaculo (Aves: Passeriformes: Rhinocryptidae) from the Andes of central Peru. *Wilson J. Orn.*125(2): 233–242.

Howell, S.N.G., 1993. A taxonomic review of the Green-fronted Hummingbird. *Bull. Brit. Orn. Club* 113(3): 179–187.

Howell, S.N.G., 2002. Additional information on the birds of Ecuador. *Cotinga* 18: 62–65.

Howell, S.N.G., 1999. *A Bird-finding Guide to Mexico.* Comstock Publishing, a Division of Cornell University Press, Ithaca, NY.

Howell, S.N.G. & Robbins, M.B., 1995. Species limits in the Least Pygmy-owl (*Glaucidium minutissimum*) complex. *Wilson Bull.* 107: 7–25.

Howell, S.N.G., Snetsinger, P.B. & Wilson, R.G., 1997. A sight record of White-fronted Swift *Cypseloides storeri* in Michoacán, Mexico. *Cotinga* 7: 23–26.

Howell, S.N.G. & Webb, S., 1995. *A Guide to the Birds of Mexico and Northern Central America.* Oxford University Press, Oxford.

Howell, T.R. & O'Neill, J.P., 1981. In Memoriam: George H. Lowery, Jr. *Auk* 98: 159–166.

Humphrey, P.S. & Livezey, B.C., 1985. Nest, eggs and downy young of the White-headed Flightless Steamer-duck. *Orn. Monog.* No 36, 944–952.

TD Humphrey, P.S. & Thompson, M.C., 1981. A new species of Steamer-duck (*Tachyeres*) from Argentina. *Occ. Papers Mus. Nat. Hist. University of Kansas.* 95: 1–12.

Hutchinson, R., Eaton, J. & Benstead, P., 2006. Observations of Cinnabar Hawk-owl *Ninox ios* in Gunung Ambang Nature Reserve, North Sulawesi, Indonesia, with a description of a secondary vocalisation. *Forktail* 22: 120–121.

Ibáñez-Hernández, G., Benítez-Díaz, H., Peterson, A.T. & Navarro S., A.G., 2003. A further specimen of the White-fronted Swift *Cyseloides storeri. Bull. Brit. Orn. Club* 104(1): 55–64.

Ikenaga, H. & Gima, T., 1993. Vocal repertoire and duetting in the Okinawa Rail *Rallus okinawae. J. Yamashina Inst. Orn.* 25: 28–39.

Imberti, S. & Casañas, H., 2010. Hooded Grebe *Podiceps gallardoi*: extinct by its 50th birthday? *Neotropical Birding* 6: 65.

Imberti, S., Casañas, H. & Roesler, I., 2011. Hooded Grebe (*Podiceps gallardoi*). Neotropical Birds Online (T.S.Schulenberg, Editor). Ithaca, Cornell Lab. of Ornithology.

Imberti, S., Sturzenbaum, S.M. & McNamara, M., 2004. Actualización de la distribución invernal del Macá Tobiano (*Podiceps gallardoi*) y notas sobre su problemática de conservación. *Hornero* 19(2): 83–89.

Inchausti, P. & Weimerskirch, H., 2001. Risks of decline and extinction of the endangered Amsterdam Albatross and the projected impact of long-line fisheries. *Biological Conservation* 100: 377–386.

TD Indrawan, M., Rasmussen, P.C. & Sunarto., 2008. A new white-eye (*Zosterops*) from the Togian Islands, Sulawesi, Indonesia. *Wilson J. Orn.* 120(1): 1–9.

TD Indrawan, M. & Somadikarta, S., 2004. A new hawk-owl from the Togian Islands, Gulf of Tomini, central Sulawesi, Indonesia. *Bull. Brit. Orn. Club* 124(3): 160–171.

Indrawan, M., Somadikarta, S., Supriatna, J., Bruce, M.D., Sunarto & Djanubudiman, G., 2006. The birds of the Togian Islands, Central Sulawesi, Indonesia. *Forktail* 22: 7–22.

Inskipp, C. & Inskipp, T., 1991. *A Guide to the Birds of Nepal.* 2nd Edition. Christopher Helm, London.

Isenmann, P. & Moali, A., 2000. *Oiseaux d'Algérie.* Société d'Études Ornithologiques de France, Paris.

TD Isler, M.L., Alonso, J.A., Isler, P.R. & Whitney, B.M., 2001. A new species of *Percnostola* antbird (Passeriformes: Thamnophilidae) from Amazonian Peru, and an analysis of species limits within *Percnostola rufifrons*. *Wilson Bull.* 113(2): 164–176.

Isler, M.L. & Isler, P.R., 1987. *The Tanagers: Natural History, Distribution and Identification.* Smithsonian Institution Press, Washington, DC.

Isler, M.L., Isler, P.R. & Whitney, B.M., 2007. Species limits in antbirds (Thamnophilidae): the Warbling Antbird (*Hypocnemus cantator*) complex. *Auk* 124(1): 11–28.

Isler, M.L., Lacerda, D.R., Isler, P.R., Hackett, S.J., Rosenberg, K.V. & Brumfield, R.T., 2006. *Epinecrophylla*, a new genus of antwrens (Aves: Thamnophilidae). *Proc. Biol. Soc. Washington* 119(4): 522–527.

Jacobs, P., Mahler, F. & Ochando, B., 1978. A propos de la couleur de la chalotte chez la Sittelle kabyle (*Sitta ledanti*). *Aves:* 15: 149–153.

Jahn, O., Palacios, B. & Valenzuela, P.M., 2007. Ecology, population and conservation status of the Choco Vireo *Vireo masteri*, a species new to Ecuador. *Bull. Brit. Orn. Club* 127(2): 161–166.

Jakosalem, P.G.C., Collar, N.J. & Gill, J.A., 2012. Habitat selection and conservation status of the endemic *Ninox* hawk-owl on Cebu, Philippines. *Bird Cons. Int.* 22 DOI:10.1017/S095927091 2000317

James, H.F. & Olson, S.L., 1991. Descriptions of 32 new species of birds from the Hawaiian Islands: Part II. Passeriformes. *Ornithol. Monogr.* 45: 1–88.

Jaramillo, A., 2010. *Serpophaga griseicapilla.* South American Classification Committee.

TD Jensen, F.P., 1983. A new species of sunbird from Tanzania. *Ibis* 125: 447–449.

Jensen, F.P. & Brøgger-Jensen, S., 1992. The forest avifauna of the Udzungwa Mountains, Tanzania. *Scopus* 15: 65–83.

Jiang, A.W., Zhou, F., Wu, Y.H. & Liu, N.F., 2013. First breeding records of Nonggang Babbler (*Stachyris nonggangensis*) in a limestone area in southern China. *Wilson J. Orn.* 125(3): 609–615.

TD Johnson, A.W. & Millie, W.R., 1972. A new species of conebill (*Conorostrum*) from northern Chile. Pp 3–8. In Johnson, A.W. *Supplement to the Birds of Chile and adjacent regions of Argentina, Bolivia and Peru.* Platt Establecimientos Gráficos S.A., Buenos Aires.

Johnson, A. & Serret, A., 1994. Búsqueda del paradero invernal del Macá Tobiano *Podiceps gallardoi. Bol. Tecnico* No 23, Fundación Vida Silvestre Argentina, Buenos Aires.

Johnson, N.K., 2002. Leapfrogging revisited in Andean birds: geographical variation in the Tody-tyrant superspecies *Poecilotriccus ruficeps* and *P. luluae. Ibis* 144: 69–84.

TD Johnson, N.K. & Jones, R.E., 2001. A new species of tody-tyrant (Tyrannidae: *Poecilotriccus*) from northern Peru. *Auk* 118(2): 334–341.

Jolliffe, T., 2000. Breeding the Lemon-breasted Canary *Serinus citrinipectus* and notes on some other *Serinus* species. *Avic. Mag.* 106(4): 178–181.

Jones, C.G. & Swinnerton, K.J., 2000. Discovery: A possible new taxon of rock thrush *Monticola* sp. from the limestone karst region of western Madagascar. *Bull. African Bird Club* 7(1): 52–53.

Jones, M.J., Linsley, M.D. & Marsden, S.J., 1995. Population sizes, status and habitat associations of the restricted-range bird species of Sumba, Indonesia. *Bird Cons. Int.* 5: 21–52.

Jønsson, K.A. & Fjeldså, J., 2006. A phylogenetic supertree of oscine passerine birds (Aves: Passeri). *Zool. Scripta* 35: 149–186.

TD Jønsson, K.A., Poulsen, M.K., Haryoko, T., Reeve, A.H. & Fabre, P-H., 2013. A new species of masked-owl (Aves: Strigiformes: Tytonidae) from Seram, Indonesia. *Zootaxa* 3635(1): 051–061.

Joseph, L., 1982. A further population of the Grey Grasswren. *Sunbird* 12(4): 51–52.

Joseph, L., 2000. Beginning an end to 63 years of uncertainty: the Neotropical parakeets known as *Pyrrhura picta* and *P. leucotis* comprise more than two species. *Proc. Acad. Nat. Sci. Philadelphia* 150: 279–292.

Joseph, L., 2002. Geographical variation, taxonomy and distribution of some Amazonian *Pyrrhura* parakeets. *Ornitología Neotropical* 13: 337–363.

Joseph, L. & Bates, J.M. in Joseph, L., 2002. Geographical variation, distribution and taxonomy of some Amazonian *Pyrrhura* parakeets. *Ornitología Neotropical* 13(4): 337–363.

Joseph, L., Slikas, B., Rankin-Baransky, K., Bazartseren, B., Alpers, D. & Gilbert, A.E., 1999. DNA evidence concerning the identities of *Crax viridirostris* Sclater, 1875, and *C. estudilloi* Allen, 1977. *Ornitología Neotropical* 10(2): 129–144.

TD Jouanin, C., 1964. Un Pétrel nouveau de la Réunion, *Bulweria baraui. Bull. Mus. natn. Hist. nat. Paris* 35(2): 593–597.

Jouanin, C. & Gill, F.B., 1967. Recherche du Petrel de Barau *Pterodroma baraui. L'Oiseau et R.F.O.*, 37(1–2): 15–19.

Jouventin, P., Martinez, J. & Roux, J.P., 1989. Breeding biology and current status of the Amsterdam Island Albatross *Diomedea amsterdamensis. Ibis* 131: 171–182.

Juiña, M. & Bonaccorso, E., 2013. Área de hogar y notas sobre la historia natural de la Grallaria Jocotoco (*Grallaria ridgelyi*). *Orn. Neotrop.* 24(1): 27–34.

Juniper, T. & Parr, M., 1998. *Parrots: a Guide to the Parrots of the World.* Pica Press, Robertsbridge, UK.

Kahindo, C., Bowie, R.C.K. & Bates, J.M., 2007. The relevance of data on genetic diversity for the conservation of Afro-montane regions. *Biological Conservation* 134: 262–270.

Kawakami, K., Eda, M., Horikoshi, K., Suzuki, H., Chiba, H. & Hiraoka, T., 2012. Bryan's Shearwaters have survived on the Bonin Islands, Northwestern Pacific. *Condor* 114(3): 507–512.

Kazmierczak, K. & Muzika, Y., 2012. A preliminary report on the apparent rediscovery of Sillem's Mountain Finch *Leucosticte sillemi. BirdingASIA* 18: 17–20.

Kelly, L & Bottomley, C., 1998. Conservation in the Cook Islands (BLI SF 2015).

TD Kemp, A.C. & Delport, W., 2002. Comments on the status of subspecies in the Red-billed Hornbill *Tockus erythrorhynchos* complex (Aves: Bucerotidae), with the description of a new taxon endemic to Tanzania. *Annals Transvaal Mus.* 39: 1–8.

Kennedy, R.S. & Miranda, H.C., 1998. In Memoriam: Dioscoro S. Rabor, 1911–1996. *Auk:* 115(1); 204–205.

TD Kennedy, R.S., Fisher, T.H., Harrap, S.C.B., Diesmos, A.C.

& Manamtam, A.S., 2001. A new species of woodcock (Aves: Scolopacidae) from the Philippines and a re-evaluation of other Asian/Papuasian woodcock. *Forktail* 17: 1–12.

TD Kennedy, R.S., Gonzales, P.C. & Miranda, H.C., 1997. New *Aethopyga* sunbirds (Aves: Nectariniidae) from the island of Mindanao, Philippines. *Auk* 114(1): 1–10.

Kennerley, P. & Pearson, D.J., 2010. *Reed and Bush Warblers*. Christopher Helm, London.

TD Kepler, C.B. & Parkes, K.C., 1972. A new species of warbler (Parulidae) from Puerto Rico. *Auk* 89: 1–18.

Kepler, C.B., Pratt, T.K., Ecton, A.M., Engilis, A. & Fluetsch, K.M., 1996. Nesting behavior of the Poo-uli. *Wilson Bull.* 108: 620–638.

Kiemann, L. Jr, & Vieira, J.S., 2013. Assessing the extent of occurrence, area of occupancy and population size of Marsh Tapaculo (*Scytalopus iraiensis*). *Animal Biodiversity and Conservation* 36(1): 47–57.

King, B.F., 2005. The song of the Cinnabar Hawk-owl *Ninox ios* in North Sulawesi, Indonesia. *Forktail* 21: 173–174.

King, B.F., 2008. Vocalisations of the Togian Boobook *Ninox burhani*. *Forktail* 24: 122–124.

King, B. & Kanwanich, S., 1978. First wild sighting of the White-eyed River Martin *Pseudochelidon sirintarae*. *Biol. Cons.* 13: 183–185.

King, B., Rostron, P., Luijendijk T., Bouwman, R. & Quispel, C., 1999. An undescribed *Muscicapa* flycatcher on Sulawesi. *Forktail* 15: 104.

King, B.F. & Yong, D., 2001. An unknown scops owl *Otus* sp. from Sumba, Indonesia. *Bull. Brit. Orn. Club* 121: 91–93.

Kirchman, J.J., Witt, C.C., McGuire, J.A. & Graves, G.R., 2009. DNA from a 100 year-old holotype confirms the identity of a potentially extinct hummingbird species. *Biology Letters* 5: 1–4.

Kirwan, G.M., Brinkhuizen, D., Calderón, D., Davis, B. & Minns, J., 2015b. Interesting and unusual records from published and unpublished sources. *Neotropical Birding* 16: 43–62.

Kirwan, G.M. & Freile, J.F., 2008. Current perspectives in Ecuadorian ornithology and conservation: a tribute to Paul Coopmans. *Cotinga* 29: 2–3.

Kirwan, G.M. & Green, G., 2011. *Cotingas and Manakins*. Christopher Helm, London.

Kirwan, G.M., Mazar Barnett, J., de Vasconcelos, M.F., Raposo, M.A., D'Angelo Neto, S. & Roesler, I., 2004. Further comments on the avifauna of the middle São Francisco Valley, Minas Gerais, Brazil. *Bull. Brit. Orn. Club* 124: 207–220.

TD Kirwan, G.M., Schweizer, M. & Copete, J.L., 2015. Multiple lines of evidence confirm that Hume's Owl *Strix butleri* (A.O.Hume, 1878) is two species, with description of an unnamed species (Aves: Non-Passeriformes: Strigidae). *Zootaxa* 3904(1): 028–050.

Kirwan, G.M., Steinheimer, F.D., Raposo, M.A. & Zimmer, K.J., 2014. Nomenclatural corrections, neotype designation and new subspecies description in the genus *Suiriri* (Aves: Passeriform; Tyrannidae). *Zootaxa* 3784 (3): 224–240.

Kirwan, G.M., Whittaker, A. & Zimmer, K.J., 2015a. Interesting bird records from the Araguaia River Valley, central Brazil, with comments on conservation, distribution and taxonomy. *Bull. Brit. Orn. Club* 135(1): 21–60.

Klicka, J., Burns, K. & Spellman, G.M., 2007. Defining a monophyletic Cardinalini: a molecular perspective. *Molecular Phylogenetics and Evolution* 45: 1014–1032.

Klop, E., Lindsell, J.A. & Siaka, A.M., 2010. The birds of Gola Forest and Tiwai Island, Sierra Leone. *Malimbus* 32(1): 33–58.

Knox, A.G., 1980. Feather protein as a source of avian taxonomic information. *Comp. Biochem. Physiol.* 65B: 45–54.

König, C., 1991a. Taxonomische und ökologische Untersuchungen an Kreischeulen (*Otus* sp.) des südlichen Südamerikas. *J. Orn.* 132: 209–214.

König, C., 1991b. Zur Taxonomie und Ökologie der Sperlingskauze (*Glaucidium* spp.) des Andenraum. *Ökol. Vögel* 13: 15–76.

TD König, C. & Straneck, R., 1989. Eine neue Eule (Aves: Strigidae) aus Nordargentinien. *Stuttgarter Beiträge zur Naturkunde, Ser. A.*, 428: 1–20.

König, C., Weick, F. & Becking, J.-H., 2008. *Owls of the World* 2nd edition. Christopher Helm, London.

Konter, A., 2008. Decline in the population of Hooded Grebe *Podiceps gallardoi*? *Cotinga* 29: 135–138.

Kotaka, N., Kudaka, M., Takehara, K. & Sato, H., 2009. Ground use pattern by forest animals and vulnerability toward invasion by *Herpestes javanicus* into Yambaru, northern Okinawa Island, southern Japan. *Jap. J. Ornithol.* 58: 28–45.

Kotaka, N. & Sawashi, Y., 2004. The Road-kill of the Okinawa Rail *Gallirallus okinawae*. *J. Yamashina Inst. Orn.* 35: 143–143.

Krabbe, N., 2008. Birding Ecuador: a tribute to Paul Coopmans. *Cotinga* 29: 12–14.

Krabbe, N., 2009. A significant northward range extension of Munchique Wood-wren (*Henicorhina negreti*) in the western Andes of Colombia. *Ornitología Colombiana* 8: 76–77.

TD Krabbe, N., Agro, D.J., Rice, N.H., Jacome, M. & Sornoza M., F., 1999. A new species of antpitta (Formicariidae: *Grallaria*) from the southern Ecuadorian Andes. *Auk* 116: 882–890.

Krabbe, N. & Ahlman, F.L., 2009. Royal Sunangel *Heliangelus regalis* at Yankuan Lodge, Ecuador. *Cotinga* 31: 132.

TD Krabbe, N., Isler, M.L., Isler, P.R., Whitney, B.M., Álvarez A., J. & Greenfield, P.J., 1999. A new species in the *Myrmotherula haematonota* superspecies (Aves: Thamnophilidae) from the western Amazonian lowlands of Ecuador and Peru. *Wilson Bull.* 111(2): 157–165.

Krabbe, N., Poulsen, B.O., Frølander, A., Hinojosa, M. & Quiroga, C., 1996. Birds of montane forest fragments in Chuquisaca Department, Bolivia. *Bull. Brit. Orn. Club* 116(4): 230–243.

TD Krabbe, N., Salaman, P., Cortes, A., Quevedo, A., Ortega, L.A. & Cadena, C.D., 2005. A new species of *Scytalopus* tapaculo from the upper Magdalena Valley, Colombia. *Bull. Brit. Orn. Club* 125(2): 93–108.

TD ×3 Krabbe, N. & Schulenberg, T.S., 1997. Species limits and natural history of *Scytalopus* tapaculos (Rhinocryptidae), with descriptions of the Ecuadorian taxa, including three new species. *Ornithological Monographs* 48: 47–88.

Krabbe, N. & Sornoza M., F., 1994. Avifaunistic results of a subtropical camp in the Cordillera del Cóndor, southeastern Ecuador. *Bull. Brit. Orn. Club* 114(1): 55–61.

Kratter, A.W., 1998. The nests of two bamboo specialists: *Celeus spectabilis* and *Cercomacra manu*. *J. Field Orn.* 69: 37–44.

Kroodsma & Brewer 2005 HBW 10.

Kühn, G., 1994. Steckbrief: Buntkopf-Papageiamadine *Erythrura coloria*. *Gefierdete Welt* 118: 114–116.

Kuroda, N., 1938. On a melanistic example of *Tribura luteoventris* from Formosa. *Tori* 10: 3–9.

Lachenaud, O., 2006. Les oiseaux du Parc National du Banco et de la forêt Classée de l'Anguededou, Côte d'Ivoire. *Malimbus* 28(2): 107–133.

TD Lafontaine, R.M. & Moulaert, N., 1998. Une nouvelle èspece de petit-duc (*Otus*: Aves) aux Comores: taxonomie et statut de conservation. *Bull. African Bird Club* 6(1): 61–65.

TD Lambert, F.R., 1998a. A new species of *Gymnocrex* from the Talaud Islands, Indonesia. *Forktail* 13: 1–6.

TD Lambert, F.R., 1998b. A new species of *Amaurornis* rail from the Talaud Islands, Indonesia, and a review of taxonomy of bush hens occurring from the Philippines to Australasia. *Bull. Brit. Orn. Club* 118(2): 67–75.

TD Lambert, F.R. & Rasmussen, P.C., 1998. A new Scops Owl from Sangihe Island, Indonesia. *Bull. Brit. Orn. Club* 204–217.

Lancelotti, J.L., Pozzi, L.M., Márquez, F., Yorio, P. & Pascual, M.A., 2009. Waterbird occurrence and abundance in the Strobel Plateau, Patagonia, Argentina. *Hornero* 24(1): 13–2.

Lane, D.F., 2008. Romancing the Stone...er, Barbet. *Neotropical Birding* 11: 27–32.

TD Lane, D., Servat, T., Valqui, H. & Lambert, F.R., 2007. A distinctive new species of tyrant flycatcher (Passeriformes: Tyrannidae: *Cnipodectes*) from south-eastern Peru. *Auk* 124(3): 762–772.

Lane, D.F., Valqui H., T., Álvarez A., J., Armenta, J. & Eckhardt, K., 2006. The rediscovery and natural history of the White-masked Antbird (*Pithys castaneus*). *Wilson Bull.* 118: 13–22.

Langrand, O., 1990. *Guide to the Birds of Madagascar.* Yale University Press, New Haven and London.

Langrand, O. & von Bechtolsheim, M., 2009. New distributional record of Appert's Tetraka (*Xanthomixis apperti*) from Salary Bay, Mikea Forest, Madagascar. *Malagasy Nature* 2.

Langrand, O. & Goodman, S.M., 1996. Current distribution and status of Benson's Rockthrush *Pseudocossyphus bensoni*, a Madagascar endemic. *Ostrich* 67: 49–54.

TD Lanyon, S.M., Stotz, D.F. & Willard, D.E., 1990. *Clytoctantes atrogularis*, a new species of antbird from western Brazil. *Wilson Bull.* 102: 571–580.

TD Lara, C.E., Cuervo, A.M., Valderrama, S.V., Calderón-F., D. & Cadena, C.D., 2012. A new species of wren (Troglodytidae: *Thryophilus*) from the dry Cauca River canyon, northwestern Colombia. *Auk* 129(3): 537–550.

Lawson, W.J., 1970. Note on the breeding of the Lemon-breasted Canary in captivity. *Ostrich* 41: 252.

TD Lawson, W.J., 1984. The West African mainland forest dwelling population of *Batis*: a new species. *Bull. Brit. Orn. Club* 104: 144–146.

Leader, P.J., 2009. Is Vaurie's Nightjar *Caprimulgus centralasicus* a valid species? *BirdingASIA* 11: 47–50.

Lebbin, D.J., 2006. Notes on birds consuming *Guadua* bamboo seeds. *Ornitol. Neotrop.* 17(4): 609–612.

Le Corre, M., Ollivier, A., Ribes, S & Jouventin, P., 2002. Light-induced mortality of petrels: a 4–year study from Réunion Island (Indian Ocean). *Biological Conservation* 105: 93–102.

LeCroy, M., 1999. Type specimens of new forms of *Lonchura. Bull. Brit. Orn. Club* 119(4): 214–220.

TD LeCroy, M. & Barker, F.K., 2006. A new species of bush-warbler from Bougainville Island and a monophyletic origin for southwest Pacific *Cettia. Amer. Mus. Novit.* 3511: 1–20.

Ledant, J-P., 1977. La Sittelle kabyle (*Sitta ledanti* Vielliard), espèce endémique montagnard récemment découverte. *Aves* 14: 83–85.

Ledant, J-P., 1978. Données comparées sur la Sittelle corse (*Sitta whiteheadi*) et la Sittelle kabyle (*Sitta ledanti*). *Aves* 15: 154–157.

Ledant, J-P., 1981. Conservation et fragilité de la fôret de Babor, habitation de la Sittelle kabyle. *Aves* 18: 1–9.

Ledant, J-P. & Jacobs, P., 1977. La Sittelle kabyle (*Sitta ledanti*): données nouvelles sur la biologie. *Aves* 14: 233–242.

Ledant, J-P., Jacobs, P., Ochando, B. & Renault, J., 1985. Dynamique de la forêt du Mont Babor et preferences écologiques de la Sitelle kabyle (*Sitta ledanti*). *Biological Conservation* 32(3): 231–254.

Lees, A., 2009. Splits, lumps and shuffles. *Neotropical Birding* 5: 20.

Lees, A., 2015. Splits, lumps and shuffles. *Neotropical Birding* 16: 4–15.

Lees, A.C., Albano, C., Kirwan, G.M., Pacheco, J.F. & Whittaker, A., 2014. The end of hope for Alagoas Foliage-gleaner *Philydor novaesi*? *Neotropical Birding* 14: 20–28.

Lees, A.C. & Peres, C.A., 2006. Rapid avifaunal collapse along the Amazonian deforestation frontier. *Biol. Cons.* 133: 198–211.

Leite, G.A., Marcelino, D.G. & Pinheiro, R.T., 2010. First description of the juvenile plumage of the Critically Endangered Kaempfer's Woodpecker *Celeus obrieni* of central Brazil. *Orn. Neotrop.* 21(3): 453–456.

Leite, G.A., Pinheiro, R.T., Marcelino, D.G., Figueira, J.E.C. & Delabie, J.H.C., 2013. Foraging behavior of Kaempfer's Woodpecker (*Celeus obrieni*), a bamboo specialist. *Condor* 115(2): 221–229.

TD Lenciono-Neto, F., 1994. Une nouvelle espèce de *Chordeiles* (Aves: Caprimulgidae) de Bahia (Brésil). *Alauda* 62(4): 242–245.

TD Lentino, M. & Restall, R., 2003. A new species of *Amaurospiza* blue seedeater from Venezuela. *Auk* 120(3): 600–606.

Lentino, M., Sharpe, C., Pérez-Emán, J.L. & Carreño, Y., 2004. *Aves registradas en la Serranía de Lajas, Serranía de Valledupar, Sierra de Perijá, Estado Zulia, en abril de 2004. Technical Report.* Colección Ornitológica Phelps, Caracas, Venezuela.

Lewis, A., 1998. Mayotte Scops Owl *Otus rutilis mayottensis. Bull. African Bird Club* 5(1): 33–34.

Lewis, C.F., 1978. Little Raven caching food. *Aust. Birds* 7: 272.

Lewison, R.L. & Crowder, L.B., 2003. Estimating Fishery Bycatch and Effects on a Vulnerable Seabird Population. *Ecological Applications* 13(3): 743–753.

TD Li, G.-Y., 1995. A new subspecies of *Certhia familiaris* (Passeriformes: Certhiidae). *Acta Zoologica Sinica* 20(3): 373–377.

Li, Z.T., Zhou, F., Lu, Z., Jiang, A.W., Yang, G. & Yu, C.X., 2013. Distribution, habitat and status of the new species Nonggang Babbler *Stachyris nonggangensis. Bird Cons. Int.* 1–8.

Linhares, K.V., Soares, F.A. & Machado, I.C.S., 2010. Nest support plants of the Araripe Manakin *Antilophia bokermanni*, a Critically Endangered endemic bird from Ceará, Brazil. *Cotinga* 32: 121–125.

TD(I) Liversidge, R., 1996. A new species of pipit in southern Africa. *Bull. Brit. Orn. Club* 116(4): 211–215.

Liversidge, R., 1998. The African pipit enigma. *Bull. African Bird Club* 5(2): 105–107.

TD(I) Liversidge, R. & Voelker, G., 2002. The Kimberley Pipit: a new African species. *Bull. Brit. Orn. Club* 122(2): 93–109.

Livezey, B.C., 1989. Feeding morphology, foraging behavior and foods of Steamer-ducks (Anatidae: *Tachyeres*). *Occ. Papers Mus. Nat. Hist. University of Kansas* 126: 1–41.

Lloyd, H., 2000. Population densities of the Black-faced Cotinga *Conioptilon mcilhennyi* in south-east Peru. *Bird Cons. Int.* 10: 277–285.

Lloyd, H., 2004. Habitat and population estimates of some threatened lowland forest bird species in Tambopata, south-eastern Peru. *Bird Cons. Int.* 14: 261–277.

Lloyd, H., 2009. Apurimac Spinetail (*Synallaxis courseni*). *Neotropical Birds online* (T. S. Schulenberg, Editor). Cornell Lab of Ornithology, Ithaca, NY.

Loaiza, J.M., Sornoza, F.A., Agreda, A.E., Aguirre, J., Ramos, R. & Canaday, C., 2005. The presence of Wavy-breasted Parakeet *Pyrrhura peruvianus* confirmed for Ecuador. *Cotinga* 23: 37–38.

TD Longmore, N.W.& Boles, W.E., 1983. Description and systematics of the Eungella Honeyeater *Meliphaga hindwoodi*, a new species of honeyeater from central eastern Queensland, Australia. *Emu* 83: 59–65.

Lopes, L.E., 2005. Field identification and new site records of Chapada Flycatcher *Suiriri islerorum. Cotinga* 24: 38–41.

Lopes, L.E., Leite, L., Pinho, J.B. & Goes, R., 2005. New bird records to the Estação Ecológica de Águas Emendales, Planaltina, Distrito Federal. *Ararajuba* 13(1): 107–108.

Lopes, L.M. & Marini, M.A., 2005. Biologia reproductiva de *Suiriri affinis* e *S. islerorum* (Aves: Tyrannidae) no cerrado do Brasil central. *Papeis Avulsos de Zoologia (São Paulo)* 45(12): 127–141.

Lopes, L.E. & Marini, M.A., 2006. Home range and habitat use by *Suiriri affinis* and *Suiriri islerorum* (Aves: Tyrannidae) in the central Brazilian Cerrado. *Studies on Neotropical Fauna and Environment* 41(2): 87–92.

López-Calleja, M.V. & Estades, C.F., 1996. Natural history of the Tamarugo Conebill (*Conirostrum tamarugense*) during the breeding period: Diet and habitat preferences. *Revista Chilena de Historia Natural* 69(3): 351–356.

López-Lanús, B. & Lowen, J.C., 1999. Observations of breeding activity in the El Oro Parakeet *Pyrrhura orcesi*. *Cotinga* 11: 46–47.

TD(I) López-Lanús, B., 2015. Una nueva especie de capuchino (Emberizidae: *Sporophila*) de los pastizales anegados del Ibera, Corrientes, Argentina. Pp 473-489 in López-Lanús, B. (2015) *Guía Audiornis de las aves de Argentina, fotos y sonidos; identificación contrapuestas y marcas sobre imágenes.* First Edition. Audiornis Producciones, Buenos Aires, Argentina. http://www.xeno-canto.org/docs/lopez-lanus-2015.pdf

López-Lanús, B., 2017. Una nueva especie de jiguero (Thraupidae: *Sicalis*) endémica de las Sierras de Ventania, pampa bonaerense, Argentina. Pp 475–497 in López-Lanús, B., (2017). *Guía Audiornis de las aves de Argentina, fotos y sonidos; identificación por características contrapuestas y marcas sobre imágines.* Second Edition. Audiornis Producciones, Buenos Aires, Argentina.

López-Ordóñez, J.P., Páez-Ortiz, C.A., Sandoval-Sierra, J.V. & Salaman, P., 2008. Una segunda localidad para *Eriocnemus mirabilis* en la Cordillera Occidental de Colombia. *Cotinga* 29: 169–171.

TD Louette, M., 1981. *Melignomon eisentrauti:* a new species of honeyguide from West Africa. *Rev. Zool. Afr.* 95(1): 131–135.

Louette, M., 1987. A new weaver from Uganda? *Ibis* 129(3): 405–406.

TD Louette, M., 1990. A new species of nightjar from Zaire. *Ibis* 132: 349–353.

Louette, M., 1992. Obituary: A. Prigogine. *Ibis* 134: 89–90.

TD Louette, M. & Benson, C.W., 1982. Swamp-dwelling weavers of the *Ploceus velatus/vitellinus* complex, with the description of a new species. *Bull. Brit. Orn. Club* 102: 24–31.

Louette, M. & Hasson, M., 2009. Rediscovery of the Lake Lufira Weaver *Ploceus ruweti*. *Bull. African Bird Club* 16: 168–173.

Louette, M., Herremans, M., Bijnens, L. & Janssens, L., 1988. Taxonomy and evolution of the brush warblers *Nesillas* on the Comoro Islands. *Tauraco* 1: 110–129.

Louette, M., Herremans, M., Stevens, J., Vangeluwe, D. & Soilih, A., 2008. Red Data Bird: Grand Comoro Scops-owl. *World Birdwatch* 13.

Lowen, J., 2002. Neotropical News. *Cotinga* 18: 11–12.

TD Lowery, G.H., Jr & O'Neill, J.P., 1964. A new genus and species of tanager from Peru. *Auk* 81: 125–131.

TD Lowery, G.H., Jr & O'Neill, J.P., 1965. A new species of cacique (Aves: Icteridae) from Peru. *Occ. Papers Mus. Zool. Univ. Louisiana.* 33: 1–4.

TD Lowery, G.H. Jr. & O'Neill, J.P.,1966. A new genus and species of cotinga from eastern Peru. *Auk* 83: 1–9.

TD Lowery, G.H. Jr. & O'Neill, J.P., 1969. A new species of antpitta from Peru and a revision of the subfamily Grallarinae. *Auk* 86(1): 1–12.

TD Lowery, G.H. & Tallman, D.A., 1976. New genus and species of nine-primaried oscine of uncertain affinities from Peru. *Auk* 93(3): 415–428.

Macedo Mestre, L.A., Barlow, J. & Thom, G., 2009. Burned Forests as a novel habitat for the Black-faced Cotinga (*Conioptilon mcilhennyi*) in the western Brazilian Amazon. *Ornitología Neotropical* 20(3): 467–470.

Macedo Mestre, L.A., Thom, G. & Cochrane, M.A. & Barlow, J., 2010. The birds of Reserva Extrativista Chico Mendes, South Acre, Brazil. *Boletim do Museu Paraense Emilio Goeldi Ciencias Naturais* 5(3): 311–333.

MacLoed, R., Ewing, S.K., Herzog, S.K., Bryce, R., Evans, K.L. & MacCormick, A., 2005. First ornithological inventory and conservation assessment for the yungas forests of the Cordilleras Cocapata and Mosetenes, Cochabamba, Bolivia. *Bird Cons. Int.* 15: 361–382.

Madge, S. & McGowan, P., 2002. *Pheasants, Partridges and Grouse.* Christopher Helm, London.

Madge, S.C. & Redman, N.J., 1989. The existence of a form of Cliff Swallow *Hirundo* sp. in Ethiopia. *Scopus* 13: 126–129.

Madika, B., Putra, D.D., Harris, J.B.C., Yong, D.L., Mallo, F.N., Rahman, A., Prawiradilaga, D.M. & Rasmussen, P.C., 2011. An undescribed *Ninox* hawk owl from the highlands of Central Sulawesi, Indonesia? *Bull. Brit. Orn. Club* 131(2): 94–102.

Mahood, S.P., Edwards, D.P., Ansell, F.A., Craik, R., 2012. An accessible site for Chestnut-eared Laughingthrush *Garrulax konkakinhensis*. *BirdingASIA* 17: 104–105.

TD Mahood, S.P., John, A.J.I., Eames, J.C., Oliveros, C.H., Moyle, R.G., Hong, Chamnan, Poole, C.M., Nielsen, H. & Sheldon, F.H., 2013. A new species of lowland tailorbird (Passeriformes: Cisticolidae: *Orthotomus*) from the Mekong floodplain of Cambodia. *Forktail* 29: 1–14.

TD Maijer, S. & Fjeldså, J., 1997. Description of a new *Cranioleuca* spinetail from Bolivia and a 'leapfrog pattern' of geographic variation in the genus. *Ibis* 139: 606–616.

Mann, N.I., Barker, F.K., Graves, J.A., Dingess-Mann, K.A. & Slater, P.J.B., 2006. Molecular data delineate four genera of '*Thryothorus*' wrens. *Molecular Phylogenetics and Evolution* 40: 750–759.

Manson-Bahr, P., 1959. Recollections of some famous British ornithologists. *Ibis* 101: 53–64.

Marín M., A., Carrión B., J.M. & Sibley, F.C., 1992. New distributional records for Ecuadorian birds. *Orn. Neotrop.* 3: 27–34.

Marin, M. & Stiles, F.G., 1992. On the biology of five species of swifts (Apodidae: Cypseloidinae) in Costa Rica. *Proc. Western Found. Vert. Zool.* 4: 287–351.

Marshall, J.T., 1978. *Systematics of Smaller Asian Night Birds based on Voice.* Orn. Monographs 25, AOU, Washington, DC.

TD Martens, J. & Eck, S., 1991. *Pnoepyga immaculata* n. sp., eine neue bodenbewohnende Timaie aus dem Nepal-Himalaya. *J. Ornithol.* 132: 179–198.

TD Martens, J., Eck, S., Päckert, M. & Sun, Y-H., 1999. The Golden-spectacled Warbler *Seicercus burkii* – a species swarm. (Aves: Passeriformes: Sylviidae) Part 1. *Zoologisches Abhandlungen, Staatliches Museum für Tierkunde Dresden.* 50(18): 281–327.

Martens, J., Eck, S. & Sun, Y-H., 2002. *Certhia tianquanensis* Li, a treecreeper with a relict distribution in Sichuan, China. *J. Ornithol.* 143(4): 440–455.

Martens, J., Eck, S. & Sun, Y-H., 2003. On the discovery of a new treecreeper in China – *Certhia tianquanensis. Bull Oriental Bird Club* 37: 65–70.

Martens, J., Sun, Y-H. & Päckert, M., 2008. Intraspecific differentiation of Sino-Himalayan bush-dwelling *Phylloscopus* leaf warblers, with description of two new taxa (*P. fuscatus, P. fuligiventer, P. affinis, P. armandi, P. subaffinis*). *Vertebrate Zool.* 58(2): 233–265.

Martens, J. & Tietze, D.T., 2006. Systematic notes on Asian birds. 65. A preliminary review of the Certhiidae. *Zoologische Meded. (Leiden)* 80(5): 273–286.

Martens, J., Tietze, D.T. & Eck, S., 2004. Radiation and species limits in the Asian Pallas's Warbler complex (*Phylloscopus proregulus* s.l.). *J. Ornithol.* 145: 206–222.

Martuscelli, P. & Yamashita, C., 1997. Rediscovery of the White-cheeked Parrot *Amazona kawalli* (Grantsau & Camargo 1989), with notes on its ecology, distribution and taxonomy. *Ararajuba* 5(2): 97–113.

Mata, H., Fontana, C.S., Mauricio, G.N., Bornschein, M.R., Vasconcelos, M.F. & Bonatto, S.L., 2009. Molecular phylogeny and biogeography of the eastern tapaculos (Aves: Rhinocryptidae: *Scytalopus: Eleoscytalopus*): cryptic diversification in Brazilian Atlantic Forest. *Molecular Phylogenetics and Evolution* 53(2): 450–462.

Mathews, G.M., 1912. (*Corvus mellori*). *Novit. Zool.* 18(3): 443.

Matthysen, E., 1998. *The Nuthatches*. T. & A.D. Poyser, London.

Mattos, J.C.F., Vale, M.M., Vecchi, M.B. & Alves, M.A.S., 2009. Abundance, distribution and conservation of the Restinga Antwren *Formicivora littoralis*. *Bird Cons. Int.* 19: 392–400.

TD Maurício, G.N., 2005. Taxonomy of southern populations in the *Scytalopus speluncae* group, with the description of a new species and remarks on the systematics and biogeography of the complex (Passeriformes: Rhinocryptidae). *Ararajuba* 13: 7–28.

TD Maurício, G.N., Belmonte-Lopes, R., Pacheco, J.F., Silveira, L.F., Whitney, B.M. & Bornschein, M.R., 2014. Taxonomy of 'Mouse-colored Tapaculos' (II): An endangered new species from the montane Atlantic Forest of southern Bahia, Brazil (Passeriformes: Rhinocryptidae: *Scytalopus*). *Auk* 131: 643–659.

Maurício, G.N., Belmonte-Lopes, R., Pacheco, J.F., Silveira, L.F., Whitney, B.M. & Bornschein, M.R., 2015. Erratum: Taxonomy of 'Mouse-colored Tapaculos' (II). An endangered new species from the montane Atlantic Forest of southern Bahia, Brazil (Passeriformes: Rhinocryptidae: *Scytalopus*). *Auk* 132: 951–952.

Maurício, G.N., Mata, H., Bornschein, M.R., Cadena, C.D., Alvarenga, H. & Bonatto, S.L., 2008. Hidden genetic diversity in Neotropical birds: molecular and anatomical data support a new genus for '*Scytalopus*' *indigoticus* species-group (Aves: Rhinocryptidae), *Molecular Phylogenetics and Evolution.* 49: 125–135.

Mayer, S., 1999. Bolivian Spinetail *Cranioleuca henricae* and Masked Antpitta *Hylopezus auricularis*. *Cotinga* 11: 71–72.

Mayr, E., 1933. Birds collected during the Whitney South Sea Expedition. XXVI. Notes on the genera *Myiagra* and *Mayrornis*. *Amer. Mus. Novit.* 651.

Mayr, E., 1946. The number of species of birds. *Auk* 63: 64–69.

Mayr, E., 1969. *Principles of Systematic Zoology*. McGraw-Hill, New York.

Mayr, E., 1971. New species of birds described from 1956 to 1965. *J. Orn.* 112(3): 302–315.

Mayr, E. & Vuilleumier, F., 1983. New species of birds described from 1966 to 1975. *J. Ornithol.* 124: 217–232.

TD Mazar Barnett, J. & Buzzetti, D.R.C., 2014. A new species of *Cichlocolaptes* Reichenbach 1853 (Furnariidae), the 'gritador-do-nordeste', an undescribed trace of the fading life in north-eastern Brazil. *Revista Brasiliera de Ornitologia* 22: 75–94.

Mazar Barnett, J., Carlos, J.C. & Roda, S.A., 2003. A new site for Alagoas endemics. *Cotinga* 20: 13.

Mazar Barnett, J., Carlos, J.C. & Roda, S.A., 2005. Renewed hope for the threatened avian endemics of northeastern Brazil. *Biodiversity and Conservation* 14: 2265–2274.

Mazar Barnett, J. & Kirwan, G.M., 1999. Neotropical Notebook. *Cotinga* 11: 103.

Mazar Barnett, J., Minns, J. Kirwan, G.M. Informaçoes adicionais sobre as aves dos estados do Paraná, Santa Catarina e Rio Grande do Sul. *Ararajuba* 12(1): 55–58.

Mazariegos H., L.A. & Salaman, P.G.W., 1999. The rediscovery of the Colourful Puffleg *Eriocnemus mirabilis*. *Cotinga* 11: 34–38.

Mazumdar, K., Gupta, A. & Samal, P.K., 2011. Documentation of avifauna in proposed Tsangyang Gyatso Biosphere Reserve, western Arunachal Pradesh, India. *CIBTech Journal of Zoology* 3(1): 74–85.

McCarthy, E.M., 2006. *Handbook of Avian Hybrids of the World*. Oxford University Press, Oxford.

McCormack, G., 2005. The status of Cook Island Birds – 1996. Cook Islands Natural Heritage Trust, Rarotonga. http://cookislands.bishopmuseum.org/showarticle.asp?id=7

Mee, A., Ohlson, J., Stewart, I., Wilson, M., Örn, P. & Diaz F., J., 2002. The Cerros del Sirá revisited: birds of submontane and montane forest. *Cotinga* 18: 46–57.

TD Mees, G.F., 1973. Description of a new member of the *Monarcha trivirgata* group from Flores, Lesser Sunda Islands (Aves: Monarchinae). *Zool. Meded. Leiden* 46(12): 180–181.

TD(I) Mees, G.F., 1974. Additions to the avifauna of Suriname. *Zool. Meded. Leiden* 48(7): 1–14.

de Melo, T.N., Olmos, F. & Quental, J., 2015. New records of *Picumnus subtilis* (Aves: Picidae), *Cnipodectes superrufus* (Aves: Tyrannidae) and *Hemitriccus cohnhafti* (Aves: Rhynchocyclidae) in Acre, Brazil. *Check List* 11: doi: http://dx.doi.org/10.15560/11.1.1519.

Mendes de Azevedo Júnior, S., Xavier do Nascimento, J.L. & Serrano do Nascimento, I., 2000. Novos registros de ocorrência de *Antilophia bokermanni* Coelho e Silva, 1999, na Chapada do Araripe, Ceará, Brasil. *Ararajuba* 8(2): 133–134.

Mestre, L.A.M., Barlow, J., Thom, G. & Cochrane, M.A., 2009. Burned forests as a novel habitat for the Black-faced Cotinga (*Conioptilon mcilhennyi*) in the western Brazilian Amazon. *Orn. Neotropical* 20: 467–470.

Mestre, L.A.M., Thom, G., Cochrane, M.A. & Barlow, J., 2010. The birds of Reserva Extrativista Chico Mendes, south Acre, Brazil. *Bol. Mus. Para. E. Goeldi Cienc. Nat.* 5: 311–333.

Meyer, D., 2016. Registros de espécies de aves ameaçadas de extinção ou raras para o Estado de Santa Catarina, sul do Brasil. *Cotinga* 38: 2–9.

Meyer de Schauensee, R., 1966. *The Species of Birds of South America with their Distribution*. Livingstone Press, Narberth, PA.

TD Meyer de Schauensee, R., 1967. *Eriocnemus mirabilis*, a new species of hummingbird from Colombia. *Notulae Naturae* 402: 1–2.

Meyer de Schauensee, R., 1970. *A Guide to the Birds of South America*. Livingston Press, Narberth. PA.

Meyer de Schauensee, R., Poole, E.L., Quinn, J.R. & Sutton, G.M., 1982. *A Guide to the Birds of South America*. Academy of Natural Sciences of Philadelphia, Philadelphia, PA.

Meyer de Schauensee, R., 1984. *The Birds of China*. Smithsonian Institution Press, Washington, DC.

Micol, T. & Jouventin, P., 1995. Restoration of Amsterdam Island South Indian Ocean, following control of feral cattle. *Biological Conservation* 73: 199–206.

Mikkola, H., 2014. *Owls of the World. A Photographic Guide* (Second Edition). Christopher Helm, London.

TD(I) Miller, A.H., 1964. A new species of warbler from New Guinea. *Auk* 81: 1–4.

Mills, M.S.L., 2010. Rock Firefinch *Lagonosticta sanguinodorsalis* and its brood parasite, Jos Plateau Indigobird *Vidua maryae* in Northern Cameroon. *Bull. African Bird Club* 17(1): 86–89.

Mills, M.S.L., Cohen, C., Francis, J. & Spottiswoode, C.N., 2015. A survey for the Critically Endangered Liben Lark *Heteromirafra archeri* in Somaliland, north-western Somalia. *Ostrich* 86(3): 291–294.

Mills, M.S.L. & Ryan, P.G., 2005. Modelling impacts of long-line fishing: what are the effects of pair-bond disruption and sex-biased mortality on albatross fecundity? *Animal Conservation* 8: 359–367.

Milot, E., Weimerskirch, H., Duchesne, P. & Bernatchez, L., 2007. Surviving with Low Genetic Diversity: the Case of Albatrosses. *Proc. Royal Soc. Biological Sciences* 274: (1611): 779–787.

TD(I) Miranda, L., Aleixo, A., Whitney, B.M., Silveira, L.F., Guilherme, E., Dantas Santos, M.P. & Schneider, M.P.C., 2013. Molecular systematics and taxonomic revision of the Ihering's Antwren complex (*Myrmotherula iheringi*: Thamnophilidae), with description of a new species from southwestern Amazonia. *Handbook of the Birds of the World* Special Volume, pp 268–271. Lynx Edicions, Barcelona.

de Miranda Ribeiro, A., 1920. Revisão dos psittacideos brasilieros. *Rev. Mus. Paulista* 12(2): 1–82.

Mittermeier, J.C., Zykowski, K., Stowe, E.S. & Lai, J.E., 2010. Avifauna of the Sipaliwini Savanna (Suriname) with insights into its biological affinities. *Bull. Peabody Mus. Nat. Hist.* 51(1): 97–122.

Mlíkovsky, J., 2009. New data on the distribution of the Marsh Tapaculo (*Scytalopus iraiensis*, Rhinocryptidae). *Ornitología Neotropical* 20: 143–146.

MMA (2014). Lista Nacional de Espécies da Fauna Ameaçadas de Extinção. Portaria No 444, Diário Oficial da União-Seção 1.

Molubah, F.P. & Garbo, M., 2010. Liberian Greenbul *Phyllastrephus leucolepis* Survey Report, July 2010. Society for the Conservation of Nature of Liberia (SCNL).

Monadjem, A., Virani, M.Z., Jackson, C & Reside, A., 2013. Rapid decline and shift in the future distribution predicted for the endangered Sokoke Scops Owl *Otus ireneae* due to climate change. *Bird Cons. Int.* 23(2): 247–258.

Monroe, B.L. Jr, 1968. *A distributional survey of the birds of Honduras.* Ornithological Monographs No 7. AOU.

Monteiro, L.R. & Furness, R.W., 1998. Speciation through temporal segregation of Madeiran Storm-petrels in Azores? *Phil. Trans. Royal Soc. Lond.* B. 353: 845–853.

Monticello, D. & Legrande, V., 2009a. Algerian Nuthatch: a photographic trip. *Dutch Birding* 31: 247–251.

Monticello, D. & Legrande, V., 2009b. Identification of Algerian Nuthatch. *Birding World* 22(8): 333–335.

Morcombe, M., 2000. *Field Guide to Australian Birds.* Steve Parish Publishing, Oxley, Queensland.

Morris, P. & Hawkins, F., 1998. *Birds of Madagascar: a Photographic Guide.* Pica Press, Robertsbridge, UK.

Morrison, J., Hack, M., Wilcove, D.S. & Greeney, H.F., 2014. A nest of the Orange-throated Tanager *Wetmorethraupis sterrhopteron. Cotinga* 36: 115–117.

Mountainspring, S., Casey, T.L.C., Kepler, C.B. & Scott, J.M., 1990. Ecology, behavior and conservation of the Poo-uli (*Melamprosops phaeosoma*). *Wilson Bull.* 102: 109–122.

Mustoe, S.H., Capper, D.R., Lowen, J.C., Leadly, J.D., Rakatomalala, D. & Randrianorivo, T., 2000. Biological surveys of Zombitse-Vohibasia National Park in south-west Madagascar. Unpublished report.

Muzika, Y., 2014. Sillem's Mountain Finch *Leucosticte sillemi* revisited. *BirdingASIA* 21: 28–33.

Naka, L., 2013. Juan Mazar Barnett (1975–2012): una vida entre plumas y amigos. *Hornero* 28(2).

Naka, L.N., 2014. The legacy of Juan Mazar Barnett (1975–2012) to Neotropical Ornithology. *Revista Brasiliera de Ornitologia* 22(2): 63–74.

Naoki, K., 2003. Notes on foraging ecology of the little-known Green-capped Tanager *Tangara meyerdeschauenseei. Orn. Neotrop.* 14: 411–414.

TD(I) Narosky, T., 1977 (*Sporophila zelichi*). *Hornero* 11: 345–348.

de Naurois, R. & Morel, G.J., 1995. Description des oeufs et du nid de la prinia aquatique *Prinia fluviatilis. Malimbus* 17(1): 28–31.

Navarro S., A.G., Benítez D., H., Sánchez B., V., García R., S. & Santana C., E., 1993. The White-faced Swift in Jalisco, Mexico. *Wilson Bull.* 105: 366–367.

TD(I) Navarro-Sigüenza, A.G., García-Hernández, M.A. & Peterson, A.T., 2013. A new species of brush-finch (*Arremon*: Emberizidae) from western Mexico. *Wilson J. Ornithol.* 125(3): 443–453.

TD Navarro-S., A.G., Peterson, A.T., Escalante P., B.P. & Benítez D., H., 1992. *Cypseloides storeri*, a new species of swift from Mexico. *Wilson Bull.* 104(1): 55–64.

Nemésio, A. & Rasmussen, C., 2009. The rediscovery of Buffon's 'Guarouba' or 'Perriche Jaune': two senior synonyms of *Aratinga pintoi* Silveira, de Lima & Höfling, 2005 (Aves: Psittaciformes). *Zootaxa* 2013: 1–16.

Nguembock, B., Fjeldsa, J., Couloux, A. & Pasquet, E., 2008. Phylogeny of *Laniarius.* Molecular data reveal *L. liberatus* synonymous with *L. erlangeri* and 'plumage coloration' as unreliable morphological characters for defining species and species groups. *Molecular Phylogenetics and Evolution* 48(2): 395–407.

TD(I) ×2 Nicolai, J., 1972. Zwei neue *Hypochera*-arten aus West-Afrika (Ploceidae, Viduinae). *J. Orn.* 113(3): 229–239.

Niemann, D., 2015. Crossing boundaries. *World Birdwatch* September 2015: 12–15.

Nores, M., 1986. Diez nuevas subespecies de aves provenientes de islas ecológicas argentinas. *Hornero* 12(4): 262–273.

TD Nores, M. & Yzurieta, D., 1979. Una nueva especie y dos nuevas subespecies de aves (Passeriformes). *Acad. Nac. de Cienc. de Córdoba, Miscelánea* 61: 4–8.

O'Donnell, C. & Fjeldså, J. 1997 *Grebes - Status Survey and Conservation Action Plan.* IUCN/SSC Grebes Scialist Group. IUCN, Gland, Switzerland and Cambridge, UK.

O'Keefe, M., Dowdall, S., Enright, K., Fahy, J., Gilligan, J., & Lillie, G., 2009. Unidentified storm-petrels off Puerto Montt, Chile, in February 2009. *Dutch Birding* 31: 223–224.

Oláh, J., 2006. Little-known Asian bird. Black-crowned Barwing *Actinodura sodangorum.* BirdingASIA 6: 73–74.

Oliveira Mafia, P., 2015. New record extends the northern limit of distribution of *Scytalopus petrophilus* (Passeriformes: Rhinocryptidae) in the Espinhaço Range, Minas Gerais, Brazil. *Check List* 11(1): 1525.

Oliveros, C.H. & Layusa, C.A.A., 2011. First description of the nest and eggs of the Calayan Rail. *J. Yamashina Inst.Orn.* 42: 143–146.

Oliveros, C.H., Reddy, S. & Moyle, R.G., 2012. The phylogenetic position of some Philippine 'babblers' spans the muscicapoid and sylvioid bird radiations. *Molecular Phylogenetics and Evolution* 65(2): 799–804.

Olmos, F., Silveira, L.F. & Benedicto, G.A., 2011. A contribution to the ornithology of Rondônia, southwest of the Brazilian Amazon. *Revista Brasiliera de Ornitologia* 19(2): 200–229.

Olrog, C.C., 1970. Adiciones a la avifauna Argentina. *Acta Zoologica Liloana* 27(17): 255–266.

Olrog, C.C., 1984. *Las aves argentinas. Una nueva guía de campo.* Administración de Parques Nacionales, Buenos Aires.

TD(I) Olrog, C.C. & Contino, F., 1966. Dos neuvos tiranidos para la fauna Argentina. *Neotropica* 12: 133–114.

Olsen, J., 2011. Chapter 44, New Species, pp 281–287, in *Australian High Country Owls.*

Olsen, J., Trost, S. & Myers, S., 2009. Owls on the Island of Sumba, Indonesia. *Australian Field Ornithology* 26: 2–14.

TD Olsen, J., Wink, M., Sauer-Gürth, H. & Trost, S., 2002. A new *Ninox* owl from Sumba, Indonesia. *Emu* 102: 223–231.

TD Olsson, U., Alström, P. & Colston, P.R., 1993. A new species of *Phylloscopus* warbler from Hainan Island, China. *Ibis* 135(1): 3–7.

Oyler-McCance, S.J., Cornman, R.S., Jones, K.L. & Fike, J.A.,

2015. Genomic single-nucleotide polymorphisms confirm that Gunnison and Greater Sage-grouse are genetically well differentiated and that the Bi-State population is distinct. *Condor* 117: 217–227.

O'Neill, J.P., 1969. Distributional notes on the birds of Peru, including twelve species previously unreported from the republic. *Occ. Pap. Mus. Zool., Louisiana State University 37.*

O'Neill, J.P., 2006. Museum expedition to northern Peru. *Museum Quarterly. LSU Museum of Natural Science* 24: 8–10.

TD O'Neill, J.P. & Graves, G.R., 1977. A new genus and species of owl (Aves: Strigidae) from Peru. *Auk* 94(3): 409–416.

TD O'Neill, J.P., Lane, D.F., Kratter, A.W., Capparella, A.P. & Joo, C.F., 2000. A striking new species of Barbet (Capitonidae: *Capito*) from the eastern Andes of Peru. *Auk* 117(3): 369–370.

TD O'Neill, J.P., Lane, D.F. & Naka, L.N., 2011. A cryptic new species of thrush (Turdidae: *Turdus*) from western Amazonia. *Condor* 113(4): 869–880.

TD O'Neill, J.P., Munn, C.A. & Francke, I.J., 1991. *Nanopsittaca dachilleae*, a new species of parrotlet from eastern Peru. *Auk* 108(2): 225–229.

Onley, D. & Scofield, P., 2007. *Albatrosses, Petrels and Shearwaters of the World.* Christopher Helm, London.

van Oosten, H., Beunen, R., van de Meulengraaf, B. & van Noort, T., 2007. White-masked Antbird *Pithys castaneus* and Orange-throated Tanager *Wetmorethraupis sterrhopteron. Cotinga* 28: 79–81.

van Oosten, H. & Cortés, O., 2009. First record of Munchique Wood-wren *Henicorhina negreti* in dpto. Chocó, Colombia. *Cotinga* 31: 128.

Orians, G.H. & Orians, E.N., 2000. Observations of the Pale-eyed Blackbird in southeastern Peru. *Condor* 102: 956–958.

Ortiz-Crespo, F.I., Greenfield, P.J. & Matheus, J.C., 1990. *Aves de Ecuador Continente y Archipiélago de Galápagos.* FEPROTUR and CECIA, Quito, Ecuador.

Ottosson, U., 2009. Två arter klippamarant? *Vår Fågelvärld* 68(3): 34.

Ozaki, K., Komeda, S., Baba, T., Toguchi, Y. & Harato, T. Declining distribution of the Okinawa Rail. Impact of introduced predators. 24th Int. Orn. Congress, Hamburg, Germany, 13–16 August 2006.

Ozaki, K., Yamamoto, Y. & Yamagishi, S., 2010. *Genes Genet.* 85: 55–63.

TD Pacheco, J.F. & Gonzaga, L.P., 1995. A new species of *Synallaxis* of the *ruficapilla/infuscata* complex from eastern Brazil (Passeriformes: Furnariidae). *Ararajuba* 3: 3–11.

TD Pacheco, J.F., Whitney, B.M. & Gonzaga, L.P., 1996. A new genus and species of furnariid (Aves: Furnariidae) from the cocoa-growing region of southeastern Brazil. *Wilson Bull.* 108: 397–433.

Päckert, M., Martens, J., Liang, W., Hsu, Y-C. & Sun, Y-H., 2013. Molecular genetic and bioacoustic differentiation of *Pnoepyga* Wren-babblers. *J. Ornithol.* 154: 329–337.

Pam, G.B. & Ottossson, U., 2006. Habitat selection in the Rock Firefinch. *J. Ornithol.*147(5): Suppl. 1: 224.

Parker, S.A., 1972. Remarks on distribution and taxonomy of the grass wrens *Amytornis textilis, modestus* and *purnelli. Emu*: 72(4): 157–166.

TD(I) Parker, S.A., 1982. A new sandpiper of the genus *Calidris. S. Aust. Nat.* 56: 63.

Parker III, T.A., 1982. Observations of some unusual rain forest and marsh birds in southeastern Peru. *Wilson Bull.* 94(4): 477–493.

Parker III, T.A., Castillo, U.A., Gell-Mann, M. & Rocha, O.O., 1991. Records of new and unusual birds from northern Bolivia. *Bull. Brit. Orn. Club* 111: 120–138.

TD Parker III, T.A & O'Neill, J.P., 1985. A new species and a new subspecies of *Thryothorus* wren from Peru. *Neotropical Ornithology, AOU Monographs* 36: 9–15.

Parker III, T.A. & Remsen, J.V. Jr., 1987. Fifty-two Amazonian bird species new to Bolivia. *Bull. Brit. Orn. Club* 107: 94–107.

Parker, T.A. & Rocha, O., 1991. La avifauna del Cerro San Simón, una localidad de campo rupestre aislado en el Depto. Beni, noreste Bolivia. *Ecol. Bolivia* 17: 15–29.

Parker III, T.A., Stotz, D.F. & Fitzpatrick, J.W., 1997. Notes on avian bamboo specialists in southwestern Amazonian Brazil. Pp 543–547 in J.V. Remsen Jr. (ed.), *Studies in Neotropical Ornithology honoring Ted Parker.* Ornithological Monographs 48. American Ornithologists' Union, Washington, DC.

Parker, V., 1999. *The Atlas of the Birds of Sul do Save, Southern Mozambique.* Avian Demography Unit and Endangered Wildlife Trust, Cape Town and Johannesburg.

Parkes, K.C., 1987. Letter: Was the 'Chinese' White-eyed River Martin an Oriental Pratincole? *Forktail* 3: 68–69.

Parkes, K.C., 1999. Review in Ornithological Literature. *Wilson Bull.* 111(1): 144–146.

Parrini, R., Raposo, M.A., Pacheco, J.F., Carvalhaes, A.M.P., Melo, T.A.J., Fonseca, P.S.H. & Minns, J., 1999. Birds of the Chapada Diamantina, Brazil. *Cotinga* 11: 86–95.

Pasquet, E., 1998. Phylogeny of the nuthatches of the *Sitta canadensis* group and its evolutionary and biogeographic implications. *Ibis* 140(1): 150–156.

Patton, J.L. & Smith, M.F., 1994. Paraphyly, polyphyly, and the nature of species boundaries in pocket gophers (genus *Thomomys*). *Syst. Biol.* 43:11–26.

Paula de Faria, I., 2007. Registros de aves globalmente ameaçadas, raras e endêmicas para a região de Vicente Pires, Distrito Federal, Brasil. *Revista Brasiliera de Ornitogia* 15(1): 117–122.

Paulian, P., 1953. Pinnipèdes, cétaces et oiseaux des îles Kerguelen et Amsterdam. *Mém. Inst. Scient. Madagascar* A 8: 111–234.

Paulian, P., 1960. Quelques données sur l'avifaune ancienne des îles Amsterdam et Saint-Paul. *L'Oiseau et R.F.O.,* 30: 18–23.

Payne, R.B., 1968a. A preliminary report on the relationships of the indigobirds. *Bull. Brit. Orn. Club* 88: 32–36.

Payne, R.B., 1968b. Mimicry and relationships in the indigobirds or combassous of Nigeria. *Nigerian Orn. Soc. Bull.* 5: 57–60.

TD Payne, R.B., 1982. Species limits in the indigobirds (Ploceidae: *Vidua*) of west Africa: mouth mimicry, song mimicry and description of new species. *Misc. Publ. Mus. Zool. University of Michigan* 162: 1–96.

TD Payne, R.B., 1998. A new species of Firefinch *Lagonosticta* from northern Nigeria and its association with the Jos Plateau Indigobird *Vidua maryae. Ibis* 140(3): 368–381.

Payne, R.B. & Barlow, C.R., 2004. Songs of Mali Firefinch *Lagonosticta virata* and their mimicry by Barka Indigobird *Vidua larvaticola* in West Africa. *Malimbus* 26: 11–18.

Payne, R.B. & Payne, L., 1994. Song mimicry and species associations of west African indigobirds *Vidua* with Quail-finch *Ortygospiza atricollis,* Goldbreast *Amandava subflava* and Brown Twinspot *Clytospiza monteiri. Ibis* 136: 291–304.

Peacock, F., 2006. *Pipits of Southern Africa.*Published by the author, Pretoria: www.pipits.co.za

Pearman, M., 1990. Behaviour and vocalisations of an undescribed Canastero *Asthenes* sp. from Brazil. *Bull. Brit. Orn. Club* 110(3): 145–153.

Pearman, M., 1993. Some range extensions and five species new to Colombia, with notes on some scarce or little known species. *Bull. Brit. Orn. Club* 113(2): 66–75.

Pedersen, T., 1997. New observations of a Zaïrean endemic: Prigogine's Greenbul *Chlorocichla prigoginei. Bull. African Bird Club* 4(2): 109–110.

Penhallurick, J. & Robson, C., 2009. The generic taxonomy of parrotbills (Aves, Timaliidae). *Forktail* 25: 137–141.

Pereira, G.A., Dantas, S.D.M., Silveira, L.F., Roda, S.A., Albano, C., Sonntag, F.A., Leal, S., Periquito, M.C., Malacco, G.B. & Lees, A.C., 2014. Status of the globally threatened forest birds of northeast Brazil. *Pap. Avulsos de Zool. (São Paulo)* 54: 177–194.

Pereira, G.A., Rodrigues, P.P., Leal, S., Periquito, M.C., Malacco da Silva, G.B., Menêzes, M., da Silva Corrêa, G., Sonntag, F.A., Nobre de Almeida, M.N. & Nunes, P.B., 2014. Important bird records from Alagoas, Pernambuco and Paraiba, north-east Brasil. *Cotinga* 36: 46–51.

Pérez-Emán, J., Sharpe, C.J., Lentino, M., Prum, R.O. & Carreño, I., 2003. New records from the summit of Cerro Guaiquinima, Estado Bolívar, Venezuela. *Bull. Brit. Orn. Club* 123(2); 78–89.

van Perlo, B., 2009. *A Field Guide to the Birds of Brazil.* Oxford University Press, Oxford.

TD(I) Peters, D.F., 1996. *Hypositta perdita* sp. nov., a new bird species from Madagascar. *Senckenbergiana Biologica* 76 (1–2): 7–14.

Peters, D.S., 2004. The third museum specimen of Stresemann's Bristlefront (*Merulaxis stresemanni* Sick, 1960). *J. Ornithol.* 145: 269–270.

Peters, J.L., 1951. *Check-list of Birds of the World, Vol 7.* Museum of Comparative Zoology, Cambridge, Massachusetts.

Peterson, A.T., 2014. Type specimens in modern ornithology are necessary and irreplaceable. *Auk* 131(3): 282–286.

Peterson, A.T., Brooks, T., Gamauf, A., Carlos T., J.C., Mallari, N.A.D., Dutson, G., Bush, S.F., Clayton, D.H. & Fernandez, R., 2008. The Avifauna of Mt. Kitanglad, Bukidnon Province, Mindanao, Philippines. *Fieldiana Zool.* 114: 1–43.

TD Phelps, W.H., 1977. Una nueva especie y dos nuevas subespecies de Aves (Psittacidae, Furnariidae) de la Sierra de Perijá cerca de la divisoria Colombo-Venezolana. *Bol. Soc. venez. Cienc. nat.* 33(134): 43–53.

Phillips, A.R., 1966. Notas sistemáticas sobre aves Méxicanas III. *Rev. Soc. méx. Hist. nat.* 25: 217–242.

Phillips, A.R., *The Known Birds of North and Middle America Part I (Hirundinidae to Mimidae; Certhiidae)* 1986; *Part II (Bombycillidae; Sylviidae to Sturnidae)* 1991. Privately published, Denver, Colorado.

Piacentino, V. de Q., Aleixo, A. & Silveira, L.F., 2009. Hybrid, subspecies or species? The validity and taxonomic status of *Phaethornis longuemareus aethopyga* Zimmer 1950 (Trochilidae). *Auk* 126(3): 604–612.

Pilgrim, J.D., Eames, J.C. & Gandy, D., 2009. The newly described Bare-faced Bulbul *Pycnonotus hualon. BirdingASIA* 12: 53–55.

Pinet, P., Jaquemet, S., Pinaud, D., Weimerskirch, H., Phillips, R.A. & Le Corre, M., 2011. Migration, wintering distribution and habitat use of an endangered tropical seabird, Barau's Petrel *Pterodroma baraui. Marine Ecology Progress Series* Vol. 423: 291–302.

Pinet, P., Jaquemet, S., Phillips, R.A. & Le Corre, M., 2012. Sex-specific foraging strategies throughout the breeding season in a tropical, sexually monomorphic small petrel. *Animal Behaviour* 83: 979–989.

Pinheiro, R.T. & Dornas, T., 2008. New records and distribution of Kaempfer's Woodpecker *Celeus obrieni. Revista Brasiliera de Ornitologia* 16(2): 167–169.

Platt, J.R., 2015. Amazing Discovery: Nearly Extinct Bird Found Breeding in Japan. https://blogs.scientificamerican.com/extinction-countdown/amazing-discovery-nearly-extinct-bird-found-breeding-in-japan/

duPont, J.E., 1971. Notes on Philippine birds. *Nemouria* 3: 1–6.

Porter, R. & Aspinall, S., 2010. *Birds of the Middle East.* 2nd Edition. Christopher Helm, London.

Porter, R.F., Dymond, J.N. & Martins, R.P., 1994. Forbes-Watson's Swift *Apus berliozi* in Socotra. *Sandgrouse* 17: 138–141.

TD Porter, R.F. & Kirwan, G., 2010. Studies of Socotran birds VI. The taxonomic status of the Socotra Buzzard. *Bull. Brit. Orn. Club* 130(2): 116–131.

TD(I) Portes, C.E.B., Aleixo, A., Zimmer, K.J., Whittaker, A., Weckstein, J.D., Gonzaga, L.P., Ribas, C.C., Bates, J.M. & Lees, A.C., 2013. A new species of *Campylorhamphus* (Aves: Dendrocolaptidae) from the Tapajós-Xingu interfluve in Amazonian Brazil. *Handbook of the Birds of the World* Special Volume, pp 258–262. Lynx Edicions, Barcelona.

Poulsen, M.K., 1995. The threatened and near-threatened birds of Luzon, Philippines, and the role of the Sierra Madre Mountains in their conservation. *Bird Cons. Int.* 5(1): 79–115.

Powell, A., 2008. *The Race to Save the World's Rarest Bird: The Discovery and Death of the Po'ouli.* Stackpole Books, Mechanicsburg, PA.

do Prado, A.D., 2006. *Celeus obrieni*: 80 anos depois. *Atualidades Ornitológicas* 4: 5.

Pratt, H.D., 1992. Is the Po'ouli a Hawaiian honeycreeper (Drepaninae)? *Condor* 94: 172–180.

Pratt, H.D., Bruner, P.L. & Berrett, D.G., 1987. *A Field Guide to the Birds of Hawaii and the Tropical Pacific.* Princeton University Press, Princeton, NJ.

Pratt, T.K. & Beehler, B.M., 2014. *Birds of New Guinea.* 2nd Edition. Princeton University Press, Princeton, NJ.

Pratt, T.K., Kepler, C.B. & Casey, T.L., 1997. Po'ouli (*Melamprosops phaeosoma*). *The Birds of North America online.* (A. Poole, Ed.). Cornell Lab. of Ornithology, Ithaca, NY.

TD Preleuthner, M. & Gamauf, A., 1998. A possible new subspecies of the Philippine Hawk-eagle (*Spizaetus philippensis*) and its future prospects. *J. Raptor Res.* 32(2): 126–135.

TD Prigogine, A., 1960. Un nouveau martinet du Congo. *Rev. Zool. Bot. Afr.* 62(1–2): 103–105.

TD(I) Prigogine, A., 1972. Description of a new bulbul from the Republic of Zaire. *Bull. Brit. Orn. Club* 92(5): 138–141.

TD(I) Prigogine, A., 1978. A new ground-thrush from Africa. *Gerfaut* 68: 482–492.

Prigogine, A., 1981. A new species of *Malimbus* from Sierra Leone? *Malimbus* 3: 55.

TD Prigogine, A., 1983. Un nouveau *Glaucidium* de l'Afrique centrale. *Rev. Zool. afr.* 97(4): 886–895.

Prigogine, A., 1989. The taxonomic status of *Zoothera kibalensis. Gerfaut* 79: 189–190.

Probst, J.-M., Le Corre, M. & Thébaud, C., 2000. Breeding habitat and conservation priorities in *Pterodroma baraui,* an endangered gadfly petrel of the Mascarene archipelago. *Biological Conservation* 93: 135–138.

Prŷs-Jones, R.P., 1979. The ecology and conservation of the Aldabra Brush Warbler *Nesillas aldabranus. Phil. Trans. Roy. Soc. London* (Ser. B*)* 286: 211–224.

Pyle, P., David, R., Eilerts, B.D., Amerson, B., McKown, M. & Booker, A., 2012. Updated information on Bryan's Shearwaters (*Puffinus bryani*) in the north Pacific Ocean, with a look toward its conservation. 'PSG 2012 Hawaii Abstracts', Pacific Seabird Group p 78, http://www.pacificseabirdgroup.org/2012mtg/PSG2012.AbstractBook.pdf

Pyle, R.L. & Pyle, P., [online], 2009. The birds of the Hawaiian Islands: occurrence, history, distribution and status. B.P.Bishop Museum, Honolulu. http://hbs.bishopmuseum.org/birds/rlpmonograph/(31 December 2009).

TD Pyle, P., Welch, A.J. & Fleischer, R.C., 2011. A new species of Shearwater (*Puffinus*) recorded from Midway Atoll, Northwestern Hawaiian Islands. *Condor* 113(3): 518–527.

Quintana, B., 1987. *Chiloé mitológico.* Temuco: Telstar Impresores.

Raffaele, H., Wiley, J., Garrido, O., Keith, A. & Raffaele, J., 1998.

A Guide to the Birds of the West Indies. Princeton University Press, Princeton NJ.

Rand, A.L. & Rabor, D.S., 1967. New Birds from Luzon, Philippine Islands. *Fieldiana Zoology* 51(6): 86.

Rainey, H., Borrow, N. & Demey, R., 2003. First recordings of vocalisations of Yellow-footed Honeyguide *Melignomon eisentrauti* and confirmed records in Ivory Coast. *Malimbus* 25(1): 31–38.

Rains, D., Weimerskirch, H. & Burg, T.M., 2011. Piecing together the global population puzzle of wandering albatrosses: genetic analysis of the Amsterdam Albatross *Diomedea amsterdamensis*. *Journal of Avian Biology* 42: 69–89.

Rajeshkumar, S., Ragunathan, C. & Rasmussen, P.C., 2012. An apparently new species of *Rallina* crake from Great Nicobar Island, India. *BirdingASIA* 17: 44–46.

Ramírez-Burbano, M.B., Sandoval-Sierra, J.V. & Gómez-Bernal, L.G., 2007. Uso de recursos florales por el Zamarrito Multicolor *Eriocnemus mirabilis* (Trochilidae) en el Parque Nacional Munchique, Colombia. *Ornitología Colombiana* 5: 64–77.

TD Rand, A.L., 1960. A new species of Babbling Thrush from the Philippines. *Fieldiana Zool.* 39(33): 377.

TD Raposo, M.A., 1997. A new species of *Arremon* (Passeriformes: Emberizidae) from Brazil. *Ararajuba* 5: 3–9.

Raposo, M.A., Tello, J.G., Dickinson, E.C. & Brito, G.R.R., 2015. Remarks on the name *Cercomacra* Sclater, 1858 (Aves: Thamnophilidae) and its type species. *Zootaxa* 3914(1): 94–96.

TD Rappole, J.H., Renner, S.C., Shwe, N.M. & Sweet, P.R., 2005. A new species of scimitar-babbler (Timaliidae: *Jabouillea*) from the sub-Himalayan region of Myanmar. *Auk* 122(4): 1064–1069.

Rappole, J.H., Rasmussen, P.C., Aung, T., Milensky, C.M. & Renner, S.C., 2008. Observations on a new species: the Naung Mung Scimitar-babbler *Jabouilleia naungmungensis*. *Ibis* 150(3): 623–627.

Rasmussen, J.F., Rabbek, C., Poulsen, B.O., Poulsen, M.K. & Bloch, H., 1996. Distributional records and natural history notes on threatened and little known birds of southern Ecuador. *Bull. Brit. Orn. Club* 116(1): 26–46.

TD Rasmussen, P.C., 1998. A new Scops-owl from Great Nicobar Island. *Bull. Brit. Orn. Club* 118(3): 141–153.

TD Rasmussen, P.C., Allen, D.N.S., Collar, N.J., Demeulemeester, B., Hutchinson, R.O., Jakosalem, P.G.C., Kennedy, R.S., Lambert, F.R. & Paguntalan, P.G.C., 2012. Vocal divergence and new species in the Philippine Hawk Owl *Ninox philippensis* complex. *Forktail* 28: 1–20.

TD Rasmussen, P.C., Round, P.D., Dickinson, E.C. & Rozendaal, F.G., 2000. A new bush-warbler (Sylviidae, *Bradypterus*) from Taiwan. *Auk* 117: 279–289.

TD Rasmussen, P., Schulenberg, T.S., Hawkins, F. & Voninavoko, R., 2000. Geographic variation in the Malagasy Scops-owl *Otus rutilis auct.*: the existence of an unrecognised species on Madagascar and the taxonomy of other Indian Ocean taxa. *Bull. Brit. Orn. Club* 120(2): 75–101.

Read, C., Tarboton, W.R., Davies, G.B.P., Anderson, M.D. & Anderson, T.A., 2014. An annotated checklist of birds of the Vilanculos Coastal Wildlife Sanctuary, southern Mozambique. *Ornithological Observations* 5: 370–408.

Redman, N., Stevenson, T. & Fanshawe, J., 2011. *Birds of the Horn of Africa* 2nd Edition. Christopher Helm, London.

Regalado, A., 2011. Feathers are flying over Colombian bird name flap. *Science* 331: 1123–1124.

Rêgo, M.A., Dantas, S., Guilherme, E. & Martuscelli, P., 2009. First records of Fine-barred Piculet *Picumnus subtilis* from Acre, western Amazonia, Brazil. *Bull. Brit. Orn. Club* 129(3): 182–191.

Reinert, B.L., Belmonte-Lopes, R., Bornschein, M.R., Sobotka, D.D., Corrêa, L., Pie, M.R. & Pizo, M.A., 2012. Nest and eggs of the Marsh Antwren (*Stymphalornis acutirostris*): the only marsh-dwelling Thamnophilid. *Wilson J. Orn.* 124(2): 286–291.

Reinert, B.L. & Bornschein, M.R., 1996. Descrição do macho adulto de *Stymphalornis acutirostris* (Aves: Formicariidae). *Ararajuba* 4(2): 103–105.

Remold, H.G. & Ramos Neto, M.B., 1995. A nest of the Restinga Tyrannulet *Phylloscartes kronei*. *Bull. Brit. Orn. Club* 115(4): 239–240.

TD Remsen, J.V., 1993. Zoogeography and geographic variation of *Atlapetes rufinucha* (Aves: Emberizinae) including a new subspecies, in southern Peru and Bolivia. *Proc. Biol. Soc. Washington* 106: 429–435.

Remsen, J. V., Jr., J. I. Areta, C. D. Cadena, S. Claramunt, A. Jaramillo, J. F. Pacheco, J. Pérez-Emán, M. B. Robbins, F. G. Stiles, D. F. Stotz, and K. J. Zimmer. Version 2016. *A classification of the bird species of South America*. American Ornithologists' Union. www.museum.lsu.edu/~Remsen/SACCBaseline.htm

Remsen, J.V. Jr & Cicero, C., 2007. A tribute to the career of Ned K. Johnson: enduring standards through changing times. *Orn. Monographs* 63: 1–17.

Remsen, J.V., Stiles, F.G. & Scott, P.E., 1986. Frequency of arthropods in stomachs of tropical hummingbirds. *Auk* 103: 436–441.

Remsen, J.V. Jr & Traylor, M.A., 1989. *An Annotated List of Birds of Bolivia*. Pp 1–57. Buteo Books, Vermilion, SD.

Renjifo M., L.M., 1994. First records of the Bay-vented Cotinga *Doliornis sclateri* in Colombia. *Bull. Brit. Orn. Club* 114(1): 101–103.

Renman, E., 1994. A possible new species of Scops Owl *Otus* sp. on Réunion? *Bull. African Bird Club* 2: 54–55.

Repenning, M., 2012. História natural, com ênfase na biologia reprodutiva, de uma população migratória de *Sporophila* aff. *plumbea* (Aves: Emberizidae) do sul do Brasil. MS dissertation, Pontifícia Universidade Católica do Rio Grande do Sul, Porto Alegre, Brazil.

TD Repenning, M. & Fontana, C.S., 2013a. A new species of Gray Seedeater (Emberizidae: *Sporophila*) from upland grasslands of southern Brazil. *Auk* 130(4): 791–803.

Repenning, M. & Fontana, C.S., 2013b. *Sporophila* aff. *plumbea*. In: Plano de ação para a conservação dos Passeriformes ameaçados dos Campos Sulinos e Espinilho. Instituto Chico Mendes e Ministério do Meio Ambiente, Brazil, Ed.

Repenning, M. & Fontana, C.S., 2016. Breeding Biology of the Tropeiro Seedeater (*Sporophila beltoni*). *Auk* 133: 484–496.

TD(I) Restall, R.L., 1996. A proposed new species of munia, genus *Lonchura* (Estrildinae). *Bull. Brit. Orn. Club* 116(3): 137–142.

Restall, R., Rodner, C. & Lentino, M., 2006. *Birds of Northern South America*. Vols 1 & 2. Christopher Helm, London.

Reynolds, M.H. & Snetsinger, T.J., 2001. The Hawai'i rare bird search 1994–1996. *Studies in Avian Biology* 22: 133–143.

Rheindt, F.E., 2004. Notes on the range and ecology of Sichuan Treecreeper *Certhia tianquanensis*. *Forktail* 20: 141–142.

Rheindt, F.E., 2006. Splits galore: the revolution in Asian leaf warbler systematics. *BirdingASIA* 5: 25–39.

Rheindt, F.E. & Hutchinson, R.O., 2013. The discovery of a new *Locustella* bush warbler on the island of Taliabu, Indonesia. *BirdingASIA* 19: 109–110.

Rheindt, F.E., Prawiradilaga, D.M., Suparno, Ashari, H. & Wilton, P.R., 2014. New and significant island records, range extensions and elevational extensions of birds in eastern Sulawesi, its nearby satellites, and Ternate. *Treubia* 41: 61–90.

Ribon, R., Whitney, B.M., Pacheco, J.F., 2002. Rediscovery of Bahia Spinetail *Synallaxis cinerea* in north-east Minas Gerais, Brazil, with additional records of some rare and threatened montane Atlantic Forest birds. *Cotinga* 17: 46–54.

Ridgely, R.S., 2012. Discovering the Jocotoco. *Neotropical Birding* 10: 4–8.

Ridgely, R.S. & Gaulin, S.J.C., 1980. The birds of Finca Merenberg, Huila Department, Colombia. *Condor* 82: 379–391.

Ridgely, R.S. & Greenfield, P., 2001. *The Birds of Ecuador.* Comstock Publishing, Ithaca, N.Y.

Ridgely, R.S. & Gwynne, J.A., 1976. *A Guide to the Birds of Panama.* Princeton University Press, Princeton, NJ.

Ridgely, R.S. & Gwynne, J.A., 1989. *A Guide to the Birds of Panama with Costa Rica, Nicaragua and Honduras.* Princeton University Press, Princeton, NJ.

Ridgely, R.S. & Tudor, G., 1989. *The Birds of South America.* University of Texas Press, Austin, TX.

Ridgely, R.S. & Tudor, G., 2009. *Field Guide to the Songbirds of South America; the Passerines.* University of Texas Press, Austin, TX.

TD Ridgely, R.S. & Robbins, M.B., 1988. *Pyrrhura orcesi*, a new parakeet from southwestern Ecuador, with systematic notes on the *P. melanura* complex. *Wilson Bull.* 100(2): 173–182.

Riley, J., 2002. Population sizes and the status of endemic and restricted-range bird species on Sangihe Island, Indonesia. *Bird Cons. Int.*, 12: 53–78.

Riley, J., 2003. Population sizes and the conservation status of endemic and restricted-range bird species on Karakelang, Talaud Islands, Indonesia. *Bird Cons. Int.* 13: 59–74.

TD(I) Ripley, S.D., 1960. Two new birds from Angola. *Postilla* 43: 1–3.

TD Ripley, S.D., 1965. Le Martinet Pâle du Socotra (*Apus pallidus berliozi*). *L'Oiseau et RFO* 35: 101–102.

TD Ripley, S.D., 1966. A notable owlet from Kenya. *Ibis* 108: 136–137.

TD(I) Ripley, S.D., 1980. A new species, and a new subspecies of bird from Tirap District, Arunachal Pradesh, and comments on the subspecies of *Stachyris nigriceps* Blyth. *Journal of the Bombay Natural History Society* 77(1): 1–5.

Ripley, S.D., 1984. A note on the status of *Brachypteryx cryptica. J. Bombay Nat. Hist. Soc.* 81(3): 700–701.

Ripley, S.D., 1985. Relationships of the Pacific warbler *Cichlornis* and its allies. *Bull. Brit. Orn. Club* 105(3): 109–112.

Ripley, S.D., 1986. In Memoriam: Rodolphe Meyer de Schauensee. *Auk* 103(1): 204–206.

TD Ripley, S.D. & Hadden, D., 1982. A new subspecies of *Zoothera* (Aves: Muscicapidae: Turdinae) from the northern Solomon Islands. *J. Yamashina Inst. Orn.* 14: 103–107.

TD Ripley, S.D. & Marshall, J.T., 1967. A new subspecies of flycatcher from Luzon, Philippine Islands. *Proc. Biol. Soc. Washington* 80: 243–244.

TD Ripley, S.D. & Rabor, D.S., 1961. The Avifauna of Mount Katanglad. *Postilla, Yale Peabody Mus. Nat. Hist.* 50: 1–20.

TD Ripley, S.D. & Rabor, D.S., 1966. *Dicaeum proprium,* new species (Aves; Family Dicaeidae). *Proc. Biol. Soc. Washington* 79: 305–306.

TD Ripley, S.D. & Rabor, D.S., 1968. Two new subspecies of birds from the Philippines and comments on the validity of two others. *Proc. Biol. Soc. Washington* 81: 31–36.

Rivalan, P., Barbraud, C., Inchausti, P. & Weimerskirch, H., 2010. Combined impacts of longline fisheries and climate on the persistence of the Amsterdam Albatross *Diomedea amsterdamensis. Ibis* 152(1): 6–18.

TD(I) Robb, M.S., van den Berg, A.B. & Constantine, M., 2013. A new species of *Strix* owl from Oman. *Dutch Birding* 35: 275–310.

Robb, M.S., Sangster, G., Aliabadian, M., van den Berg, A.B., Constantine, M., Irestedt, M., Khani, A., Musavi, S.B., Nunes, J.M.G., Willson, M.S. & Walsh, A.J., 2015. The rediscovery of *Strix butleri* (Hume, 1878) in Oman and Iran, with molecular resolution of the identity of *Strix omanensis* Robb, van den Berg and Constantine, 2013. bioRxiv preprint posted online Aug 20, 2015; doi: http//dx.doi.org/10.1101/025122.

TD Robbins, M.B. & Howell, S.N.G., 1995. A new species of Pygmy-owl (Strigidae: *Glaucidium*) from the Eastern Andes. *Wilson Bull.* 107(1): 1–6.

Robbins, M.B. & Ridgely, R.S., 1990. The avifauna of an upper tropical cloud forest in southwestern Ecuador. *Proc. Acad. Nat. Sci. Philadelphia* 142: 59–71.

TD Robbins, M.B., Rosenberg, G.H. & Sornoza M., F., 1994. A new species of Cotinga (Cotingidae, *Doliornis*) from the Ecuadorian Andes, with comments on plumage sequences in *Doliornis* and *Ampelion. Auk* 111(1): 1–7.

TD Robbins, M.B. & Stiles, F.G., 1999. A new species of Pygmy-Owl (Strigidae: *Glaucidium*) from the Pacific slope of the Northern Andes. *Auk* 116(2): 305–315.

Robel, D., 2000. Unidentified green turaco in Ethiopia. *Bull. African Bird Club* 7(12): 56.

Roberson, D. & Caratello, R., 1997. Updates to the avifauna of Oaxaca, Mexico. *Cotinga* 7: 21–22.

Robertson, C.J.R. & Nunn, G.B., 1998. Towards a new taxonomy for albatrosses. In: Robertson, G. & Gales, R., (eds), *Albatross Biology and Conservation*, pp 13–19. Surrey Beatty & Sons, Chipping Norton, Australia.

TD Robertson, C.J.R. & Warham, J., 1992. Nomenclature of the New Zealand Wandering Albatrosses (*Diomedea exulans*). *Bull. Brit. Orn. Club* 112: 74–81.

Robertson, H.A. & de Monchy, P.J.M., 2912. Varied success from the landscape-scale management of kiwis *Apteryx spp.* in five sanctuaries in New Zealand. *Bird Cons. Int.* 22: 429–444.

Robertson, I., 1992. New information on birds in Cameroon. *Bull. Brit. Orn. Club* 112(1): 36–42.

Robertson, I.S., 1995. First field observations of the Sidamo Lark *Heteromirafra sidamoensis. Bull. Brit. Orn. Club* 115(4): 241–243.

Robertson, J.S., 1962. Mackay Report. *Emu* 61: 270–274.

Robiller, F., 2009. Ein ungewolter Mischlung – Buntkopf-Papageiamadine-Blaugrüne Papageiamadine. *Gefierdete Welt* 133(7): 14–15.

Robson, C., 2002. *Birds of Thailand.* Princeton University Press, Princeton, NJ.

Robson, C., 2005. *A Field Guide to the Birds of South-East Asia.* Princeton University Press, Princeton, NJ.

Robson, C. & Davidson, P., 1995. Some recent records of Philippine birds. *Forktail* 11: 162–167.

Robson, C., Roddis, S. & Loseby, T., 2016. From the Field. *BirdingASIA* 25: 122–128.

Robson, N.F., 1990. First recorded nest of the Lemonbreasted Canary in the field. *Ostrich* 61: 84–85.

Roda, S.A., Carlos, C.J. & Costa Rodrigues, R., 2003. New and noteworthy records for some endemic and threatened birds of the Atlantic forest of north-eastern Brazil. *Bull. Brit. Orn. Club* 123(4): 227–236.

Roda, S.A., Pereira, G.A. & Dantas, S.D., 2009. Alagoas Antwren *Myrmotherula snowi*: a new locality and remarks on its conservation. *Cotinga* 31: 144–146.

TD Rodrigues, E.L., Aleixo, A., Whittaker, A. & Naka, L.N., 2013. Molecular systematics and taxonomic revision of the Lineated Woodcreeper complex (*Lepidocolaptes albolineatus*: Dendrocolaptidae), with the description of a new species from southwestern Amazonia. *Handbook of the Birds of the World* Special Volume, pp 248–252. Lynx Edicions, Barcelona.

Rodríguez, G., López-Velasco, D. & Geale, D., 2012. Unidentified barred owl *Ciccaba* sp. at Manu National Park, Peru. *Neotropical Birding* 10: 9–13.

Rodríguez, J.V., 1982. *Aves del Parque Nacional Los Kátios.* INDERENA, Bogotá.

Rodríguez-Mojica, R., 2004. First report of cavity-nesting in Elfin-woods Warbler (*Dendroica angelae*) at Maricao State Forest, Puerto Rico. *Cotinga* 22: 21–23.

Rodwell, S.P., Sauvage, A., Rumsey, S.J.R. & Braunlich, A., 1996. An annotated check-list of the birds occurring at the Parc National des Oiseaux du Djoudi in Senegal, 1984–1994. *Malimbus* 18: 74–111.

Roesler, I., Casañas, H. & Imberti, S., 2011. Final countdown for the Hooded Grebe? *Neotropical Birding* 9: 3–7.

Roesler, I., Imberti, S., Casañas, H., Mahler, B. & Reboreda, J.C., 2012. Hooded Grebe *Podiceps gallardoi* population decreased by eighty per cent in the last twenty-five years. *Bird Cons. Int.*22: 371–384.

Roesler, I., Imberti, S., Casañas, H. & Volpe, N., 2012. A new threat to the globally Endangered Hooded Grebe *Podiceps gallardoi*: the American Mink *Neovison vison*. *Bird Cons. Int.* 22: 383–387.

Roesler, I., Kirwan, G.M., Agostini, M.G., Beadle, D., Shirihai, H. & Binford, L.C., 2009. First sight record of White-chested Swift *Cypseloides lemosi* and White-chinned Swift *C. cryptus* in Peru. *Bull. Brit. Orn. Club* 129(4): 222–228.

Rohwer, S., Filardi, C.E., Bostwick, K.S. & Peterson, A.T., 2000. A critical examination of Kenyon's Shag (*Phalacrocorax [Stictocarbo] kenyoni*). *Auk* 117(2): 308–320.

TD de Roo, A., 1967. A new species of *Chlorocichla* from north-eastern Congo (Aves: Pycnonotidae). *Rev. Zool. Bot. afr.* 75: 392–395.

TD Roselaar, C.S., 1992. *Leocosticte sillemi nov. spec.*, a new species of mountain finch from western Tibet. *Bull. Brit. Orn. Club* 112: 225–231.

Roselaar, C.S., 1994. Notes on Sillem's Mountain-finch, a recently described species from western Tibet. *Dutch Birding* 16: 20–26.

TD Roux, J-P., Jouventin, P., Mougin, J-L., Stahl, J-C. & Weimerskirch, H., 1983. Un nouvel albatros *Diomedea amsterdamensis* n. sp. découvert sur l'île Amsterdam (37°50' S., 77°35' E). *L'Oiseau et RFO* 53: 1–10.

Rowe, S. & Empson, R., 1996a. Distribution and abundance of the Tanga'eo or Mangaia Kingfisher *Halcyon tuta ruficollaris*. *Notornis* 43(1): 35–42.

Rowe, S. & Empson, R., 1996b. Observations on the breeding behaviour of the Tanga'eo or Mangaia Kingfisher *Halcyon tuta ruficollaris*. *Notornis* 43(1): 43–48.

TD Rowley, I., 1967. A fourth species of Australian corvid. *Emu* 66(3): 191–210.

Rowley, I., 1973. The comparative ecology of Australian corvids. 4. Nesting and rearing of young to independence. *CSIRO Wild. Res. Ann. Rep.* 18: 91–130.

Rowley, I. & Russell, E.M., 1997. *Fairywrens and Grasswrens*. Oxford University Press, Oxford.

TD Rowley, J.S, & Orr, R.T., 1964. A new hummingbird from southern Mexico. *Condor* 66(2): 81–84.

TD Rozendaal, F.G., 1987. Description of a new species of bush warbler of the genus *Cettia* Bonaparte, 1834 (Aves: Sylviidae) from Yamdena, Tanimbar Islands, Indonesia. *Zool. Meded.* 61(14): 177–202.

Ruiz-Esparza, J., Adriano da Rocha, P., de Souza Ribeiro, A., Ferrari, S.F. & Araujo, H.F.P., 2011. Expansion of the known range of Tawny Piculet *Picumnus fulvescens* including the south bank of the São Francisco River in north-east Brazil. *Bull. Brit. Orn. Club* 217–221.

TD Rumboll, M.A.E., 1974. Una nueva especie de Macá (Podicipedidae). *Comunicaciones Mus. Argentino Cien. Nat. Buenos Aires Zool.* 4(5): 33–35.

TD(I) Ruschi, A., 1972. Uma nova espécie de Beija-flor do E.E.Santo. *Phaethornis margarettae* Ruschi. *Bol. Mus. Biol. Prof. Mello Leitão,* Sér. Zool. No 35.

TD(I) Ruschi, A., 1973a. Uma nova espécie de Beija-flor de E.E.Santo. *Bol. Mus. Biol. Prof. Mello Leitão,* Sér. Zool. No 36.

TD(I) Ruschi, A., 1973b. Uma nova espécie de *Threnetes* (Aves: Trochilidae). *Bol. Mus. Biol. Prof. Mello Leitão, Santa Teresa,* Sér. Zool. No 37.

Ruschi, A., 1973c. *Beija-flores*. Museu de Biologia 'Prof. Mello Leitão'.

TD(I) Ruschi. A., 1975. *Threnetes cristinae* n. sp. *Bol. Mus. Biol. Prof. Mello Leitão,* Sér. Zool. No 38.

TD(I) Ruschi, A., 1982. Uma nova espécie de beija-flor do Brasil: *Amazilia rondoniae n. sp.* e a chave para determinar as espécies de *Amazilia* que ecortem no Brasil. *Bol. Mus. Melio Leitão* Sér. Zool. 100: 1–2.

Ruwet, J.-C., 1962. Aspects de la vie ornithologique au Lac de Retenue de la Lufira (Katanga). *Gerfaut* 52: 448–456.

Ruwet, J.-C., 1963. Notes zoologiques et éthologiques sur les oiseaux des plaines de la Lufira Supérieure (Katanga). *Rev. Zool. Bot. afr.* 68: 49–55.

Ruwet, J.-C., 1965. Notes écologiques et éthologiques sur les oiseaux des plaines de la Lufira Supérieure (Katanga). *Rev. Zool. Bot. afr.* 72: 389–427.

Safford, R.J., Ash, J.S. & Duckworth, J.W., 1997. Reply to Forero and Tella. *Ibis* 139: 410–411.

TD Safford, R.J., Ash, J.S., Duckworth, J.W., Telfer, M.G. & Zewdie, C., 1995. A new species of nightjar from Ethiopia. *Ibis* 137: 301–307.

Safford, R. & Hawkins, F., 2013. *The Birds of Africa Vol. VIII: the Malagasy Region*. Christopher Helm, London.

Saguindang, F.J. & Nuneza, O.M., 2004. The endemic and threatened birds on Mt Kimangkil Range, Bukidnon Province, Mindanao Island. *Philippine Scientist* 41: 176–196.

Saguindang, F.J., Nuneza, O.M. & Tabaranza, B.R., 2002. The avifauna of Mt Kimangkil Range, Bukidnon Province, Mindanao Island, Philippines. *Asia Life Sciences* 11(1): 9–28.

Salaman, P., 1996. The uncovered treasures of the Río Ñambí Paradise. *World Birdwatch* 18(4): 16–19.

TD Salaman, P., Coopmans, P., Donegan, T.M., Mulligan, M., Cortés, A., Hilty, S.L. & Ortega, L.A., 2003. A new species of wood-wren (Troglodytidae: *Henicorhina*) from the western Andes of Colombia. *Ornitología Colombiana* 1: 4–21.

Salaman, P., Negret, R.R. & Ortega, L.A., 1999. In Memoriam: Alvaro José Negret, 1949–1998.

TD Salaman, P.G.W. & Stiles, F.G., 1996. A distinctive new species of vireo (Passeriformes: Vireonidae) from the western Andes of Colombia. *Ibis* 138(4): 610–619.

Salamolard, M., Ghistemme, T., Couzi, F.-X., Minatchy, N. & Le Corre, M., 2007. Impacts des éclairage urbains sur les Petrels de Barau sur l'Îsle la Réunion et mesures pour reduire ces impacts. *Ostrich* 78(2): 449–452.

Salewski, V., Schmaljohann, H. & Herremans, M., 2005. New bird records from Mauretania. *Malimbus* 27(1): 19–32.

Sallaberry, A.,M., Hirsimaki, H., Olausson, A., 2010. Range extension of Puna Ibis *Plegadis ridgwayi* and new observations of Tamarugo Conebill *Conirostrum tamarugense* in northern Chile. *Cotinga* 32: 157.

TD Salomonsen, F., 1961. A new Tit-Babbler (*Stachyris hypogrammica*, sp. nov.) from Palawan, Philippine Islands. *Dansk Ornitologisk Forenings Tidsskrift* 55: 219–221.

TD Salomonsen, F., 1962(1963). Whitehead's Swiftlet (*Collocalia whiteheadi* Ogilvie-Grant) in New Guinea and Melanesia. *Vidensk. Medd. fra Dansk naturh. Foren.* 115: 205–281.

Sánchez-González, L.A., Oliveros, C., Puna, N & Moyle, R.G., 2010. Nests, nest placement and eggs of three Philippine endemic birds. *Wilson J. Orn.* 122(2): 587–591.

Sangster, G., 1998. New species of scops-owl. *Dutch Birding* 20: 322–323.

Sangster, G., 1999. Cryptic species of storm-petrels in the Azores? *Dutch Birding* 21: 101–106.

TD Sangster, G., King, B.F., Verbelen, P. & Trainor, C.R., 2013. A new owl species of the genus *Otus* (Aves: Strigidae) from Lombok, Indonesia. *PLoS ONE* 8(2): e53712.

Sangster, G., Roselaar, C.S., Irestedt, M. & Ericson, P.G.P., 2016. Sillem's Mountain Finch is a valid species of rosefinch (*Carpodacus*, Fringillidae). *Ibis* 158: 184–189.

TD Sangster, G. & Rozendaal, F.G., 2004. Systematic notes on Asiatic birds 41. Territorial songs and species-level taxonomy of nightjars of the *Caprimulgus macrurus* complex, with the description of a new species. *Zool. Verh. Leiden* 350: 7–45.

Santos, M.P.D., Cerqueira, P.V. & Soares, L.M. dos M., 2010. Avifauna in six localities in south-central state of Maranhão, Brazil. *Ornitologia* 4(1): 49–65.

Santos, M.P.D., Silveira, L.F. & Cardoso da Silva, J.M., 2011. Birds of Serra do Cachimbo, Pará State, Brazil. *Rev. Bras. Orn.* 19(2): 244–259.

Santos, M.P.D. & Vasconcelos, M.F., 2007. Range extension for Kaempfer's Woodpecker *Celeus obrieni* in Brazil, with the first male specimen. *Bull. Brit. Orn. Club* 127(3): 249–252.

Schifter, H., 2000. A further specimen of the Fine-barred Piculet (*Picumnus subtilis*) in the Museum of Natural History, Vienna, Austria. *Ornitología Neotropical* 11(3): 247–248.

TD Schmidt, B.K., Foster, J.T., Angehr, G.R., Durrant, K.L. & Fleischer, R.C., 2008. A new species of African Forest Robin from Gabon (Passeriformes: Muscicapidae: *Stiphrornis*). *Zootaxa* 1850: 27–42.

Schodde, R., 1982. *The Fairy Wrens – A Monograph of the Maluridae.* Melbourne: Lansdowne.

Schodde, R., 1984. First specimens of Campbell's Fairy-wren, *Malurus campbelli*, from New Guinea. *Emu* 84: 249–250.

Schodde, R., 1985. The Rusty-tailed Flyeater *Gerygone ruficauda* Ford & Johnstone – a case of mistaken identity. *Emu* 85: 49–50.

TD Schodde, R. & Christidis, L., 1987. Genetic differentiation and subspeciation in the Grey Grasswren *Amytornis barbatus* (Maluridae). *Emu* 87: 188–192.

Schodde, R. & Holyoak, D.T., 1977. Application of *Halcyon ruficollaris* Holyoak and *Alcyone ruficollaris* Banker. *Bull. Brit. Orn. Club* 97(1): 32.

Schodde, R. & Weatherly, R.G., 1982, in Schodde, R., 1982. *The Fairy-wrens: a monograph of the Maluridae.* Landsdowne, Melbourne.

TD Schodde, R. & Weatherly, R.G., 1983. Contributions to Papuasian ornithology. 8. Campbell's Fairy-wren (*Malurus campbelli*) a new species from New Guinea. *Emu* 82: 308–309.

Schuchmann, K.-L., Weller, A.-A. & Heynen, I., 2001. Systematics and biogeography of the Andean genus *Eriocnemus* (Aves: Trochilidae). *J. Orn.* 142: 431–481.

Schulenberg, T.S., 1987. Observations of two rare birds, *Upucerthia albigula* and *Conirostrum tamarugense*, from the Andes of south-western Peru. *Condor* 89(3): 654–658.

Schulenberg, T.S., Albujar, C. & Rojas, J.I., 2006. Pages 86–98, 185–196, 263–273 in Vriesendorp, C., Schulenberg, T.S., Alverson, W.S., Moskovits, D.K. & Rojas Moscoso, J.I. (eds). Peru. Sierra del Divisor Rapid Biological Inventories Report 17. The Field Museum, Chicago.

Schulenberg, T.S. & Awbrey, K., 1997. *The Cordillera del Cóndor region of Ecuador and Peru: a biological assessment.* Conservation International, Washington, DC.

TD Schulenberg, T.S. & Binford, L.C., 1985. A new species of tanager (Emberizidae: Thraupinae, *Tangara*) from southern Peru. *Wilson Bull.* 97: 413–420.

Schulenberg, T.S. & Jaramillo, A., 2012. São Francisco Sparrow (*Arremon franciscanus*). *Neotropical Birds Online* (T.S. Schulenberg, Editor), Cornell Lab. of Ornithology, Ithaca, NY.

Schulenberg, T.S. & Kirwan, G.M., 2011. Acre Antshrike (*Thamnophilus divisorius*) Neotropical Birds Online (T.S. Schulenberg, Editor), Cornell Lab. of Ornithology, Ithaca, NY.

Schulenberg, T.S. & Kirwan, G.M., 2012a. Orange-eyed Flycatcher (*Tolmomyias traylori*). Neotropical Birds online (T.S.Schulenberg, Editor), Cornell Lab of Ornithology, Ithaca, NY.

Schulenberg, T.S. & Kirwan, G.M., 2012b. Pale-billed Antpitta (*Grallaria carrikeri*). Neotropical Birds online (T.S.Schulenberg, Editor), Cornell Lab of Ornithology, Ithaca, NY.

Schulenberg, T.S. & Kirwan, G.M., 2012c. Chestnut Antpitta (*Grallaria blakei*). Neotropical Birds online (T.S.Schulenberg, Editor), Cornell Lab of Ornithology, Ithaca, NY.

TD Schulenberg, T.S. & Parker, T.A. III, 1997. A new species of tyrant-flycatcher (Tyrannidae: *Tolmomyias*) from the western Amazonian basin. *Orn. Monographs* 48: 723–731.

Schulenberg, T.S., Stotz, D.F., Lane, J.P., O'Neill, J.P. & Parker, T.A. III, 2007., Revised Edition 2010. *Birds of Peru.* Princeton University Press, Princeton, NJ.

TD Schulenberg, T.S. & Williams, M.D., 1982. A new species of antpitta (*Grallaria*) from northern Peru. *Wilson Bull.* 94: 105–113.

Sclater, P.L., 1886. *Cat. Birds. Brit. Mus.* 11.

Scott, D. & Brooke, M.de L., 1993. The rediscovery of the Grey-winged Cotinga *Tijuca condita* in south-eastern Brazil. *Bird Cons. Int.* 3: 1–12.

Scott, J.M., Mountainspring, S., Ramsey, F.L. & Kepler, C.B., 1986. Forest bird communities in the Hawaiian Islands: Their dynamics, ecology and conservation. *Studies in Avian Biology* 9. Cooper Ornithological Society. Allen Press, Lawrence, KS.

Seddon, N., Barnes, R., Butchart, S.H.M., Davies, C.W. & Fernández, M., 1996. Recent observations and notes on the ecology of the Royal Sunangel *Heliangelus regalis*. *Bull. Brit. Orn. Club* 116(1): 46–49.

Seeholzer, G.F., Justiniano, M.A., Harvey, M.G. & Smith, B.T., 2015. Ornithological inventory along an elevational gradient in the río Cotacajes Valley, dptos. La Paz and Cochabamba, Bolivia. *Cotinga* 37: 87–101.

TD Seeholzer, G.F., Winger, B.M., Harvey, M.G., Cáceres, A.D. & Weckstein, J.D., 2012. A new species of barbet (Capitonidae: *Capito*) from the Cerros del Sira, Ucayali, Peru. *Auk* 129(3): 1–9.

Serle, W., 1959. Note on the immature plumage of the Honey-guide *Melignomen zenkeri* Reichenow. *Bull. Brit. Orn. Club* 79: 65.

Shany, N., Alván, J.D. & Alonso, J.A., 2007. Mishana Tyrannulet *Zimmerius villarejoi*. *Neotropical Birding* 66.

Shany, N., Díaz A., J. & Álvarez A., J., 2007a. Finding white-sand forest specialists in Allpahuayo-Mishana Reserve, Peru. *Neotropical Birding* 2: 60–68.

Shany, N., Díaz A., J. & Álvarez A., J., 2007b. Iquitos Gnatcatcher *Polioptila clementsi*. *Neotropical Birding* 67.

Sharpe, C., 2015. *Lipaugus weberi*. *Neotropical Birding* 17: 28–31.

Sharpe, C.J., Spencer, A., Pieplow, N. & Gibbons, B., 2017. Photographs and status of White-fronted Swift *Cypseloides storeri*, a poorly-known Data Deficient Mexican endemic. *Neotropical Birding* 20: 15–20.

Shirihai, H., 1996. *The Birds of Israel.* Academic Press, London & San Diego.

Shirihai, H., 2007. *A Complete Guide to Antarctic Wildlife.* 2nd Edition. A&C Black, London.

Shirihai, H. & Bretagnolle, V., 2010. First observations at sea of Vanuatu Petrel *Pterodroma* (*cervicalis*) *occulta*. *Bull. Brit. Orn. Club* 130(2): 132–140.

TD(I) Shirihai, H., Sinclair, I. & Colston, P.R., 1995. A new *Puffinus* shearwater from the western Indian Ocean. *Bull. Brit. Orn. Club* 115(2): 75–87.

Shimelis, A., 1999. A range extension for the Ankober Serin *Serinus ankoberensis*. *Bull. African Bird Club* 6(2): 135–136.

TD Short, L.L., 1969. A new species of blackbird (*Agelaius*) from Peru. *Occ. Papers Mus. Zool. Louisiana State University* 36: 1–8.

Short, L.L., 1973. A new race of *Celeus spectabilis* from eastern Brazil. *Wilson Bull.* 85: 465–467.

Short, L.L. & Horne, J.F.M., 1985. Social behavior and systematics of African barbets (Aves: Capitonidae). Pp 255–278 in *Proc. Intern. Symp. Afr. Vertebr.* Mus. A. Koenig, Bonn.

Short, L.L. & Horne, J.F.M., 1988. Capitonidae: barbets and tinkerbirds. In Fry, C.H., Keith, S. & Urban, E.K. (eds). *The Birds of Africa* Vol. 3. Academic Press, London.

Short, L.L. & Horne, J.F.M., 2001. *Toucans, Barbets and Honeyguides.* Oxford University Press, Oxford.

Shu, X.L., Zhou, L.F., Li, Y.L. & Du, Y., 2009. The current situation and conservation of the threatened animals in limestone region in southwestern China. *Genomics and Applied Biology* 28: 828–834.

Sibley, C.G. & Monroe, B.L. Jr., 1990. *Distribution and Taxonomy of Birds of the World.* Yale University Press, New Haven, CT.

TD Sick, H., 1960. Zur Systematik und Biologie der Bürzelstelzer (Rhinocryptidae), speziell Brasiliens. *J. Ornithol.* 101: 141–174.

Sick, H., 1961. *Tukani.* Labor, Barcelona.

TD Sick, H., 1969. Über eine Töpfervögel (Furnariidae) aus Rio Grande do Sul, Brasilien, mit Beschreibung eines neuen *Cinclodes. Beitrage zur Neotropischen Fauna* 61: 63–79.

Sick, H., 1973. Nova contribução ao conhecimento de *Cinclodes pabsti* Sick, 1969 (Furnariidae, Aves). *Rev. Bras. Biol.* 33: 109–117.

Sick, H., 1993. *Birds in Brazil.* Princeton University Press, Princeton, NJ.

TD(I) Siegel-Causey, D., 1991. Systematics and biogeography of North Pacific shags, with a description of a new species. *Occ. Papers Mus. Nat. Hist. Univ. Kansas* 140: 1–17.

Sillem, J.A., 1934. Ornithological results of the Netherlands Karakorum Expedition 1929/1930. *Org. Club Nederl. Vogelk.*7: 1–48.

Sillem, J.A., 1935. Aves. Pp 452–499 in Visser, P.C. (ed), *Wissenschaftliche Ergebnisse der niederlandischen Expeditionen in den Karakorum und die angrenzenden Gebeite 1922, 1925 und 1929/1930.* Leipzig.

TD(I) da Silva, J.M.C., Novaes, F.C. & Oren, D.C., 1995. (*Hylexetastes [perroti] brigidai*). *Bull. Brit. Orn. Club* 115(4): 200–202.

TD da Silva, J.M., Coelho, G. & Gonzaga, L.P., 2002a. Discovered on the brink of extinction: a new species of Pygmy-owl (Strigidae: *Glaucidium*) from the Atlantic Forest of north-eastern Brazil. *Ararajuba* 10(2): 123–130.

TD(I) da Silva, J.M., Novaes, F.C. & Oren, D.C., 2002b. Differentiation of *Xiphocolaptes* (Dendrocolaptidae) across the river Xingu, Brazilian Amazonia: recognition of a new phylogenetic species and biogeographic implications. *Bull. Brit. Orn. Club* 122(3): 185–193.

da Silva, M., Pichorim, M., Cardoso, M.Z., 2008. Nest and egg description of threatened *Herpsilochmus* spp. from coastal forest habitats in Rio Grande do Norte, Brazil (Aves: Thamnophilidae). *Revista Brasiliera de Zoologia* 25(3): 570–572.

da Silva, M., de Albuquerque França, B.R., de Lima Hagi, L.Y.G., Neto, M.R., Valdenor de Oliveira, D. & Pichorim, M., 2011. New sites and range extensions for endemic and endangered birds in extreme north-east Brazil. *Bull. Brit. Orn. Club* 131(4): 234–240.

Silveira, L.F., 2010. *Formicivora* aff. *acutirostris,* p 209. In: Bressan, P.M., Kierulff, M.C.M. & Sugieda, A.M., (eds). *Fauna Ameaçada de extinção no Estado de São Paulo – Vertebrados.* São Paulo Fundação Parque Zoológico de São Paulo e Secretaria do Meio Ambiente.

TD Silveira, L.F., de Lima, F.C.T. & Hofling, E., 2005a. A new species of *Aratinga* (Psittaciformes: Psittacidae) from Brazil, with taxonomic remarks on the *Aratinga solstitialis* complex. *Auk* 122(1): 292–305.

Silveira, L.F., Develey, P.F., Pacheco, J.F. & Whitney, B.M., 2005b. Avifauna of the Serra das Lontras-Javi montane complex, Bahia, Brazil. *Cotinga* 24: 45–54.

Silveira, L.F., Olmos, F. & Long, A.J., 2003. Birds in Atlantic Forest fragments in north-east Brazil. *Cotinga* 20: 32–46.

Sim, I.M.W. & Zefania, S., 2002. Extension of the range of the Red-shouldered Vanga *Calicalia rufocarpalis* in southwest Madagascar. *Bull. Brit. Orn. Club* 122(3): 194–1.

Simon, J.E. & Magnano, G.R., 2013. Rediscovery of the Cryptic Forest-Falcon *Micraster mintoni* Whittaker, 2002 (Falconidae) in the Atlantic forest of southeastern Brazil. *Revista Brasiliera de Ornitologia* 21(4): 257–262.

Sinclair, I. & Langrand, O., 2013. *Birds of the Indian Ocean Islands.* Third Edition. Struik Nature, Cape Town, South Africa.

TD(I) Smith, E.F.G., Arctander, P., Fjeldså, J. & Amir, O.G., 1991. A new species of shrike (Laniidae: *Laniarius*) from Somalia, verified by DNA-sequence data from the only known individual. *Ibis* 133(3): 227–235.

Smith, F.T.H., 1984. Victorian records of a sandpiper new to science. *Aust. Bird Watcher* 10: 264–265.

Smythies, B.E., 2000. *Birds of Borneo:* 4th Edition. Natural History Publications, Kota Kinabalu, Borneo.

Snethlage, E., 1914. Catalogo das aves amazonicas contendo todas as especies descriptas e mencionadas até 1913. *Bol. Mus. Goeldi (Museu Paraense)* 8: 1–500.

TD Snow, D.W., 1980. A new species of cotinga from southeastern Brazil. *Bull. Brit. Orn. Club* 100(4): 213–215.

Snow, D.W., 1982. *The Cotingas: Bellbirds, Umbrellabirds and Other Species.* British Museum and Comstock Publishing.

Socolar, S.J., González, O. & Forero-Medina, G., 2013. Noteworthy bird records from the northern Cerros del Sirá, Peru. *Cotinga* 35: 24–36.

Sophasan, S. & Dobias, R., 1984. The fate of the 'Princess Bird' or White-eyed River Martin (*Pseudochelidon sirintarae*). *Nat. Hist. Bull. Siam Soc.* 32(1): 1–10.

Sornoza Molina, F., 2000. Fundación Jocotoco: conservation action in Ecuador. *World Birdwatch* 22: 14–17.

de Sousa Azevedo, L.A., Alexandre, S., Santos, M.P.D. 2013. New molecular evidence supports the species status of Kaempfer's Woodpecker. *Genetics and Molecular Biology* 36(2): 192–200.

de Souza, R.D.R., 2013. Primeiro registro do tapaculo-serrano *Scytalopus petrophilus* para el estado de São Paulo. *Atualidades Ornitologicas* 175: 12–13.

Spottiswoode, C.N., 2010. Finding southern Ethiopia's endemic birds. *Bull. African Bird Club* 17(1): 106–113.

Spottiswoode, C.N., Wondafrash, M., Gabremichael, M.N., Dellelegn Abebe, Y., Mwangi, M.A.K., Collar, N.J. & Dolman, P.M., 2009. Rangeland degradation is poised to cause Africa's first recorded avian extinction. *Animal Conservation* 12: 249–257.

Spottiswoode, C. *et al.*, 2013. Rediscovery of a long-lost lark reveals the conspecificity of endangered *Heteromirafra* populations in the Horn of Africa. *J. Ornithol.* DOI 10.1007/s10336-013-0948-1.

TD Stager, K.E., 1961. A new bird of the genus *Picumnus* from eastern Brazil. *Contrib. Sci. Los Angeles Co. Mus.* 46: 1–4.

TD Stager, K.E., 1968. A new piculet from southeastern Peru. *Contrib. Sci. Los Angeles Co. Mus.* 153: 1–4.

Stahl, J.-C. & Bartle, J.A., 1991. Distribution, abundance and aspects of the pelagic ecology of Barau's Petrel *Pterodroma baraui* in the South-West Indian Ocean. *Notornis* 38: 211–225.

Stanley, T.R., Aldridge, C.L., Saher, D.J. & Childers, T.M., 2015. Daily nest survival rates of Gunnison Sage-grouse (*Centrocercus minimus*): assessing local- and landscape-scale drivers. *Wilson J. Orn.* 127(1): 59–71.

Stap, D., 1990. *A Parrot without a Name.* Alfred A. Knopf (Random House, NY).

Statius Müller, P.L., 1776. *Natursyst* (Suppl.) 74–79, 80–81, 90. (*Psittacus maculatus*).

TD Stepanyan, L.S., 1972. Novi vid roda *Locustella* (Sylviidae: Aves) iz vostochnoi Palearktiki [A new species of the genus *Locustella* (Aves: Sylviidae) from the eastern Palearctic]. *Zoologicheskiy Zhurnal* 51: 1896–1897.

Stepanyan, L.S., 1974. *Paradoxornis heudei polivanovi* Stepanyan ssp. n. (Paradoxornithidae, Aves) iz basseina Ozera Khanka [New subspecies from the Lake Khanka Basin]. *Zoologicheskiy Zhurnal* 53(8); 1270–1272.

Stepanyan, L.S., 1990. Novaya versiya proiskhozhdeniya Pesochnika Koksa [A new hypothesis of the origin of Cox's Sandpiper] *Calidris paramelanotos* (Scolopacidae, Aves). *Zoologicheskiy Zhurnal* 69: 148–151.

Stepanyan, L.S., 1998. O vidovoi samostoyatel'nosti *Paradoxornis polivanovi* (Paradoxornithidae, Aves). [On the status of *Paradoxornis polivanovi* as an independent species]. *Zoologicheskiy Zhurnal* 77(10): 1158–1161.

Stevenson, T. & Fanshawe, J., 2002. *Birds of East Africa.* Christopher Helm, London.

Stiles, F.G., 1990. Un encuentro con el mosquerito antioqueño *Phylloscartes lanyoni* Graves. *Boletin SAO* 1(2): 12–13.

TD Stiles, F.G., 1992. A new species of antpitta (Formicariidae; *Grallaria*) from the Eastern Andes of Colombia. *Wilson Bull.* 104: 389–399.

TD Stiles, F.G., 1996. A new species of Emerald Hummingbird (Trochilidae, *Chlorostilbon*) from the Sierra de Chiribiquete, southeastern Colombia, with a review of the *C. mellisugus* complex. *Wilson Bull.* 108(1): 1–27.

TD Stiles, F.G., Laverde-R., O. & Cadena, C.D., 2017. A new species of tapaculo (Rhinocryptidae: *Scytalopus*) from the Western Andes of Colombia. *Auk* 134(2): 377–392.

Stiles, F.G., Rosselli, L. & Bohórquez, C.I., 1999. New and noteworthy records of birds from the middle Magdalena valley of Colombia. *Bull. Brit. Orn. Club* 119: 113–129.

Stoeckle, M. & Hebert, P., 2008. Barcode of Life. *Scientific American* 299(4): 82.

Stopiglia, R. & Raposo, M.A., 2006. The name *Synallaxis whitneyi* Pacheco and Gonzaga, 1995, is not a synonym of *Synallaxis cinereus* Wied, 1831 (Aves: Passeriformes: Furnariidae). *Zootaxa* 1166: 49–55.

Stopiglia, R., Raposo, M.A. & Teixeira, D.M., 2013. Taxonomy and geographic variation of the *Synallaxis ruficapilla* Vieillot, 1819 species-complex (Aves: Passeriformes: Furnariidae). *J. Ornithol.* 154: 191–207.

Storer, R.W., 1981. The Hooded Grebe on Laguna de los Escarchados: ecology and behavior. *Living Bird* 19: 51–67.

Storer, R.W., 1982. A hybrid between Hooded and Silvery Grebes (*Podiceps gallardoi* and *Podiceps occipitalis*). *Auk* 99: 632–636.

Stotz, D.F., Fitzpatrick, J.W., Parker, T.A. & Moskovits, D.K., 1996. *Neotropical Birds: Ecology and Conservation.* University of Chicago Press, Chicago.

Stotz, D.F., Lanyon, S.M., Schulenberg, T.S., Willard, D.E., Peterson, A.T. & Fitzpatrick, J.W., 1997. An avifaunal survey of two tropical forest localities on the middle Rio Jiparaná, Rondônia, Brazil. *Ornithological Monographs* 48: 763–781.

Stranek, R.J., 1993. Aportes para la unificación de *Serpophaga subcristata* y *Serpophaga munda*, y la revalidación de *Serpophaga griseiceps* (Aves: Tyrannidae). *Revista del Museo Argentino de Ciencias Naturales "Bernardo Rivadavia", Zoología.* 16: 51–63.

TD Stranek, R., 2008. Una nueva especie de *Serpophaga* (Aves: Tyrannidae). *Revista FAVE Ciencias Veterinarias* 6(1–2): 31–42.

Stranek, R. & Johnson, A.,1985. Vocalizaciones en relación al comportamiento del Macá Tobiano (*Podiceps gallardoi* Rumboll). *Revista del Museo Argentino de Ciencias Naturales 'Bernado Rivadavia' e Instituto Nacional de Investigación de las Ciencias Naturales: Zoología* 13: 177–188.

Stresemann, E., 1948. Der Naturforscher Friedrich Sellow (1831) und sein Beitrag zur Kenntnis Brasiliens. *Zoologischen Jahrbücher* 77: 401–425.

Stuart, S.N., Jensen, F.P. & Brøgger-Jensen, S., 1987. Altitudinal zonation of the avifauna in Mwanihana and Magombera Forests, eastern Tanzania. *Gerfaut* 77: 163–186.

Sun, Y-H., Jiang, Y-X., Martens, J. & Bi, Zh-L., 2009. Notes on the breeding biology of the Sichuan Treecreeper (*Certhia tianquanensis*). *J. Ornithol.* 150(4): 909–913.

Sykes, B.R., 2006. New liocichla found in India at Arunachal Pradesh's Eaglenest Sanctuary. *BirdingASIA* 6: 72.

Sykes, B., 2013. Rinjani Scops Owl *Otus jolandae* revealed from Lombok, Lesser Sundas, Indonesia. *BirdingASIA* 19: 111–120.

Tallman, D.A., Parker, T.A. III, Lester, G.D. & Hughes, R.A., 1978. Notes on two species of birds previously unreported from Peru. *Wilson Bull.* 90: 445–446.

Tarburton, M.K., 1990. Breeding Biology of the Atiu Swiftlet. *Emu* 909: 175–179.

Taylor, B. & van Perlo, B., 1998. *Rails: a Guide to the Rails, Crakes, Gallinules and Coots of the World.* Pica Press, Sussex, UK.

Tebb, G., Morris, P. & Los, P., 2008. New and interesting bird records from Sulawesi and Halmahera, Indonesia. *BirdingASIA* 10: 67–76.

TD Teixeira, D.M., 1987. A new tyrannulet *Phylloscartes* from northeastern Brazil. *Bull. Brit. Orn. Club* 107(1): 37–41.

Teixeira, D.M., 1992. Obituary: Helmut Sick. *Ibis* 134: 90.

Teixeira, D.M. & Carnevalli, N., 1989. Nova espécie de *Scytalopus* Gould 1837, do noreste do Brasil. *Boletim Museu Nacional, Nova Série, Rio de Janeiro, Zoologia* 331: (20 November) 1–11.

TD Teixeira, D.M. & Gonzaga, L.P., 1983a. A new antwren from northeastern Brazil. *Bull. Brit. Orn. Club* 103(4): 133–135.

TD Teixeira, D.M. & Gonzaga, L.P., 1983b. Um novo Furnariidae para o nordeste do Brasil: *Philydor novaesi* sp. nov. *Bol. Mus. Paraense Emilio Goeldi* (n. ser. Zool.) 124: 1–22.

Teixeira, D.M. & Gonzaga, L.P., 1985. Uma nova subespecie de *Myrmotherula unicolor* do noreste do Brasil. *Bol. Museu Nacional* (n. ser. Zool.) 310: 1–16.

Teixeira, D.M., Nacinovic, J.B. & Luigi, G., 1988. Notes on some birds of northeastern Brazil. *Bull. Brit. Orn. Club* 108(2): 75–79.

TD Tello, J.G., Degner, J.F., Bates, J.M. & Willard, D.E., 2006. A New Species of Hanging-parrot (Aves: Psittacidae: *Loriculus*) from Camiguin Island, Philippines. *Fieldiana: Zoology* 106: 49–57.

Tello, J.G., Raposo, M., Bates, J.M., Bravo, G.A., Cadena, C.D. & Maldonado-Coelho, M., 2014. Reassessment of the systematics of the widespread Neotropical genus *Cercomacra* (Aves: Thamnophilidae). *Zool. J. Linn. Soc.* 170(3): Special Issue: S1: 546–565.

TD Tennyson, A.J.D., Palma, R.L., Robertson, H.A.,Worthy, T.H. & Gill, R.J., 2003. A new species of kiwi (Aves: Apterygiformes) from Okarito, New Zealand. *Rec. Auckland Mus.*40: 55–64.

Thiebot, J.-B., Delord, K., Marteau, C. & Weimerskirch, H., 2014a. Stage-dependent distribution of the Critically Endangered Amsterdam Albatross in relation to Economic Exclusive Zones. *Endangered Species Research* 23(3): 263–276.

Thiebot, J.-B., Barbraud, C., Delord, K., Marteau, C. & Weimerskirch, H., 2014b. Do introduced mammals chronically impact the breeding success of the world's rarest albatross? *Ornithological Science* 13(1): 41–46.

TD Thonglongya, K., 1968. A new martin of the genus *Pseudochelidon* from Thailand. *Thai National Scientific Papers, Fauna Series no 1.* Applied Scientific Research Corporation of Thailand, Bangkok.

Thonglongya, K., 1969. Report of an expedition in northern Thailand to look for breeding sites of *Pseudochelidon sirintarae* (21 May–27 June). Applied Scientific Research Corporation of Thailand, Bangkok.

Tibbett, S., 2006. Forbes-Watson's Swifts nesting in Dhofar. *Phoenix* 22: 23.

Tietze, D.T. & Martens, J., 2009. Morphometric characterisation of treecreepers (genus *Certhia*). *J. Ornithol.* 150(2): 431–457.

Tietze, D.T., Martens, J. & Sun, Y-H., 2006. Molecular phylogeny of treecreepers (*Certhia*) detects hidden diversity. *Ibis* 148(3): 477–488.

Tobias, J., 2000. A little-known oriental bird: White-eyed River-martin *Eurochelidon sirintarae. Bull. Oriental Bird Club* 31: 45–48.

Tobias, J., 2003a. Notes on breeding behaviour in Black-faced Cotinga *Conioptilon mcilhennyi. Cotinga* 19: 80–81.

Tobias, J., 2003b. Further sightings of Selva Cacique *Cacicus koepckeae* in Manu National Park, Peru. *Cotinga* 19: 79–80.

Tobias, J.A., Butchart, S.H.M. & Collar, N.J., 2006. Lost and found: a gap analysis for the Neotropical avifauna. *Neotropical Birding* 1: 4–22.

Tobias, J.A., Lebbin, D.J., Aleixo, A., Andersen, M.J., Guilherme, E., Hosner, P.A. & Seddon, N., 2008. Distribution, behavior and conservation status of the Rufous Twistwing *Cnipodectes superrufus. Wilson J. Orn.* 120(1): 38–49.

Tobias, J.A., Seddon. N., Spottiswoode, C.N., Fishpool, L.D.C. & Collar, N.J., 2010. Quantitative criteria for species delimitation. *Ibis* 152: 724-746.

Todd, F.S., 1979. *Waterfowl: Ducks, Geese and Swans of the World.* Harcourt-Brace Jovanovich, New York.

Töpfer, T., 2006. Systematic notes on Asian birds. 60. Remarks on the systematic position of *Ficedula elisae* (Weigold, 1922). *Zoologische Meded. (Leiden)* 80: 203–212.

Totterman, S.L., 2009. Vanuatu Petrel (*Pterodroma occulta*) discovered breeding on Vanua Lava, Banks Islands, Vanuatu. *Notornis* 56: 57–62.

Totterman, S.L., 2012. Sexual differences in vocalisations and playback-response behaviour of the Vanuatu Petrel (*Pterodroma occulta*). *Notornis* 59: 97–104.

Trai, Le Trong & Craik, R.C., 2008. Mekong Wagtail *Motacilla samveasnae* – resident breeder in Vietnam? *BirdingASIA* 9: 68–69.

Trainor, C.R., Verbelen, P & Johnstone, R.L., 2012. The avifauna of Alor and Pantar, Lesser Sundas, Indonesia. *Forktail* 28: 77–92.

TD Traylor, M.A., 1967. A new species of *Cisticola. Bull. Brit. Orn. Club* 87: 45–48.

Traylor, M.A., Jr., 1982. Notes on tyrant flycatchers (Aves: Tyrannidae). *Fieldiana Zoology* 13: 1–22.

Turner, A. & Rose, C., 1989. *A Handbook of the Swallows and Martins of the World.* Christopher Helm, London, UK.

Uitamo, E., 1999. Modelling deforestation caused by the expansion of subsistence farming in the Philippines. *J. Forest Economics* 5(1): 99–122.

Ujihara, M., 2002. An apparent juvenile Cox's Sandpiper in Japan. *Birding World* 15(8): 346–347.

Urban, E.K., Fry, C.H. & Keith, S., (eds) 1997. *The Birds of Africa.* Vol 5. Academic Press, London.

Uy, J.A.C. & Borgia, G., 2000. Sexual selection drives rapid divergence in bowerbird display traits. *Evolution* 54: 273–278.

TD(I) Valdés-Velásquez, A. & Schuchmann, K.-L., 2009. A new species of hummingbird (*Thalurania*: Trochilidae) from the western Colombian Andes. *Ornithol. Anz.* 48: 143–149.

Vallotton, L., 2016. The juvenile plumage of Tamarugo Conebill *Conirostrum tamarugense. Cotinga* 38: 60–63.

TD Valqui, T. & Fjeldså, J., 1999. New brush-finch *Atlapetes* from Peru. *Ibis* 141: 194–198.

Valqui, T. & Fjeldså, J., 2002. *Atlapetes melanopsis* nom. nov. for the Black-faced Brushfinch. *Ibis* 144: 347.

van den Berg, A.B., 1982. Plumages of the Algerian Nuthatch. *Dutch Birding* 4: 98–100.

van den Berg, A.B., Smeenk, C., Bosman, C.A.W., Haase, B.J.M., van der Niet, A.M. & Cadée, G.C., 1991. Barau's Petrel *Pterodroma baraui*, Jouanin's Petrel *Bulweria fallax* and other seabirds in the northern Indian Ocean in June–July 1984 and 1985. *Ardea* 79:1–14.

van Eijk, P., 2013. Presumed second locality for Omani Owl. *Dutch Birding* 35(6): 387–388.

de Vasconcelos, M.F., D'Angelo Neto, S., Kirwan, G.M., Bornschein, M.R., Diniz, M.G. & Francisco da Silva, J., 2006. Important ornithological records from Minas Gerais state, Brazil. *Bull. Brit. Orn. Club* 126: 212–238.

TD Vaurie, C., 1960. Systematic notes on Palearctic birds. No 39. Caprimulgidae. A new species of *Caprimulgus. Amer. Mus. Novit.* 1985: 1–10.

Vaurie, C., 1971. *Classification of the Ovenbirds (Furnariidae).* H.F. & G. Witherby, London.

Vaurie, C., 1980. Taxonomy and geographical distribution of the Furnariidae (Aves: Passeriformes). *Bull. Amer. Mus. Nat. Hist.* 166: 1–357.

TD Vaurie, C., Weske, J.S. & Terborgh, J.W., 1972. Taxonomy of *Schizoeca fuliginosa* (Furnariidae), with description of two new subspecies. *Ibis* 95(2): 142–144.

Vecchi, M.B. & Alves, M.A.S., 2008. New records of the Restinga Antwren *Formicivora littoralis* Gonzaga and Pacheco (Aves, Thamnophilidae) in the state of Rio de Janeiro, Brazil: inland extended range and threats. *Brazilian Journal of Biology* 68(2): 1–6.

Verbelen, P. & Demeulemeester, B., 2012. Field observations of an unidentified frogmouth *Batrachostomus* on Siberut, Mentawai Islands, West Sumatra, Indonesia. *BirdingASIA* 17: 106–108.

Vermeulen, J., 2000. Trip report: Ethiopia, October 8–27, 1999. Unpublished. www.camacdonald.com/birding/tripreports/Ethiopia99.html

Vickery, P.D., Finch, D.W. & Donahue, P.K., 1987. Juvenile Cox's Sandpiper (*Calidris paramelanotos*) in Massachusetts, a first New World occurrence and a hitherto undescribed plumage. *American Birds* 41(5): 1366–1369.

TD Vielliard, J., 1976a. La Sittelle Kabyle. *Alauda Suppl. Spec.* 44(3): 351–352.

Vielliard, J., 1976b. New living fossil showing emergence of species in Mediterranean basin: *Sitta ledanti* sp. nov. (Aves: Sittidae). *Comptes Rendus Hebdom. Séance. Acad. Scien. Serie D* 283(10): 1193–1195.

Vielliard, J., 1978. Le Djebel Babor et sa Sittelle *Sitta ledanti* Vielliard 1976. *Alauda* 46: 1–42.

Vielliard, J., 1980. Remarques complémentaires sur la Sittelle kabyle *Sitta ledanti* Vielliard 1976. *Alauda* 48: 139–150.

TD Vielliard, J., 1989. Uma nova espécie de *Glaucidium* (Aves: Strigidae) da Amazonia. *Rev. Bras. Zool.* 6: 685–693.

TD(I) Vielliard, J., 1990. Uma nova espécie de *Asthenes* da serra do Cipó, Minas Gerais, Brasil. *Ararajuba* 1: 121–122.

Virani, M., 1995. Sokoke Scops Owl in Tanzania. *Swara* 18: 34.

Virani, M., 2000. Home range and movement patterns of Sokoke Scops Owl *Otus ireneae*. *Ostrich* 71(1&2): 139–142.

Virani, M.Z., Njoroge, P. & Gordon, I., 2010. Disconcerting trends in populations of the endangered Sokoke Scops Owl *Otus ireneae* in the Arabuko-Sokoke Forest, Kenya. *Ostrich* 81(2): 155–158.

TD(I) Vo Quy, 1973. *Chim Viet Nam* (Birds of Vietnam). In Vietnamese.

Voelker, G., 1999. Molecular evolutionary relationships in the avian genus *Anthus* (Pipits: Motacillidae) *Molecular Phylogenetics and Evolution* 11: 84–94.

Voelker, G., Outlaw, R.K. & Bowie, R.C.K., 2010a. Pliocene forest dynamics as a primary driver of African bird speciation. *Global Ecol. Biogeogr.* 19: 111–121.

TD Voelker, G., Outlaw, R,K., Reddy, S., Tobler, M., Bates, J.M., Hackett, S.J., Kahindo, C., Marks, B.D., Peterhans, J.C.K. & Gnoske, T.P., 2010b. A new species of boubou (Malaconotidae: *Laniarius*) from the Albertine Rift. *Auk* 127(3): 678–689.

Vuilleumier, F., 1995. Five great Neotropical ornithologists. *Orn. Neotrop.* 6: 97–111.

Vuilleumier, F., 1998. The need to collect birds in the Neotropics. *Orn. Neotrop.* 9: 201–203.

Vuilleumier, F. & Mayr, E., 1987. New species of birds described from 1976 to 1980. *J. Orn.* 128(2): 137–150.

Vuilleumier, F., LeCroy, M. & Mayr, E., 1992. New species of birds described from 1981 to 1990. *Bull. Brit. Orn. Club Centenary Suppl.* 112A: 267–307.

Walker, B., Stotz, D.F., Pequeño, Y. & Fitzpatrick, J.W., 2006. Birds of the Manu Biosphere Reserve. Pp 23–49 in B.D.Patterson, D.F.Stotz & S.Solari (eds). *Mammals and Birds of the Manu Biosphere Reserve. Fieldiana Zoology*, new series, 110.

Walker, K. & Elliott, G., 1999. Population changes and biology of the Wandering Albatross *Diomedea exulans gibsoni* at the Auckland Islands. *Emu* 99: 239–247.

Walker, K. & Elliott, G., 2005. Population changes and biology of the Antipodean Wandering Albatross (*Diomedea antipodensis*). *Notornis* 52: 206–214.

Walker, K. & Elliott, G., 2006. At sea distribution of Gibson's and Antipodean Wandering Albatrosses, and relationships with long-line fisheries. *Notornis* 53: 265–290.

Walker, K., Elliott, G., Nicholls, D., Murray, D. & Diks, P., 1995. Satellite tracking of Wandering Albatross (*Diomedea exulans*) from the Auckland Islands: preliminary results. *Notornis* 42: 127–137.

Walpole, P., 2010. *Figuring the Forest Figures: Understanding Forest Cover Data in the Philippines and where we might be proceeding.* Environmental Science for Social Change, Quezon City, Philippines.

Warakagoda, D., 2001a. Discovery of a new species of owl in Sri Lanka. *Ceylon Bird Club Notes*, January–February: 1–4.

Warakagoda, D., 2001b. The new species of scops owl in Sri Lanka. *Tyto* 6(2): 54–59.

Warakagoda, D., 2001c. The discovery of Serendib Scops Owl *Otus* sp. in Sri Lanka. *Sri Lanka Naturalist* 4(4): 57–59.

Warakagoda, D., 2001d. The discovery of a new owl. *Loris* 22(5): 45–47.

Warakagoda, D., 2006. Sri Lanka's Serendib Scops Owl. *Birding Asia* 6: 68–71.

TD Warakagoda, D. & Rasmussen, P., 2004, A new species of scops-owl from Sri Lanka. *Bull. Brit. Orn. Club* 124: 85–105.

Warburton, T., 2009. The Philippine Owl Conservation Programme; why is it needed? *Ardea* 97(4): 429.

Wardill, J.C., 2001. Notes on the Talaud Rail *Gymnocrex talaudensis* from Karakelang Island, North Sulawesi, Indonesia. *Forktail* 17: 116–118.

Waterhouse, J.H.L., 1949. *A Roviana and English Dictionary*, rev. ed. Epworth, Sydney.

Waugh, D., 2004. Der Schutz des bedrohten El-Oro-Sittichs. *Gefierdete Welt* 128(4): 113–115.

Webster, J.D., 1988. Skeletons and genera of tanagers. *Proc. Indiana Acad. Sci.* 98: 581–593.

Webster, J.D., 1992. The manubrium-sternum bridge in songbirds (Oscines). *Proc. Indiana Acad. Sci.* 101: 299–308.

Wege, D., 1996. Threatened birds of the Darién highlands, Panama; a reassessment. *Bird Cons. Int.* 6: 175–179.

Weick, F., 2006. *Owls (Strigiformes): Annotated and Illustrated Checklist.* Springer-Verlag, Berlin and Heidelberg.

Weimerskirch, H., 2004. Diseases threaten Southern Ocean albatrosses. *Polar Biol.* 27: 374–379.

Weimerskirch, H., Brothers, N. & Jouventin, P., 1997. Population dynamics of Wandering Albatross *Diomedea exulans* and Amsterdam Albatross *D. amsterdamensis* in the Indian Ocean and their relationships with long-line fisheries: conservation implications. *Biological Conservation* 79: 257–270.

Welch, G. & Welch, H., 1998. Mystery birds from Djibouti. *Bull. African Bird Club* 5(1): 45–50.

TD(I) Weller, A.-A., 2011. Geographic and age-related variation in the Violet-throated Sunangel (*Heliangelus viola*, Trochilidae): evidence for a new species and subspecies. *Ornitología Neotropical* 22(4): 601–614.

TD Weske, J.S. & Terborgh, J.W., 1974. *Hemispingus parodii*, a new species of tanager from Peru. *Wilson Bull.* 86(2): 97–103.

TD Weske, J.S. & Terborgh, J.W., 1977. *Phaethornis koepckeae*, a new species of Hummingbird from Peru. *Condor* 79: 143–147.

TD Weske, J.S. & Terborgh, J.W., 1981. *Otus marshalli*, a new species of Screech-owl from Perú. *Auk* 98(1): 1–7.

TD Wetmore, A., 1963. Additions to records of birds known from the Republic of Panamá. *Smithsonian Misc. Coll.* 145(6): 1–14.

TD Wetmore, A., 1964. A Revision of the American Vultures of the genus *Cathartes*. *Smithsonian Misc. Coll.* 146(6), (4539): 14–18.

Wetmore, A., 1965, 1968, 1973. The Birds of the Republic of Panama, Parts 1–3. *Smiths. Misc. Coll.*, Vol 150, pts 1–3.

Wheatley, N., 1994. *Where to Watch Birds in South America.* Christopher Helm/A&C Black, London.

Wheatley, N. & Brewer, D., 2001. *Where to Watch Birds in Central America, Mexico and the Caribbean.* Christopher Helm/A&C Black, London.

White, A.W., 2010. Finding 'South Hills Crossbills'. *Winging It* 2010: 7.

Whitehouse, A.J. & Ribon, R., 2010. Finding Stresemann's Bristlefront in Minas Gerais, Brazil. *Neotropical Birding* 6: 36–39.

TD Whitney, B.M., 1994. A new *Scytalopus* tapaculo (Rhinocryptidae) from Bolivia, with notes on other Bolivian members of the genus and the *magellanicus* complex. *Wilson Bull.* 106(4): 585–614.

Whitney, B.M., 1997. Birding the Alta Floresta region, northern Mato Grosso, Brazil. *Cotinga* 7: 64–68.

Whitney, B.M., 2005. *Clytoctantes (atrogularis?)* in Amazonas, Brazil, and its relationship to *Neoctantes niger* (Thamnophilidae). *Bull. Brit. Orn. Club* 125: 108–113.

TD Whitney, B.M. & Álvarez A., J., 1998. A new *Herpsilochmus* antwren (Aves: Thamnophilidae) from northern Amazonian Peru and adjacent Ecuador: the role of edaphic heterogeneity of *terra firme* forest. *Auk* 115: 559–576.

TD Whitney, B.M. & Álvarez A., J., 2005. A new species of gnatcatcher from white-sand forests of northern Amazonian Peru, with revision of the *Polioptila guianensis* complex. *Wilson Bull.* 117: 113–127.

TD Whitney, B.M., Cohn-Haft, M., Bravo, G.A., Schunck, F. & Silveira, L.F., 2013. A new species of *Herpsilochmus* antwren from the Aripuanã-Machado interfluvium in central Amazonian Brazil. *Handbook of the Birds of the World* Special Volume, pp 277–281. Lynx Edicions, Barcelona.

TD Whitney, B.M., Ferreira de Vasconcelos, M.F, Silveira, L.F. & Pacheco, J.F., 2010. *Scytalopus petrophilus* (Rock Tapaculo): a new species from Minas Gerais, Brazil. *Revista Brasiliera de Ornitologia* 18(2): 73–88.

TD(I) Whitney, B.M., Isler, M.L., Bravo, G.A., Aristizábal, N., Schunck, F., Silveira, L.F. & Piacentini, V. de Q., 2013. A new species of *Epinecrophyla* antwren from the Aripuanã-Machado interfluvium in central Amazonian Brazil with a revision of the 'stipple-throated antwren' complex. *Handbook of the Birds of the World* Special Volume, pp 263–267. Lynx Edicions, Barcelona.

TD Whitney, B.M., Isler, M.L., Bravo, G.A., Aristizábal, N., Schunck, F., Silveira, L.F., Piacentini, V. de Q., Cohn-Haft, M. & Rêgo, M.A., 2013. A new species of antbird in the *Hypocnemus cantator* complex from the Aripuanã-Machado interfluvium in central Amazonian Brazil. *Handbook of the Birds of the World* Special Volume, pp 282–285. Lynx Edicions, Barcelona.

Whitney, B.M. & Oren, D.C., 2001. Primeiro registro de *Nanopsittaca dachilleae* no Brasil. *Natteria* 2: 26.

TD Whitney, B.M., Oren, D.C. & Brumfield, R.T., 2004. A new species of *Thamnophilus* antshrike (Aves: Thamnophilidae) from the Serra do Divisor, Brazil. *Auk* 121: 1031–1039.

Whitney, B.M. & Pacheco, J.F., 1995. Distribution and conservation status of four *Myrmotherula* antwrens (Formicariidae) in the Atlantic Forest of Brazil. *Bird Cons. Int.* 5: 421–439.

Whitney, B.M. & Pacheco, J.F., 1997. Behavior, vocalisations and relationships of some *Myrmotherula* antwrens (Thamnophilidae) in eastern Brazil, with comments on the 'plain-winged' group. *Ornithological Monographs* 48: 809–819.

Whitney, B.M. & Pacheco, J.F., 2001. *Synallaxis whitneyi* Pacheco and Gonzaga, 1995, is a synonym of *Synallaxis cinerea* Wied, 1831. *Natteria* 2: 34–35.

TD Whitney, B.M., Pacheco, J.F., Buzzetti, D.R.C. & Parrini, R., 2000. Systematic revision and biogeography of the *Herpsilochmus pileatus* complex, with description of a new species from northeastern Brazil. *Auk* 117(4): 869–891.

Whitney, B.M., Pacheco, J.F., Moreira de Fonseca, P. & Barth, R.H., Jr, 1996. The nest and nesting ecology of *Acrobatornis fonsecai* (Furnariidae), with implications for intrafamilial relationships. *Wilson Bull.* 108: 434–448.

Whitney, B.M., Pacheco, J.F., Moreira de Fonseca, P.S., Webster, R.E., Kirwan, G.M. & Mazar Barnett, J., 2003. Reassignment of *Chordeiles vielliardi* Lenciono-Neto, 1994, to *Nyctiprogne* Bonaparte, 1857, with comments on the latter genus and some presumably related chordeilines (Caprimulgidae). *Bull. Brit. Orn. Club* 123(2): 103–112.

TD Whitney, B.M., Piacentini, V. de Q., Schunk, F., Aleixo, A., de Sousa, B.R.S., Silveira, L.F. & Rêgo, M.A., 2013. A name for Striolated Puffbird west of the Rio Madeira with revision of the *Nystalus striolatus* complex. *Handbook of the Birds of the World* Special Volume, pp 240–244. Lynx Edicions, Barcelona.

TD Whitney, B.M., Schunck, F., Rêgo, M.A. & Silveira, L.F., 2013a. A new species of *Zimmerius* tyrannulet from the upper Madeira-Tapajós interfluve in central Amazonian Brazil: Birds don't always occur where they 'should'. *Handbook of the Birds of the World* Special Volume, pp 286–291. Lynx Edicions, Barcelona.

TD(I) Whitney, B.M., Schunck, F., Rêgo, M.A. & Silveira, L.F., 2013b. A new species of flycatcher in the *Tolmomyias assimilis* radiation from the lower Sucunduri-Tapajós interfluvium in central Amazonian Brazil heralds a new chapter in Amazonian biogeography. *Handbook of the Birds of the World* Special Volume, pp 297–305. Lynx Edicions, Barcelona.

Whittaker, A., 2001. Alagoas reserve threatened. *Cotinga* 16: 10.

TD Whittaker, A., 2003. A new species of forest-falcon (Falconidae: *Micrastur*) from southeastern Amazonia and the Atlantic Rainforests of Brazil. *Wilson Bull.* 114(4): 421–445.

Whittaker, A., 2004. Noteworthy ornithological records from Rondônia, Brazil. *Bull. Brit. Orn. Club* 124(4): 239–271.

Whittaker, A., 2009. Pousada Rio Roosevelt; a provisional avifaunal inventory in south-western Amazonian Brazil, with information on life-history, new distributional data and comments on taxonomy. *Cotinga* 31: 20–43.

Whittaker, A., Aleixo, A. & Poletto, F., 2008. Corrections and additions to an anotated checklist of birds of the upper Rio Urucu, Amazonas, Brazil. *Bull. Brit. Orn. Club* 128: 11.

TD Whittaker, A., Aleixo, A., Whitney, B.M., Smith, B.T. & Klicka, J., 2013. A distinctive new species of gnatcatcher in the *Polioptila guianensis* complex (Aves: Polioptilidae) from western Amazonian Brazil. *Handbook of the Birds of the World* Special Volume, pp 301–305. Lynx Edicions, Barcelona.

Whittaker, A. & Oren, D.C., 1999. Important ornithological records from the Rio Juruá, western Amazonia, including twelve additions to the Brazilian avifauna. *Bull. Brit. Orn. Club* 119(4): 235–260.

Whittingham, M.J. & Atkinson, P.W., 1996. A species split in Mexico: Sumichrast's and Nava's Wren *Hylorchilus sumichrasti* and *H. navai*. *Cotinga* 6: 20–22.

Wiedenfeld, D.A., 1982. A nest of the Pale-billed Antpitta (*Grallaria carrikeri*) with comparative remarks on antpitta nests. *Wilson Bull.* 94: 580–582.

TD(I) Williams, J.G., 1966. A new species of swallow from Kenya. *Bull. Brit. Orn. Club* 86: 40.

Williams, J., 1980. *A Field Guide to the Birds of East Africa*. Collins, Glasgow.

Williams, M.D (ed.), 1986. Report on the Cambridge Ornithological Expedition to China, 1985.

Willis, E.O., 1987. Primeiros registros de *Geotrygon saphirina* (Aves: Columbidae) e *Grallaria* sp. cf. *eludens* (Aves: Formicariidae) no oeste do Brasil. *Resumos XIV Congr. Bras. Zool.*, Juiz de Fora, MG p 153.

TD Willis, E.O. & Oniki, Y., 1992. A new *Phylloscartes* (Tyrannidae) from southeastern Brazil. *Bull. Brit. Orn. Club* 112(3): 158–165.

TD Wolters, H.E., 1974. Aus den ornithologischen Sammlung des Museums Alexander Koenig III. Ein neuer *Malimbus* (Ploceidae, Aves) von der Elfenbeinküste. *Bonn. Zool. Beitr.* 25: 290–291.

TD Woxfold, I.A., Duckworth, J.W. & Timmins, R.J., 2009. An unusual new bulbul (Passeriformes: Pycnonotidae) from the limestone karst of Lao PDR. *Forktail* 25: 1–12.

Wright, D. & Jones, P., 2005. Population densities and habitat associations of the range-restricted Rock Firefinch *Lagonosticta sanguinodorsalis* on the Jos Plateau, Nigeria. *Bird Cons. Int.*15(3): 287–295.

TD Yamashina, Y. & Mano, T., 1981. A New Species of Rail from Okinawa Island. *J. Yamashina Inst. Org.* 13: 1–6.

TD Young, J.R., Braun, C.E., Oyler-McCance, S.J., Hupp, J.W. & Quinn, T.W., 2000. A new species of sage-grouse (Phasianidae: *Centrocercus*) from southwestern Colorado. *Wilson Bull.* 112(4): 445–453.

Yuang, H.-W., Ding, T.-S. & Hsieh, H.-I., 2005. Short-term Responses of Animal Communities to Thinning in a *Cryptomeria japonica* (Taxodiaceae) Plantation in Taiwan. *Zool. Studies* 44(3): 393–402.

Zhang, Y.-Y., Wang, N., Zhang, J. & Zheng, G.-M., 2006. Acoustic difference of Narcissus Flycatcher complex. *Acta Zoologica Sinica* 52: 648–654.

TD(I) Zheng Guangmei, Song Jie, Zhang Zhengwang, Zhang Yanyun & Guo Dongsheng, 2000. A new species of flycatcher (*Ficedula*) from China (Aves: Passeriformes: Muscicapidae). *Journal of Beijing Normal University (Natural Science)* 36(3): 405–409.

TD Zhou, F. & Jiang, A.W., 2008. A new species of babbler (Timaliidae: *Stachyris*) from the Sino-Vietnamese border region of China. *Auk* 125: 420–424.

Zimmer, J.T., 1939. Studies of Peruvian birds. No 32. *Amer. Mus. Novit.* 1044.

Zimmer, K.J., Parker, T.A. III, Isler, M.L. & Isler, P.R.,1997. Survey of a southern Amazonian avifauna, the Alta Floresta region, Mato Grosso, Brazil. *Ornithological Monographs* 48: 887–918.

TD(I) Zimmer, K.J., Whittaker, A. & Oren, D.C., 2001. A cryptic new species of flycatcher (Aves: *Suiriri*) from the cerrado region of central South America. *Auk* 118(1): 56–78.

TD(I) Zimmer, K.J., Whittaker, A., Sardelli, C.H., Guilherme, E. & Aleixo, A., 2013. A new species of *Hemitriccus* tody-tyrant from the state of Acre, Brazil. *Handbook of the Birds of the World* Special Volume, pp 292–296. Lynx Edicions, Barcelona.

Zimmerman, D.A., Turner, D.J. & Pearson, D.J., 1996. *Birds of Kenya and Northern Tanzania.* Christopher Helm/A&C Black, London.

Zink, R.M. & McKitrick, M.C., 1995. The debate over species concepts and its implications for ornithology. *Auk* 112(3): 701–719.

Ziswiler, V., Güttinger, H.R. & Bregula, H., 1972. *Monographie der Gattung* Erythrura *Swainson, 1837 (Aves: Passeres: Estrildidae).* Bonner Zoologische Monographien No 2.

Zuccon, D., Prŷs-Jones, R., Rasmussen, P.C. & Ericson, P.G.P., 2012. The phylogenetic relationships and generic limits of finches (Fringillidae). *Molecular Phylogenetics and Evolution* 62(2): 581–596.

Zusi, R.L., 1978. Remarks on the generic allocation of *Pseudochelidon sirintarae. Bull. Brit. Orn. Club* 98(1): 13–15.

PHOTOGRAPHIC CREDITS

Bloomsbury Publishing would like to thank those listed below for providing photographs and for permission to reproduce copyright material within this book. Whilst every effort has been made to trace and acknowledge all copyright holders, we would like to apologise for any errors or omissions, and invite readers to inform us so that corrections can be made in any future editions.

Front cover: top – Scarlet-banded Barbet *Capito wallacei* (Dubi Shapiro); bottom left – Jocotoco Antpitta (*Grallaria ridgelyi*) (Nigel Voaden); bottom centre – Okinawa Rail (*Gallirallus okinawae*) (feathercollector/www.shutterstock.com); bottom right – Mayotte Scops Owl (*Otus mayottensis*) (Ken Behrens). Back cover: Lulu's Tody-Flycatcher (*Poecilotriccus luluae*) (Miguel Lezama).

Copyright information for the photographs in this book: T = top; B = bottom; L = left; R = right.

Adam Riley 75R, 119
Adrian Constantino/www.birdingphilippines.com 312
Agustin Carrasco 192L
Alexis Guevara and Euclides Campos 23
Amir Ben Dov 80L,R, 368T
André Grassi Corrêa 159T
Andre Moncrieff 61
Andres Cuervo 193
Andrew Spencer 99, 109
Andrew Whittaker/Birding Brazil Tours 205, 222
Arthur Grosset 41T, 62, 125T
Ashley Banwell 251
Barry Wright 7B
Bob Gress 82
Bram Demeulemeester 50, 52, 92, 343
Brian Schmidt/Smithsonian Institution 298L
Bruce Beehler 249, 296
Bruno Rennó 41B, 63L, 107, 142, 149, 159B, 186
Buttu Madika 342B
Carl Downing 112, 237
Carlos Calle 214, 329
Carlos Figuerero 326L,R
Carlos Gussoni 145, 146
Carlton Ward Jr 298R
Chris Tzaros 225
Christian Artuso 75L, 342TL
Christian Nunes 332
Christopher L. Wood 285
Christopher Milensky/Smithsonian Institution 270
Ciro Albano 95, 125B, 128R, 131, 144, 162, 163T,BL, 169, 170L,R, 173L, 180, 181, 183, 185, 196B, 212, 213, 219, 220, 327, 367T
Claudia Hermes 192R
Daniel J. Lebbin 152, 176B
Daniel Lane 58, 148, 150, 153L,R, 297, 323, 324, 336
Daniel Uribe 175T, 190
Dante Buzzetti 239
David Ascanio 215, 216, 218
David Beadle 111L, 363TR, 363B
David Cook 231, 240
David Hoddinott 234
David Monticelli 293L,R
Don Geoff Tabaranza 46
Doug Wechsler 371
Dubi Shapiro 87, 89, 121, 122, 189, 238, 262, 362, 364BR
Dušan Brinkhuizen 60, 133, 143, 160, 334, 342TR
Eduardo Carrión L. 194
Edward Vercruysse 275
Elis Simpson 172T,B
Fabio Schunck 167L,R, 178L,R
Fabrice Schmitt 30T, 77, 84, 135, 140, 179T,B, 207, 217, 289, 331, 338
Fernando Angulo Pratolongo 115T
Francesco Rovero/MUSE-Museo delle Scienze 27
Gerald McCormack 102, 103, 116, 117, 259T,B, 260T,B, 364TL
Gerrit Vyn 24, 25
Graham Ekins 208
H. E. McClure 242
Hanne and Jens Eriksen 104L,R
Ian Montgomery 226R
Igor Castillo/Proyecto AmaurospizaVE 339, 340
Jack Jeffrey Photography 319
Jacob and Tini Wijpkema 78, 79, 129
Jainy Kuriakose 69, 367B
James Eaton 2-3, 10, 90, 94, 97, 247, 250, 263, 264, 266, 271, 274L, 281, 283, 284, 315B, 348, 349
Jan F. Rasmussen 244
János Oláh/Birdquest 282
João Quental 56, 101, 127, 128L, 134, 147, 163BR, 168, 176T, 187, 364TR
Johannes Ferdinand 232R
John C. Mittermeier 236T,B
John Caddick 313
Jon Hornbuckle 111R, 206, 370
José Álvarez Alonso 141, 166, 290
Josh Beck 59L,R
Julian Heavyside 191
Kazuto Kawakami 36
Ken Behrens 34, 35, 45, 73, 235, 252, 317B, 363TL
Kim Westerskov/www.gettyimages.co.uk 30B
Kirk Zufelt 33, 49, 416

Knud Andreas Jønsson 65
Kurt Hennige 28T,B
Lars Petersson 156B
Laval Roy 233R
Len Robinson 224
Li Liwei 294
Liao Pen-shing 254T,B, 256
Lindsay Hansch 226L
Louis A. Hansen 8T, 26L,R, 305R
Luis A. Mazariegos 113
Luis Eduardo Urueña 209
Mami Ikehata 47
Marc Chrétien 55
Marc Guyt/www.agami.nl 12, 38
Márcio Repenning 325L,R
Mark Beaman 318L,R
Markus Lagerqvist 232L, 258, 299, 305L, 307R
Markus Lilje 118, 261
Matthias Dehling 83
Michael G. Harvey 120, 337L,R
Michael Kearns 277
Michel Hasson 308, 309L,R
Michelle & Peter Wong 267, 268
Miguel Lezama 151, 201, 210, 333, 335T,B, 344B, 366
Mike Morel 322L,R
Niall Perrins 317TL,TR
Nick Athanas 8B, 115B, 156T, 188, 196T, 200, 202, 286
Nigel Voaden 1, 9B, 108, 175B, 204, 301, 314, 364BL
Nik Borrow 124, 300, 310
Nitin Srinivasa Murthy 274R
Pablo Florez 344T
Paul Oliver 307L
Per Alström 257, 295
Pete Morris 39, 71, 72, 255
Peter Hosner 273
Philippe Verbelen 70, 74, 91
Ray Plowman 29
Rémi Bigonneau 9T, 31, 32
Ricardo Plácido 157, 158
Richard Greenhalgh 138T
Richard Hoyer 85T
Richard Porter 43L,R
Robert Hutchinson 7T, 44, 68, 265, 279, 303, 347
Robert Lewis 105, 106L,R, 288
Robert Martin 51
Robert Prŷs-Jones 253
Robson Czaban 21, 63R, 126, 132, 136
Rodrigo Tapia Jiménez 37L,R
Ross Gallardy 93, 138B, 165, 198, 245, 276
S. Rajeshkumar 341
Samantha Klein 182
Sergio Gregorio 184
Stephen Jones 173R
Steve Bird 287
Sylvia Ramos 302, 304
Tanguy Deville 85B
Tasso Leventis 66, 315T
Ted Shimba 48
Thang Nguyen 280
Thiago Laranjeiras 365
Thierry Nogaro 155
Thomas Arndt 53L,R, 57
Tim Avery 100
Tim Laman/Getty Images 229
Tonji Ramos 278, 306T,B
Tui de Roy/www.naturepl.com 22
Uditha Hettige 67, 368B
Vik Dunis 228
Werner Suter 311
Yann Kolbeinsson 233L

INDEX

The following pages comprise species names from the 'Species Accounts' (including alternative names) and 'Invalid Species' chapters.

▲ Vanuatu Petrel. Vanua Lava, Banks Island, Vanuatu, April 2014 (*Kirk Zufelt*).